Terms	Meaning	Section
	Formal Logic	
$\neg p$	Not p	2.1
$p \wedge q$	p and q	2.1
$p \vee q$	p or q	2.1
$p \rightarrow q$	p implies q	2.1
$p \leftrightarrow q$	p is equivalent to q	2.1
$S \models \chi$	S logically implies χ	2.3.3
$\mathcal{P} \neq \mathcal{NP}$	Conjecture about complexity	2.5.6
$(\forall x) P(x)$	For all x, $P(x)$	2.7.2
$(\exists x) P(x)$	There exists an x such that $P(x)$	2.7.2
$(\forall x \in V) P(x)$	For all $x \in V$, $P(x)$	2.7.3
$(\exists x \in V) P(x)$	There exists an $x \in V$ such that $P(x)$	2.7.3
$A[i \mathrel{.\,.} j]$	Array with elements $A[i], \ldots, A[j]$	2.7.3
$\mid$	Sheffer stroke	2.4
$\veebar$	Exclusive or	2.4
$\downarrow$	Pierce arrow	2.9
$(x, y) \in R$ or $x R y$	x is R-related to y	3.1
R^{-1}	The inverse of the relation R	3.2.1
$R \circ S$	Composition of relations R and S	3.2.2
R^{+}	$\cup_{i=1}^{\infty} R^i$	3.4.4
R^{*}	$\cup R_{i=0}^{\infty} R^i$	3.4.4
$n \equiv m (mod\, p)$	$n - m = kp$ for some $k \in \mathbb{N}$	3.6
Id_X	Identity relation	3.1
Le_X	Less than or equal relation	3.1
Gt_X	Greater than relation	3.1
Ge_X	Greater than or equal relation	3.1
$[x]$	Equivalence class of x	3.6
$m \mid n$	m divides n	3.8.1
$R \bowtie S$	Equijoin of relations R and S	3.10.2

www.brookscole.com

www.brookscole.com is the World Wide Web site for
Brooks/Cole and is your direct source to dozens of
online resources.

At www.brookscole.com you can find out about
supplements, demonstration software, and student
resources. You can also send email to many of our
authors and preview new publications
and exciting new technologies.

www.brookscole.com
Changing the way the world learns®

Discrete Mathematics
for Computer Science

Discrete Mathematics for Computer Science

Gary Haggard
Bucknell University

John Schlipf
University of Cincinnati

Sue Whitesides
McGill University

THOMSON
BROOKS/COLE

Australia · Canada · Mexico · Singapore · Spain
United Kingdom · United States

THOMSON

™

BROOKS/COLE

Publisher: *Bob Pirtle*

Assistant Editor: *Stacy Green*

Editorial Assistant: *Katherine Cook*

Technology Project Manager: *Earl Perry*

Marketing Manager: *Tom Ziolkowski*

Marketing Assistant: *Erin Mitchell*

Advertising Project Manager: *Bryan Vann*

Signing Representative: *Stephanie Shedlock*

Project Manager, Editorial Production:
Cheryll Linthicum

Art Director: *Vernon Boes*

Print/Media Buyer: *Doreen Suruki*

Permissions Editor: *Chelsea Junget*

Production Service: *Hearthside Publishing Service;
Anne Seitz*

Text Designer: *Roy Neuhaus*

Copy Editor: *Hearthside Publishing Service;
Wesley Morrison*

Illustrator: *Hearthside Publishing Service;
Jade Myers*

Cover Designer: *Roy R. Neuhaus*

Cover Image: *DigitalVision*

Cover Printer: *Phoenix Color Corp*

Compositor: *ATLIS*

Printer: *Phoenix Color Corp*

Printed in the United States of America

1 2 3 4 5 6 7   09 08 07 06 05

For more information about our products,
contact us at:
Thomson Learning Academic Resource Center
1-800-423-0563

For permission to use material from this text or product, submit a request online at
http://www.thomsonrights.com.
Any additional questions about permissions can be submitted by email to thomsonrights@thomson.com.

Library of Congress Control Number: 2004113828

ISBN 0-534-49501-X

Thomson Higher Education
10 Davis Drive
Belmont, CA 94002-3098
USA

Asia
Thomson Learning
5 Shenton Way #01-01
UIC Building
Singapore 068808

Australia/New Zealand
Thomson Learning
102 Dodds Street
Southbank, Victoria 3006
Australia

Canada
Nelson
1120 Birchmount Road
Toronto, Ontario M1K 5G4
Canada

Europe/Middle East/Africa
Thomson Learning
High Holborn House
50/51 Bedford Row
London, WC1R 4LR
United Kingdom

Latin America
Thomson Learning
Seneca, 53
Colonia Polanco
11560 Mexico D.F.
Mexico

Spain/Portugal
Paraninfo
Calle Magallanes, 25
28015 Madrid
Spain

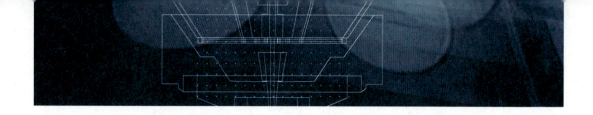

Contents

CHAPTER 2

Formal Logic   89

CHAPTER **3**

Relations   157

CHAPTER 4

CHAPTER **6**

Graph Theory   331

CHAPTER **7**

Counting and Combinatorics   421

CHAPTER 8

CHAPTER **9**

Recurrence Relations   549

APPENDIX

Preface

As the discipline of computer science has matured, it has become clear that a study of discrete mathematical topics is an essential part of the computer science major. The course in discrete structures has two primary aims. The first is to introduce students to the rich mathematical structures that naturally describe much of the content of the computer science discipline, including many structures that are frequently used in modeling and implementing solutions to problems. The second is to help students develop the skills of mathematical reasoning to learn new concepts and material in computer science. This learning takes place not only while they are students but also after graduation and throughout their professional life.

During the past few years, researchers in areas of computer science as diverse as the analysis of algorithms, database systems, and artificial intelligence have made ever-increasing use of discrete mathematical structures to clarify and explain key concepts and problems. As a reflection of this emphasis, careful discussions of applications such as a relational database system, the complexity of a computation, and normal forms of propositions are included in this text. The discussions of these topics build on a strong, focused development of fundamental ideas about sets, logic, relations, and functions as well as graph theory and combinatorics.

The diagram that follows gives an indication of the order in which the material can be covered. The six chapters referred to in the box contain the fundamental topics. These chapters are used to guide students in learning how to express mathematically precise ideas in the language of mathematics.

The two chapters dealing with graph theory and combinatorics are also core material for a discrete structures course, but this material always seems more intuitive to students than the formalism of the first four chapters. Topics from the first four chapters are freely used in these later chapters. The chapter on discrete probability builds on the chapter on combinatorics. The chapter on the analysis of algorithms uses notions from the core chapters but can be presented at an informal level to motivate the topic without spending a lot of time with the details of the chapter. Finally, the chapter on recurrence relations primarily uses the early material on induction and an intuitive understanding of the chapter on the analysis of algorithms.

PREFACE

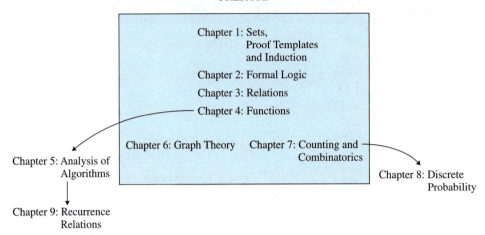

The material in Chapters 1 through 4 deals with sets, logic, relations, and functions. This material should be mastered by all students. A course can cover this material at different levels and paces depending on the program and the background of the students when they take the course. Chapter 6 introduces graph theory, with an emphasis on examples that are encountered in computer science. Undirected graphs, trees, and directed graphs are studied. Chapter 7 deals with counting and combinatorics, with topics ranging from the addition and multiplication principles to permutations and combinations of distinguishable or indistinguishable sets of elements to combinatorial identities.

Enrichment topics such as relational databases, languages and regular sets, uncomputability, finite probability, and recurrence relations all provide insights regarding how discrete structures describe the important notions studied and used in computer science. Obviously, these additional topics cannot be dealt with along with the all the core material in a one-semester course, but the topics provide attractive alternatives for a variety of programs. This text can also be used as a reference in courses. The many problems provide ample opportunity for students to deal with the material presented.

To the Student

A major aim of this book is to help you develop mathematical maturity—elusive as this objective may be. We interpret this as preparing you to understand how to do proofs of results about discrete structures that represent concepts you deal with in computer science. A correct proof can be viewed as a set of reasoned steps that persuade another student, the course grader, or the instructor about the truth of the assertion. Writing proofs is hard work even for the most experienced person, but it is a skill that needs to be developed through practice. We can only encourage you to be patient with the process. Keep trying out your proofs on other students, graders, and instructors to gain the confidence that will help you in using proofs as a natural part of your ability to solve problems and understand new material.

Solutions for the odd numbered Exercises are included on the CD that comes with the text. These solutions provide models for solving problems.

Outline for One-Semester Course

This text contains much more material than can be covered in a typical one-semester course. This diversity of material, however, allows a much broader range of courses to use the text. For a program that requires a one semester (13–14 weeks) study of discrete topics, the following outline provides coverage of the fundamental material:

Chapter 1: Sets, Proof Templates, and Induction   *(8 lectures)*
    Basic Definitions
    Operations on Sets
    The Principle of Inclusion-Exclusion
    Mathematical Induction
    A Second Form of Induction

Chapter 2: Formal Logic   *(4 lectures)*
    Introduction to Propositional Logic
    Truth and Logical Truth
    Predicates and Quantification

Chapter 3: Relations   *(5 lectures)*
    Definitions and Operations
    Special Types of Relations
    Equivalence Relations
    Ordering Relations

Chapter 4: Functions   *(4 lectures)*
    Basic Definitions
    Operations on Functions
    The Pigeon-Hole Principle

Chapter 5: Analysis of Algorithms   *(2 lectures)*
    Comparing Growth Rates of Functions
    Complexity of Programs

Chapter 6: Graph Theory   *(4 lectures)*
    Definitions
    Connected Graphs
    The Königsberg Bridge Problem
    Trees
    Spanning Trees
    Directed Graphs (Optional)

Chapter 7: Counting and Combinatorics   *(4–5 lectures)*
    Counting Principles
    Permutations and Combinations
    Permutations and Combinations with Repetitions
    Combinatorial Identities (Optional)
    Pascal's Triangle (Optional)

With a semester comprising about 40 lectures, this schedule provides time for exams and additional time to modify the course to respond to particular curricular and/or student needs. The one chapter that is quite often left to other courses is Chapter 5. If time permits, however, this material gives a good overview of the relationship between programs and their complexity.

Many variations can be made based on what other courses are included in the program. In some programs, topics in Chapters 1 through 4, particularly basic properties of sets and functions, will be covered in prerequisite courses and may be reviewed quickly in a discrete mathematics course. The sections on Induction, the Principle of Inclusion-Exclusion, and the Pigeon-Hole Principle, however, should normally be covered. In other programs, if material on the analysis of algorithms has already been discussed in computer science courses, then Chapter 4 might be a review, to a certain extent, and take less time. Optionally, material on directed graphs might be eliminated. Depending on the needs of the program, the lectures saved above may be spent on other material on the book.

Outline for a One-Quarter Course

With only 30 lectures in a one-quarter course, the syllabus presented earlier needs to be cut to about 27 lectures.

Provided the material of Chapter 5 is covered in other computer science courses, this chapter can be omitted without difficulty. If other mathematics courses explain the idea of a function, the only necessary material in Chapter 4 is the Pigeon-Hole Principle, which can save at least one lecture. Finally, eliminating the material on directed graphs should allow the basic ideas of graph theory to be covered in four lectures. In addition, the nine lectures scheduled for Chapters 1 and 2 may be shortened one or two lectures.

Incorporating these suggestions, the following is a possible syllabus for a one-quarter course (10 weeks):

Chapter 1: Sets   *(7 lectures)*
    Basic Definitions
    Operations on Sets
    The Principle of Inclusion-Exclusion
    Mathematical Induction
    A Second Form of Induction

Chapter 2: Formal Logic   *(3 lectures)*
    Introduction to Propositional Logic
    Truth and Logical Truth
    Predicates and Quantification

Chapter 3: Relations   *(4 lectures)*
    Definitions and Operations
    Special Types of Relations
    Equivalence Relations
    Ordering Relations

Chapter 4: Functions   *(3 lectures)*
    Basic Definitions
    Operations on Functions
    The Pigeon-Hole Principle

Chapter 6: Graph Theory   *(4 lectures)*
    Definitions

In both sample syllabi the number of lectures committed to material should leave time for two or three exams and for review days. In addition, instructors should find time to spend a full day on problems of special interest without being forced to give up material from the outline.

Help Requested

The authors have tried their best to make the text as error-free as possible. Needless to say, we are not perfect and likely have missed some problems that really need to be corrected to improve the text. We would appreciate it very much if any errors would be brought to our attention. (We intend to provide a small reward for the first notice of any problem brought to our attention.) Send comments to haggard@bucknell.edu along with your snail-mail address. We will acknowledge any help we receive and let you know if anyone else has already noticed the problem you uncovered. We will be very grateful for any help we receive as we intend to make this text the best learning tool we can. A collection of the changes we make will be posted at http://www.eg.bucknell.edu/~discrete/errorfile.pdf.

Gary Haggard
John Schlipf
Sue Whitesides

CHAPTER 1

Sets, Proof Templates, and Induction

The concept of a *set* underlies most of modern mathematics and much of computer science. To use sets as a foundation for all the other structures in this text, we first need to understand both the language used to describe sets and the operations normally associated with sets. The language of sets is very precise. When we use this language carefully, we gain precision in expressing problems and describing solutions to problems. Understanding basic operations on sets and the properties of these operations is a model for the approach that is used to introduce most other discrete structures in this text. In extending our understanding of operations on sets, we will learn proof techniques to explore other discrete mathematical topics, such as relations, functions, and graphs. We will use these proof techniques, for example, to prove that algorithms are correct and to determine how well we have chosen an algorithm for a given task.

This chapter has five main sections. The first introduces the notion of a set and the language for describing collections of elements. In addition, this section introduces several proof templates that are guides to both understanding and constructing proofs. The second deals with the common operations on sets: unions, intersections, complements, products, and the power set of a set. Some additional proof templates are introduced that are drawn from proofs in this section. The third provides a way to count the number of elements in a collection of sets in which some of the sets may contain some of the same elements that the other sets contain. The fourth and fifth deal with important proof techniques called the *Principle of Mathematical Induction* and the *Strong Form of Mathematical Induction*. We use induction to find the set of elements for which a statement about the integers is true.

An important application of the Principle of Mathematical Induction, in both its forms, is to show how algorithms can be proven to be correct without any execution by a computer.

1.1 Basic Definitions

The idea of a set is simple: A **set** is a collection of elements. The set {white, red, green} contains the names of the colors white, red, and green and nothing else. The set {0, 1, 2, 3, 4, 5, 100345679231} contains seven integers. The set {red, yellow, blue} contains the names of primary colors. A set of stamps stored in loose-leaf notebooks on a shelf is usually called a stamp collection. The set of past presidents of the United States consists of

$$\{\text{George Washington, John Adams, Thomas Jefferson, \ldots}\}$$

The "..." is called an ellipsis and indicates that the list contains other elements.

What is the basic characteristic of a set? For any set A and any element b, either b is in A or b is not in A. If you ask whether an element is in a set, the answer is either yes or no.

Is 0 in $\{1, 2\}$? No.
Is 0 in $\{0, 1, 2, 3, 4, 5, 100345679231\}$? Yes.
Is New York in $\{$Liverpool, London, Los Angeles $\}$? No.
Is green in $\{$red, yellow, blue$\}$? No.
Is New York in $\{$England, France, United States$\}$? No.

In mathematical terminology, 0 **is an element of** $\{0, 1, 2, 3, 4, 5, 100345679231\}$, and green **is not an element of** $\{$red, yellow, blue$\}$.

The expression "is an element of" is denoted by the symbol $\in$, a form of the Greek letter epsilon. For example, we write

$$0 \in \{0, 1, 2, 3, 4, 5, 100345679231\}$$

and

$$\text{green} \notin \{\text{red, blue, yellow}\}$$

The slash through the $\in$ symbol means *not,* just as it does in $\neq$. **Is a member of, is contained in,** or simply **is in** means the same as "is an element of." Mathematics, like ordinary language, is full of synonyms.

Despite its frequent use, the term *set* is not defined in terms of other concepts. Like the terms *point* and *line* in plane geometry, *set* is a primitive concept. Just assume there are elements, there are sets, and that for a set A and an element b, the assertion $b \in A$ is either true or false.

An important distinction needs to be made between 1 and $\{1\}$. They are not the same. By itself, 1 is a number but not a set, and $\{1\}$ is the set containing the element 1. Similarly, $\{1\}$ and $\{\{1\}\}$ are not the same thing: $1 \in \{1\}$, but $\{1\} \notin \{1\}$. So, also, $\{1\} \in \{\{1\}\}$, but $1 \notin \{\{1\}\}$. Similarly, the set $\{1, 2\}$ has two elements, 1 and 2, but $\{\{1, 2\}\}$ has only one element, $\{1, 2\}$.

The number of times an element is listed and the order in which elements are listed are both unimportant. For example, the elements of $\{2, 3\}$ are 2 and 3. The elements of $\{3, 2, 2\}$ are also 2 and 3. Consequently, these two sets contain the same elements. That is, these two sets are *equal,* as we shall see later.

1.1.1   Describing Sets Mathematically

We present three different methods to describe a set. The first is by a list of all the elements. The second is by a description of some property the elements have. The third is by a description based on some other sets. In all these methods, we use the symbols { and } to indicate that a set is being defined. The "language" used to describe sets is called **set-theoretic notation.**

Set-Theoretic Notation

Three methods to describe the set with elements 0, 1, 2, 3, 4, 5, 6, 7, 8, and 9 are:

1. List the elements in braces: $\{0, 1, 2, 3, 4, 5, 6, 7, 8, 9\}$. We can also abbreviate this list as $\{0, 1, 2, \ldots, 9\}$.
2. Describe the elements in terms of some property they satisfy:

$$\{x : x \text{ is an integer and } x > -1/2 \text{ and } x < 19/2\}$$

This notation is read as "the set of (all) x such that x is an integer and x is greater than minus one-half and x is less than nineteen halves." The colon is read as "such that." The description following the colon tells what property these x's have.
3. Describe the elements as the set of all elements *in some other set* that satisfy some property. Here, if $\mathbb{Z}$ denotes the set of integers, then the set can be defined as

$$\{x \in \mathbb{Z} : x > -1/2 \text{ and } x < 19/2\}$$

Methods 2 and 3 are almost the same. Method 3 is preferred, however, because in some really peculiar circumstances, method 2 can cause trouble.[1]

There is a particular disadvantage to using ellipses with method 1. If someone writes

$$A = \{0, 1, 2, \ldots, 7\}$$

it is assumed everyone will understand what is intended—that is, the set A contains the elements 0, 1, 2, 3, 4, 5, 6, and 7. Frequently, however, the intended pattern is not as obvious as the person using the ellipsis thinks. Suppose

$$A = \{2, 4, \ldots, 65536\}$$

What are the other elements? Guessing what was meant requires understanding the pattern that gives rise to the elements listed. Since $65536 = 2^{16}$, one conjecture might be

$$A = \{2^1, 2^2, 2^3, 2^4, 2^5, 2^6, 2^7, 2^8, 2^9, 2^{10}, 2^{11}, 2^{12}, 2^{13}, 2^{14}, 2^{15}, 2^{16}\}$$

It could just as well be conjectured that

$$A = \{2, 2^2, 2^{2^2}, 2^{2^{2^2}}\} = \{2, 4, 16, 65536\}$$

There are endless other possibilities, with no real way to choose among them. (Nobody said the pattern had to be simple.) This notation should *only* be used when it is obvious from the context exactly what is meant.

---

[1] After Cantor defined set theory, researchers found some paradoxes. The most famous is Russell's paradox, which is similar to the so-called "liar's paradox": "This sentence is a lie." Work through it: If it is false, then it is true, and if it is true, then it is false.

Russell's paradox is this. Let x be the set of all sets that are not elements of themselves. Now, is x an element of itself?

Work through it: If it is, then it is not, and if it is not, then it is. What's wrong? Most modern set theorists assert that using definition method 2 is at fault—note that Bertrand Russell (English mathematician and philosopher, 1872–1970) used that form in defining x. The set of all sets which are not elements of themselves is deemed "too big" to be a set. By using definition method 3, we avoid constructing sets which are "too big."

Because we are not going into axiomatic set theory, however, we will be unable to avoid method 2 entirely in this book.

We often list the elements of a set in a way that shows an obvious association between the natural numbers and the elements of the set. For example, $1, 2, 2^2, 2^3, 2^4, \ldots$. We call such a set a *sequence*. We can refer to a sequence by $a_0, a_1, a_2, \ldots$. In the above example, we have $a_0 = 1$, $a_1 = 2$, $a_2 = 2^2, \ldots$, $a_n = 2^n, \ldots$. The notion of a sequence will be examined more carefully in Section 4.4. At this time, we just need to have a way to refer to sets of this form.

Special Sets

There are special names for certain common sets of numbers. Some of them are listed here.

Special Sets

$\mathbb{N}$: the set of **natural numbers,** or the set of non-negative integers $\{0, 1, 2, 3, 4, \ldots\}$.

$\mathbb{Z}$: the set of **integers,** or $\{\ldots, -3, -2, -1, 0, 1, 2, 3, \ldots\}$.

$\mathbb{Q}$: the set of **rational numbers,** or the set of fractions of integers with nonzero denominator, such as $\frac{1}{3}$ or $\frac{2357}{9731}$.

$\mathbb{R}$: the set of **real numbers,** or the set of numbers written with a decimal point, such as $\pi = 3.14159\ldots$, -2.715, or $2.35353535\ldots$.

$\emptyset$: the **empty set,** or the set $\{\ \}$ with no elements.

In some circumstances, it is convenient to write the set of squares of natural numbers as $\{x^2 : x \in \mathbb{N}\}$ rather than as $\{x : x \in \mathbb{N} \text{ and for some } k \in \mathbb{N}, x = k^2\}$.

1.1.2   Set Membership

To prove an element is a member of a set, you must prove that the element shares the property that defines membership. For example, we can define the notion of a number being a prime without knowing that any particular number is a prime. We must then show that any number we think is a prime has the defining property. First, we need to know what a *divisor* is before we can define a prime. For integers m and n, we say m is a divisor of n, denoted as $m|n$, if there is a natural number k such that $n = m \cdot k$. A natural number p is prime if $p \neq 1$ and its only divisors are 1 and p. Let $P = \{n : n \in \mathbb{N} \text{ and } n \text{ is a prime}\}$. In Example 1, we will show that P is nonempty.

Example 1.   Prove that 3 is a prime—that is, that $3 \in P$.

Solution.   We must show 3 has the property that its only divisors are 1 and 3. Since the only other possibility is 2 and 2 does not divide 3, 3 is therefore a prime.   ■

A divisor of an integer is also called a **factor.**

1.1.3   Equality of Sets

In mathematics, precise language is important if we are all to understand the same meaning for a statement. For example, what does it mean for two sets to be equal?

Definition 1.   Let A and B be sets. Then, $A = B$ or A is **equal** to B if both A and B have the same elements.

The word *if* has a special meaning when used in definitions. Definition 1 states that $A = B$ if A and B have the same elements. Since the word *if* is inside a definition, it is implied that $A \neq B$ whenever the condition is not satisfied. Thus, "$A = B$" is just a short way to say "A and B have the same elements."

Example 2.

(a) $\{n \in \mathbb{Z} : n^2 - n - 2 = 0\} = \{n \in \mathbb{Z} : (n - 2)(n + 1) = 0\} = \{2, -1\}.$
(b) $\{n \in \mathbb{N} : n^2 - n - 2 = 0\} = \{2\}$ because $-1 \notin \mathbb{N}.$
(c) $\{x \in \mathbb{N} : (x + 1)^2 - (x - 1)^2 - 4x = 0\} = \mathbb{N}$ because

$$(x + 1)^2 - (x - 1)^2 - 4x = x^2 + 2x + 1 - x^2 + 2x - 1 - 4x$$
$$= 0$$

is an **algebraic identity** (true for all x).

The empty set $\emptyset$ can be described in several ways; for example,

$\{x \in \mathbb{N} : x < x\}.$
The set of continents south of Antarctica.
The set of round squares.

Why is $\{x \in \mathbb{N} : x < x\}$ equal to the set of round squares? We know that if two sets are equal, then they must have the same elements. So, if

$$\{x \in \mathbb{N} : x < x\} \neq \text{ the set of round squares}$$

then there is an element in one of these sets that is not in the other set. This cannot be true, however, since neither set has any elements at all.

1.1.4   Finite and Infinite Sets

Some sets, like $\{0, 1, 2, 3\}$, have the property that a person could list their elements and finish listing them. We can describe this condition a little more formally: Either the set has no elements, or its elements can be matched with the elements of some subset $\{1, 2, \ldots, n\}$ of the natural numbers. Such sets are called a **finite set.** So, {Liverpool, London, Los Angeles} is a finite set—that is, elements could be matched as

$$1 \text{ (Liverpool)}, 2 \text{ (London)}, \text{ and } 3 \text{ (Los Angeles)}$$

The empty set $\emptyset$ has zero elements, so it is also finite.

Some sets are **infinite sets,** or **not finite sets,** like $\mathbb{Z}$, $\mathbb{R}$, and $\mathbb{N}$. There is no way to match all the elements of $\mathbb{Z}$ with a set $\{1, 2, \ldots, n\}$ for any fixed n.

1.1.5   Relations Between Sets

Besides equality, another important relation between sets occurs when all the elements of one set are also elements of a second set.

Definition 2.   Let A and B be sets. A is a **subset** of B, written as $A \subseteq B$, if every element of A is also an element of B. A is a **proper subset** of B, written $A \subset B$, if $A \subseteq B$ but $A \neq B$.

"Is not a subset" is denoted with $\not\subseteq$, whereas "is not a proper subset" is denoted with $\not\subset$. For example, $\{1, 2, 3\} \not\subseteq \{1, 3, 4\}$, since $2 \in \{1, 2, 3\}$ and $2 \notin \{1, 3, 4\}$. Similarly, $\{1, 4\} \not\subset \{1, 2, 3\}$, since $4 \in \{1, 4\}$ and $4 \notin \{1, 2, 3\}$. Also, $\{1, 2, 3, 2, 1\} \subseteq \{1, 2, 3\}$, but $\{1, 2, 3, 2, 1\} \not\subset \{1, 2, 3\}$.

We now state formally two facts that follow immediately from the definitions.

Theorem 1.   Let A be a set.

(a) $A \subseteq A$.
(b) $\emptyset \subseteq A$.

Proof.

(a) To say that $A \subseteq A$, according to Definition 2, means that each element of A is an element of A, which is clearly true.
(b) Since $\emptyset$ has no elements, the statement "for every element x, if $x \in \emptyset$, then $x \in A$" cannot be false, because $\emptyset$ has no elements. In this case we say the statement is **vacuously true**. ∎

We use the filled box that appears at the end of a Proof for a theorem or the end of Solution for an example as a separator. In some instances, when an example includes a discussion, we also use this to separate the example from the following text.

The idea behind proving that one set is a subset of a second set involves proving that every element of the first set is an element of the second set. It would not be very convenient if each element of the first set had to have its own proof of membership in the second set. Example 3 uses a proof that each element of the first set is an element of the second set by simply proving the result for a completely arbitrary element of the first set. A completely arbitrary element is one that has no property to use in the proof except that it is a member of the first set. An "arbitrary element of a set" is a (hypothetical) element whose *only* property is that it belongs to that set. In mathematics, the phrase "let $x \in A$" means "x is my name for an arbitrary element of A." Assuming that we are dealing with a completely arbitrary element allows us to prove the membership of every element with a single proof.

Example 3.   Prove that the sets $A = \{2^1, 2^2, 2^3, 2^4, \ldots\}$ and $B = \{2, 4, 6, 8, \ldots\}$ satisfy $A \subseteq B$.

Solution.   An arbitrary element of A is of the form 2^i for some $i \in \{1, 2, 3, \ldots\}$. An arbitrary element of B is of the form $2 \cdot j$ for some $j \in \{1, 2, 3, \ldots\}$. Clearly, $2^i = 2 \cdot j$ for the integer $j = 2^{i-1}$. Since an arbitrary element of A is an element of B, we conclude $A \subseteq B$. ∎

If and Only If

Mathematical statements about how facts are related, including many mathematical theorems, are **implications**. For example, "if you eat your carrots, you will grow big and strong," or "if Sally is in the science lab, then she is doing her chemistry experiment," or "if $x > 1$, then $x^2 > x$" are all implications. An implication starts with a hypothesis that is assumed to be true and then uses various means to prove a conclusion. We denote an implication as $a \Rightarrow b$, where a is the hypothesis and b is the conclusion. Two implications are used in the standard mathematical expression **if and only if.** The statement

$$a \text{ if and only if } b$$

means that *if a is true then b is true* ($a \Rightarrow b$) and that *if b is true then a is true* ($b \Rightarrow a$). Equivalently, it means that *a* and *b* are either both true or both false.

In a proof of an *if and only if* statement, a proof of "if *a*, then *b*" is usually labeled ($\Rightarrow$), whereas a proof of "if *b*, then *a*" is usually labeled ($\Leftarrow$). The if and only if statement is often denoted by $\Leftrightarrow$. The arrow notation is used in Theorem 2.

Theorem 2.   Let *A* and *B* be sets. Then, $A = B$ **if and only if** $A \subseteq B$ **and** $B \subseteq A$.

(What the Proof entails:)   We must prove two things. The first that $A = B$ implies that $A \subseteq B$ and $B \subseteq A$. The second is that if $A \subseteq B$ and $B \subseteq A$, then $A = B$.

Proof.

($\Rightarrow$) Prove that if $A = B$, then $A \subseteq B$ and $B \subseteq A$. Suppose $A = B$. Then, $A \subseteq B$ and $B \subseteq A$ by Theorem 1.

($\Leftarrow$) Prove that if $A \subseteq B$ and $B \subseteq A$, then $A = B$. To prove this, begin by supposing that $A \subseteq B$ and $B \subseteq A$. Then, for any *x*, if $x \in A$, $x \in B$, since $A \subseteq B$. Furthermore, if $x \in B$, then $x \in A$, since $B \subseteq A$. Therefore, the sets *A* and *B* have the same elements. By Definition 1, $A = B$.   ∎

1.1.6   Venn Diagrams

In most discussions, attention is limited to elements and subsets of a fixed set. For example, elementary arithmetic is usually limited to elements and subsets of $\mathbb{Z}$ (the integers) or of $\mathbb{Q}$ (the rationals). In a study of some period of history, attention may be limited to the set of all persons living at that time. In computer science, it may be the set of all file names on a hard disk. Such sets are called **universal sets,** or **universes.** They are the "universes of discourse" for a time.

There is a very convenient type of diagram, called a **Venn diagram,** for illustrating set-theoretic relationships. Start with a rectangle, and let the points in the rectangle represent the elements of a universal set, as shown in Figure 1.1.

Figure 1.1   Venn diagram of a universal set *U*.

Subsets of the universal set are represented by circles or ovals in the rectangle, as shown in Figure 1.2. For example, suppose that *A*, *B*, and *C* are subsets of the universal set *U*. The region within the circle for *A* represent the elements of *A*, and similarly for *B* and *C*. Figure 1.2 shows *A*, *B*, and *C* where $A \subset B$, *A* and *C* have no elements in common, and *B* and *C* have elements in common but neither is a subset of the other.

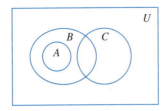

Figure 1.2 Sample Venn diagram.

Venn diagrams are frequently used to build intuition for proofs. The diagrams are designed to present fairly general pictures of what is known, and these pictures can often help a person to see set-theoretic relationships. A good Venn diagram can be very useful, but a Venn diagram itself is not a proof. In particular, if a mistake is made in drawing the Venn diagram, it is often possible to think that a property is true when it really is not. In especially complicated cases, it may be very difficult to see whether the picture is correct. The picture may be vague on certain points as well. For example, Figure 1.2 suggests that there are elements of B that are in neither A nor C. This may or may not be true. Nevertheless, a *good* Venn diagram is valuable both in suggesting whether a statement could be true and in motivating and illustrating the proof.

Theorem 3.   Let A, B, and C be sets. If $A \subseteq B$ and $B \subseteq C$, then $A \subseteq C$.

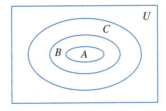

Figure 1.3 $A \subseteq B$ and $B \subseteq C$.

(The Venn diagram in Figure 1.3 is drawn so that $A \neq B$ and $B \neq C$, but this is not necessarily true. The Venn diagram suggests that if you start with an element of A, then that element is in B. Then, if the element is in B, it suggests that it is also in C. The proof will proceed using these two steps.)

Proof.   We must prove that if $A \subseteq B$ and $B \subseteq C$, then $A \subseteq C$. Let $x \in A$, and prove that $x \in C$. (Think of this as starting the proof by picking x arbitrarily from A.) Then, since $A \subseteq B$, $x \in B$. We now have $x \in B$, and we are given that $B \subseteq C$. So, it follows that $x \in C$. Since every element of A is an element of C, it follows that $A \subseteq C$. ∎

1.1.7   Templates

The proofs in this section use very typical techniques that you will see throughout the book. When you try to construct a proof, getting started is always a bit daunting. The templates shown here will give you ideas about what you need to do in a proof. The templates will describe what is needed to prove that an element is in a set, that one set is or is not a subset of another set, and that two sets are or are not equal. First, we state a template for proving that an element is a member of a set.

Template 1.1   Element Membership in a Set

Let A be a given set. To prove $x \in A$, show that x has the property that defines membership in A.

Example 4.   Let $A = \{n : n \in \mathbb{N}$ and $n = 3k + 5$ for some $k \in \mathbb{N}\}$. Is $23 \in A$?

Solution.   To show that $23 \in A$, we must find a natural number k_0 such that

$$23 = 3k_0 + 5$$

since every element of A has the form $3k + 5$ for some $k \in \mathbb{N}$. To find out if there is such a k, we simply solve the equation for k_0 and see if the solution is an integer.

$$3k_0 + 5 = 23$$
$$3k_0 = 23 - 5$$
$$k_0 = 18/3 = 6$$

Since 6 is a natural number, we know $3 \cdot 6 + 5 = 23 \in A$.   ∎

We use the template for element membership in a set to develop a template for proving that one set is a subset of another set.

Template 1.2   Set Inclusion

To prove that one set is a subset of another ($A \subseteq B$), show that every element x of A is also in B.

Example 5.   Let

$$A = \{n : n = 2k + 5 \text{ for some } k \in \mathbb{N}\}$$

and

$$B = \{n : n = 2j + 1 \text{ for some } j \in \mathbb{N}\}$$

Is $A \subseteq B$?

Solution.   By writing out a few of the elements in each of these sets, we can at least get an idea about whether we think $A \subseteq B$. The first six elements of A are 5, 7, 9, 11, 13, and 15. The first six elements of B are 1, 3, 5, 7, 9, and 11. The difference between the two sets seems to be the initial values. To show that $A \subseteq B$, we must take an arbitrary element of A, say, $n = 2k_0 + 5$ for some $k_0 \in \mathbb{N}$, and show that this can be written as $2j + 1$ for some

$j \in \mathbb{N}$, which would prove that $2k_0 + 5 = 2j + 1 \in B$. The algebra needed to see if this is possible involves solving for j in terms of k_0. This computation

$$2j + 1 = 2k_0 + 5$$
$$2j = 2k_0 + 5 - 1 = 2k_0 + 4$$
$$j = k_0 + 2$$

shows that for any k_0, the needed j is $k_0 + 2$, which is clearly an element of $\mathbb{N}$. This says that we can write an element of A as $2k_0 + 5 = 2(k_0 + 2) + 1$ for $k_0 + 2 \in \mathbb{N}$. Therefore, since an arbitrary element of A is an element of B, we have $A \subseteq B$. ∎

If the condition in Template 1.2 is not satisfied—that is, for two sets A and B, we have $A \nsubseteq B$—this means that there is an element in A that is not an element of B. We summarize this observation in Template 1.3.

Template 1.3 Set Non-Inclusion

To prove that one set is not a subset of another ($A \nsubseteq B$), show that some element x of A is not in B.

Example 6. Let

$$A = \{n \in \mathbb{N} : n = 2k^2 - 3 \text{ for some } k \in \mathbb{N}\}$$

and

$$B = \{n : n \in \mathbb{N} \text{ and } n = j^2 + 3 \text{ for some } j \in \mathbb{N}\}$$

Prove that $A \nsubseteq B$.

Solution. We must find some element of A that is not an element of B. If we list the first few elements of A and B, perhaps a candidate element will appear. $A = \{-3, -1, 5, 15, 47, \ldots\}$, and $B = \{3, 4, 7, 12, 19, \ldots\}$. An obvious candidate is -3, since $-3 \in A$ and $-3 \notin B$. We need to show that there is some fixed integer of the form $2k_0 - 3$, for $k_0 \in \mathbb{N}$, that can be written as $j^2 + 3$ for some choice of $j \in \mathbb{N}$. If such a j existed, we would have $2k_0 - 3 = j^2 + 3$. In the case $k_0 = 0$, the element j would have to satisfy $-3 = j^2 + 3$ or $j^2 + 6 = 0$ for -3 to be in B. Since no such j exists, $-3 \notin B$. ∎

One last possibility for set inclusion: One set can be a subset of a second subset, but not every element of the first set need be an element of the second set. This proper inclusion is formalized in the next template.

Template 1.4 Proper Set Inclusion

To prove that one set is a proper subset of another ($A \subset B$), first prove that $A \subseteq B$, and then show that some element x of B is not in A.

Example 7. Let

$$A = \{n \in \mathbb{N} : n \geq 2 \text{ and } n = 4j - 5 \text{ for some } j \in \mathbb{N}\}$$

and

$$B = \{n \in \mathbb{N} : n \geq 0 \text{ and } n = 2k + 1 \text{ for some } k \in \mathbb{N}\}$$

Prove that $A \subset B$.

Solution. To show that $A \subseteq B$, we must show that every element of A is an element of B. Let $n = 4j_0 - 5$ be an arbitrary element of A for some fixed $j_0 \in \mathbb{N}$. To show that $n \in B$, we must show that $n = 2k + 1$ for some $k \in \mathbb{N}$. We see if this is possible by solving for k:

$$2k + 1 = 4j_0 - 5$$
$$2k = 4j_0 - 6$$
$$k = 2j_0 - 3$$

Now, $2j_0 - 3 \geq 0$, since $j_0 \geq 2$. Since $2(2j_0 - 3) + 1 = 4j_0 - 5$, every element of A is an element of B, and $A \subseteq B$.

For $0 \in \mathbb{N}$, $2 \cdot 0 + 1 = 1 \in B$. If $1 \in A$, then $1 = 4j - 5$ for some $j \in \mathbb{N}$. By solving for j, we find that j must be equal to 3/2, which is not a natural number. Therefore, no j exists. Therefore, $1 \in B$, and $1 \notin A$. It follows that $A \subset B$.   ∎

The next step is to deal with set equality and set inequality.

In the case that we have two set descriptions and want to know that the sets are equal, there are actually two things to prove. The proof that two sets are equal follows Template 1.5.

Template 1.5   Set Equality

To prove that $A = B$ for sets A and B, prove that $A \subseteq B$ and $B \subseteq A$.

Example 8. Let

$$A = \{n : n = 2j \text{ for some } j \in \mathbb{N}\}$$

and

$$B = \{n : n = 2k + 2 \text{ for some } k \in \mathbb{Z} \text{ and } k \geq -1\}$$

Prove that $A = B$.

Solution. To show that an arbitrary element of A, say, $2j_0$ for some $j_0 \in \mathbb{N}$, is an element of B, we must find a $k \in \mathbb{Z}$ such that $k \geq -1$ and $2j_0 = 2k + 2$. Solving for k gives $k = j_0 - 1$. Since $j_0 \geq 0$, $k = j_0 - 1 \geq -1$, and $2k + 2 \in B$. To show that an arbitrary element of B, say, $2k_0 + 2$ for some $k_0 \in \mathbb{Z}$ and $k_0 \geq -1$, is an element of A, we must find a $j \in \mathbb{N}$ such that $2j = 2k_0 + 2$. This implies that j must satisfy $j = k_0 + 1$ if $2k_0 + 2$ is an element of A. Since $k_0 \geq -1$, we have $k_0 + 1 \geq 0$, and $k_0 + 1$ defines an element of A. Because $A \subseteq B$ and $B \subseteq A$, it follows that $A = B$. ■

There are often many different descriptions for a single set. One problem is to make sure that the set description that is given in fact describes the set intended. The idea of a proof to show that two sets are not equal is given in the next template.

Template 1.6   Set Inequality

To prove that $A \neq B$ for sets A and B, prove that $A \nsubseteq B$ or $B \nsubseteq A$.

The way to show that $A \subseteq B$ is false is to find an element x such that $x \in A$ and $x \notin B$. Showing that $B \subseteq A$ is false is done analogously, but only one of these implications needs to be shown to prove that $A \neq B$.

Example 9.   Let

$$A = \{n : n \in \mathbb{N} \text{ and } n = 4j^2 - 3 \text{ for some } j \in \mathbb{N}\}$$

and

$$B = \{n \in \mathbb{N} : n = 2k^2 - 3 \text{ for some } k \in \mathbb{N}\}$$

Prove that $A \neq B$.

Solution.   To show that $A \neq B$, it suffices to find an element in A that is not an element of B. We will show that 1 is such an element.

We can write $1 = 4(1)^2 - 3$. Therefore, $1 \in A$. If $1 \in B$, then $1 = 2k^2 - 3$ for some $k \in \mathbb{N}$. If this were so, then the element k would satisfy the equation $k^2 = 2$. Since no such k exists in $\mathbb{N}$, $1 \notin B$. ■

Other Proofs

In Theorem 2 (Section 1.1.5) we needed to prove two things to conclude the result. This kind of a proof is typical when you are trying to prove that two statements are simply different ways of saying the same thing. In Theorem 2 we found that set equality could be stated either in terms of element membership or in terms of the subset relation. The proof is formalized in Template 1.7.

> ### Template 1.7   Implications and If and Only If
>
> To prove "if a, then b" and "a if and only if b" results, use one of the two forms:
>
> **Form 1:** To prove "if a, then b," assume a and derive b.
>
> **Form 2:** To prove "a if and only if b," prove "if a, then b," and then prove "if b, then a."
>
> In a proof of an if and only if statement, a proof of "if a, then b" is usually labeled ($\Rightarrow$), whereas a proof of "if b, then a" is usually labeled ($\Leftarrow$). The if and only if statement is often written using $\Leftrightarrow$.

1.2   Exercises

1. Let X be the set of all students at a university. Let A be the set of students who are first-year students, B the set of students who are second-year students, C the set of students who are in a discrete mathematics course, D the set of students who are international relations majors, E the set of students who went to a concert on Monday night, and F the set of students who studied until 2 AM on Tuesday. Express in set theoretic notation the following sets of students:

   (a) All second-year students in the discrete mathematics course.

   Sample Solution. $\{x \in X : x \in B \text{ and } x \in C\}$.

   (b) All first-year students who studied until 2 AM on Tuesday.
   (c) All students who are international relations majors and went to the concert on Monday night.
   (d) All students who studied until 2 AM on Tuesday, are second-year students, and are not international relations majors.
   (e) All first- and second-year students who did not go to the concert on Monday night but are international relations majors.
   (f) All students who are first-year international relations majors or who studied until 2 AM on Tuesday.
   (g) All students who are first- or second-year students who went to a concert on Monday night.
   (h) All first-year students who are international relations majors or went to a concert on Monday night.

2. Find at least two different ways to fill in the ellipses in the set descriptions given. For example, $\{2, 4, \ldots, 12\}$ could be written either $\{2n : 1 \le n \le 6 \text{ and } n \in \mathbb{N}\}$ or $\{n + 1 : n \in \{1, 3, 5, 7, 11\}\}$.

   (a) $\{1, 3, \ldots, 31\}$
   (b) $\{1, 2, \ldots, 26\}$
   (c) $\{2, 5, \ldots, 32\}$

3. Write three descriptions of the elements of the set $\{2, 5, 8, 11, 14\}$.

4. How many elements does each of the following sets have?

   (a) $A = \emptyset$
   (b) $B = \{\emptyset\}$
   (c) $C = \{\{0, 1\}, \{1, 2\}\}$
   (d) $D = \{0, 1, 2, \{0, 1\}, \{1, 2\}, \{0, 1, 2\}, A\}$
   (e) $E = \{0, \{\{1, \{3, 5\}, \{4, 5, 7\}, 8\}\}\}$

5. Which of the following pairs of sets are equal? For each pair that is unequal, find an element that is in one but is not in the other.

   (a) $\{0, 1, 2\}$ and $\{0, 0, 1, 2, 2, 1\}$
   (b) $\{0, 1, 3, \{1, 2\}\}$ and $\{0, 1, 2, \{2, 3\}\}$
   (c) $\{\{1, 3, 5\}, \{2, 4, 6\}, \{5, 5, 1, 3\}\}$ and $\{\{3, 5, 1\}, \{6, 4, 4, 4, 2\}, \{2, 4, 4, 2, 6\}\}$
   (d) $\{\{5, 3, 5, 1, 5\}, \{2, 4, 6\}, \{5, 1, 3, 3\}\}$ and $\{\{1, 3, 5, 1\}, \{6, 4, 2\}, \{6, 6, 4, 4, 6\}\}$
   (e) $\emptyset$ and $\{x \in \mathbb{N} : x > 1 \text{ and } x^2 = x\}$
   (f) $\emptyset$ and $\{\emptyset\}$

6. This problem concerns the following six sets:

$$A = \{0, 2, 4, 6\} \quad B = \{1, 3, 5\} \quad C = \{0, 1, 2, 3, 4, 5, 6, 7\}$$
$$D = \emptyset \quad E = \mathbb{N} \quad F = \{\{0, 2, 4, 6\}\}$$

   (a) What sets are subsets of A?
   (b) What sets are subsets of B?
   (c) What sets are subsets of C?
   (d) What sets are subsets of D?
   (e) What sets are subsets of E?
   (f) What sets are subsets of F?

7. Let $A = \{n : n \in \mathbb{N} \text{ and } n = 2k + 1 \text{ for some } k \in \mathbb{N}\}$, $B = \{n : n \in \mathbb{N} \text{ and } n = 4k + 1 \text{ for some } k \in \mathbb{N}\}$, and $C = \{m \in \mathbb{N} : m = 2k - 1 \text{ and } k \in \mathbb{N} \text{ and } k \geq 1\}$. Prove the following:

   (a) $35 \in A$
   (b) $35 \in C$
   (c) $35 \notin B$
   (d) $A = C$
   (e) $B \subseteq A$
   (f) $B \subseteq C$
   (g) $B \subset A$
   (h) $B \subset C$

8. Let $A = \{n : n \in \mathbb{N} \text{ and } n = 3k + 2 \text{ for some } k \in \mathbb{N}\}$, $B = \{n : n \in \mathbb{N} \text{ and } n = 5k - 1 \text{ for some } k \in \mathbb{N} \text{ such that } k \geq 5\}$, and $C = \{m \in \mathbb{N} : m = 6k - 4 \text{ and } k \in \mathbb{N} \text{ and } k \geq 1\}$. Prove the following:

   (a) $C \subseteq A$
   (b) $A \neq B$
   (c) $B \neq C$
   (d) $A \neq C$
   (e) $C \subset A$

9. Describe in words the difference between $\emptyset$ and $\{\emptyset\}$.
10. Let A, B, and C be sets.

    (a) Prove that if $A \subset B$ and $B \subseteq C$, then $A \subset C$.
    (b) Prove that if $A \subseteq B$ and $B \subset C$, then $A \subset C$.
    (c) Prove that if $A \subseteq B$ and $A \not\subseteq C$, then $B \not\subseteq C$.

1.3   Operations on Sets

In many areas of computer science and mathematics, from formal logic to object-oriented programming, the operations to be performed must be considered in the context of a specific set. For example, familiar operations, such as addition, subtraction, multiplication, and division, are performed within a specific set of numbers, such as the integers, rationals, or reals. This section discusses operations on sets and introduces the most common operations: union, intersection, difference, complement, product, and power set of a set. We study the laws these operations satisfy as well as how they interact with one another. We then extend our discussion to lattices and boolean algebras. Lattices and boolean algebras have two operations defined on their elements such that a set of special axioms for these operations holds. An example of a lattice is a family of sets with the operations defined as set union and set intersection.

1.3.1   Union and Intersection

The two simplest operations on sets involve combining two sets into one (union) and finding common elements in two sets (intersection). These operations obey many of the general rules that addition and multiplication with real numbers also satisfy. The first operation consists of combining two sets into a set containing the elements of both sets.

Definition 1.   Let A and B be sets. The **union** of A and B, denoted $A \cup B$, is

$$\{x : x \in A \text{ or } x \in B\}$$

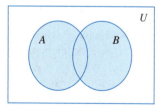

Figure 1.4 $A \cup B$.

    The Venn diagram for set union (shown in Figure 1.4) illustrates what was stated in the definition. We do, however, need to clarify the meaning of the word *or* in the definition. When mathematicians say $x \in A$ *or* $x \in B$, they generally mean $x \in A$ or $x \in B$ or *both*. This interpretation is called the **inclusive or** because it includes the possibility that both may be true.

Example 1.

(a) $\{1, 2, 3\} \cup \{3, 4, 5\} = \{1, 2, 3, 3, 4, 5\} = \{1, 2, 3, 4, 5\}$.
(b) $\{1, 2, \{1, 2, 3\}\} \cup \{1, 2, 3, \{1, 2\}\} = \{1, 2, 3, \{1, 2\}, \{1, 2, 3\}\}$.
(c) $\mathbb{N} \cup \mathbb{Z} = \mathbb{Z}$.
(d) For any set A, $A \cup \emptyset = A$.

Why was the definition of the union of three or more sets not given? The more general union operation, for any finite number of sets, is handled by the assumption that $A \cup B \cup C$ means $(A \cup B) \cup C$. Therefore, it is only necessary to find unions of two sets at a time, and that has already been defined. The shaded region in Figure 1.5 shows the union of three sets.

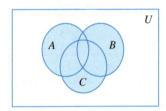

Figure 1.5 $A \cup B \cup C$.

The next theorem proves some fundamental results about set union.

Theorem 1. Let A, B, and C be sets. Then:

(a) $A \cup A = A$.
(b) $A \subseteq A \cup B$ and $B \subseteq A \cup B$.
(c) $A \cup B = B \cup A$.   **(Commutative Law for Union)**
(d) $A \cup (B \cup C) = (A \cup B) \cup C$.   **(Associative Law for Union)**

(What the proof entails.)   Parts (a) and (b) follow directly from the definition of union. The proof of (c) will be given. Since the proof of (d) uses an argument similar to the one used in (c), it will be left as an exercise for the reader. Part (c) says that the order in which the union of two sets is formed does not matter. Part (d) states that $A \cup B \cup C$ makes sense even without parentheses.

Proof.   (c) Follow the template for set equality to prove that (i) $A \cup B \subseteq B \cup A$ and (ii) $B \cup A \subseteq A \cup B$. For (i), use the template for set inclusion to prove that for any $x \in A \cup B$, it follows that $x \in B \cup A$.

Suppose that $x \in A \cup B$. Then (ia) $x \in A$ or (ib) $x \in B$. In case (ia), since $x \in A$, by Definition 1 we have $x \in B \cup A$. In case (ib), since $x \in B$, by Definition 1 we have $x \in B \cup A$. This completes the proof of (i).

The proof of (ii) is analogous.   ∎

What do we mean when we say that one proof is **analogous** to another? In this context, it means that the two proofs have essentially the same logic. Here, for example, one can form the proof of part (ii) from the proof of part (i) by interchanging A and B.

The second important set operation, intersection, forms a set from the elements common to two sets.

Definition 2.   Let A and B be sets. The **intersection** of A and B, denoted by $A \cap B$, is

$$\{x : x \in A \text{ and } x \in B\}$$

The intersection of A and B is shaded in Figure 1.6.

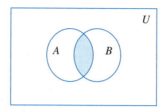

Figure 1.6 $A \cap B$.

Example 2.

(a) $\{1, 2, 3\} \cap \{3, 4, 5\} = \{3\}$.
(b) $\{1, 2, 3\} \cap \{4, 5, 6\} = \emptyset$.
(c) $\mathbb{N} \cap \mathbb{Z} = \mathbb{N}$.
(d) For any set A, $A \cap \emptyset = \emptyset$.
(e) $\{1, 2, 3\} \cap \{\{1, 2, 3\}\} = \emptyset$. (The first set has three elements, 1, 2, and 3, whereas the second set has only one element, $\{1, 2, 3\}$.)

Theorem 2 proves some fundamental results about set intersection. Like set union, set intersection satisfies the commutative and associative laws.

Theorem 2.   Let A, B, and C be sets.

(a) $A \cap A = A$.
(b) $A \cap B \subseteq A$ and $A \cap B \subseteq B$.
(c) $A \cap B = B \cap A$.   (**Commutative Law for Intersection**)
(d) $A \cap (B \cap C) = (A \cap B) \cap C$.   (**Associative Law for Intersection**)

(What the proof entails.)   Parts (a) and (b) follow directly from the definition of intersection. Part (c) says that the order in which the intersection of two sets is formed does not matter. Part (d) states that $A \cap B \cap C$ makes sense even without parentheses.

Proof.   (c) Again, follow Template 1.5 (Set Equality). Prove that (i) $A \cap B \subseteq B \cap A$ and (ii) $B \cap A \subseteq A \cap B$. For (i), follow the template for proving one set is a subset of another. That is, assume $x \in A \cap B$, and show $x \in B \cap A$.

Suppose $x \in A \cap B$. Then, $x \in A$ and $x \in B$. Equivalently, $x \in B$ and $x \in A$, since no order is implied by the word *and*. Therefore, $x \in B \cap A$. The proof of (ii) is analogous.
(d) This part is left as an exercise for the reader.   ■

The distributive laws for addition and multiplication for real numbers have analogues with the operations of union and intersection with sets, as Theorem 3 shows.

Theorem 3.   (**Set Distributivity**)   Let A, B, and C be sets. Then:

(a) $A \cup (B \cap C) = (A \cup B) \cap (A \cup C)$.   (**Distributive Law for Union**)
(b) $A \cap (B \cup C) = (A \cap B) \cup (A \cap C)$.   (**Distributive Law for Intersection**)

Proof. The proofs are left as an exercise for the reader. ∎

The intersection of two sets may not contain any elements. If there are no elements in a set intersection, we call the sets disjoint.

Definition 3. Let A and B be sets. Then, A and B are **disjoint sets** if $A \cap B = \emptyset$.

Example 3.

(a) Verify that $\{1, 2, 3\}$ and $\{4, 5, 6\}$ are disjoint.
(b) Verify that $\{1, 2, 3\}$ and $\{\{1, 2, 3\}\}$ are disjoint.
(c) For any set A, verify that A and $\emptyset$ are disjoint.

The reader may be asking whether there is any reason or need to prove additional theorems about the union and intersection operations on sets. There are two reasons to prove additional theorems. First (and most obviously), the results will be needed later. Second, proofs of these results are fairly easy examples of proofs, and they provide good models for constructing other proofs. Additional opportunities to write proofs will be given in the exercises.

Theorem 4 shows how set inclusion and the operations of union and intersection are related.

Theorem 4. Let A, B, and C be sets. Then:

(a) If $A \subseteq B$ or $A \subseteq C$, then $A \subseteq B \cup C$.
(b) If $B \subseteq A$ and $C \subseteq A$, then $B \cup C \subseteq A$.
(c) If $A \subseteq B$ and $A \subseteq C$, then $A \subseteq B \cap C$.
(d) If $B \subseteq A$ or $C \subseteq A$, then $B \cap C \subseteq A$.

(Motivation for the proof.) For part (a), there are two cases: (i) $A \subseteq B$, and (ii) $A \subseteq C$. The Venn diagrams (see Figure 1.7) illustrate both parts of (a) and will help in understanding the proof.

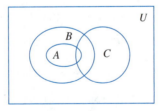 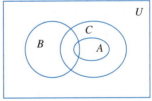

Figure 1.7 A. $A \subseteq B \Rightarrow A \subseteq B \cup C$. B. $A \subseteq C \Rightarrow A \subseteq B \cup C$.

For part (b), one Venn diagram suffices (see Figure 1.8).

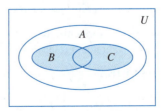

Figure 1.8 $B \subseteq A$ and $C \subseteq A \Rightarrow B \cup C \subseteq A$.

Proof.   (a) Suppose that $A \subseteq B$ or $A \subseteq C$.

Case 1: $A \subseteq B$. Follow Template 1.2 (Set Inclusion) for proving that one set is a subset of another. Show that every element of A is also an element of $B \cup C$.

Let $x \in A$. The goal is to show that $x \in B \cup C$. Since $x \in A$ and $A \subseteq B$, we have $x \in B$. But,

$$B \cup C = \{x : x \in B \text{ or } x \in C\}$$

Therefore, $x \in B \cup C$.

Case 2: $A \subseteq C$. The proof is analogous to that in part (a).
(b) Suppose $x \in B \cup C$. Then, either $x \in B$ or $x \in C$.

Case 1: $x \in B$. Since $B \subseteq A$, it follows that $x \in A$.

Case 2: $x \in C$. Since $C \subseteq A$, it follows that $x \in A$. Therefore, $B \cup C \subseteq A$.
(c)–(d) Exercises for the reader.   ∎

The proof of Theorem 4 shows how a template can be used. It also demonstrates another proof technique: **proof by cases.** The assumption was that $A \subseteq B$ or $A \subseteq C$. The proof breaks down into the two ways that this could happen: (1) $A \subseteq B$, or (2) $A \subseteq C$. Each case was handled separately. This is a general approach: If there are relatively few ways that some assumption can be met, then one can handle them separately.

Let's review what Theorem 4 asserts. Contrast the following two statements:

(a) "If A is a subset of both B and C, then A is a subset of their union."

$$\text{If } A \subseteq B \text{ or } A \subseteq C, \text{ then } A \subseteq B \cup C$$

(b) "If A is a subset of the union of B and C, then A is a subset of either B or C."

$$\text{If } A \subseteq B \cup C, \text{ then } A \subseteq B \text{ or } A \subseteq C$$

Theorem 4 asserts that statement (a) is true. Theorem 4 does not assert statement (b). In fact, statement (b) is false in general. How would it be shown that statement (b) is false? Statement (b) asserts that some relationship is true for *all* sets A, B, and C. To prove it to be false, then, we must find just *one* example where it is false. That is, we must find three sets A, B, and C such that $A \subseteq B \cup C$ but (i) $A \not\subseteq B$ and (ii) $A \not\subseteq C$ (see Exercise 9 in Section 1.4).

The theorems proved so far can easily be used to show something that is not entirely obvious.

Theorem 5.   (An Absorption Law)   Let A and B be sets. Then,

$$A \cup (A \cap B) = A$$

Proof.   As usual, prove that $A \cup (A \cap B) \subseteq A$ and $A \subseteq A \cup (A \cap B)$. For the first part, we have $A \subseteq A$ by Theorem 1 in Section 1.1.5 and $A \cap B \subseteq A$ by Theorem 2(b) in Secton 1.3.1. These two conditions imply that $A \cup (A \cap B) \subseteq A$ by Theorem 4(b) in Section 1.3.1. For the second part, start with $A \subseteq A$, which gives $A \subseteq A \cup (A \cap B)$ by Theorem 4(a) in Section 1.3.1.   ∎

This result is one that is needed later. When we discuss boolean algebras and their relation to electrical circuits, this **Absorption Law** is particularly useful.

Generalized Unions and Intersections

The definitions of union and intersection make good sense for any finite number of sets, because the operations are associative. There are occasions, however, when one would like to express the idea of the union of an infinite collection of sets. This leads to the generalization of the notion of set union and intersection given in the next definition.

Definition 4.   Let $\mathcal{X}$ be a *set of sets*. Then,

$$\cup\mathcal{X} = \{x : x \text{ is contained in some set in } \mathcal{X}\}$$

and

$$\cap\mathcal{X} = \{x : x \text{ is contained in every set in } \mathcal{X}\}$$

If

$$\mathcal{X} = \{X_0, X_1, \ldots, X_n, \ldots\}$$

That is, the elements of $\mathcal{X}$ are indexed with the natural numbers, the union of sets $\cup\mathcal{X}$ is usually written as

$$\cup_{i=0}^{\infty} X_i = X_0 \cup X_1 \cup X_2 \cup \cdots \cup X_n \cup \cdots$$

and the *intersection of sets* $\cap\mathcal{X}$ is usually written as

$$\cap_{i=0}^{\infty} X_i = X_0 \cap X_1 \cap X_2 \cap \cdots \cap X_n \cap \cdots$$

Example 4.

(a) $U_i = (-1/i, 1/i) \subseteq \mathbb{R}$ where $i \in \mathbb{N} - \{0\}$. Then, $\cup_{i=1}^{\infty} U_i = (-1, 1)$, $\cap_{i=1}^{\infty} U_i = \{0\}$.

(b) $V_i = (i, i+2) \subseteq \mathbb{R}$, where $i \in \mathbb{N} - \{0\}$. Then, $\cup_{i=1}^{\infty} V_i = (1, \infty)$, $\cap_{i=-\infty}^{\infty} V_i = \emptyset$.

1.3.2   Set Difference, Complements, and DeMorgan's Laws

In Venn diagrams, pairs of sets are often drawn so that it appears as if there are elements in each of the sets that are not in the other set. Often, it is important to find these elements. This operation on sets is called *set difference*. In other instances, we are interested in the elements that are not in a set. The operation of finding these elements is called **complementation.** Finally, we would like to understand how union and intersection interact with the operation of set difference and complementation. The relationships are described by **DeMorgan's Laws.** We start this section by defining set difference.

Definition 5.   Let A and B be sets. The **set difference** of A and B, denoted $A - B$, is

$$\{x : x \in A \text{ and } x \notin B\}$$

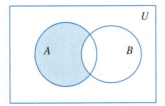

Figure 1.9 $A - B$.

Example 5.

(a) Let $A = \{1, 2, \ldots, 10\}$ and $B = \{3, 5, 7, 9\}$. Then, $A - B = \{1, 2, 4, 6, 8, 10\}$.
(b) Let $A = \mathbb{N}$ and $B = \{2i : i \in \mathbb{N}\}$. Then, $A - B = \{2i + 1 : i \in \mathbb{N}\}$.

The difference $A - B$ is also sometimes call the **relative difference.** The Venn diagram (shown in Figure 1.9) gives an intuitive understanding of this notion. Remember that Venn diagrams suggest relations between or among sets but are not actually proofs of relationships between or among sets. Theorem 6 proves some key relationships involving the difference of two sets, $A - B$ and $B - A$.

Theorem 6.   Let A and B be sets. Then:

(a) $A - B$ and $B - A$ are disjoint, $A - B$ and $A \cap B$ are disjoint, and $A \cap B$ and $B - A$ are disjoint.
(b) $A = (A - B) \cup (A \cap B)$.
(c) $A \cup B = (A - B) \cup (A \cap B) \cup (B - A)$.
(d) $A \subseteq B$ if and only if $A - B = \emptyset$.

Proof.   If you look at a Venn diagram for two sets and identify $A - B$, $B - A$, and $A \cap B$, it looks like the sets are disjoint. This theorem says that your intuition from the diagram is correct. The proofs of (a)–(d) are left as exercises for the reader.   ■

Complement of a Set

Recall that a *universal set* is a set that contains as a subset every set currently being discussed. In a context in which there is a universal set, another set theoretic operation can be defined.

Definition 6.   Let U be a universal set and A be a subset of U. The **complement** of A, denoted $\overline{A}$, is

$$\{x : x \in U \text{ and } x \notin A\}$$

Sometimes, to emphasize that U is a universal set, $\overline{A}$ is also called the **absolute difference.**

With this definition, we can restate Definition 5 as $A - B = A \cap \overline{B}$. Some important identities concern complements, especially how they interact with other set-theoretic operations.

Theorem 7.   Let U be a universal set and A and B be subsets of U. Then:

(a) $\overline{\overline{A}} = A$.   ($\overline{\overline{A}}$ is the complement of $\overline{A}$.)
(b) $A \subseteq B$ if and only if $\overline{B} \subseteq \overline{A}$.
(c) $A = B$ if and only if $\overline{A} = \overline{B}$.

(What the proof entails.)   Part (a) tells us that the complement only produces something new the first time it is applied. Part (b) says that if A is a subset of B, then set inclusion goes the other way for the complements; that is, the complement of B is contained in the complement of A. In part (c), we prove that if two sets are equal, then their complements are equal.

Proof.   (a) Show that (i) $\overline{\overline{A}} \subseteq A$ and (ii) $A \subseteq \overline{\overline{A}}$. To prove (i), suppose $x \in \overline{\overline{A}}$. Then, $x \in U$ and $x \notin \overline{A}$. But, then $x \in A$. To prove (ii), suppose $x \in A$. Then, $x \in U$, but $x \notin \overline{A}$. So, $x \in \overline{\overline{A}}$.

(b) ($\Rightarrow$) Show that if $A \subseteq B$, then $\overline{B} \subseteq \overline{A}$. Prove the result by contradiction (see Template 1.3). Assume that for *some* subsets A and B of U, $A \subseteq B$ and $\overline{B} \not\subseteq \overline{A}$, and derive a contradiction. Since $\overline{B} \not\subseteq \overline{A}$, there is some $x \in \overline{B} - \overline{A}$. Pick such an x. Since $x \notin \overline{A}$, it follows that $x \in A$. Since $A \subseteq B$, we have $x \in B$. But, it was assumed that $x \in \overline{B}$; hence, x has the property that $x \notin B$. Since both $x \in B$ and $x \notin B$ were proved, this gives a contradiction.
($\Leftarrow$) The proof is analogous to the proof of ($\Rightarrow$) using (a).
(c) Exercise 11 in Section 1.4. ■

A Computer Representation for Sets

Let $U = \{1, 2, 3, 4, 5, 6\}$ be a set, and let $X \subseteq U$. A **bit representation** for X is a six-digit binary number $x_1x_2x_3x_4x_5x_6$ with bit x_i for $1 \le i \le 6$ defined as

$$x_i = \begin{cases} 1 & \text{if } i \in X \\ 0 & \text{for } i \notin X \end{cases}$$

For example, if $B = \{2, 3, 6\}$, then $B = 011001$. The operations of union, intersection, and complement can be carried out using operators *UNION, INTER, DIFF,* and *COMP* that operate on binary numbers bit-by-bit. Let $B, C \subseteq U$ with $B = b_1b_2b_3b_4b_5b_6$ and $C = c_1c_2c_3c_4c_5c_6$. Define the union as $UNION(B, C) = x_1x_2x_3x_4x_5x_6$ where for $1 \le i \le 6$,

$$x_i = \begin{cases} 1 & \text{if } b_i = 1 \text{ or } c_i = 1 \\ 0 & \text{otherwise} \end{cases}$$

Define the intersection as $INTER(B, C) = x_1x_2x_3x_4x_5x_6$, where for $1 \le i \le 6$,

$$x_i = \begin{cases} 1 & \text{if } b_i = 1 \text{ and } c_i = 1 \\ 0 & \text{otherwise} \end{cases}$$

Define the complement as $COMP(B) = x_1x_2x_3x_4x_5x_6$, where for $1 \le i \le 6$,

$$x_i = \begin{cases} 1 & \text{if } b_i = 0 \\ 0 & \text{otherwise} \end{cases}$$

Define the relative difference as $DIFF(B, C) = x_1x_2x_3x_4x_5x_6$, where for $1 \le i \le 6$,

$$x_i = \begin{cases} 1 & \text{if } b_i = 1 \text{ and } c_i = 0 \\ 0 & \text{otherwise} \end{cases}$$

Example 6.   Let $B = \{1, 2, 3, 4, 5\}$ and $C = \{3, 4, 5, 6, 7, 8\}$ be subsets of the universal set $U = \{1, 2, \ldots, 9\}$. Find $UNION(B, C)$, $INTER(B, C)$, $COMP(C)$, and $DIFF(B, C)$.

Solution. $B \cup C = \{1, 2, 3, 4, 5, 6, 7, 8\}$. $B \cap C = \{3, 4, 5\}$. $\overline{C} = \{1, 2, 9\}$. $B - C = \{1, 2\}$. Therefore,

$UNION(B, C) = 111111110$
$INTER(B, C) = 001110000$
$COMP(C) = 110000001$
$DIFF(B, C) = 110000000$ ■

DeMorgan's Laws

DeMorgan's Laws are among the most important and useful results about sets. These laws describe how union, intersection, and complement are related. Figure 1.10 indicates what the laws tell us.

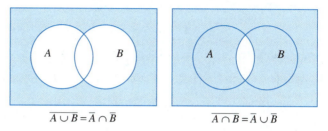

$$\overline{A \cup B} = \overline{A} \cap \overline{B} \qquad\qquad \overline{A \cap B} = \overline{A} \cup \overline{B}$$

Figure 1.10   DeMorgan's Laws.

Theorem 8.   (DeMorgan's Laws)   Let U be a universal set, and let A and B be subsets of U. Then:

(a) $\overline{(A \cup B)} = \overline{A} \cap \overline{B}$.   **(DeMorgan's Law for Union)**
(b) $\overline{(A \cap B)} = \overline{A} \cup \overline{B}$.   **(DeMorgan's Law for Intersection)**

Proof.
(a) Show that (i) $\overline{(A \cup B)} \subseteq \overline{A} \cap \overline{B}$ and (ii) $\overline{A} \cap \overline{B} \subseteq \overline{(A \cup B)}$.
   (i) Pick an arbitrary $x \in \overline{(A \cup B)}$. Since $x \in U - (A \cup B)$, it follows that $x \notin A \cup B$. For x not to be in this union means it may not be in either of the sets. So, $x \notin A$ and $x \notin B$. Hence, since $x \in U - A = \overline{A}$ and $x \in U - B = \overline{B}$, it follows that $x \in \overline{A} \cap \overline{B}$.
   (ii) Pick an arbitrary $x \in \overline{A} \cap \overline{B}$. Then, $x \in \overline{A}$, so $x \notin A$. Also, $x \in \overline{B}$, so $x \notin B$. Therefore, $x \notin (A \cup B)$, and consequently, $x \in \overline{(A \cup B)}$.
(b) The proof is left for the reader.   ■

   Theorem 8 resembles the ways that *and* and *or* interact with *not* (which are also called DeMorgan's Laws in logic). For example, "not $(x$ is greater than 3 or x is odd)" is equivalent to "x is not greater than 3, and x is not odd." A more thorough study of logic is given in *Chapter 2*.

Example 7.   Verify DeMorgan's Laws for the sets $A = \{1, 2, 3, 4\}$ and $B = \{3, 5, 6, 8\}$ when the universal set is $U = \{1, 2, 3, 4, 5, 6, 7, 8\}$.

Solution. $\overline{A \cup B} = \overline{\{1, 2, 3, 4, 5, 6, 8\}} = \{7\}.$ $\overline{A} = \{5, 6, 7, 8\}.$ $\overline{B} = \{1, 2, 4, 7\}.$ $\overline{A} \cap \overline{B} = \{7\}.$ It now follows that $\overline{A \cup B} = \overline{A} \cap \overline{B}.$ ■

   DeMorgan's Laws are important tools for proving results about how union, intersection, and complementation interact. The notion of the symmetric difference is a particular instance of this. Symmetric difference identifies the elements of two sets that are not in their intersection. This set $(A \cup B) - (A \cap B)$ is shown in Figure 1.11.

Figure 1.11   Elements in two sets that are not in the intersection: $A \cup B - A \cap B$.

We can define the elements of two sets that are not in their intersection in terms of unions, intersections, and complements of the sets. After the definition of this set, we will show that the operation of forming this set satisfies both the commutative and the associative law.

Definition 7.   Let A and B be sets. The set

$$A \oplus B = (A - B) \cup (B - A)$$

is the **symmetric difference** of A and B.

Example 8.   Let $A = \{1, 2, 3, 4\}$ and $B = \{3, 4, 5, 6\}$. Then, $A \oplus B = \{1, 2, 5, 6\}$.

Some obvious facts about the symmetric difference are collected in Theorem 9.

Theorem 9.

(a) For any set A, we have $A \oplus \emptyset = A$.
(b) For any set A, we have $A \oplus A = \emptyset$.
(c) For any two sets A and B it follows that $A \oplus B = B \oplus A$.

Proof.   (a) and (b) follow directly from the definition.

(c) Since $A \oplus B = (A - B) \cup (B - A) = (B - A) \cup (A - B) = B \oplus A$

the result follows.   ■

In Theorem 9(c), it is shown that $\oplus$ is a commutative operation. The next theorem shows how you prove that symmetric difference is also an associative operation.

Theorem 10. $A \oplus (B \oplus C) = (A \oplus B) \oplus C$.

Proof.

$$(A \oplus B) \oplus C = ((A \oplus B) - C) \cup (C - (A \oplus B))$$
$$= (((A - B) \cup (B - A)) - C) \cup (C - ((A - B) \cup (B - A)))$$

To simplify the proof, we will reduce the two terms on the right side separately. When we have reduced these two terms, we can combine the reductions to complete a reduction of $(A \oplus B) \oplus C$.

The first step will be to replace various expressions of the form $X - Y$ with $X \cap \overline{Y}$ where X and Y represent any pair of the sets A, B, and C:

$$((A - B) \cup (B - A)) - C = ((A \cap \overline{B}) \cup (B \cap \overline{A})) \cap \overline{C}$$
$$= (A \cap \overline{B} \cap \overline{C}) \cup (\overline{A} \cap B \cap \overline{C}) \quad \text{(Distributive Law)}$$

The second term involves a few more steps than the first term:

$$C - ((A - B) \cup (B - A)) = C \cap \overline{((A \cap \overline{B}) \cup (B \cap \overline{A}))}$$
$$= C \cap (\overline{(A \cap \overline{B})} \cap \overline{(B \cap \overline{A})}) \quad \text{(DeMorgan's Law)}$$
$$= C \cap ((\overline{A} \cup B) \cap (\overline{B} \cup A)) \quad \text{(DeMorgan's Law and } \overline{\overline{A}} = A)$$
$$= C \cap (((\overline{A} \cup B) \cap \overline{B}) \cup ((\overline{A} \cup B) \cap A)) \quad \text{(Distributive Law)}$$
$$= C \cap (((\overline{A} \cap \overline{B}) \cup (B \cap \overline{B})) \cup ((\overline{A} \cap A) \cup (B \cap A))) \quad \text{(Distributive Law)}$$

$$= C \cap ((\overline{A} \cap \overline{B}) \cup (B \cap A)) \quad (A \cap \overline{A} = B \cap \overline{B} = \emptyset)$$

$$= (C \cap \overline{A} \cap \overline{B}) \cup (C \cap A \cap B) \quad \text{(Distributive Law)}$$

Putting the reduced form of these two terms together gives a new description of $(A \oplus B) \oplus C$.

$$((A \oplus B) \oplus C) = (A \cap \overline{B} \cap \overline{C}) \cup (\overline{A} \cap B \cap \overline{C}) \cup (\overline{A} \cap \overline{B} \cap C) \cup (A \cap B \cap C)$$

By similar steps, the term $A \oplus (B \oplus C)$ can be reduced to these same expression. We leave this reduction to the reader. After this second reduction, we can conclude

$$A \oplus (B \oplus C) = (A \oplus B) \oplus C$$ ■

The Logic of Statements

Theorem 8(b) is closely tied to an issue in the logic of sentences. The issue is the relationship between an if-then statement and its **converse,** its **inverse,** and its **contrapositive.** A statement such as "if a, then b" can be rewritten as "if b, then a," and you might wonder if the first statement is true whether or not you can deduce anything about the truth of the second. We start with a statement such as "if a, then b." The obvious variants of this statement are "if b, then a," "if not a, then not b," and "if not b, then not a." What we would like to understand is whether any one of these statements being true (or false) implies that any other of these statements is true (or false). Consider the statement

"If George is a horse, then George is an animal."

The **inverse** of this statement is

"If George is not a horse, then George is not an animal."

The **converse** of this statement is

"If George is an animal, then George is a horse."

And, finally, the **contrapositive** of the statement is

"If George is not an animal, then George is not a horse."

The statement and its contrapositive are both true, whereas the inverse and converse are probably false (depending on who George is).

As another example, consider the following:

Statement: "If my cat is a horse, then my cat is an animal."
Inverse: "If my cat is not a horse, then my cat is not an animal."
Converse: "If my cat is an animal, then my cat is a horse."
Contrapositive: "If my cat is not an animal, then my cat is not a horse."

As the two examples illustrate, the if-then statements are **equivalent statements** to their contrapositives. It can be shown that in general, based on logic alone, a statement is true if and only if its contrapositive is true. In writing a proof, it may be easier to use the contrapositive of a statement than to use the statement itself. A proof of the contrapositive of your objective is called an **indirect proof.** In the cat/horse example, the statement and its contrapositive are vacuously true, but the inverse and converse are false. An if-then statement is normally not equivalent to its inverse or to its converse, but the converse and inverse are always equivalent to each other.

1.3.3   New Proof Templates

The first new proof idea was used in Theorem 4 (Section 1.3.1). In Theorem 4(a) there were two possibilities in the hypothesis. We needed to prove that regardless of which possibility was true, the conclusion followed. The proof was actually two proofs! In general, there can be any number of cases. The proof technique is outlined in Template 1.8.

Template 1.8   Proof by Cases

To prove a theorem by cases:

1. List all possible cases that will cover every circumstance in which the hypothesis might hold.
2. For each possible case, prove the conclusion separately.

The proof of Theorem 4 is a simple proof by cases; we will present more complicated examples later. As you proceed, be aware of the following recommendations in using a proof by cases:

1. Make sure you need to use a proof by cases. If you break a proof into cases, you must normally treat each case separately, which tends to make your proof long. If you don't need to break the proof into cases, your proof will often be shorter. If only one step of your proof needs to be broken into cases, then break only that step into cases. Even more risky is breaking cases into subcases. Suppose you write a proof breaking into four cases, and each case breaks into four subcases, and each subcase breaks into four sub-subcases. That gives you $4 \cdot 4 \cdot 4 = 64$ cases in all to prove, and that almost inevitably makes your proof longer than it otherwise might be.
2. Make sure you list all possible cases. The proof of Theorem 4(a) in Section 1.3.1, consisted of only two possible cases, and they were obvious from the problem. However, problems sometimes break down into more than two cases, and when they do, it is easy to miss some cases.
3. You need to prove that your list of cases covers all possible cases.
4. When claiming that two cases are analogous, make sure that one case is truly analogous to another. There may well be logical subtleties that arise in one case that didn't arise in an earlier case. (Indeed, that is often why we break a problem into cases in the first place!) Before saying that two cases are analogous, think carefully through the details to make sure they are!

The discussion following Theorem 4 pointed out another proof technique. In discussing the statement of Theorem 4(a), it was pointed out that it is useful to understand what a theorem does not say. Quite often, it is not true that what seems intuitively to be quite reasonable is, in fact, true. In the case discussed following Theorem 4, a **counterexample** would give a concrete instance of sets that satisfy the hypothesis of the alternate statement, whereas the same sets do not satisfy the conclusion. This proof idea is shown in Template 1.9.

Template 1.9   Disproof by Counterexample

To disprove results starting "for every $x \in A$," find an x that can be proven to be in A and for which the result fails.

Theorem 7 in Section 1.3.2 proved that if $A \subseteq B$, then $\overline{B} \subseteq \overline{A}$ by assuming that $A \subseteq B$ and $\overline{B} \not\subseteq \overline{A}$. We then showed this led to a contradiction. The format for this proof is summarized in Template 1.10.

Template 1.10   Proof by Contradiction

To prove an assertion a by contradiction, use one of the following two forms:

Form 1: Assume assertion a is false, and prove that some other assertion b is false where assertion b is known to be true.

Form 2: Assume assertion a is false. For some assertion b, prove that *both* assertion b is true and assertion b is false.

The statement of Theorem 7(b) can be construed as saying that a statement and its contrapositive have the same truth value. For example, think of "$A \subseteq B$" as being translated "if $x \in A$, then $x \in B$." Similarly, think of "$\overline{B} \subseteq \overline{A}$" as "if $x \notin B$, then $x \notin A$." How would an if-then statement be proved to be true or false just when its contrapositive is? The answer is almost exactly the way that Theorem 7(b) was proved. The idea of this proof technique is summarized in Template 1.11.

Template 1.11   Indirect Proof

To prove a theorem using an indirect proof, prove "if p, then q" by proving "if not q, then not p."

1.3.4   Power Sets and Products

We started by introducing you to thinking about sets of objects and not just individual objects. After introducing sets, we made precise what it means for one set to be a subset of another. We can also take one more step, however, and think of the set consisting of all subsets of a set.

Definition 8.   Let A be a set. The **power set** of A, denoted $\mathcal{P}(A)$, is

$$\mathcal{P}(A) = \{X : X \subseteq A\}$$

Example 9.

(a) $\mathcal{P}(\emptyset) = \{\emptyset\}$. Even though $\emptyset$ has no elements, $\mathcal{P}(\emptyset)$ has the one element $\emptyset$.
(b) $\mathcal{P}(\mathcal{P}(\emptyset)) = \mathcal{P}(\{\emptyset\}) = \{\emptyset, \{\emptyset\}\}$.
(c) $\mathcal{P}(\{1, 2\}) = \{\emptyset, \{1\}, \{2\}, \{1, 2\}\}$.
(d) $\mathcal{P}(\{1, 2, 3\}) = \{\emptyset, \{1\}, \{2\}, \{3\}, \{1, 2\}, \{2, 3\}, \{1, 3\}, \{1, 2, 3\}\}$.
(e) $\mathcal{P}(\{1, 2, \{3\}\}) = \{\emptyset, \{1\}, \{2\}, \{\{3\}\}, \{1, 2\}, \{2, \{3\}\}, \{1, \{3\}\}, \{1, 2, \{3\}\}\}$.
(f) $\mathcal{P}(\{\{1, 2, 3\}\}) = \{\emptyset, \{1, 2, 3\}\}$. This is true, because the set $\{\{1, 2, 3\}\}$ has only one element, $\{1, 2, 3\}$. So, there are only two subsets of $\{\{1, 2, 3\}\}$, one that contains $\{1, 2, 3\}$ and one that does not.

Products of Sets

The next operation on sets is familiar, because it is the formalism behind the way we are used to seeing points in two-dimensional space represented as ordered pairs.

Definition 9.   For any sets X and Y, the **product** $X \times Y$ is the set of all ordered pairs (a, b) such that $a \in X$ and $b \in Y$. When $X = Y$, this set is also denoted X^2. Similarly, the product of n sets $X_1, \ldots, X_n$ is the set of all ordered n-tuples $(x_1, \ldots, x_n)$ of elements such that $x_1 \in X_1, \ldots$, and $x_n \in X_n$. When n copies of the same set X are used, the resulting Cartesian product $X \times \ldots \times X$ is the set of all ordered n-tuples of elements in X, denoted X^n.

Example 10.   Let $X = \{0, 1\}$ and $C = \{a, b\}$. Then, $X \times C = \{(0, a), (0, b), (1, a), (1, b)\}$, and $C \times C = \{(a, a), (a, b), (b, a), (b, b)\}$. The product of two sets is sometimes referred to as the Cartesian product.

1.3.5   Lattices and Boolean Algebras

The design of computer chips involves very complex interactions of very small components or building blocks called *gates*. The complete design that is of a computer chip is called a *combinatorial circuit*. The mathematical structure we will introduce here can be used to design, represent, and optimize *combinatorial circuits*. We will look more closely at gates and *combinatorial circuits* in Chapter 2, but we first need to understand the underlying mathematical structure.

Definition 10.   A **lattice** is a set X with two operations, called **meet**, denoted as $\wedge$, and **join**, denoted as $\vee$, that satisfy the following properties for all $x, y, z \in X$:

$x \wedge y = y \wedge x$	**Commutative Law for Meet**
$x \vee y = y \vee x$	**Commutative Law for Join**
$x \wedge (y \wedge z) = (x \wedge y) \wedge z$	**Associative Law for Meet**
$x \vee (y \vee z) = (x \vee y) \vee z$	**Associative Law for Join**
$x \wedge (x \vee y) = x$	**Absorption Law for Meet**
$x \vee (x \wedge y) = x$	**Absorption Law for Join**

To say that something is a lattice, we must explicitly say (i) what the set of objects is and (ii) what the meet and join operations are. After specifying these, we must show that meet and join so interpreted satisfy all the required axioms. Meet and join can be any operations on the set so long as the axioms are satisfied. The operations of meet and join can be as simple as union and intersection defined on a set of sets. Whatever the operations are defined to be, however, the first task is to show that the operations satisfy the Commutative and Associative Laws.

Example 11. Let X be a set, and let $L = \mathcal{P}(\mathcal{X})$. Let join be defined as the union of two subsets of X and meet as the intersection of two subsets of X. Then, L together with union and intersection is a lattice.

Solution. By Theorems 1 and 2 in Section 1.3.1, the Commutative and Associative Laws hold. To prove that the Absorption Law holds, we use the result of Theorem 5 in Section 1.3.1. ∎

The next example shows that the interpretation of meet and join can be rather different from unions and intersections.

Example 12. Let $X \subseteq \mathbb{R}$. Let meet be defined as the minimum of two elements of X and join as the maximum of two elements of X. Then, X together with the minimum and the maximum operations is a lattice.

Solution. The Commutative Law for Meet in this context says that the minimum of two real numbers is the same regardless of the order in which you consider the elements. The remainder of the Commutative and Associative Laws for Meet and Join are straightforward to verify. The Absorption Law for Meet says that the minimum of an element x together with the maximum of the two elements x and y where y is any other element is just x. This just says that either x is the minimum of $\{x, x\}$ or the minimum of $\{x, y\}$ where $y \geq x$. In either case, the result follows. The remaining details are left as Exercise 21 in Section 1.4. ∎

There are two additional properties that are used to distinguish different kinds of lattices. The first of these laws, the Distributive Law, is familiar in the context of union and intersection.

Definition 11. Let X with the operations meet ($\wedge$) and join ($\vee$) be a lattice. X is a **distributive lattice** if the following two properties are satisfied for all $x, y, z \in X$:

$x \wedge (y \vee z) = (x \wedge y) \vee (x \wedge z)$	**Distributive Law for Meet**
$x \vee (y \wedge z) = (x \vee y) \wedge (x \vee z)$	**Distributive Law for Join**

The Distributive Laws for Meet and Join are proved for the interpretation of meet as intersection and join as union in Theorem 3 (Section 1.3.1).

The final property we need is stated abstractly in terms of two special elements that must be identified in the set of elements forming a lattice. The usual way to prove this result is to assume that the lattice has this property and then determine what these special elements must be.

Definition 12. Let X together with the operations meet ($\wedge$) and join ($\vee$) be a lattice. X is a **complemented lattice** if

1. There are two (unequal) elements, one called the **minimum element,** denoted $\perp$ (read **bottom**), and the other called the **maximum element,** denoted $\top$ (read **top**), such that for every $x \in X$,

$$x \wedge \top = x, \quad x \wedge \perp = \perp, \quad x \vee \top = \top, \text{ and } x \vee \perp = x$$

2. For each $x \in X$, there is an element $\neg x \in X$ such that $x \wedge \neg x = \perp$ and $x \vee \neg x = \top$.

Example 13. Let A be a set, and let $X = \mathcal{P}(A)$. The lattice on X with meet defined as intersection and join defined as union is a complemented lattice.

Solution. Let $\top = A$ and $\perp = \emptyset$. Since for any $B \in X$ we have $B \cap \top = B \cap A = B$, $B \cap \perp = B \cap \emptyset = \emptyset$, $B \cup \top = B \cup A = A$, and $B \cup \perp = B \cup \emptyset = B$, X is a complemented lattice. ∎

The definition does not tell you what elements of a lattice should be $\perp$ and $\top$ or what the relationship between $\neg x$ and x is. For the lattice of subsets of a set A, we can use $\emptyset = \perp$ and A itself as $\top$. We also define $\neg x$ as the complement of x. With these definitions of $\top$, $\perp$, and $\neg x$ for this lattice, you can show that the lattice is complemented. The details are left as Exercise 23 in Section 1.4.

The mathematical structure that is of importance in computer science can now be defined.

Definition 13. A **boolean algebra** is a complemented, distributive lattice.

The boolean algebra used by computer scientists to model combinatorial circuits is based on the set of elements $\{0, 1\}$ and the operations shown in Table 1.1 where $\vee$ is the meet and $\wedge$ is the join.

Table 1.1 Operations for a Boolean Algebra

$\vee$	0	1
0	0	1
1	1	1

$\wedge$	0	1
0	0	0
1	0	1

Example 14. Let B be a set of elements assigned values from the set $\{0, 1\}$, and let the operations $\vee$ and $\wedge$ be defined on B as described in Table 1.1. B together with $\vee$ as meet and $\wedge$ as join forms a boolean algebra.

Here, it turns out that there is only one possible choice for each of $\top$, $\bot$, and $\neg$:

$$\top = 1, \quad \bot = 0, \quad \neg 0 = 1, \text{ and } \neg 1 = 0$$

Indeed, there is always only one possible choice (see Exercise 24 in Section 1.4). Thus, in any boolean algebra, we may refer to $\top$, $\bot$, and each $\neg x$ without ambiguity.

Solution. The proof requires showing that no matter what value $x, y \in B$ have, the axioms of a boolean algebra hold. As an example, we will show that the operation meet is commutative. Let $x, y \in B$. Then,

		y	
	$\vee$	0	1
x	0	0	1
	1	1	1

		x	
	$\vee$	0	1
y	0	0	1
	1	1	1

We can simply check that for all possible pairs, $x \vee y = y \vee x$ for the meet operation. Similar proofs are needed for the other axioms and will be left for the reader. ■

In Section 2.1.4, we will consider this boolean algebra by another name. In place of 1 and 0, we will call the values of the elements *TRUE* and *FALSE*. The operations will be *or* and *and*. Then, for example, we can interpret a variable x to be *TRUE* if there is a current flowing in a wire X—and similarly, y to be *TRUE* if there is a current flowing in wire Y. This turns out to be a very natural way to look at computer circuits.

1.4 Exercises

1. Let $A = \{1, 2, 3, \ldots, 10\}$, $B = \{2, 3, 6, 8\}$, and $C = \{3, 5, 4, 8, 2\}$. Find the following:
   (a) $B \cup C$
   (b) $B \cap C$
   (c) $B - C$
   (d) $A - B$
   (e) $A - C$
2. Let $U = \{0, 1, 2, 3, 4, 5, 6, 7, 8, 9\}$, $A = \{0, 1, 2, 3\}$, $B = \{0, 2, 4\}$, and $C = \{0, 3, 6, 9\}$.
   (a) Find $A \cup B$, $A \cap B$, $\overline{A}$, $\overline{(A \cap B)}$, and $(B \cup C) - A$.
   (b) Find $\mathcal{P}(A)$, $\mathcal{P}(B)$, $\mathcal{P}(A \cap B)$, $\mathcal{P}(A) \cap \mathcal{P}(B)$.
   (c) Is $\mathcal{P}(A \cup B) = \mathcal{P}(A) \cup \mathcal{P}(B)$? Prove your answer.
   (d) Why doesn't $\overline{\mathcal{P}(A)}$ make sense?
3. Let $A = \{0, 3\}$ and $B = \{x, y, z\}$. Find the following:
   (a) $A \times B$
   (b) $A \times A \times B$
   (c) $B \times A$
   (d) $B \times A \times B$

4. Let $X = \{2, 4\}$, $Y = \{1, 4\}$, and $Z = \{0, 4, 8\}$. Construct the following sets:

   (a) $X \times Y$
   (b) $X \times Y \times Z$
   (c) $Y \times Z$
   (d) $Z \times Y \times X$
   (e) $Z \times X \times Y$

5. Prove Theorem 1(d).
6. Prove Theorem 2(d).
7. (a) Draw Venn diagrams to illustrate Theorems 3(a) and 3(b).
   (b) Prove Theorem 3(a).
   (c) Prove Theorem 3(b).
8. (a) Draw Venn diagrams to illustrate Theorems 4(c) and 4(d).
   (b) Prove Theorem 4(c).
   (c) Prove Theorem 4(d).
9. Find three sets A, B, and C where $A \subseteq B \cup C$ but $A \nsubseteq B$ and $A \nsubseteq C$.
10. (a) Draw Venn diagrams illustrating the four parts of Theorem 6.
    (b) Prove Theorem 6(a).
    (c) Prove Theorem 6(b).
    (d) Prove Theorem 6(c).
    (e) Prove Theorem 6(d).
11. Prove Theorem 7(c).
12. (a) Prove Theorem 9(b) using as a model the proof of Theorem 9(a).
    (b) Prove Theorem 9(b) using Theorem 7(c).
13. Let $A = \{1, 2, \{\{1, 2\}\}\}$.

    (a) How many elements does A have? How many elements does $\mathcal{P}(A)$ have? How many elements does $\mathcal{P}(\mathcal{P}(A))$ have?
    In parts (b)–(m) determine, whether each of the following is true, and if not, explain why not.
    (b) $1 \in A$
    (c) $\{1, 2\} \in A$
    (d) $\{\{1, 2\}\} \in A$
    (e) $\emptyset \in A$
    (f) $1 \in \mathcal{P}(A)$
    (g) $\{1, 2\} \in \mathcal{P}(A)$
    (h) $\{\{1, 2\}\} \in \mathcal{P}(A)$
    (i) $\emptyset \in \mathcal{P}(A)$
    (j) $1 \in \mathcal{P}(\mathcal{P}(A))$
    (k) $\{1, 2\} \in \mathcal{P}(\mathcal{P}(A))$
    (l) $\{\{1, 2\}\} \in \mathcal{P}(\mathcal{P}(A))$
    (m) $\emptyset \in \mathcal{P}(\mathcal{P}(A))$

14. For each of the following statements, find the corresponding inverse, converse, and contrapositive.

    (a) If the stars are shining, then it is the middle of the night.
    (b) If the Wizards won, then they scored at least 100 points.
    (c) If the exam is hard, then the highest grade is less than 90.

15. Which of the following statements are correct? Prove each correct statement. Disprove each incorrect statement by finding a counterexample.

    (a) A and B are disjoint if and only if B and A are disjoint. (Read the statement carefully—the order in which the sets are listed might matter!)

    (b) $A \cup B$ and C are disjoint if and only if both the following are true: (i) A and C are disjoint and (ii) B and C are disjoint.

    (c) $A \cap B$ and C are disjoint if and only if both the following are true: (i) A and C are disjoint and (ii) B and C are disjoint.

    (d) $A \cup B$ and C are disjoint if and only if one of the following is true: (i) A and C are disjoint or (ii) B and C are disjoint.

    (e) $A \cap B$ and C are disjoint if and only if one of the following is true: (i) A and C are disjoint or (ii) B and C are disjoint.

    (f) Let U be a universal set with $A, B \subseteq U$. A and B are disjoint if and only if $\overline{A}$ and $\overline{B}$ are disjoint.

16. For (a) and (b), prove the stated result. For (c) and (d), find a counterexample to show that these conjectures are false.

    (a) $A \oplus B = (A \cup B) - (A \cap B)$

    (b) $A \cap (B \oplus C) = (A \cap B) \oplus (A \cap C)$

    (c) $(A \cap B) \oplus (C \cap D) \subseteq (A \oplus C) \cap (B \oplus D)$

    (d) $(A \cup B) \oplus (C \cup D) \subseteq (A \cup C) \oplus (B \cup D)$

17. Given any four integers $x_1, x_2, x_3,$ and x_4, none of which is even and none of which is a multiple of 5, prove that some consecutive product of these integers ends in the digit 1. A consecutive product is one term, two terms in a row, three terms in a row, or all four terms in a row using the order in which the integers appear in the list x_1, x_2, x_3, x_4. (*Hint:* Use a proof by cases.)

18. Prove by contradiction that 7 is a prime number.

19. Prove by contradiction that $\sqrt{2}$ is not a rational number.

20. Prove by contradiction that $\mathbb{Z}$ has no smallest element.

21. Complete the proof of Example 12.

22. For parts (a) and (b), let U be any set, and let $X = \mathcal{P}(U)$.

    (a) Prove that X with the operations $\cap$ for meet and $\cup$ for join is a distributive lattice.

    (b) Prove that X with the operations $\cup$ for meet and $\cap$ for join is a distributive lattice.

23. Let U be any set, and let $X = \mathcal{P}(U)$. Prove that X with the operations $\cup$ for meet and $\cap$ for join is a complemented lattice.

24. Recall that in the definition of a boolean algebra, we did not require that $\top$, $\bot$, and each $\neg x$ *be specified;* we merely said they must exist. So, it is natural to ask whether there might be several elements that could equally well be chosen as $\top$ or $\bot$ or, for some element x of the boolean algebra, several different possible choices for $\neg x$. Show that in a complemented lattice:

    (a) There is only one possible choice of elements $\top$ and $\bot$ satisfying the definition of a complemented lattice. (*Hint:* Suppose there were two possible choices for $\top$, say, $\top_1$ and $\top_2$. Evaluate $\top_1 \wedge \top_2$ in two different ways.)

    (b) For each element x of a complemented, distributive lattice, there is only one possible choice for $\neg x$ that satisfies the definition of $\neg x$. (*Hint:* Suppose there were two choices, say, $\neg x_1$ and $\neg x_2$, for $\neg x$. Find two ways to evaluate $\neg x_1 \wedge x \vee \neg x_2$.)

25. Prove that in a boolean algebra
$$a \vee (b \wedge c) = (a \vee b) \wedge c$$

if and only if
$$a \vee (b \wedge (a \vee c)) = (a \vee b) \wedge (a \vee c)$$

and
$$a \wedge (b \vee (a \wedge c)) = (a \wedge b) \vee (a \wedge c)$$

This property of a boolean algebra is called *modularity*.

26. Prove that in a boolean algebra, DeMorgan's Laws hold; that is,
$$\neg(x \vee y) = \neg x \wedge \neg y$$
$$\neg(x \wedge y) = \neg x \vee \neg y$$

27. Let $U = \{1, 2, 3, 4, 5, 6, 7, 8, 9, 10\}$ be a universal set. Let $A, B, C \subseteq U$ such that $A = \{1, 3, 4, 8\}$, $B = \{2, 3, 4, 5, 9, 10\}$, and $C = \{3, 5, 7, 9, 10\}$. Use bit representations for A, B, and C together with *UNION, INTER, DIFF*, and *COMP* to find the bit representation for the following:

(a) $A \cup B$

(b) $A \cap B \cap C$

(c) $(A \cup C) \cap B$

(d) $(A - B) \cup C$

(e) $A \cap (B - (C \cap B))$

(f) $A - (B - C)$

(g) $(A \cup B) \cup (C - B)$

1.5   The Principle of Inclusion-Exclusion

A great deal can often be learned just by counting the elements in a set. Unfortunately, it turns out that even though counting is sometimes very easy, it is sometimes very difficult, especially if the set whose elements are being counted has a very complicated description. As we show later, the Principle of Inclusion-Exclusion is a widely used method for counting the number of elements in the union or the intersection of sets.

1.5.1   Finite Cardinality

Before we focus on counting elements in unions of sets that are not disjoint, we need to make clear some fundamental ideas about how we compute the number of elements in a set.

Definition 1.   **(Informal)**   For a finite set, the **cardinality** of A is the number of elements in A. If A is infinite, then the cardinality of A is infinite. The cardinality of A is denoted by $|A|$.

Example 1. $|\{1, 2, 3\}| = 3$. $|\emptyset| = 0$. $|\mathcal{P}(\emptyset)| = 1$. $|\{\{1, 2, 3\}\}| = 1$. $|\mathbb{Z}|$ is infinite.

This definition should be viewed as a temporary one. The topic of cardinality will be dealt with in Chapter 4, in which this informal definition will be replaced with a more formal one. In Chapter 4, the idea of two sets *having the same cardinality* ($|X| = |Y|$) will be extended to include sets with infinitely many elements. The informal definition of cardinality suffices for finite sets; and in this section, only finite sets are considered.

Theorem 1. (Basic Counting Theorem) Let A and B be subsets of a finite universal set U.

(a) Let $B \subseteq A$. Then:
  i. $|B| \leq |A|$.
  ii. $|A - B| = |A| - |B|$.
  iii. $|B| = |A|$ if and only if $B = A$.

(b) Let A and B be disjoint finite sets. Then, $|A \cap B| = 0$, and $|A \cup B| = |A| + |B|$.

(c) $|\overline{A}| = |U| - |A|$.

Proof. The proof is left to the reader. ∎

Example 2. The population of Atlanteas is 830, of which 250 are adult females and 380 are children.

(a) How many adults live in Atlanteas?
(b) How many adult males live in Atlanteas?
(c) How many "females and children" live in Atlanteas?

Solution. The universal set is $U = \{$residents of Atlanteas$\}$. The subsets of interest are

$$A = \{\text{adults}\}, \ F = \{\text{adult females}\}, \ M = \{\text{adult males}\}, \text{ and } C = \{\text{children}\}$$

(a) $|A| = |U| - |C| = 830 - 380 = 450$ (part (c))
(b) $|M| = |A| - |F| = 450 - 250 = 200$ (since $M \subseteq A$ use part (a))
(c) $|F \cup C| = |F| + |C| - |F \cap C| = 250 + 380 - 0 = 630$ (since $F \cap C = \emptyset$, use part (b)) ∎

The results in parts (a) and (b) of Theorem 1 are very special, because they make strong assumptions about A and B. If these assumptions fail, the conclusions are generally incorrect.

Example 3. Let $U = \{0, 1, 2\}$, $A = \{0, 1\}$, and $B = \{1, 2\}$.

(a) Since A is not a subset of B, the hypothesis of part (a) does not hold. Neither does the conclusion. $|A| = |B|$, but $A \neq B$, and $|A - B| = 1 \neq 0 = |A| - |B|$.
(b) The sets A and B are not disjoint sets, so the hypothesis of part (b) does not hold. We have $|A \cap B| = |\{1\}| = 1 \neq 0$ and $|A \cup B| = |\{0, 1, 2\}| = 3 \neq 4 = |A| + |B|$, so neither conclusion holds.

A more interesting question is the cardinalities of $|A \cap B|$ and $|A \cup B|$ when neither A nor B is a subset of the other and the sets are not disjoint. There are, of course, some trivial truths, such as $0 \leq |A \cap B|$ and $|A \cap B| \leq |A|$. What is interesting, however, is the relationship between $|A \cap B|$ and $|A \cup B|$. The reader should study the two examples below and then, *before reading any further,* try to identify a pattern.

Example 4. Let $A = \{0, 1, 2, 3, 4, 5, 6, 7\}$ and $B = \{4, 5, 6, 7, 8, 9, 10, 11, 12, 13\}$. So, $|A| = 8$, $|B| = 10$, $|A \cap B| = 4$, and $|A \cup B| = 14$.

Example 5. The population of Atlantis is 834, of which 500 are females. There are 175 people who are at least two meters tall, and only 10 of the females are at least two meters tall. How many males are less than two meters tall?

Solution.   Of the 834 people, 500 are females, so 334 are males. Of the 175 people at least two meters tall, 10 are females. This says that there are at $175 - 10 = 165$ males who are at least two meters tall.

Since there are 334 males in total and 165 of them are at least two meters tall, there are $169 = 334 - 165$ males who are less than two meters tall. The situation is shown in Figure 1.12.   ■

	Males	Females	
At least two meters tall	165	10	
Less than two meters tall	169	490	

Figure 1.12   Population characteristics.

In Example 5 we first counted how many males were at most two meters tall by dividing the set of all Atlanteans up into two sets, one consisting of all the females and the other consisting of all the males. We did this correctly, because we knew the number of Atlanteans and the number of females. We then made another such count to find the number of males who were at least two meters tall. We knew the total number of Atlanteans who were at least two meters tall and the number of females who were at least two meters tall. A simple subtraction gave the number of males at least two meters tall. We see that in computing the size of both sets, we knew the size of two of the sets, and the two subsets were disjoint. This result is an application of Theorem 1(b) in Section 1.5.1. We next deal with the case in which A and B are not disjoint.

1.5.2   Principle of Inclusion-Exclusion for Two Sets

Example 6.   A deck of cards has four suits: Clubs, Diamonds, Hearts, and Spades. Diamonds and Hearts are called red suits; Clubs and Spades are called black suits. Each suit contains 13 cards with values Ace(1), 2, 3, 4, 5, 6, 7, 8, 9, 10, Jack, Queen, and King. How many cards are black or have the value of 3?

Solution.   Let A be the set of black cards and B the set of 3's. The example asks for the size of $|A \cup B|$. Clearly, $|A| = 26$, and $|B| = 4$. The problem is that two of the 3's are also black. In this case, $|A \cap B| \neq \emptyset$. The count $|A| + |B|$ overcounts by $|A \cap B|$. The answer is

$$|A \cup B| = |A| + |B| - |A \cap B| = 26 + 4 - 2 = 28$$ ■

The card problem in Example 6 is a special example of the **Principle of Inclusion-Exclusion** that we prove in more generality next.

Theorem 2.   (Principle of Inclusion-Exclusion for Two Sets)   Let A and B be finite sets. Then,

$$|A \cup B| = |A| + |B| - |A \cap B|$$

The number of elements in the union of two finite sets is the sum of the number of elements in each of the sets minus the number of elements in their intersection. A Venn diagram for these sets is shown in Figure 1.13.

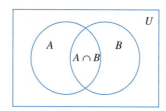

Figure 1.13 $|A \cup B|$.

(What the proof entails.)   What procedure could be used to count the elements in $A \cup B$? First, count all the elements of A. Then, count all the elements of B. In the process, all the elements of $A \cap B$ have been counted twice, so subtract $|A \cap B|$ to compensate.

Proof.   The set $A \cup B = (A - B) \cup (A \cap B) \cup (B - A)$, and any pair of $(A - B)$, $(A \cap B)$, and $(B - A)$ are disjoint (Theorem 6 in Section 1.3.2). It follows immediately that $(A - B)$ and $(A \cap B) \cup (B - A)$ are also disjoint. Hence, by using Theorem 1(b) of this section, we get

$$|A \cup B| = |A - B| + |(A \cap B) \cup (B - A)|$$
$$= |A - B| + |A \cap B| + |B - A|$$

By Theorem 6(b)

$$|A| = |A - B| + |A \cap B|$$
$$|B| = |A \cap B| + |B - A|$$

Putting the last two equations together gives

$$|A| + |B| = |A - B| + 2 \cdot |A \cap B| + |B - A|$$
$$|A| + |B| - |A \cap B| = |A - B| + |A \cap B| + |B - A|$$

Now, substituting this into the equation for $|A \cup B|$, we get the required result:

$$|A \cup B| = |A| + |B| - |A \cap B| \qquad \blacksquare$$

1.5.3   Principle of Inclusion-Exclusion for Three Sets

Figure 1.14 $|A \cup B \cup C|$.

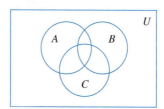

The decomposition of two sets into disjoint subsets is fairly obvious. For three or more sets, however, this is not as obvious a step. Figure 1.14 will help you to understand the next theorem if you identify each of the regions of $A \cup B \cup C$.

The sets of interest are identified as A, B, C, $A \cap B$, $A \cap C$, $B \cap C$, and $A \cap B \cap C$.

Theorem 3.   (Principle of Inclusion-Exclusion for Three Sets)   Let A, B, and C be finite sets. Then,

$$|A \cup B \cup C| = |A| + |B| + |C| - |A \cap B| - |A \cap C| - |B \cap C| + |A \cap B \cap C|$$

Proof.   The same style of proof as used in Theorem 2 could be used, but in this case, there would be seven pieces to keep track of instead of three. A clearer way to proceed is to use Theorem 2 in Section 1.5.2.

$|A \cup B \cup C| = |(A \cup B) \cup C|$ (by the definition of the union of three sets)

$= |A \cup B| + |C| - |(A \cup B) \cap C|$ (by Theorem 2 in Section 1.5.2)

$= |A \cup B| + |C| - |(A \cap C) \cup (B \cap C)|$ (by Distributive Law for Intersection)

$= |A \cup B| + |C| - (|A \cap C| + |B \cap C| - |(A \cap C) \cap (B \cap C)|)$ (by Theorem 2 in Section 1.5.2 on $(A \cap C) \cup (B \cap C)$)

$= |A \cup B| + |C| - |A \cap C| - |B \cap C| + |(A \cap C) \cap (B \cap C)|$

(removing parentheses)

$= |A \cup B| + |C| - |A \cap C| - |B \cap C| + |A \cap B \cap C|$

(simplifying $|(A \cap C) \cap (B \cap C)|$)

$= |A| + |B| - |A \cap B| + |C| - |A \cap C| - |B \cap C| + |A \cap B \cap C|$

(by Theorem 2 in Section 1.5.2 again)

$= |A| + |B| + |C| - |A \cap B| - |A \cap C| - |B \cap C| + |A \cap B \cap C|$ ∎

When the Principle of Inclusion-Exclusion is applied in the next example, the solution becomes straightforward.

Example 7.   A particular political campaign mailing is expected to appeal to three groups of people: liberals, people earning more than \$45,000 a year, and people with children under five years of age. The mailing list includes 30,000 people, including 15,000 conservatives and 15,000 liberals. Of the 30,000 on the mailing list, 17,500 earn more than \$45,000 a year, including 10,001 of the liberals. In the set of people, 3500 have children under five years of age, including 1000 conservatives, 2500 liberals, and 900 of those who earn more than \$45,000 a year. Only one of the liberals earns more than \$45,000 a year and also has children under the age of five. How many people on the mailing list are liberals, or earn more than \$45,000 a year, or have children under five years of age? (As usual, by *or* we mean the *inclusive or.*)

Solution.   Among people on the mailing list, let L be the set of liberals, E the set of people who earn more than \$45,000 a year, and C the set of people who have children under five years of age (see Figure 1.15).

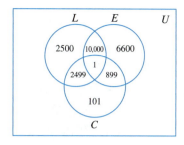

Figure 1.15 Counting liberals and children.

The Principle of Inclusion-Exclusion for Three Sets says that

$$|L \cup E \cup C| = |L| + |E| + |C| - |L \cap E| - |L \cap C| - |E \cap C| + |L \cap E \cap C|$$
$$= 15{,}000 + 17{,}500 + 3500 - 10{,}001 - 2500 - 900 + 1$$
$$= 22{,}600 \qquad \blacksquare$$

Here, as often happens, there is a different way to count this collection of elements. First, note that of the 900 people who have children under five years of age and who earn more than \$45,000 a year, only one is a liberal; the other 899 are conservatives. So, among the 1000 conservatives with children under five years of age, $1000 - 899 = 101$ do not earn more than \$45,000. There are 15,000 liberals, plus $17{,}500 - 10{,}001 = 7499$ conservatives making more than \$45,000 a year, plus 101 conservatives with children under five years of age. Therefore, the answer is

$$15{,}000 + 7{,}499 + 101 = 22{,}600$$

The Principle of Inclusion-Exclusion is also used to solve problems in number theory. Before we explain that example, we need to remind ourselves of one fact from number theory: For 2, 5, and 30, it is clear that $2|30$ and $5|30$. Moreover, it is clear that $2 \cdot 5 = 10|30$. It is not always true, however, that the product of two divisors of a number is again a divisor of the number. For example, 5, 10, and 30 have the property that $5|30$ and $10|30$, but $5 \cdot 10 = 50|30$ is false. What is true is that if m is an integer and both p and q are primes such that $p|m$ and $q|m$, then $p \cdot q|m$.

Example 8. How many natural numbers between 1 and 30,000,000 (including 1 and 30,000,000) are divisible by 2, 3, or 5?

Solution. Let

$$D_i = \{n \in \mathbb{N} : 1 \leq n \leq 30{,}000{,}000 \text{ and } n \text{ is divisible by } i\}$$

What is $|D_2 \cup D_3 \cup D_5|$? The number is difficult to count directly, so we use the Principle of Inclusion-Exclusion. $|D_2| = 15{,}000{,}000$, $|D_3| = 10{,}000{,}000$, and $|D_5| = 6{,}000{,}000$. How about $|D_2 \cap D_3|$? Since 2 and 3 are both prime, an integer n is divisible by both 2 and 3 if and only if n is divisible by $2 \cdot 3 = 6$. So $D_2 \cap D_3 = D_6$, and $|D_6| = 5{,}000{,}000$. Similarly,

$$|D_2 \cap D_5| = |D_{10}| = 3{,}000{,}000$$
$$|D_3 \cap D_5| = |D_{15}| = 2{,}000{,}000$$

and

$$| D_2 \cap D_3 \cap D_5 | = | D_{30} | = 1,000,000$$

Now, by the Principle of Inclusion-Exclusion for Three Sets,

$$| D_2 \cup D_3 \cup D_5 | = | D_2 | + | D_3 | + | D_5 | - | D_2 \cap D_3 | - | D_2 \cap D_5 |$$
$$- | D_3 \cap D_5 | + | D_2 \cap D_3 \cap D_5 |$$
$$= 22,000,000 \qquad \blacksquare$$

Often, a problem is posed in terms of finding how many objects do not have one or more of a set of properties. For example, suppose we were asked to find the number of integers between 1 and 30,000,000 that are not divisible by any of the integers 2, 3, or 5. The solution is $| \overline{D_2 \cup D_3 \cup D_5} |$ where D_2, D_3, and D_5 are defined as in Example 8. The answer is

$$| \overline{D_2 \cup D_3 \cup D_5} | = 30,000,000 - | D_2 \cup D_3 \cup D_5 |$$

In Example 8 we have shown that

$$| D_2 \cup D_3 \cup D_5 | = 22,000,000$$

so

$$|\overline{D_2 \cup D_3 \cup D_5}| = 8,000,000$$

Next, we study an example that looks quite different from counting the number of values having some set of properties.

Example 9.   (The Hat Check Problem)   Three Victorian gentlemen, called $G_1, G_2,$ and G_3, arrive at a restaurant and check their top hats. The cloakroom attendant loses the numbers on the three hats and doesn't know which hat is whose. Rather than admitting the error, the attendant gives the three hats back to the three gentlemen at random. Let h_i represent the hat that belongs to gentleman G_i where $1 \le i \le 3$. The notation $h_i h_j h_k$ represents hat h_i being given to G_1 by the attendant, h_j being given to G_2 by the attendant, and h_k being given to G_3 by the attendant.

How many random assignments of hats result in at least one gentleman receiving his own hat?

Solution.   There are six ways the three hats can be handed back. The first gentleman to request his hat back may be given any of the three hats. The second gentleman may be given either of the two remaining hats. The third gentleman must get the last hat. Multiply $3 \times 2 \times 1 = 6$ to get the number of possible ways.

Of those six ways to hand back the hats, obviously only one gets each hat back to its owner. How many ways get *at least one* hat back to its owner? This question can be answered using the Principle of Inclusion-Exclusion.

Let U be the set of all six ways the attendant can give the three top hats back. Let H_i, for $1 \le i \le 3$, be the set of all the ways where G_i gets his own hat back. Now, $| H_1 | = 2$, for if G_1 gets his own hat back, then there are two hats to return to G_2 and G_3. These two hats can be returned to these two gentlemen in two different ways. By symmetry, $|H_1| = |H_2| = |H_3| = 2$. If G_1 and G_2 get their own hats returned, then there is one hat left to be given to G_3. There is one way to return this hat to G_3. Therefore, $|H_1 \cap H_2| = 1$. By symmetry, $|H_1 \cap H_2| = |H_1 \cap H_3| = |H_2 \cap H_3| = 1$. Finally, $|H_1 \cap H_2 \cap H_3| = 1$.

By the Principle of Inclusion-Exclusion, we compute $|H_1 \cup H_2 \cup H_3|$ as

$$
\begin{aligned}
|H_1 \cup H_2 \cup H_3| &= |H_1| + |H_2| + |H_3| - |H_1 \cap H_2| - |H_1 \cap H_3| \\
&\quad - |H_2 \cap H_3| + |H_1 \cap H_2 \cap H_3| \\
&= 2 + 2 + 2 - 1 - 1 - 1 + 1 = 4
\end{aligned}
$$

That is, of the six possible ways to hand back the hats, in four of them at least one gentleman gets his own hat back. ■

1.5.4 Principle of Inclusion-Exclusion for Finitely Many Sets

Notice the alternating plus and minus signs in the Principle of Inclusion-Exclusion. For the union of three sets, add the sizes of all the individual sets (intersections of one set), subtract the sizes of the intersections of two sets, and add the size of the intersection of all three sets. This alternation continues for computing the size of the union of more than three sets.

To state the next theorem neatly, we define two terms. Neither term is commonly used, but each is quite understandable in the context of the Principle of Inclusion-Exclusion.

Definition 2. Let $A_1, A_2, \ldots, A_n$ be sets. An **odd intersection** from

$$
A_1, A_2, \ldots, A_n
$$

is an intersection of an odd number of the A_i's. An **even intersection** is an intersection of an even, positive number of A_i's.

Example 10. Let $A_1, A_2, A_3, A_4,$ and A_5 be sets. Odd intersections are:

$n = 1 : A_1, A_2, A_3, A_4, A_5$
$n = 3 : A_1 \cap A_2 \cap A_3, A_1 \cap A_2 \cap A_4, A_1 \cap A_2 \cap A_5, A_1 \cap A_3 \cap A_4,$
  $A_1 \cap A_3 \cap A_5, A_1 \cap A_4 \cap A_5, A_2 \cap A_3 \cap A_4, A_2 \cap A_3 \cap A_5,$
  $A_2 \cap A_4 \cap A_5, A_3 \cap A_4 \cap A_5$
$n = 5 : A_1 \cap A_2 \cap A_3 \cap A_4 \cap A_5$

Even intersections are:

$n = 0 : \emptyset$
$n = 2 : A_1 \cap A_2, A_2 \cap A_3, A_3 \cap A_4, A_4 \cap A_5,$
  $A_1 \cap A_3, A_2 \cap A_4, A_3 \cap A_5,$
  $A_1 \cap A_4, A_2 \cap A_5,$
  $A_1 \cap A_5$
$n = 4 : A_1 \cap A_2 \cap A_3 \cap A_4, A_1 \cap A_2 \cap A_3 \cap A_5,$
  $A_1 \cap A_3 \cap A_4 \cap A_5, A_2 \cap A_3 \cap A_4 \cap A_5$ ■

Theorem 4. (Principle of Inclusion-Exclusion For Finitely Many Sets) Let $A_1, A_2,$ $\ldots, A_n$ be finite sets ($n \geq 1$). Then, $|A_1 \cup A_2 \cup \cdots \cup A_n|$ equals the sum of the cardinalities of all odd intersections from $A_1, A_2, \ldots, A_n$ (including single sets) minus the sum of the cardinalities of all even intersections from $A_1, A_2, \ldots, A_n$.

1.6   **Exercises**

1. In a class of 35 students who are either biology majors or have blonde hair, there are 27 biology majors and 21 blondes. How many biology majors must be blonde?

2. A film class had 33 students who liked Hitchcock movies, 21 students who liked Spielberg movies, and 17 students who liked both kinds of films. How many students were in the class if every student is represented in the survey?

3. A tennis camp has 39 players. There are 25 left-handed players and 22 players who have a two-handed back stroke. How many left-handed players have a two-handed back stroke if every player is represented in these two counts?

4. A car manufacturer determines that automatic transmission, power steering, and a CD player are the three most important features in generating sales. The production schedule for the next day has these features incorporated in cars as shown in the following table:

Car	Automatic Transmission	Power Steering	CD Player
A	x	x	
B	x	x	x
C		x	
D		x	x
E			x
F		x	x
G	x		x
H		x	x

   (a) How many cars have at least one of these features? Even though you can see the answer, use the Principle of Inclusion-Exclusion to derive it.

   (b) How many cars have two or more of these features? Again, use the Principle of Inclusion-Exclusion to derive the answer.

5. A marketing class did a survey of the number of fast-food outlets near campus. The results of the survey showed the following:

Type of Food Sold	No. of Outlets
Hamburgers	15
Tacos	25
Pizza	21
Hamburgers and tacos	11
Hamburgers and pizza	10
Tacos and pizza	14
Hamburgers and tacos and pizza	9
Served none of these items	5

How many fast food outlets are there near campus?

6. At the beginning of the semester, an instructor of a music appreciation class wants to find out how many of the 250 students had heard recordings of the music of Mozart, Beethoven, Haydn, or Bach. The survey showed the following:

Composer Listened to by Students	No. of Students
Mozart	125
Beethoven	78
Haydn	95
Bach	62
Mozart and Beethoven	65
Mozart and Haydn	50
Mozart and Bach	48
Beethoven and Haydn	49
Beethoven and Bach	39
Haydn and Bach	37
Mozart, Beethoven, and Haydn	22
Mozart, Beethoven, and Bach	19
Mozart, Haydn, and Bach	18
Beethoven, Haydn, and Bach	13
Mozart, Beethoven, Haydn, and Bach	9

How many students had listened to none of the composers?

7. A marketing class did a sample survey to find out how many of a class of 125 people owned CDs of the Beatles, Alabama, or Bob Marley. The results of the survey showed the following:

Recording Artist	No. of Students Owning CDs
Beatles	65
Alabama	46
Bob Marley	29
Beatles and Alabama	18
Beatles and Bob Marley	21
Bob Marley and Alabama	12
Beatles, Bob Marley, and Alabama	9

How many of the students owned no CD featuring these performers?

8. The language department wanted to know how many of the 2000 students at the university were not studying a language. Class rosters showed the number of students studying some combination of French, German, and Spanish, as recapped in the following table:

Language	No. of Students
French	75
German	68
Spanish	199
French and German	32
French and Spanish	41
German and Spanish	11
French and German and Spanish	7

How many students were not studying a language?

9. How many integers between 1 and 250 are divisible by 3 or 5?

10. In the game of tic-tac-toe, every game ends with one player winning or with a draw. In a tic-tac-toe tournament, the players merely count the number of times they win or draw. The match winner is the player with the larger total. If a match between two players A and B consists of 25 games, player A has a score of 19, and player B has a score of 23, how many draws were there?

11. There are 76 students enrolled in Anth229, Intermediate Anthropology. Each of these students is also required to enroll in either one or both of Biol313, Physiology, and Engl218, Victorian Poets. Of these 76 students, there are 35 in Biol313 and 49 in Engl218. How many students are enrolled in all three classes?

12. The enrollment for the four courses Biol212, Poli115, Econ313, and Fina215 is 108, 203, 315, and 212, respectively. No student is in all four of these courses. No student is in the three courses Biology 212, Fina215, and Poli115. No student takes Econ313 and Fina215 in the same semester. Poli115 and Fina215 are not allowed in the same term. There are 39 students in both Biol212 and Poli115, and 48 students in both Poli115 and Econ313 as well as in the two courses Biol212 and Econ313. Biol212, Poli115, and Econ313 have a common enrollment of 73. Biol212 and Fina215 have a common enrollment of 67. How many different students are enrolled in these four courses?

13. How many numbers between 1 and 1000 are not divisible by 3, 7, or 9?

14. How many integers between 500 and 10,000 are divisible by 5 or 7?

15. (a) How many numbers between 1 and 70,000,000, including both 1 and 70,000,000, are divisible by 2, 5, or 7?
    (b) How many numbers between 1 and 6,000,000, including both 1 and 6,000,000, are divisible by 4, 5, or 6?

16. Determine how many numbers between 1 and 21,000,000,000, including 1 and 21,000,000,000, are divisible by 2, 3, 5, or 7.

17. How many numbers between 1 and 21,000,000, including both 1 and 21,000,000, are divisible by 2, 3, or 5 but *not* by 7?

18. Find the number of integers between 1 and 1000, including both 1 and 1000, that are not divisible by any of 5, 6, or 8.

19. Find the number of integers between 1 and 1000, including 1 and 1000, that are not divisible by any of 4, 5, or 6.
20. Find the number of integers between 1 and 1000, including 1 and 1000, that are not divisible by any of 4, 6, 7, or 10.
21. (a) Extend Example 9 to cover four Victorian gentlemen and four top hats. With four gentlemen, there are $4 \times 3 \times 2 \times 1 = 24$ ways to give the hats back.
    (b) Modify part (a) to ask the number of ways, with four gentlemen and four hats, that *at least two* gentlemen can get their own hats back.
    (c) Solve Example 9 using an alternative proof that counts the number of ways that *no* gentleman gets his own hat back and subtracts that value from the total number of ways for the hats to be given back.
    (d) *Challenge:* Solve part (b) using the same methods as for part (c).

1.7   Mathematical Induction

Mathematical induction is a powerful and fundamental technique for proving results about *all* natural numbers. It is most important when it is possible to write down a proof for each individual natural number but difficult—or even impossible—to give a single direct proof that works for all natural numbers. This proof technique also often is used to prove that algorithms are correct and to determine expressions for the complexity of algorithms.

1.7.1   A First Form of Induction

One of the easiest methods (**algorithms**) for sorting a list of numbers into increasing order is called **selection sort.** This algorithm first finds the smallest element in the list and then interchanges it with the first element. After removing the smallest element from further consideration, the algorithm finds and removes from consideration the smallest element remaining (those elements other than the element now first in the list). This process is repeated until the list has just one element remaining. Since finding a smallest element in a set with n elements requires $n - 1$ comparisons, a selection sort, operating on $n + 1$ numbers, always makes

$$n + (n - 1) + (n - 2) + \cdots + 1$$

comparisons.

Example 1.   Carry out a selection sort on the list 2, 1, 4, 3, 5.

Solution.   In step i of the process, the ith smallest element is found among the elements in positions $i, i + 1, \ldots, 5$ and is interchanged with the element in position i where $1 \leq i \leq 4$. (See Selection Sort Steps on page 46.)

To appreciate how many comparisons are needed, it is necessary to find a simpler way to write the expression for the total number of comparisons. ∎

How do you go about adding up all the natural numbers from 0 to n where n can be 5, or 500, or 5000, or any other number? We all know how to do it in a tedious fashion for any particular n, but that brute force method does not give an easy way to appreciate the size of the sum for arbitrary n. (Nor does it give a way to compute the sum quickly.) The problem is to find a simpler way to express the sum.

Selection Sort Steps						
Initial order	2	1	4	3	5	
Step One	2	1	4	3	5	Identify smallest (nonboxed element) in four comparisons
	[1]	2	4	3	5	Swap 2 with 1
Step Two	[1]	2	4	3	5	Identify smallest (nonboxed element) in three comparisons
	[1]	[2]	4	3	5	No swap needed
Step Three	[1]	[2]	4	3	5	Identify smallest (nonboxed element) in two comparisons
	[1]	[2]	[3]	4	5	Swap 4 with 3
Step Four	[1]	[2]	[3]	4	5	Identify smallest (nonboxed element) in one comparison
	[1]	[2]	[3]	[4]	5	No swap needed
Step Five	[1]	[2]	[3]	[4]	5	Identify smallest (nonboxed element) in zero comparisons
Final Order	1	2	3	4	5	Number of comparisons $= 4 + 3 + 2 + 1 + 0$

One way to proceed is to try to find a pattern for small instances of the problem: Add up, say, the natural numbers from 0 to n for $n = 0, 1, 2, 3, 4$, and try to find a pattern. Patterns can be very misleading, however, because a pattern that may look correct for the first few numbers may very easily fail later on. If a possible pattern is found, it is necessary to prove whether it works in general. Consider the sums for the first few integers:

$$0 = 0$$
$$0 + 1 = 1$$
$$0 + 1 + 2 = 3$$
$$0 + 1 + 2 + 3 = 6$$
$$0 + 1 + 2 + 3 + 4 = 10$$
$$0 + 1 + 2 + 3 + 4 + 5 = 15$$

To find a different form for the problem often requires an idea that is not particularly obvious. In this case, if you multiply each of the sums by two and then factor the doubled value, you can hopefully see a pattern emerging. This transformation of the sums gives

$$2 \cdot 0 = \ \ 0 = 0 \cdot 1$$
$$2 \cdot (0 + 1) = \ \ 2 = 1 \cdot 2$$
$$2 \cdot (0 + 1 + 2) = \ \ 6 = 2 \cdot 3$$
$$2 \cdot (0 + 1 + 2 + 3) = 12 = 3 \cdot 4$$
$$2 \cdot (0 + 1 + 2 + 3 + 4) = 20 = 4 \cdot 5$$

The pattern that seems to be emerging is

$$2 \cdot (0 + 1 + 2 + \cdots + n) = n \cdot (n + 1)$$

It is not obvious that this formula is true for *all n*. It is true for $n = 0, 1, 2, 3$, and 4, but as yet, we have no reason to believe it is true for, say, $n = 12$, or 347, or any of the integers for which we have not shown it to be true.

What is needed is a method to *prove* that the conjectured formula is correct for all $n \in \mathbb{N}$. The standard method of proof for a result claimed to hold for every natural number is called **mathematical induction.** Such proofs use an axiom of arithmetic called the **Principle of Mathematical Induction.** This is not like Template 1.2 (Set Inclusion), since we are not proving the same thing for every element of $\mathbb{N}$. For example, for the sum of the first *n* integers, suppose we want to prove the sum is 6 for $n = 3$ but 15 for $n = 5$. Before stating the general principle, we present an example showing how the principle is used to prove that our conjectured formula for adding the natural numbers from 0 to *n* is true for all natural numbers *n*.

Theorem 1. For any natural number *n*,

$$0 + 1 + 2 + \cdots + n = \frac{n \cdot (n + 1)}{2}$$

Proof.   Step 1: **(Base step)** Prove the result for $n = 0$, the smallest natural number. The sum on the left-hand side of the equals sign is just the sum of all the natural numbers starting at 0 and going up to 0—that is, it is just 0. The number on the right-hand side is $0 \cdot (0 + 1)/2$, which is also 0. Therefore, the two sides are equal, and the result is true for $n = 0$.

Step 1: **(Inductive step)** Let *n* be any natural number for which the result is true. Prove the result is also true for $n + 1$. The assumption that the result is true for *n* is called the **inductive hypothesis** or **inductive assumption.** Assuming the result is true for *n* means that

$$0 + 1 + 2 + \cdots + n = \frac{n \cdot (n + 1)}{2}$$

Use this assumed-correct result to prove the required result for $n + 1$—that is, to prove that

$$0 + 1 + 2 + \cdots + n + (n + 1) = \frac{(n + 1) \cdot (n + 2)}{2}$$

To prove this, we start by regrouping the terms on the left-hand side:

$$0 + 1 + 2 + \cdots + n + (n + 1) = (0 + 1 + 2 + \cdots + n) + (n + 1)$$

By the inductive hypothesis, the result is true for *n*, so we can substitute $n(n + 1)/2$ for the terms in the first pair of parentheses on the right-hand side. We get

$$(0 + 1 + 2 + \cdots + n) + (n + 1) = \frac{n \cdot (n + 1)}{2} + (n + 1) \quad \text{(using the inductive hypothesis)}$$

$$= \frac{n \cdot (n + 1) + 2 \cdot (n + 1)}{2} \quad \text{(simplifying the algebra)}$$

$$= \frac{(n + 1) \cdot (n + 2)}{2}$$

This means the formula is true for $n + 1$.

Since we have proved that the formula is true for $n = 0$ and is true for $n + 1$ whenever it is true for n, we can conclude that the formula is valid for all natural numbers. This reasoning is call the Principle of Mathematical Induction. ■

Let $T = \{n \in \mathbb{N} : 0 + \cdots + n = n(n+1)/2\}$:

1. Since $0 \in T$ by the base step, by the inductive step, $0 + 1 = 1 \in T$.
2. Apply the inductive step again: since $1 \in T$, $1 + 1 = 2 \in T$.
3. And again: since $2 \in T$, $2 + 1 = 3 \in T$.

To prove that $100 \in T$, apply the inductive step 100 times. To prove that $10{,}000 \in T$, apply it 10,000 times. For any specific natural number n, one can show that $n \in T$ by showing that $0 \in T$ and then applying the inductive step n times. An inductive proof is often visualized as an infinite line of dominoes, with the dominoes being pushed over one at a time starting with the first one. Figure 1.16 gives another way of thinking about what happens in an inductive proof.

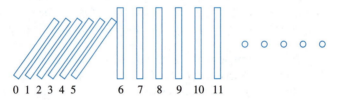

0 1 2 3 4 5   6 7 8 9 10 11

Figure 1.16   Falling dominoes.

A First Form of the Principle of Mathematical Induction

The Principle of Mathematical Induction gives a method for writing a *single* proof that proves *all* natural numbers are in T. Sometimes, this statement of the Principle of Mathematical Induction is called its **first form.**

Principle of Mathematical Induction

Let T be a subset of the natural numbers (that is, $T \subseteq \mathbb{N}$), and let $n_0 \in \mathbb{N}$. Suppose

(Base step) $n_0 \in T$, and

(Inductive step)   for all natural numbers n such that $n \geq n_0$, if $n \in T$, then $n + 1 \in T$.

Then, every natural number greater than or equal to n_0 is in T. That is,

$$T = \{n : n \in \mathbb{N} \text{ and } n \geq n_0\}$$

In the proof of Theorem 1, T was defined to be the set of all natural numbers for which the formula

$$\sum_{i=0}^{n} i = n(n+1)/2$$

is true. So, in that case, we choose $n_0 = 0$. The base step of that proof showed that $n_0 = 0 \in T$. The inductive step showed that if $n \in T$, then $n + 1 \in T$. Now, by the Principle of Mathematical Induction, $T = \{n \in \mathbb{N} : n \geq n_0 = 0\} = \mathbb{N}$.

We give a picture of what is involved in an inductive proof in Figure 1.17.

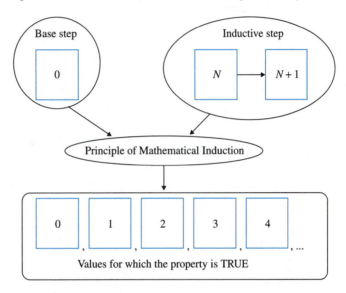

Figure 1.17   The parts of an inductive proof.

1.7.2   A Template for Constructing Proofs by Induction

Template 1.12 should help you to understand and construct a proof by induction.

Template 1.12   Using the Principle of Mathematical Induction

To construct a proof using the **Principle of Mathematical Induction,** choose an $n_0 \in \mathbb{N}$ appropriate to the problem. Let $T = \{n \in \mathbb{N} : n \geq n_0$ and property P holds for $n\}$:

- **(Base step)**   Prove that $n_0 \in T$.
- **(Inductive step)**   Let $n \in T$, and prove that $n + 1 \in T$. The assumption that $n \in T$ is called the inductive hypothesis.
- Infer by the Principle of Mathematical Induction that every natural number $n \geq n_0$ is in T.

The examples that follow show the power of this proof method. Some of the inequalities verified here by induction will appear again in later chapters when we consider the complexity of programs.

Example 2.   For any natural number n such that $n \geq 2$, show that $n + 1 < n^2$. Since we wish to prove our result for every n such that $n \geq 2$, we must choose $n_0 = 2$ and let

$$T = \{n \in \mathbb{N} : n \geq 2 \text{ and } n + 1 < n^2\}$$

According to our template, the proof now has three essential parts: (i) a base step, (ii) an inductive step, and (iii) an application of the Principle of Mathematical Induction.

For the base step, we must prove that $n_0 \in T$. In this case, we must prove for $n_0 = 2$ that $n_0 + 1 < n_0^2$. When the proof of the base step is complete, we know $T \neq \emptyset$, because $n_0 \in T$. We would then like to know what elements greater than n_0 are also in T. The elements of T other than n_0 are found using the inductive step and the Principle of Mathematical Induction.

The inductive step begins by picking an arbitrary element n of T. We then write out property P for n to see what this assumption tells us. Here, it means $n \geq 2$ and $n + 1 < n^2$.

To complete the inductive step, we must show that $n + 1 \in T$. We write out property P for $n + 1$ to see what we need to prove. In this case, it means that $n + 1 \geq 2$ and $(n + 1) + 1 < (n + 1)^2$. We must then figure out how to prove that property P holds for $n + 1$ knowing that property P is true for n. When we complete this proof, the Principle of Mathematical Induction tells us that for all n such that $n \geq n_0$, we have $n \in T$.

Solution.   Let $n_0 = 2$. Let $T = \{n \in \mathbb{N} : n \geq 2 \text{ and } n + 1 < n^2\}$. Prove by induction that $n \in T$ provided $n \geq 2$.

(Base step)   To show that $n_0 \in T$, show that $2 \geq 2$ and $2 + 1 < 2^2$. Both are obviously true. Therefore, $2 \in T$.

(Inductive step)   Let $n \geq n_0$. Show that if $n \in T$, then $n + 1 \in T$. That is, assume $n \geq 2$ and $n + 1 < n^2$, and prove that (i) $n + 1 \geq 2$ and (ii) $(n + 1) + 1 < (n + 1)^2$. To prove (i), observe that since $n \in T$, we have $n \geq 2$. Therefore, $n + 1 \geq 2$. To prove (ii), use the following chain of equalities and inequalities:

$$(n + 1)^2 = n^2 + 2n + 1$$
$$> (n + 1) + 2n + 1 \quad \text{(using the inductive hypothesis: } n^2 > n + 1)$$
$$> (n + 1) + 1 \quad \text{(using the inductive hypothesis: } n \geq 2 > 0)$$

Therefore, $n + 1 \in T$.

By the Principle of Mathematical Induction, $T = \{n \in \mathbb{N} : n \geq 2\}$. ■

You can see the parts of the template being used as you study Example 3. Identify the steps of the template as they appear in this example.

Example 3.   Recall that $n! = n \cdot (n - 1) \cdot (n - 2) \cdots 2 \cdot 1$. For any natural number n such that $n \geq 4$, prove that $n! > n^2$.

Solution.   Let $n_0 = 4$. Let $T = \{n \in \mathbb{N} : n \geq 4 \text{ and } n! > n^2\}$. Prove by induction for every natural number n that $n \in T$ provided that $n \geq 4$.

(**Base step**)   Show that $4 \in \mathcal{T}$. Since $4! = 24$ and $4^2 = 16$, we have $4! > 4^2$, so $4 \in \mathcal{T}$.

(**Inductive step**)   Let $n \in \mathcal{T}$, then show that $n + 1 \in \mathcal{T}$. As before, it is trivial to show that $n + 1 \geq 4$, so it only remains to show that $(n + 1)! > (n + 1)^2$. To prove this, use the following chain of equalities and inequalities:

$$
\begin{aligned}
(n + 1)! &= (n + 1) \cdot n! &&\text{(definition of } n!)\\
&> (n + 1) \cdot n^2 &&\text{(using the inductive hypothesis)}\\
&> (n + 1) \cdot (n + 1) &&\text{(use Example 2 in Section 1.7.2 with } n \geq 4 \geq 2)\\
&= (n + 1)^2
\end{aligned}
$$

Therefore, $n + 1 \in \mathcal{T}$.

By the Principle of Mathematical Induction, $\mathcal{T} = \{n \in \mathbb{N} : n \geq 4\}$. ∎

You should now show that n_0 could not be chosen smaller.

In Examples 2 and 3 in Section 1.7.2 we did not prove the results were true for all $n \in \mathbb{N}$. It is quite typical that important relations may not be true for some finitely many small integers and, instead, are only true for all integers greater than or equal to some "large" integer.

1.7.3   Application: Fibonacci Numbers

A famous and often-studied sequence of numbers, called the **Fibonacci numbers,** was defined by Leonardo Fibonacci (1170–1250, born in Italy).[2] The first few numbers in this sequence are

$$1, 1, 2, 3, 5, 8, 13, 21, \ldots$$

Denote the nth Fibonacci number by F_n, and let the first element of the sequence be denoted as F_0. The defining rule for the elements of this sequence is $F_0 = F_1 = 1$ and $F_n = F_{n-1} + F_{n-2}$ for $n \geq 2$. After the initial values given for F_0 and F_1, the following Fibonacci numbers can be found by adding together the two previous Fibonacci numbers; for example, F_2 is the sum of F_1 and F_0. The first six Fibonacci numbers are

$$
\begin{aligned}
F_0 &= F_1 = 1\\
F_2 &= F_1 + F_0 = 2\\
F_3 &= F_2 + F_1 = 3\\
F_4 &= F_3 + F_2 = 5\\
F_5 &= F_4 + F_3 = 8
\end{aligned}
$$

We computed the nth Fibonacci number for $n \geq 2$ by adding together the preceding two Fibonacci numbers. A definition of this sort is called a **recursive definition,** because the value we want is given in terms of previously computed values. (We could not compute a value for F_4 directly from the value 4 as we could if the sequence were defined as $G(n) = 4 \cdot n$.) The resulting sequence is called a **recursively defined sequence.**

The Fibonacci numbers are probably best known as a source of recreational mathematics but are also the source of inspiration for searching and sorting methods. Many results concerning Fibonacci numbers are proved by induction. Example 4 shows a typical proof.

---

[2] We will abbreviate *born* to *b.* for other famous persons.

The Fibonacci numbers were defined by Leonardo of Pisa, *filius* (son of) Bonacci, who lived around 1200. Leonardo developed the sequence in predicting the size of a population of rabbits. F_n is the number of pairs of rabbits he predicted one would have n months after buying a pair of baby rabbits under the assumption that a pair of rabbits matured in one month and produced a pair of offspring each month thereafter.

Month (n)	Old Pairs	New Pairs	F_n
0			1
1			1
2			2
3			3
4			5
5			8

Furthermore, he assumed that the rabbits always produced a male and a female as each pair of offspring. So, $F_0 = 1$ for the pair just purchased. $F_1 = 1$, because after one month, the original pair has just matured and are only now ready to start breeding. $F_2 = 2$, because the original pair has just had one pair of offspring. $F_3 = 3$, because the original pair has had another pair of offspring and the first offspring have just matured and are only now ready to start breeding. What happens during month n? All the rabbits alive during month $n - 1$ are still alive. In addition, all the rabbits alive during month $n - 2$ have matured, and each pair has had one pair of offspring. Hence, $F_n = F_{n-1} + F_{n-2}$.

Example 4. Show that the identity $F_1 + F_3 + F_5 + \cdots + F_{2n-1} = F_{2n} - 1$ is true for all $n \geq 1$.

Solution. Let $n_0 = 1$. Prove the identity by induction on n. Let

$$T = \{n \in \mathbb{N} : n \geq 1 \text{ and } F_1 + F_3 + \cdots + F_{2n-1} = F_{2n} - 1\}$$

Prove that $T = \{n \in \mathbb{N} : n \geq 1\}$.

(Base step) Prove the result for $n = 1$. The left-hand side in this case is just the sum of all the Fibonacci numbers starting with F_1 and ending with $F_{(2 \cdot 1 - 1)}$. There is just one such Fibonacci number, F_1, and the value of the left-hand side is 1. The right-hand side is $F_{2 \cdot 1} - 1 = F_2 - 1 = 2 - 1 = 1$. So, the two sides are equal, and $1 \in T$.

(Inductive step)   Let $n \geq n_0$. Show that if $n \in \mathcal{T}$, then $n + 1 \in \mathcal{T}$. Since $n \geq 1$, $n + 1 \geq 1$. Assume for n that

$$F_1 + F_3 + \cdots + F_{2n-1} = F_{2n} - 1$$

and prove that

$$F_1 + F_3 + \cdots + F_{2n-1} + F_{2(n+1)-1} = F_{2(n+1)} - 1$$

The required computation is

$$
\begin{aligned}
&F_1 + F_3 + \cdots + F_{2n-1} + F_{2(n+1)-1} \\
&= (F_1 + F_3 + \cdots + F_{2n-1}) + F_{2(n+1)-1} && \text{(making the formula for } n \text{ clear)} \\
&= F_{2n} - 1 + F_{2(n+1)-1} && \text{(using the inductive hypothesis)} \\
&= (F_{2n} + F_{2(n+1)}) - 1 && \text{(rearranging terms)} \\
&= F_{2n+2} - 1 && \text{(using the definition of } F_{2n+2}) \\
&= F_{2(n+1)} - 1
\end{aligned}
$$

Therefore, $n + 1 \in \mathcal{T}$.

By the Principle of Mathematical Induction, $\mathcal{T} = \{n \in \mathbb{N} : n \geq 1\}$.   ■

1.7.4   Application: Size of a Power Set

The next result was referred to in the discussion of computer switches in Section 1.3.4 and will be proved several times in the book using several different ideas. Recall that $\mathcal{P}(X)$, the power set of X, is the set of all subsets of X.

Theorem 2.   (Size of a Power Set)   Let X be any finite set with n elements. Then, $\mathcal{P}(X)$ has 2^n elements.

The proof of Theorem 2 can be proved by induction on the number of elements in X. First, we prove an auxiliary result called a **lemma**. A lemma is the same as a theorem, except that the result is not particularly important in its own right but only gives a step in another proof. Just as procedures divide programs into manageable parts, lemmas are tools for dividing a proof into smaller, more comprehensible pieces.

Lemma 1:   Let X be any set, and let $b \notin X$. If X has (exactly) n subsets, then $X \cup \{b\}$ has exactly $2n$ subsets.

Proof.   List the subsets of X:

$$S_1, S_2, S_3, \ldots, S_n$$

Each of these is also a subset of $X \cup \{b\}$. Now, create n more subsets of $X \cup \{b\}$:

$$S_1 \cup \{b\}, S_2 \cup \{b\}, \ldots, S_n \cup \{b\}$$

Obviously, each $S_i \cup \{b\}$ is also a subset of $X \cup \{b\}$. Now, we have a list of $2n$ subsets of $X \cup \{b\}$:

$$S_1, S_2, \ldots, S_n, S_1 \cup \{b\}, S_2 \cup \{b\}, \ldots, S_n \cup \{b\}$$

Show that (i) no subset of $X \cup \{b\}$ appears twice in this list and (ii) every subset of $X \cup \{b\}$ appears in this list.

Once these two assertions have been proven, it will follow that these $2n$ subsets are all the subsets of $X \cup \{b\}$, so $X \cup \{b\}$ has $2n$ subsets. We prove (ii) and leave the proof of (i) as Exercise 32 in Section 1.9 for the reader.

To prove (ii), follow Template 1.1 (Element Membership in a Set). Let S be an arbitrary subset of $X \cup \{b\}$. If $b \notin S$, then $S \subseteq X$, so S is one of the S_i's. If $b \in S$, let $S' = S - \{b\}$. Then, $S' \subseteq X$, so S' is some S_i, and then S is $S_i \cup \{b\}$ (for the same i). In either case, S is on the list. ∎

Proof of Theorem 2.   Let $\mathcal{T} = \{n \in \mathbb{N} : \text{for every finite set } X \text{ with } n \text{ elements}, \mathcal{P}(X) \text{ has } 2^n \text{ subsets}\}$. We will prove by induction that $\mathcal{T} = \mathbb{N}$.

(Base step)   Let $n_0 = 0$. The only set with zero elements is $\emptyset$. The only subset of $\emptyset$ is $\emptyset$, so $\mathcal{P}(\emptyset)$ has $1 = 2^0$ elements. Therefore, $0 \in \mathcal{T}$.

(Inductive step)   Let $n \geq 0$. Show that if $n \in \mathcal{T}$, then $n + 1 \in \mathcal{T}$. Using the hypothesis that every set with n elements has 2^n subsets, prove that every set with $n + 1$ elements has 2^{n+1} subsets. Let X be an arbitrary set with $n + 1$ elements. Pick one element $y \in X$, and let $Z = X - \{y\}$. Then, Z has n elements, so by the hypothesis, Z has 2^n subsets. By Lemma 1, $X = Z \cup \{y\}$ has $2 \cdot 2^n = 2^{n+1}$ subsets. Therefore, $n + 1 \in \mathcal{T}$.

By the Principle of Mathematical Induction, $\mathcal{T} = \mathbb{N}$. ∎

1.7.5   Application: Geometric Series

A **finite geometric series**, or just a **geometric series**, is the sum of terms of the form $a \cdot r^i$ where $a, r \in \mathbb{R} - \{0\}$, $r \neq 1$, and $0 \leq i \leq n$. For example,

$$\sum_{i=0}^{n} a \cdot r^i = a + a \cdot r + a \cdot r^2 + \cdots + a \cdot r^n$$

is a geometric series. As another example, let $a = 5$, $r = -3$, and $n = 5$, giving

$$5 + 5 \cdot (-3) + 5 \cdot (-3)^2 + 5 \cdot (-3)^3 + 5 \cdot (-3)^4 + 5 \cdot (-3)^5$$

as a geometric series. Although

$$\sum_{i=5}^{20} 3 \cdot 2^i$$

does not look like a geometric series, it is easy to transform this finite geometric series into a more familiar looking expression:

$$\sum_{i=5}^{20} 3 \cdot 2^i = 3 \cdot 32 + 3 \cdot 64 + \cdots + 3 \cdot 2^{20}$$

$$= 96 \cdot 1 + 96 \cdot 2 + \cdots + 96 \cdot 2^{15}$$

$$= \sum_{i=0}^{15} 96 \cdot 2^i$$

A very useful feature of a geometric series is that we can find a closed form for its sum. Here, we focus on the sum of a finite geometric series. The sum of an infinite geometric series is usually studied in a calculus course, since the limiting process is needed. Although it seems to be unrelated at first, we will begin by proving that for any $n \in \mathbb{N}$, $1 - x^{n+1}$ has $1 - x$ as a factor. After proving this by induction, we will apply the result to summing the finite geometric series.

Theorem 3.   For any natural number n and for any real number x, prove that

$$(1 - x)(1 + x + x^2 + \cdots + x^n) = 1 - x^{n+1}$$

Solution.   This result is just a familiar factoring rule. The ellipses usually suggest that a proof by induction is needed. Fix an arbitrary $x \in \mathbb{R}$. Let $n_0 = 0$ and

$$T = \{n \in \mathbb{N} : \text{for any } x \in \mathbb{R}, (1 - x)(1 + x + x^2 + x^3 + \cdots + x^n) = 1 - x^{n+1}\}$$

(Base step)   Show that $0 \in T$. Substituting 0 for n gives $(1 - x)(1) = 1 - x^1$ as required.

(Inductive step)   Let $n \geq 0$. Show that if $n \in T$, then $n + 1 \in T$. Since $n \in T$, it is assumed that

$$(1 - x)(1 + x + x^2 + x^3 + \cdots + x^n) = 1 - x^{n+1}$$

We must prove that $n + 1 \in T$ or that

$$(1 - x)(1 + x + x^2 + x^3 + \cdots + x^{n+1}) = 1 - x^{n+2}$$

Use the following chain of equalities to complete the proof:

$(1 - x)(1 + x + x^2 + x^3 + \cdots + x^n + x^{n+1})$
$\quad = (1 - x)(1 + x + x^2 + x^3 + \cdots + x^n) + (1 - x)x^{n+1}$ (making the formula for n clear)
$\quad = 1 - x^{n+1} + (1 - x)x^{n+1}$ (using the inductive hypothesis)
$\quad = 1 - x^{n+1} + x^{n+1} - x^{n+2}$ (simplifying the expression)
$\quad = 1 - x^{n+2}$

Therefore, $n + 1 \in T$.

By the Principle of Mathematical Induction, $T = \mathbb{N}$. ∎

Corollary 1:   For $r \in \mathbb{R}$ with $r \neq 1$,

$$\sum_{i=0}^{n} a \cdot r^i = a \cdot \frac{1 - r^{n+1}}{1 - r}$$

Proof. $\sum_{i=0}^{n} a \cdot r^i = a \sum_{i=0}^{n} r^i = a \cdot \frac{1 - r^{n+1}}{1 - r}$ ∎

Corollary 1 gives a formula for finding the sum of a finite geometric series.

Example 5.

(a) $1 + 3 + 3^2 + 3^3 + \cdots + 3^n = (1 - 3^{n+1})/(-2) = (3^{n+1} - 1)/2.$
(b) $2 + 10 + 50 + \cdots + 1250 = 2 + 2 \cdot 5 + 2 \cdot 5^2 + 2 \cdot 5^3 + 2 \cdot 5^4$
$$= 2(1 - 5^5)/(-4)$$
$$= 1562$$ ∎

The next example shows how to compute the sum when the first term does not clearly correspond to what is expected for a term of the form $a \cdot r^0$.

Example 6.   Find the sum of

$$3 \cdot 2 + 3 \cdot 2^2 + 3 \cdot 2^3 + \cdots + 3 \cdot 2^n$$

Solution.   Rewrite the expression with $3 \cdot 2 = 6$ as a factor of each term:

$$(6 \cdot 2^0 + 6 \cdot 2^1 + \cdots + 6 \cdot 2^{n-1})$$

We now have a geometric series with n terms with $a = 6$ and $r = 2$. The sum is

$$\sum_{i=0}^{n-1} 6 \cdot 2^i = 6 \cdot (1 - 2^n)/(-1) = 6 \cdot (2^n - 1)$$ ∎

1.8   Program Correctness

An important problem in computer science is to prove that a program executes correctly for all possible data sets. There is no simple way to do this. In fact, it is impossible in principle to prove **correctness** for very complicated programs. Many techniques, however, are useful for proving the correctness of a wide variety of programs.

One method for checking the correctness of a program is to test the program on lots of data to make sure it comes up with the right answers in each case. Obviously, this technique is useful, but it can be used only to find errors, not to establish correctness. The problem is that if no errors are found by running a program on test data, the only conclusion one can draw with any assurance is that the program works correctly on all the data tested.

Another useful technique is to prove mathematically that the algorithms (the principles behind the program) are correct. Such proofs are often proofs by induction.

Before we examine some algorithms, we need to explain how we will present the steps of an algorithm. The language we use is called a **pseudocode,** because it is a mixture of normal language and the precise syntax of a programming language.

1.8.1   Pseudocode Conventions

A *variable* will simply be a name that will represent a place in a computer to store a value. When we use a variable, like X, we are referring to the value that is stored in the location the machine assigns to X. A simple assignment statement of the form

$$variable = expression$$

computes the value of the expression on the right-hand side of the equal sign and then stores the result in the location indicated by the name on the left-hand side. To cause branching in the code, we use a condition test of the form

$$\text{if } condition \text{ then}$$
$$S_1$$
$$\text{else}$$
$$S_2$$

When this code is executed, the **condition** is evaluated to be either *TRUE* or *FALSE.* If the condition is *TRUE,* then the code represented by S_1 is executed, the code represented by S_2 is not executed, and the execution then continues at the first command following S_2. If the condition is *FALSE,* then the code represented by S_2 is executed, the code represented by S_1 is not executed, and the execution then continues at the first command following S_2.

For a statement that can cause repetition of a block of code, we generally use a *for* construct. The code

$$\text{for } i = 1 \text{ to } n \text{ do}$$
$$S$$

starts by initializing i to the value 1. If the value of i is less than or equal to n, the commands represented by S will then be executed. The code S may or may not use i as a variable. At the end of executing this indented code, i will be incremented by 1 and then tested to see if it is still less than or equal to n. If this condition for the current value of i is evaluated as *TRUE,* the loop is executed again using the new value of i. When the condition is tested with a value of i for which the condition is evaluated as *FALSE,* the program continues at the next line following the code represented by S. We often refer to such code as a **for loop.** To display a result, we use the word *print* followed by a list of the names of the storage locations whose values are to be displayed (think of print on the screen or of output to a printer). Comments in the code will appear as /* *any text as a comment* */. Comments are skipped over when the program is executed.

With just these four instructions (assignment, condition, repetition, and printing) for pseudocode, we can write instructions that could easily be turned into valid code in some programming language.

An additional way to cause repetition of a block of code is with use of a **while loop:**

$$\text{while } condition$$
$$S$$

A while statement is a command to execute the code indented below the while statement over and over again as long as the condition written just after the word *while* is evaluated as *TRUE.* If this condition is evaluated as *FALSE* when the loop is first reached, the indented statements are executed zero times, that is, they are not executed.

Many authors use the word *algorithm* to describe only strategies for programs that will ultimately stop. Others would say there is no "output" unless it stops. We include our apologies for our use of the word and present the following program as algorithm. The algorithm will use a while loop to "repeat forever" a block of code, since the condition in the while statement can never be false. This is just an instruction to execute the while loop without stopping—or until someone turns off the computer. In this case, it is called an **infinite loop,** since it could go on forever!

1.8.2   An Algorithm to Generate Perfect Squares

We now demonstrate how you can generate all the perfect squares. A **perfect square** is any integer n that is equal to k^2 for some integer k. For example, 1 is a perfect square, because it is equal to 1^2. In addition 9 is a perfect square, because 9 equals 3^2. For the **Perfect Squares algorithm,** we give an intuitive argument that the program is correct.

Algorithm: Perfect Squares

INPUT:
OUTPUT: List of perfect squares

Counter $= 0$
while (*TRUE*) /* repeat forever */
    Counter $=$ *Counter* $+ 1$
    print *Counter* $\cdot$ *Counter*

To understand the Perfect Squares algorithm, trace its execution for the first few values of *Counter.* The algorithm starts with *Counter* equal to 0 and then repeats the last two instructions forever. The first time through, the algorithm adds 1 to *Counter,* giving *Counter* the value of 1, and prints $1 \cdot 1$. The second time through, it adds 1 to *Counter,* increasing *Counter* from 1 to 2, and prints $2 \cdot 2$. The third time, it adds 1 to *Counter,* increasing *Counter* from 2 to 3, and prints out $3 \cdot 3$. And so forth. It is obvious that the algorithm works. In fact, for any natural number k, after the kth time through the loop, *Counter* is set to k and the first k perfect squares have been printed.

1.8.3   Two Algorithms for Computing Square Roots

There are many algorithms for finding the square root of an integer. The two algorithms presented here use different strategies for finding a better and better approximation of a square root. The first was found on ancient Babylonian cuneiform tablets. The second is a variant of one that has been taught in schools. A fundamental problem with any approximation algorithm is to have a bound on how far an approximation is from the true value. For both algorithms presented here, we can prove a result about the bound using induction.

Square Root I

The **Square Root I algorithm** provides a method of finding an approximation to the square root of an integer. In this case, the first approximation is a value less than the desired result. Each iteration of the procedure gives a larger value than the previous one.

Algorithm: Square Root I

RESULT: Approximation of $\sqrt{17}$

$Root = 4$
$DecimalPlaceValue = 1$
for $i = 1$ to 8 do
    $DecimalPlaceValue = DecimalPlaceValue/10$
        /* Search for the digit at the decimal place.*/
    $Digit = 9$
        /* 9 is the largest possible value for *Digit*. */
    $AddOn = Digit \cdot DecimalPlaceValue$
    while$((Root + AddOn) \cdot (Root + AddOn) > 17)$ do
        /* *Digit* is too big, so try a smaller value. */
        $Digit = Digit - 1$
        $AddOn = Digit \cdot DecimalPlaceValue$
        /* At the end of the **while loop** the next digit is found. */
    $Root = Root + AddOn$
print $Root$

The code starts by approximating $\sqrt{17}$ by 4. The variable *DecimalPlaceValue* is used to keep track of which decimal digit is being added to the approximation. When i is equal to n, *DecimalPlaceValue* will be equal to 10^{-n}. The first value added to the previous approximation is $Digit \cdot 10^{-n}$ where *Digit* is 9. The while loop sees if adding $Digit \cdot 10^{-n}$ gives a new value for the approximation by computing

$$(Root + Digit \cdot 10^{-n})^2 < 17$$

If the value of *Digit* gives

$$(Root + Digit \cdot 10^{-n})^2 > 17$$

a new, smaller value of *Digit* is tried. At some point, *Digit* will take on a first value, say $Digit_{first}$, for which

$$(Root + Digit_{first} \cdot 10^{-n})^2 \leq 17$$

The new approximation for $\sqrt{17}$ will be formed by adding $Digit_{first} \cdot 10^{-n}$ to *Root* to form the next approximation. The first iteration of the for loop gives the value 1 for $Digit_{first}$. Consequently,

$$Root = 4 + 1 \cdot 10^{-1} = 4.1$$

for $i = 1$. Now, in the second iteration, $Digit_{first}$ takes the value 2, so

$$Root = 4.1 + 2 \cdot 10^{-2} = 4.12$$

The process continues to add decimal digits to the intial approximation for as many iterations of the for loop as are required by the code. For the code shown, the final approximation is $Root = 4.12310562$ and $Root \cdot Root = 16.9999999536$.

For Square Root I, we can find an explicit formula for a bound on the error after n iterations. Let R_n denote the value of $Root$ after n iterations of the for loop. The error term is defined as $\epsilon_n = \sqrt{17} - R_n$ for $n \in \mathbb{N}$.

Theorem 4.   Prove that for Square Root I, the error bound ϵ_n for R_n satisfies the inequality $\epsilon_n < 10^{-n}$ for each $n \in \mathbb{N}$.

Proof.   Let $n_0 = 0$. Let $\mathcal{T} = \{n \in \mathbb{N} : R_n \leq \sqrt{17} < R_n + 10^{-n}\}$.

(Base step)   For $n = 0$, the result follows, since $Root$ is 4 and the for loop is not executed. Clearly, $4 < \sqrt{17} < 5$, so $\epsilon_0 = \sqrt{17} - 4 < 1 = 10^{-0}$. Therefore, $0 \in \mathcal{T}$.

(Inductive step)   Choose $n \geq n_0$ such that $n \in \mathcal{T}$. Now, prove that $n + 1 \in \mathcal{T}$. That is, assume $R_n \leq \sqrt{17} < R_n + 10^{-n}$, and prove that $R_{n+1} \leq \sqrt{17} < R_{n+1} + 10^{-(n+1)}$. By the inductive hypothesis,

$$R_n \leq \sqrt{17} < R_n + 10^{-n} = R_n + 10 \cdot 10^{-(n+1)}$$

The search for $Digit$ finds the largest integer $Digit$ where $R_{n+1} + Digit \cdot 10^{-(n+1)} \leq \sqrt{17}$. Since $Digit$ is the largest such integer,

$$R_n + Digit \cdot 10^{-(n+1)} \leq \sqrt{17} < (R_n + Digit \cdot 10^{-(n+1)}) + 10^{-(n+1)}$$

Since $Digit + 1$ has the property that

$$(R_n + (Digit + 1) \cdot 10^{-(n+1)})^2 > 17$$

then

$$R_n + Digit \cdot 10^{-(n+1)} = R_{n+1} \leq \sqrt{17} < R_n + Digit \cdot 10^{-(n+1)} + 10^{-(n+1)}$$
$$= R_{n+1} + 10^{-(n+1)}$$

as desired, and $n + 1 \in \mathcal{T}$.

Therefore, $\mathcal{T} = \mathbb{N}$ by the Principle of Mathematical Induction.   ∎

Square Root II

The **Square Root II algorithm** produces an approximation of the square root of an integer by generating approximations that are alternately larger than the square root and then smaller than the square root. Each iteration of the procedure, however, brings the value of the approximation closer to the true value of the square root.

Algorithm: Square Root II

RESULT: Approximation of $\sqrt{17}$

$Root = 4$
for $i = 1$ to 4 do
    $Root = (Root + 17/Root)/2$
print $Root$

The computation starts by assigning 4 to *Root*. The value in *Root* at any time will represent the current approximation to $\sqrt{17}$. For each iteration of the for loop, the current approximation is improved by evaluating the expression

$$(Root + 17/Root)/2.$$

and storing the "better" approximation in *Root*. This process continues until the for loop has been executed four times. The value of *Root* after each of the first four iterations is shown in Table 1.2.

Table 1.2   Output from Square Root II

Values of *Root* for I = 1, 2, 3, 4	
I	*Root*
1	4.125
2	4.12310606060606
3	4.12310562561768
4	4.12310562561766

With any iterative algorithm, it is important to know with each iteration that the error gets smaller. Let R_n denote the value of *Root* after the for loop has been executed n times. Then, as before $\epsilon_n = \sqrt{17} - R_n$ is the error in the calculation after n executions of the for loop. The error can be either positive or negative.

Theorem 5.   Prove that for Square Root II, the error bound ϵ_n for R_n satisfies the inequality $|\epsilon_n| < (1/2)^{6 \cdot 2^n - 3}$ for each $n \in \mathbb{N}$.

Proof.   Let $n_0 = 0$. Let $\mathcal{T} = \{n \in \mathbb{N} : |\epsilon_n| < (1/2)^{6 \cdot 2^n - 3}\}$.

(Base step)   Since $(4.1)^2 = 16.81$ and $(4.125)^2 = 17.015625$, it follows that

$$4.1 < \sqrt{17} < 4.125$$

Now, $R_0 = 4$, so

$$4.1 - R_0 < \sqrt{17} - R_0 < 4.125 - R_0$$
$$\epsilon_0 < 0.125 \qquad (R_0 = 4)$$
$$\epsilon_0 < (1/2)^{6 \cdot 2^0 - 3}$$

Therefore, $0 \in T$.

(Inductive step)   The remainder of the proof is left as an exercise for the reader.   ■

Exercise 39 in Section 1.9 explores other properties of this algorithm.

1.9   Exercises

Assume that all variables not given an explicit domain are elements of $\mathbb{N}$.

1. Show that for $n = 0, 1, 2$ the following is true:
$$1^2 + 2^2 + 3^2 + \cdots + n^2 = n(n+1)(2n+1)/6$$

2. Find all the elements of $\{0, 1, 2, 3\}$ that, when substituted for n, satisfy:
$$\frac{1}{1 \cdot 2} + \frac{1}{2 \cdot 3} + \cdots + \frac{1}{n(n+1)} = \frac{n}{n+1}$$

3. Write out the information that describes what the inductive step assumes and what the step must prove for proving
$$1^2 + 2^2 + 3^2 + \cdots + n^2 = n(n+1)(2n+1)/6$$
with n_0 given.

4. Write out the information that describes what the inductive step assumes and what the step must prove for proving
$$1^5 + 2^5 + 3^5 + \cdots + n^5 = \frac{1}{6}n^6 + \frac{1}{2}n^5 + \frac{5}{12}n^4 - \frac{1}{12}n^2$$
with n_0 given.

5. Write out the information that describes what the inductive step assumes and what the step must prove for proving that 6 divides $n^3 + 5n$ with n_0 given.

6. Write out the information that describes what the inductive step assumes and what the step must prove for proving that 120 divides $n^5 - 5n^3 + 4n$ with n_0 given.

7. Show for $n = 0, 1, 2$ that
$$(n+1)(2n+1)(2n+3)/3 + (2n+3)^2 = (n+2)(2n+3)(2n+5)/3$$

8. Show that
$$(n+1)(2n+1)(2n+3)/3 + (2n+3)^2 = (n+2)(2n+3)(2n+5)/3$$

9. Show that
$$n^2 + n + 2(n+1) = (n+1)^2 + (n+1)$$

10. Show that
$$\sum_{i=0}^{n} F_{2i+1} = F_{2n+2} - 1$$
for $n = 1, 2, 3, 4$.

11. For which elements $n \in \{0, 1, 2, 3, 4, 5\}$ does 6 divide $n^3 + 5n$?

12. Show that 8 divides $k^2 - 1$ for $k \in \{1, 3, 5, 7\}$.

13. Find the smallest $n \in \mathbb{N}$ such that $2n^2 + 3n + 1 < n^3$.
14. Prove by induction for $n \geq 0$:

$$2 + 4 + 6 + \cdots + 2n = n^2 + n$$

15. Prove by induction:

   (a) $1^2 + 2^2 + 3^2 + \cdots + n^2 = n(n+1)(2n+1)/6$ for $n \geq 0$
   (b) $1^3 + 2^3 + 3^3 + \cdots + n^3 = (1 + 2 + 3 + \cdots + n)^2$ for $n \geq 0$
   (c) $1^4 + 2^4 + 3^4 + \cdots + n^4 = n(n+1)(2n+1)(3n^2+3n-1)/30$ for $n \geq 0$
   (d) $1^5 + 2^5 + 3^5 + \cdots + n^5 = \frac{1}{6}n^6 + \frac{1}{2}n^5 + \frac{5}{12}n^4 - \frac{1}{12}n^2$ for $n \geq 0$

16. Prove by induction:

   (a) $0 \cdot 2^0 + 1 \cdot 2^1 + 2 \cdot 2^2 + 3 \cdot 2^3 + \cdots + n \cdot 2^n = (n-1)2^{n+1} + 2$ for $n \geq 0$
   (b) $1^2 + 3^2 + 5^2 + \cdots + (2n+1)^2 = (n+1)(2n+1)(2n+3)/3$ for $n \geq 0$
   (c) $1^2 - 2^2 + 3^2 + \cdots + (-1)^{n-1}n^2 = (-1)^{n-1}n(n+1)/2$ for $n \geq 0$
   (d) $1 \cdot 2 + 2 \cdot 3 + 3 \cdot 4 + \cdots + n \cdot (n+1) = n(n+1)(n+2)/3$ for $n \geq 0$
   (e) $1 \cdot 2 \cdot 3 + 2 \cdot 3 \cdot 4 + 3 \cdot 4 \cdot 5 + \cdots + n \cdot (n+1) \cdot (n+2) = n(n+1)(n+2)$
       $(n+3)/4$ for $n \geq 0$

17. Prove by induction:

   (a) $\frac{1}{1 \cdot 2} + \frac{1}{2 \cdot 3} + \cdots + \frac{1}{n(n+1)} = \frac{n}{n+1}$ for $n \geq 1$
   (b) $\frac{1}{2} + \frac{2}{2^2} + \frac{3}{2^3} + \cdots + \frac{n}{2^n} = 2 - \frac{n+2}{2^n}$ for $n \geq 1$

18. Prove by induction that 8 divides $(2n+1)^2 - 1$ for all $n \in \mathbb{N}$.
19. Prove by induction for $n \geq 0$:

   (a) 3 divides $n^3 + 2n$
   (b) 5 divides $n^5 - n$
   (c) 6 divides $n^3 - n$
   (d) 6 divides $n^3 + 5n$

20. Prove by induction for all $n \in \mathbb{N}$:

   (a) 7 divides $n^7 - n$
   (b) 11 divides $n^{11} - n$
   (c) 13 divides $n^{13} - n$
   (d) 120 divides $n^5 - 5n^3 + 4n$

21. Prove by induction: The sum of the cubes of any three consecutive natural numbers is divisible by 9.
22. Show that any integer consisting of 3^n identical digits is divisible by 3^n. Verify this for 222; 777; 222,222,222; and 555,555,555. Prove the general statement for all $n \in \mathbb{N}$ by induction.
23. Prove by induction that the following identities are true for the Fibonacci numbers:

   (a) $\sum_{i=0}^{n} F_{2i+1} = F_{2n+2} - 1$ for $n \geq 0$
   (b) $\sum_{i=1}^{n} F_i^2 = F_n \cdot F_{n+1} - 1$ for $n \geq 1$
   (c) $\sum_{i=0}^{n} F_i = F_{n+2} - 1$ for $n \geq 0$

24. Find the Fibonacci numbers F_8 through F_{15}. Prove the following results for the Fibonacci numbers:

   (a) F_{3n} and F_{3n+1} are odd, and F_{3n+2} is even for $n \geq 0$
   (b) $F_0 + F_2 + \cdots + F_{2n} = F_{2n+1}$ for $n \geq 0$

(c) $F_0 + F_3 + \cdots + F_{3n} = F_{3n+2}/2$ for $n \geq 0$

(d) $F_{n+1}^2 = F_n \cdot F_{n+2} - (-1)^n$ for $n \geq 0$

25. The *Lucas numbers* are defined as $L_0 = 2$, $L_1 = 1$, and $L_n = L_{n-1} + L_{n-2}$ for $n \geq 2$. Prove the following identities for Lucas numbers.

(a) $L_1 + L_2 + \cdots + L_n = L_{n+2} - 3$ for $n \geq 1$

(b) $L_1^2 + L_2^2 + L_3^2 + \cdots + L_n^2 = L_n \cdot L_{n+1} - 2$ for $n \geq 2$

(c) $L_2 + L_4 + \cdots + L_{2n} = L_{2n+1} - 1$ for $n \geq 2$

26. Find the value of the following sums:

(a) $2 + \frac{2}{3} + \frac{2}{9} + \cdots + \frac{2}{3^n}$

(b) $1 - \frac{1}{2} + \frac{1}{4} - \frac{1}{8} + \cdots + (\frac{-1}{2})^n$

(c) $-2 + 4 - 8 + 16 + \cdots + (-2)^{11}$

(d) $1.03 + (1.03)^2 + (1.03)^3 + \cdots + (1.03)^n$

27. Find a rational number representing each of the following repeating decimals:

(a) $0.537537537537537537537537537\ldots$

(b) $31.254696969696969696969696\ldots$

28. A fixed dose of a given drug increases the concentration of that drug above normal levels in the bloodstream by an amount C_0 (measured in percent). The effect of the drug wears off over time such that the concentration at some time t is $C_0 e^{-kt}$ where k is the known rate at which the concentration of the drug in the bloodstream declines.

(a) Find the residual concentration R, the accumulated amount of the drug above normal levels in the bloodstream, at time t after n doses given at intervals of t_0 hours starting with the first dose at $t = 0$.

(b) If the drug is alcohol and 1 oz. of alcohol has $C_0 = 0.05\%$, how often can a "dose" be taken so that the residual concentration is never more than 0.15%? Assume $k = (1/3)\ln(2)$.

29. (a) Prove by induction that $2^n > n$ for all $n \geq 0$.

(b) Prove that $2^n > n$ directly from Theorem 2 in Section 1.7.4, without explicit use of induction. (That is, Theorem 2 in Section 1.7.4 itself was proved using induction, but you should not have to do any additional induction.)

(c) Prove by induction that $2^n > n^3$ for $n \geq 10$.

30. Prove by induction:

(a) There is a natural number k such that $n! > n^3$ for all $n \geq k$. (Try to find the least such number k.)

(b) $n! > n^4$ for $n \geq 7$.

31. Let $T = \{n \in \mathbb{N} : \sin(n \cdot \pi) = 0\}$. Prove that $T = \mathbb{N}$. (*Hint:* $\sin(a + b) = \sin(a) \cdot \cos(b) + \cos(a) \cdot \sin(b)$.)

32. Prove assertion 1 from Lemma 1.

33. (a) Suppose you take out a mortgage for A dollars at a monthly interest rate I and a monthly payment P. (To calculate I: if the annual interest rate is 12%, divide by 12 to get a monthly rate of 1%, then replace the percentage with the decimal fraction 0.01.) Let A_n denote the amount you have left to pay off after n months. So, $A_0 = A$ by definition. At the end of each month, you are first charged interest

on all the money you owed during the month, and then your payment is subtracted. So,

$$A_{n+1} = A_n(1 + I) - P$$

Prove by induction that

$$A_n = \left(A - \frac{P}{I}\right)(1 + I)^n + \frac{P}{I}$$

(b) Use this to calculate the monthly payment on a 30-year loan of $100,000 at 12% interest per year. (Note that the formula is inexact, since money is always rounded off to a whole number of cents. The derivation here does not do that. We use 12% to make the arithmetic easier. You should consult a local bank to find a current value.)

34. Sometimes, induction is not necessary for a proof, but an inductive proof can be simpler than a noninductive proof. This is true for Examples 2 and 3 of Section 1.7.2.

   (a) Find proofs of Examples 2 and 3 using familiar algebra but no explicit induction.[3]
   (b) *Optional:* Find proofs of Examples 2 and 3 using calculus. (To some students calculus may be more familiar than induction, but it is certainly more complicated theoretically!)

35. Prove Theorem 4 of Section 1.5.4 in full generality. You may use Theorem 3 of Section 1.5.3, since it has already been proven. (*Hint:* Use induction on the number of sets).

36. For natural number exponents and nonzero bases, most of the familiar laws of exponents can be proved by induction on the exponents using the facts that $b^0 = 1$ (for $b \neq 0$) and $b^{n+1} = b \cdot b^n$. Assuming that m and n are natural numbers and both r and s are nonzero real numbers, prove the following:

   (a) $r^{m+n} = r^m \cdot r^n$.
   (b) $r^{mn} = (r^m)^n$.
   (c) If $r > 1$, then $r^m > r^n$ if and only if $m > n$.
   (d) If $n, r, s > 0$, then $r^n > s^n$ if and only if $r > s$.

37. A common use of induction is to prove various facts that seem to be fairly obvious but are otherwise awkward or impossible to prove. These frequently involve expressions with ellipses. Use induction to show that:

   (a) $X \cup (X_1 \cap X_2 \cap X_3 \cap \cdots \cap X_3^n) = (X \cup X_1) \cap (X \cup X_2) \cap \cdots \cap (X \cup X_n)$
   (b) $X \cap (X_1 \cup X_2 \cup X_3 \cup \cdots \cup X_n) = (X \cap X_1) \cup (X \cap X_2) \cup \cdots \cup (X \cap X_n)$
   (c) $\overline{(X_1 \cap X_2 \cap \cdots \cap X_n)} = \overline{X_1} \cup \overline{X_2} \cup \cdots \cup \overline{X_n}$
   (d) $\overline{(X_1 \cup X_2 \cup \cdots \cup X_n)} = \overline{X_1} \cap \overline{X_2} \cap \cdots \cap \overline{X_n}$

38. (a) Prove that $x \in X_0 \cap X_1 \cap \cdots \cap X_n$ if and only if $x \in X_i$ for *every* i such that $0 \leq i \leq n$.
   (b) Prove that $x \in X_0 \cup X_1 \cup \cdots \cup X_n$ if and only if $x \in X_i$ for *some* i such that $0 \leq i \leq n$.
   (c) Use part (a) to give another proof of Exercise 37(a).

---

[3] We say *explicit* induction since, in the development of arithmetic from the foundations, almost everything about + and · is proved by induction, including the familiar algebra needed for this problem.

39. Refer to the Square Root II algorithm.

(a) Finish the proof of Theorem 5.

(b) Show that $\epsilon_{n+1} = -\epsilon_n^2/(2R_n)$. (*Hint:* Simplify $\sqrt{17} - (R_n + (17/R_n))/2$.)

(c) How close do you think the value printed is to the actual value of $\sqrt{17}$? Approximately how many decimal digits in accuracy is that?

40. *Challenge:* Exactly where is the mistake in the following proof that all personal computers are the same brand? Let $\mathcal{T} = \{n \in \mathbb{N} : n \geq 1$ and in every set of n personal computers, all the personal computers are the same brand$\}$. Prove by induction that for every natural number n such that $n \geq 1$ is in $\mathcal{T}$.

(**Base step**) $1 \in \mathcal{T}$, since, trivially, if a set of personal computers contains only one computer, then every (one) computer in the set has the same brand.

(**Inductive step**)   Suppose $n \in \mathcal{T}$. We need to show $n + 1 \in \mathcal{T}$. So, let P be any set of $n + 1$ personal computers. Pick any computer $c \in P$; we need to show that every computer in P is the same brand as c. So, let d be any computer in P. If $d = c$, then, trivially, d and c are the same brand. Otherwise, $c \in P - \{d\}$. The set $P - \{d\}$ contains n computers, so by inductive hypothesis, all the computers in $P - \{d\}$ are the same brand. Furthermore, $d \in P - \{c\}$, and, also by inductive hypothesis, all the computers in $P - \{c\}$ are the same brand. Now, let e be a computer in both $P - \{c\}$ and $P - \{d\}$. Then, d is the same brand as e, and c is the same brand as e. Therefore, d is the same brand as c.

41. Using the Principle of Mathematical Induction, prove each of the following different forms of the principle:

(a) *Induction with a possibly negative starting point:* Suppose that $S \subseteq \mathbb{Z}$, that some integer $n_0 \in S$, and that for every $n \in \mathbb{Z}$, if $n \in S$ and $n \geq n_0$, then $n + 1 \in S$. Then, for every integer $n \geq n_0$, we have $n \in S$.

(b) *Induction downward:* Suppose that $S \subseteq \mathbb{Z}$, that some integer $n_0 \in S$, and that for every $n \in \mathbb{Z}$, if $n \in S$ and $n \leq n_0$, then $n - 1 \in S$. Then, for every integer $n \leq n_0$, we have $n \in S$.

(c) *Finite induction upward:* Let $n_0, n_1 \in \mathbb{Z}$, $n_0 \leq n_1$. Suppose that $S \subseteq \mathbb{Z}$, $n_0 \in S$, and for every $n \in \mathbb{Z}$, if $n \in S$, $n \geq n_0$, and $n < n_1$, then $n + 1 \in S$. Then, every integer n where $n_0 \leq n \leq n_1$ is in S.

(d) Suppose $S \subseteq \mathbb{N}$ is infinite, and suppose that for every $n \in \mathbb{N}$, if $n + 1 \in S$, then $n \in S$. Prove that $S = \mathbb{N}$.

1.10   Strong Form of Mathematical Induction

The **Fundamental Theorem of Arithmetic** states some familiar results about factoring integers. Part of the Fundamental Theorem of Arithmetic is the result that every integer $n > 1$ can be factored as a product

$$n = p_1 \cdot p_2 \cdots p_k$$

for some **prime numbers** $p_1, p_2, \ldots, p_k$. The p_i's are not required to be distinct, and k simply denotes the number of factors needed to express p. For example, $4 = 2 \cdot 2$ is a factorization of 4 into two primes. If $k = 1$, then n is a prime, and $n = n$ is a factorization into primes. We just *define* the term *factorization into primes* to include the one-prime case.

The proof that every integer $n > 1$ can be factored into primes goes as follows: If n is prime, then $n = n$ is a factorization of n into primes. Otherwise, if n is not a prime, then n can be factored as $n = k \cdot m$ for some integers m and k where $n > m$, $k > 1$. Since k and m are both less than n, we can conclude that m and k can be factored. We would now use the factorizations of m and k to form a factorization of n.

This is *not* an application of an inductive hypothesis *as induction has been presented so far*. The problem is that the Principle of Mathematical Induction only uses the result for $n - 1$ to prove the result for $n = (n - 1 + 1)$. Here, the result for n has to be proved from the same result for *two* smaller numbers k and m, *neither* of which (it turns out) is $n - 1$. In fact, $k, m \leq n/2$.

The **Strong Form of Mathematical Induction** has a somewhat different form of inductive hypothesis: It assumes the result for *all* natural numbers k where $n_0 \leq k < n$— with $n_0 \in \mathbb{N}$ just as before—and then proves the result for n. This was what we needed for factoring—whatever k, m are, we get to apply the inductive hypothesis to both of them.

We now give a formal statement of this new form of induction and then complete the proof that every integer can be written as a product of primes.

Strong Form of Mathematical Induction

Let $\mathcal{T} \subseteq \mathbb{N}$ and $n_0 \in \mathbb{N}$. Suppose that for all natural numbers $n \geq n_0$, if $n_0, n_0 + 1, \ldots, n - 1 \in \mathcal{T}$, then $n \in \mathcal{T}$. Then, every natural number $n \geq n_0$ is in $\mathcal{T}$.

If n_0 is equal to zero, then the Strong Form of Mathematical Induction proves that $\mathcal{T} = \mathbb{N}$.

The Strong Form of Mathematical Induction is also sometimes called *Complete Induction* or *Course of Values Induction*. It is the inductive hypothesis which is "stronger," not the principle itself. Indeed, any theorem provable with the strong form of induction is also provable with the first form, *but* such proofs may require some awkward complications.

To use the strong form of induction, one must prove the if-then statement that

$$\text{if } n_0, n_0 + 1, \ldots, n - 1 \in \mathcal{T}, \text{ then } n \in \mathcal{T}$$

Virtually always, the proof is broken into cases. For some values of n, *including* n_0, the result is proven directly; this set of cases is sometimes called the **base step.** For the other values of n, the result is proved using the assumption that $n_0, n_0 + 1, \ldots, n - 1 \in \mathcal{T}$. This is called the **inductive step,** and that assumption is called the **inductive hypothesis.**

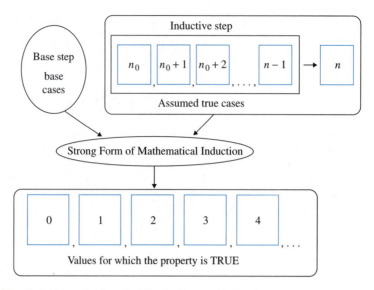

Figure 1.18   Typical proof using the Strong Form of Induction.

Using Figure 1.18 as a guide, we now return to proving the result about factoring integers. As noted above, the proof breaks into two cases: one case for prime numbers n, and one case for nonprimes.

Theorem 1.   (Part of the Fundamental Theorem of Arithmetic)   Every natural number n such that $n > 1$ can be factored into a product of one or more primes.

Proof.   The proof will use the Strong Form of Mathematical Induction. Let $n_0 = 2$, and let

$$T = \{n \in \mathbb{N} : n > 1 \text{ and } n = p_1 \cdot p_2 \cdots p_k \text{ for some prime numbers } p_1, p_2, \ldots, p_k.$$

Let n be any natural number greater than or equal to 2.

(Base step)   The base cases deal with any n that is a prime. Since n is prime, $n = n$ is a factorization of n into the product of one prime.

(Inductive step)   In this step, we will prove the result for any n that is not a prime. Assume that for all m where $2 \le m < n$, $m \in T$. Now, prove $n \in T$.

Since n is not prime, n can be factored as $n = k \cdot m$ where $k \ne 1$ and $m \ne 1$. It follows easily that $1 < k < n$ and that $1 < m < n$. Hence, by the inductive hypothesis, $k, m \in T$. So, k and m can be factored into products of primes:

$$k = p_1 \cdot p_2 \cdots p_i \quad \text{and} \quad m = q_1 \cdot q_2 \cdots q_j$$

Then,

$$n = p_1 \cdot p_2 \cdots p_i \cdot q_1 \cdot q_2 \cdots q_j$$

so n can be factored into a product of primes. Therefore, $n \in \mathcal{T}$.

By the Strong Form of Mathematical Induction, $\mathcal{T} = \{n \in \mathbb{N} : n > 1\}$. ∎

1.10.1   Using the Strong Form of Mathematical Induction

The Strong Form of Mathematical Induction is often used to prove a **closed form** for the elements of a **recursively defined sequence** like the Fibonacci sequence. A closed form for the elements is a representation for each term that can be computed without knowing any other element(s) of the sequence. Exercise 16 in Section 1.11 is to show that the nth Fibonacci number can be computed as

$$F_n = \frac{1}{\sqrt{5}} \cdot \left(\frac{1 + \sqrt{5}}{2} \right)^{n+1} - \frac{1}{\sqrt{5}} \cdot \left(\frac{1 - \sqrt{5}}{2} \right)^{n+1}$$

for each $n \in \mathbb{N}$. This expression is a closed form for the Fibonacci sequence.

The next example is similar to the result about Fibonacci numbers, but the computations are less complex. The verification of the closed form for the Fibonacci numbers is left as an exercise.

Example 1.   The terms of a sequence are given recursively as

$$a_0 = 0, a_1 = 2, \quad \text{and} \quad a_n = 4(a_{n-1} - a_{n-2}) \text{ for } n \geq 2$$

Prove by induction that $b_n = n \cdot 2^n$ is a closed form for the sequence. That is, prove that $a_n = b_n$ for every $n \in \mathbb{N}$.

Solution.   Let $n_0 = 0$ and $\mathcal{T} = \{n \in \mathbb{N} : b_n = a_n\}$. In this case, two elements of the sequence, a_0 and a_1, are defined directly. As is fairly typical, these special cases constitute the base cases for the proof.

(Base step)   The two base cases are $n = 0$ and $n = 1$ Evaluating b_0 and b_1 gives $b_0 = 0$ and $b_1 = 2$. Thus, $a_0 = b_0$ and $a_1 = b_1$, so $0, 1 \in \mathcal{T}$.

(Inductive step)   We now deal with any n such that $n \geq 2$. Assume that for all k where $0 \leq k < n$, $k \in \mathcal{T}$. Prove that $n \in \mathcal{T}$ by showing $a_n = b_n$. Since $n \geq 2$, $n - 1, n - 2 \geq 0$, so $n - 1, n - 2 \in \mathcal{T}$.

$$
\begin{aligned}
a_n &= 4(a_{n-1} - a_{n-2}) \quad \text{(by definition of } a_n\text{)} \\
&= 4((n-1)2^{n-1} - (n-2)2^{n-2}) \quad \text{(by inductive hypothesis)} \\
&= 4(n \cdot 2^{n-1} - 2^{n-1} - n \cdot 2^{n-2} + 2 \cdot 2^{n-2}) \\
&= 4(n(2^{n-1} - 2^{n-2}) - (2^{n-1} - 2 \cdot 2^{n-2})) \\
&= 4(n(2 \cdot 2^{n-2} - 2^{n-2}) - (2 \cdot 2^{n-2} - 2 \cdot 2^{n-2})) \\
&= 4 \cdot n \cdot 2^{n-2} \ = \ n \cdot 2^n
\end{aligned}
$$

Therefore, $b_n = a_n$ and $n \in \mathcal{T}$.

By the Strong Form of Mathematical Induction, $\mathcal{T} = \mathbb{N}$. That is, $b_n = n \cdot 2^n$ is a closed form for the terms of the recursively defined sequence. ∎

Constructing a proof by induction using the Strong Form of Induction requires a different template than the one for the first Principle of Mathematical Induction. This new template makes clear what is being done at each step, but be careful: There is more variety in the form of proofs using the Strong Form of Induction than in proofs using the ordinary Principle of Mathematical Induction.

Template 1.13   Using the Strong Form of Mathematical Induction

To construct a proof using the Strong Form of Mathematical Induction, choose an $n_0 \in \mathbb{N}$ appropriate to the problem. Let

$$\mathcal{T} = \{n \in \mathbb{N} : n \geq n_0 \text{ and property } P \text{ holds of } n\}$$

(Base step)   Show explicitly that property P holds for certain numbers n, called the base cases. n_0 should be one of those values; the choice of the other values depend on the problem.

(Inductive step)   For all $n \geq n_0$ not covered in the base case, assume that property P holds for all $k = n_0, n_0 + 1 \ldots, n - 1$, and prove that property P holds for n.

Infer by the Strong Form of Mathematical Induction that

$$\mathcal{T} = \{n \in \mathbb{N} : n \geq n_0\}$$

Using the Strong Form of Mathematical Induction

As in an ordinary inductive proof, an inductive proof using the strong form of induction has three essential parts: (i) a base step, (ii) an inductive step, and (iii) an application of the Strong Form of Mathematical Induction.

Translating the problem includes specifying n_0 and *clearly defining* the set $\mathcal{T}$ whose elements the inductive proof will determine—that is, *clearly stating* the property P to be verified. This *definition* does not tell us that any number is in $\mathcal{T}$.

The first step of the proof is called the base step, and it involves proving the result for the base case(s). Identify *one or more* values for which property P can be verified directly. Often, one might verify it directly for values $n_0, n_0 + 1, n_0 + 2, \ldots, n_1$ for some $n_1 \geq n_0$. In the base step of the proof, prove directly that $n_0, n_0 + 1, \ldots, n_1 \in \mathcal{T}$. As in Example 1, the base cases often correspond to the initial conditions specified in the problem.

The inductive step is usually quite different in the Strong Form of Mathematical Induction from the inductive step in the Principle of Mathematical Induction. Begin by letting $n \geq n_0$ be an arbitrary natural number *that is not covered in the base case.* Assume that $n_0, n_0 + 1, \ldots, n - 1 \in \mathcal{T}$. To complete the inductive step, use that assumption to show that $n \in \mathcal{T}$. Again, start by writing out property P for n to see what is to be proved. There is no real formula for the next part of the inductive proof. Figure out how to prove property P holds for n knowing that property P holds for $n_0, n_0 + 1, \ldots, n - 1$. When that is done, use the Strong Form of Mathematical Induction to infer that for all $n \geq n_0, n \in \mathcal{T}$.

In practice, you may often try to work out the Inductive step first. You will then see certain values—and you may as well assume that n_0 *must* be one of them—for which the argument doesn't use the inductive hypothesis. These values are identified as the **base cases.**

Example 2. The terms of a sequence are given recursively as

$$a_0 = 1, \quad a_1 = 1, \quad \text{and} \quad a_n = 2 \cdot a_{n-1} + 3 \cdot a_{n-2} \text{ for } n \geq 2$$

Prove by induction that $b_n = \frac{1}{2} \cdot 3^n + \frac{1}{2} \cdot (-1)^n$ is a closed form for the sequence.

Solution. Let $n_0 = 0$ and $T = \{n \in \mathbb{N} : b_n = a_n\}$.

(Base step) Identify $n = 0, 1$ as the base cases. The defined values in such a definition often are the base cases. Evaluate b_0 and b_1 directly:

$$b_0 = \tfrac{1}{2}(3^0 + (-1)^0) = \tfrac{1}{2}(1 + 1) = 1 = a_0$$
$$b_1 = \tfrac{1}{2}(3^1 + (-1)^1) = \tfrac{1}{2}(3 - 1) = 1 = a_1$$

So, $0, 1 \in T$.

(Inductive step) Now, let $n \geq 2$, and assume for $k = 0, 1, \ldots, n - 1$ that $k \in T$. Prove that $n \in T$ by showing that $a_n = b_n$:

$$a_n = 2a_{n-1} + 3a_{n-2} \quad \text{(by definition of } a_n)$$

$$= 2 \cdot \frac{1}{2}(3^{n-1} + (-1)^{n-1}) + 3 \cdot \frac{1}{2}(3^{n-2} + (-1)^{n-2}) \quad \text{(by inductive hypothesis)}$$

$$= 3^{n-1} + (-1)^{n-1} + \frac{3}{2} \cdot 3^{n-2} + \frac{3}{2}(-1)^{n-2}$$

We know that $(-1)^{n-2} = (-1)^n$, $(-1)^{n-1} = -(-1)^n$, and $3^{n-1} = 3 \cdot 3^{n-2}$. So,

$$a_n = 3 \cdot 3^{n-2} + \frac{3}{2} \cdot 3^{n-2} - (-1)^n + \frac{3}{2}(-1)^n$$

$$= \frac{9}{2} \cdot 3^{n-2} + \frac{1}{2}(-1)^n$$

$$= \frac{1}{2} \cdot 3^n + \frac{1}{2}(-1)^n$$

$$= b_n$$

as desired. Therefore, $n \in T$.

By the Strong Form of Mathematical Induction, $T = \mathbb{N}$. That is, $b_n = \frac{1}{2} \cdot 3^n + \frac{1}{2} \cdot (-1)^n$ is a closed form for the terms of the recursively defined sequence. ■

Unlikely as it might seem, we can use the Strong Form of Mathematical Induction to show which amounts of postage can be made from a fixed number of several denominations of stamps.

Example 3. The country of Oz issues only 3-cent and 8-cent stamps. What amounts of postage are possible with just these two kinds of stamps?

Solution. Obviously, some packages will require a lot of surface area to affix all the required postage! By experimentation, we can find out that all of 0, 3, 6, 8, 9, 11, 12, 14, 15, 16, 17, 18, 19, 20, and 21 cents are possible. Since we are getting all amounts of 14 cents or greater, we conjecture that all amounts except 1, 2, 4, 5, 7, 8, 10, and 13 cents are possible.

We conjectured that all values starting at 14 are possible, so we handle all $n < 14$ separately. We noted that 0, 3, 6, 8, 9, 11, and 12 cents are all possible. Amounts of 1, 2, 4, 5, 7, 8, 10, or 13 cents are impossible: To get any of those amounts, one would need to use, at most, 4 stamps (why?), and we can list all the possible combinations of 0–4 stamps to show that none add up to 1, 2, 4, 5, 7, 8, 10, or 13 cents.

Let

$$\mathcal{T} = \{n \in \mathbb{N} : n \geq 14 \text{ and } n = k \cdot 3 + l \cdot 8 \text{ for some } k, l \in \mathbb{N}\}$$

We must then prove that every natural number $n \geq 14$ is in $\mathcal{T}$.

(Base step)   After some experimentation, we decide the base cases are 14, 15, and 16. Since $14 = 2 \cdot 3 + 1 \cdot 8$, $15 = 5 \cdot 3 + 0 \cdot 8$, and $16 = 0 \cdot 3 + 2 \cdot 8$, we have $14, 15, 16 \in \mathcal{T}$.

(Inductive step)   Let $n \geq 14$, and assume that $14, 15, 16, \ldots, n - 1 \in \mathcal{T}$. Now, prove that $n \in \mathcal{T}$.

Since 14, 15, and 16 are base cases, every possible value for n that is not a base case and is greater than or equal to 14 is also greater than or equal to 17. For $n \geq 17$, we have $n - 3 \geq 14$. So, by the inductive hypothesis, for some $k, l \in \mathbb{N}$, $n - 3 = k \cdot 3 + l \cdot 8$. Then,

$$n = (n - 3) + 3 = k \cdot 3 + l \cdot 8 + 3 = (k + 1) \cdot 3 + 8 \cdot l$$

So, $n \in \mathcal{T}$, as desired.

By the Strong Form of Mathematical Induction, $\mathcal{T} = \{n \in \mathbb{N} : n \geq 14\}$.   ■

There are some other values for which Oz can make postage—for example, 3, 6, 8, 9, 11, and 12. When we looked carefully at the inductive step, we saw we would have to be able to go back three from any n for which we were proving the postage amount could be made. We were then more clear on what the base cases would need to be. Consequently, the base step proved postage can be made for $n = 14, 15,$ and 16. It is not unusual that the base cases are identified by trying the **inductive step** of the proof. Note that in the proof of the inductive case above, before applying the inductive hypothesis to $n - 3$, we checked that $n - 3 \geq n_0$. Not making that check is a very easy way to make an error. In this case, had we not made that check, we might have started with $n = 12$, asserted that $n - 3 = 9$ was in $\mathcal{T}$, and proceeded as with the inductive case above—and we would have "proved" something that was actually false.

1.10.2   Application: Algorithm to Compute Powers

Suppose you want to compute x^n for some nonzero real number x and some natural number n. One way is to multiply together n copies of x, a task that requires $n - 1$ multiplications. Are there faster ways to complete this computation? We will prove that the following algorithm computes x^n using far fewer multiplications for large values of n.

Algorithm: Compute Powers

INPUT: A nonzero real number x and a natural number n
OUTPUT: The value of x^n

FastPower(x, n) /* The initial call */

FastPower $(base, expont)$ /* The recursive procedure */
    if $(expont = 0)$ then
        return 1
    else
        if $(expont$ is odd) then
            return $base \cdot$ *FastPower*$(base \cdot base, (expont - 1)/2)$
        else
            return *FastPower*$(base \cdot base, expont/2)$

The algorithm presented uses a programming feature called **recursion.** In this algorithm, a call to the algorithm *FastPower* is part of its own code. In a programming language that supports this feature, the compiler will keep track of which version of *FastPower* is being executed and which values should be used for the arguments. For more details about how recursion is implemented in a programming language, the reader should consult a manual for a language such as Java, C, or C^{++}.

The reader should trace through the algorithm by hand for some sample values of *base* and *expont*. For example, a computer executing this algorithm to compute 2^5 will go through the following steps:

FastPower$(2, 5)$ identifies 5 as odd and computes

$$2 \cdot FastPower(2 \cdot 2, (5 - 1)/2) = 2 \cdot FastPower(4, 2)$$

To execute *FastPower*$(4, 2)$ requires the execution of

$$FastPower(4 \cdot 4, 2/2) = FastPower(16, 1)$$

Now, $expont = 1$ is odd, so the program computes

$$16 \cdot FastPower(16 \cdot 16, (1 - 1)/2) = 16 \cdot FastPower(256, 0)$$

When *FastPower*$(256, 0)$ is executed, the program starts the return process. *FastPower*$(256, 0)$ returns 1 to *FastPower*$(16, 1)$. The returning value using *FastPower*$(16, 1)$ is $16 \cdot FastPower(16, 1) = 16$. This value is *FastPower*$(4, 2)$, which must be multiplied by 2 before that value is returned to *FastPower*$(2, 5)$. Thus, *FastPower*$(2, 5) = 32$.

The flow of control for this example is shown in Figure 1.19 on page 74.

Even though the example computation for 2^5 works correctly, it is, however, not quite obvious that the *FastPower* algorithm correctly calculates powers for every nonzero base

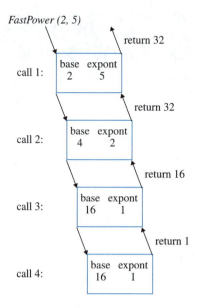

FastPower (2, 5)

return 32

call 1:
base	expont
2	5

return 32

call 2:
base	expont
4	2

return 16

call 3:
base	expont
16	1

return 1

call 4:
base	expont
16	1

Figure 1.19   Flow of control for *FastPower (2, 5)*.

and every exponent. Using the Strong Form of Mathematical Induction, we now prove that the algorithm is correct for all cases.

Theorem 2.   The *FastPower* algorithm returns the value $base^n$ for $base \in \mathbb{R} - \{0\}$, and $n \in \mathbb{N}$.

Proof.   The proof is by induction on the value of n. Let $n_0 = 0$ and

$$\mathcal{T} = \{n \in \mathbb{N} : \text{ for every } base \ \in \mathbb{R} - \{0\}, FastPower(base, n) = base^n\}$$

Prove by the Strong Form of Mathematical Induction that $\mathcal{T} = \mathbb{N}$.

(Base step)   For $n = 0$, the algorithm returns 1, as required. So, $0 \in \mathcal{T}$.

(Inductive step)   Let $n > 0$. Assume that for all k such that $0 \le k < n$, $k \in \mathcal{T}$. Now, prove that $n \in \mathcal{T}$.
   This case breaks into two subcases:

Case 1: n is odd. So, $n = 2k + 1$ for some $k \in \mathbb{N}$. Clearly, $0 \le k < n$. By familiar properties of exponentiation,

$$base^{2k+1} = base \cdot base^{2k}$$
$$= base \cdot (base^2)^k$$

By the inductive hypothesis, since $k < n$, the algorithm correctly computes b^k for any b. In particular, it computes $(base^2)^k$; thus, $base \cdot (base^2)^k = base^{2k+1}$.

Case 2: n is even. The proof is analogous to the proof of Case 1. In either case, $n \in \mathcal{T}$.
   By the Strong Form of Mathematical Induction, $\mathcal{T} = \mathbb{N}$.   ∎

   FastPower is actually used in many computer science applications when the exponent is known to be an integer. Special computer chips are used in cryptography for doing arithmetic of numbers up to approximately 300 digits. These chips essentially compute powers this way, with one modification: *FastPower*, as written, makes a **recursive call**—it invokes (another copy of) itself. To calculate 2^5, for example, the procedure was called four times (the original call and three recursive calls). There is computer overhead in each of these calls. It turns out that the special chips have had the recursive calls replaced with a loop, producing the program actually used. Interested readers should try writing this algorithm nonrecursively.

1.10.3   Application: Finding Factorizations

The Fundamental Theorem of Arithmetic was proved at the beginning of this section. As important as the result is, however, it does not provide any insight regarding how one goes about finding such a factorization. The two algorithms here explore factoring integers. The first looks for the largest odd divisor. In a theorem we will prove later, the proof does not provide a method for finding the largest odd divisor but, instead, uses the Fundamental Theorem of Arithmetic to guarantee the existence of such a factor. When you actually want to find the elements that the theorem only says will exist, you can use the first algorithm as a method for doing this step of the proof. The second algorithm takes the guarantee of the Fundamental Theorem of Arithmetic that a factorization exists and actually finds it. Later, you will be asked to prove that these algorithms are correct. At this point it, however, is important to understand what the algorithms are doing.

Largest Odd Divisor

A **while loop** controls the iterations in **Largest Odd Divisor** algorithm, because each iteration reduces the number being considered by a factor of 2 until only an odd number remains.

Algorithm: Largest Odd Divisor

INPUT: Integer value $N > 0$
OUTPUT: Largest odd divisor of N

```
LargeOdd (N)
    while (Mod(N, 2) = 0)
        N = N/2
    print N
```

In this code, the condition $mod(N, 2) = 0$ returns *TRUE* when N is divisible by 2 (N is even). The code returns *FALSE* when N is not divisible by 2 (N is odd). The first test of the condition simply asks if the original number is odd. If the number is odd, it is certainly the largest odd factor, and N is printed. If the condition is *TRUE* and N is even, then the code controlled by the while loop divides N by a factor of 2. The resulting value ($N/2$) is used in the condition the next time the while statement is executed. If the condition is *TRUE,* the division by 2 is repeated. Eventually, the condition in the while statement with the value $N/2^k$, where 2^k is the highest power of 2 that is a factor of N, will be evaluated as *FALSE,* because the value tested is odd. In this case, the process terminates by printing the final value of $N/2^k$. For example, if $N = 78$, the condition $Mod(78, 2) = 0$ is *TRUE* and N is replaced by $78/2 = 39$. Now, when the condition $Mod(39, 2) = 0$ is tested, the condition is *FALSE.* The while loop is exited, and the value of $N/2 = 39$ is printed.

Theorem 3.   Prove that the Largest Odd Divisor algorithm is correct.

Proof.   Exercise for the reader.   ■

Factorization

Often, a small insight that does not seem particularly significant can make a big difference in developing an algorithm. In the code for *PrintFactors,* the idea is that if an integer n can be factored as $j \cdot k$ where $1 < j, k < n$, then either j or k is in the range 1 to $\sqrt{n}$. To find a factor of n, we can focus on finding a value between 2 and $\sqrt{n}$ rather than a value from 2 to $n - 1$.

Algorithm: Print a Prime Factorization of an Integer

INPUT: Integer $N > 1$
OUTPUT: Factors of N

PrintFactors (N) /* Initial call */

PrintFactors (n) /* The recursive procedure */
    $RootN = \sqrt{n}$
    $TrialFactor = \lfloor RootN \rfloor$
    while $(mod(n, TrialFactor) \neq 0)$ do
        /* If *TrialFactor* is a divisor of n, the loop
        will be executed zero times. */
        $TrialFactor = TrialFactor - 1$
    if $(TrialFactor \leq 1)$ then
        print n
    else
        PrintFactors$(TrialFactor)$
        PrintFactors$(n/TrialFactor)$

The procedure *PrintFactors* is designed to display the factors of any integer. For example, we know that the factors of 12 are 2, 2, and 3. The value of *RootN* is initially assigned the value $\lfloor\sqrt{12}\rfloor = 3$. Therefore, the first time through *PrintFactors,* we set *TrialFactor* equal to 3, and we test $Mod(n, TrialFactor) \neq 0$. The condition is *FALSE,* which means that 3 is a factor of n. *TrialFactor* is greater than 1, so we call *PrintFactors*(3) and *PrintFactors*(12/3). *PrintFactors*(3) prints the factor 3. *PrintFactors*(4) starts by setting *TrialFactor* equal to 2. Because now *Mod(n, TrialFactor)* ≠ 0 is *FALSE,* we call *PrintFactors*(2) and *PrintFactors*(4/2). These two calls to *PrintFactors* both print a 2, completing the factorization of 12.

When you trace the execution of a procedure, some visual help to see how control passes from one step to another can be valuable. In Figure 1.20 we show how 376 is factored. The while loop determines whether there is a factor for *n* starting with $\sqrt{n}$ and working down to 1. The figure displays the flow of control after the while loop has been executed. Each time the while loop identifies a factor, it prints the factor and terminates. This is seen when *PrintFactors*(2) is executed. If the while loop identifies a factor of *n* such that

$$n = TrialFactor \cdot (n/TrialFactor)$$

and *TrialFactor* is greater than 1, then it executes *PrintFactors* again on both *TrialFactor* and *n/TrialFactor*. This is indicated, for example, in the case of *PrintFactors*(4) that must execute both *PrintFactors*(2) and *PrintFactors*(4/2) when the while loop identifies 2 as a factor.

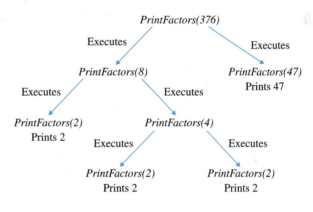

Factors: 2, 2, 2, 47

Figure 1.20   Flow of control for *PrintFactors(376)*.

Theorem 4.   Prove that the algorithm *PrintFactors* is correct.

Proof.   Exercise for the reader.   ∎

1.10.4   Application: Binary Search

If you think about how you look for a name in a phone book, you will have a good idea of what the code in the *BinarySearch* algorithm does. A common process is the following: You open a phone book to about the page where you think the name should appear. If you have turned past the name you want, you continue this process with the first part of the phone book. Otherwise, you have not gone far enough in the phone book, so you continue this process using the pages from that point forward to the end of the phone book. More mechanically, you could think of a program always choosing a page halfway through those that could possibly contain the name. If the name is not on the middle page, the search continues either in the first half of the pages being considered or in the last half of the pages being considered. This strategy is just what *BinarySearch* does by repeatedly halving the range of pages that it thinks could contain the name. Eventually, the process comes to a page that must either contain the name or the process knows that the name does not occur in the phone book.

Algorithm: Binary Search of Phone Directory

INPUT: *Name* to be found in the phone directory *City*
OUTPUT: Message indicating whether or not *Name* was found

BinarySearch(Name, City)
    FirstPage = the page number of the first page of *City*
    LastPage = the page number of the last page of *City*
    PageFound = FALSE
    NameFound = FALSE
    while (*FirstPage* ≤ *LastPage* and *PageFound* = FALSE) do
        MiddlePage = $\lfloor (FirstPage + LastPage)/2 \rfloor$
        if (*Name* falls between the first name on page *MiddlePage*
            and the last name on page *MiddlePage*) then
            PageFound = TRUE
      else
            if (*Name* is alphabetically less than
                the first name on page *MiddlePage*) then
                LastPage = *MiddlePage* − 1
            else
                FirstPage = *MiddlePage* + 1
    if (*PageFound* = TRUE) then
        Examine all names on page *MiddlePage*
        if (*Name* is found on *MiddlePage*) then
            NameFound = TRUE
        else
            NameFound = FALSE
    if (*NameFound* = TRUE) then
        Print a message saying *Name* is on *MiddlePage*
    else
        Print a message saying *Name* is not in *City*

Example 4.   Determine whether Joe Smith is in a phone book with 521 pages. For this problem, suppose Joe Smith appears on page 326.

Solution. We start with *FirstPage* $= 1$ and *LastPage* $= 521$. *MiddlePage* $= \lfloor (1 + 521)/2 \rfloor = 261$. Since Joe Smith should appear after page 261, we let *FirstPage* $= 262$. Now, *MiddlePage* $= \lfloor (262 + 521)/2 \rfloor = 391$. Since Joe Smith is not on page 391 and we are beyond the page we want, we let *LastPage* $= 390$ and compute *MiddlePage* $= \lfloor (262 + 390)/2 \rfloor = 326$. We find Joe Smith on this page and return an appropriate message.

Theorem 5. Prove that the algorithm Binary Search of Phone Directory is correct.

Proof. Exercise for the reader.   ■

1.11   Exercises

Assume that all variables not given an explicit domain are elements of $\mathbb{N}$.

1. The terms of a sequence are given recursively as $a_0 = 2$, $a_1 = 6$, and $a_n = 2 a_{n-1} + 3 a_{n-2}$ for $n \geq 2$. Find the first eight terms of this sequence.
2. The terms of a sequence are given recursively as $p_0 = 3$, $p_1 = 7$, and $p_n = 3 p_{n-1} - 2 p_{n-2}$ for $n \geq 2$. Find the first eight terms of this sequence.
3. The terms of a sequence are given recursively as $a_0 = 0$, $a_1 = 4$, and $a_n = 8 a_{n-1} - 16 a_{n-2}$ for $n \geq 2$. Find the first eight terms of this sequence.
4. Prove that with just 3-cent and 5-cent stamps, you can make any amount of postage less than 35 cents (any natural number of cents) except 1 cent, 2 cents, 4 cents, and 7 cents.
5. The terms of a sequence are given recursively as $p_0 = 1$, $p_1 = 2$, and $p_n = 2 p_{n-1} - p_{n-2}$ for $n \geq 2$. Write out the information that the inductive step assumes and what the step must prove in proving $b_n = 2 \cdot 3^n$ is a closed form for the sequence. Suppose $n_0 = 0$ and the base cases are 0 and 1.
6. The terms of a sequence are given recursively as $p_0 = 3$, $p_1 = 7$, and $p_n = 3 p_{n-1} - 2 p_{n-2}$ for $n \geq 2$. Write out the information that the inductive step assumes and what the step must prove in proving $b_n = 2^{n+2} - 1$ is a closed form for the sequence. Suppose $n_0 = 1$ and the base cases are 0 and 1.
7. The terms of a sequence are given recursively as $a_0 = 0$, $a_1 = 4$, and $a_n = 8 a_{n-1} - 16 a_{n-2}$ for $n \geq 2$. Write out the information that the inductive step assumes and what the step must prove in proving $b_n = n \, 4^n$ is a closed form for the sequence. Suppose $n_0 = 1$ and the base cases are 0 and 1.
8. Given that $b_{n-1} = 2 \cdot 3^{n-1}$ and $b_{n-2} = 2 \cdot 3^{n-2}$, prove that if $b_n = 2b_{n-1} + 3b_{n-2}$, then $b_n = 2 \cdot 3^n$ provided $n \geq 2$.
9. Given that $b_{n-1} = 2^{n+1} - 1$ and $b_{n-2} = 2^n - 1$, prove that if $b_n = 3b_{n-1} - 2b_{n-2}$, then $b_n = 2^{n+2} - 1$ provided $n \geq 2$.
10. Given that $b_{n-1} = (n-1)4^{n-1}$ and $b_{n-2} = (n-2)4^{n-2}$, prove that if $b_n = 8b_{n-1} - 16b_{n-2}$, then $b_n = n4^n$ provided $n \geq 2$.
11. The terms of a sequence are given recursively as $a_0 = 2$, $a_1 = 6$, and $a_n = 2 a_{n-1} + 3 a_{n-2}$ for $n \geq 2$. Prove by induction that $b_n = 2 \cdot 3^n$ is a closed form for the sequence.
12. The terms of a sequence are given recursively as $p_0 = 3$, $p_1 = 7$, and $p_n = 3 p_{n-1} - 2 p_{n-2}$ for $n \geq 2$. Prove by induction that $b_n = 2^{n+2} - 1$ is a closed form for the sequence.

13. The terms of a sequence are given recursively as $a_0 = 0$, $a_1 = 4$, and $a_n = 8\,a_{n-1} - 16\,a_{n-2}$ for $n \geq 2$. Prove by induction that $b_n = n\,4^n$ is a closed form for the sequence.

14. The terms of a sequence are given recursively as $p_0 = 1$, $p_1 = 2$, and $p_n = 2\,p_{n-1} - p_{n-2}$ for $n \geq 2$. Prove by induction that $b_n = 1 + n$ is a closed form for the sequence.

15. (a) Prove that with just 3-cent and 5-cent stamps, you can make any amount of postage (any natural number of cents) except 1 cent, 2 cents, 4 cents, and 7 cents.
    (*Hint:* That you can make 0-cent postage is obvious. You need to prove two things: (i) that you can assemble any amount of postage except 1 cent, 2 cents, 4 cents, and 7 cents; and (ii) that you cannot assemble these four amounts. Be careful about whether you use the Principle of Mathematical Induction or the Strong Form of Mathematical Induction.)
    (b) What amounts of postage can be assembled with 4-cent and 7-cent stamps only?
    (c) What amounts of postage can be assembled with 8-cent and 10-cent stamps only?
    (d) What amounts of postage can be assembled with 7-cent, 8-cent, and 10 cent stamps only?
    (e) What amounts of postage can be assembled with 2-cent and 5-cent stamps only?

16. Prove by induction that

$$F_n = \frac{1}{\sqrt{5}}\left(\frac{1+\sqrt{5}}{2}\right)^{n+1} - \frac{1}{\sqrt{5}}\left(\frac{1-\sqrt{5}}{2}\right)^{n+1}$$

is a closed form for the Fibonacci sequence.

17. Prove that $F_{n+m} = F_n \cdot F_m + F_{m-1} \cdot F_{n-1}$ for $m \geq 1$. Prove the following corollaries:

    (a) $F_{n-1} \mid F_{2n-1}$.
    (b) $F_{n-1} \mid F_{3n-1}$.
    (c) $F_n^2 + F_{n+1}^2$ is a Fibonacci number.

18. In how many ways can you climb a ladder with n rungs if at each step you can go up either one or two rungs? The terms of a sequence are given recursively as $a_1 = 1$, $a_2 = 2$, and $a_n = a_{n-1} + a_{n-2}$ for $n \geq 2$. Prove by induction that $b_n = F_{n+1}$ gives the terms of this sequence where F_{n+1} is the $(n + 1)$st Fibonacci number.

19. The Lucas numbers are defined as $L_0 = 2$, $L_1 = 1$, and $L_n = L_{n-1} + L_{n-2}$ for $n \geq 2$. Prove that $L_{n+1} = F_{n-1} + F_{n+1}$ for $n \geq 2$.

20. Trace through the execution of the procedure *FastPower* on the following inputs:

    (a) *base* = 3, *expont* = 9.
    (b) *base* = 2, *expont* = 10.
    (c) *base* = 5, *expont* = 6.
    (d) Count the number of multiplications needed in (a)–(c).

21. What *exactly* is wrong with the following "proof" that for every real number $x \geq 0$, $x = 2x$:
    Suppose the result is true for all real numbers y where $0 \leq y < x$.
    Case 1: $x = 0$. Then, $2x = 2 \cdot 0 = 0 = x$.
    Case 2: $x > 0$. Then, $0 < x/2 < x$. So, by hypothesis, $x/2 = 2(x/2) = x$. Doubling both sides, deduce that $x = 2x$. So, the result holds for every real number $x \geq 0$ by the Strong Form of Mathematical Induction.

22. *Challenge:* There is a third principle related to induction, the Principle of Well-Ordering for the Natural Numbers. It is the following: If $\mathcal{T} \subseteq \mathbb{N}$ and $\mathcal{T} \neq \emptyset$, then $\mathcal{T}$ contains a minimum element; that is, there is a natural number $n_0 \in \mathcal{T}$ such that for all natural numbers $k < n_0$, we have $k \notin \mathcal{T}$.

   (a) Use the Principle of Well-Ordering for the Natural Numbers instead of the Strong Form of Mathematical Induction to prove that

$$0 + 1 + 2 + \cdots + n = \frac{n \cdot (n + 1)}{2}$$

   (*Hint:* Let $\mathcal{T} = \{n \in \mathbb{N} : 0 + 1 + 2 + \cdots + n \neq n \cdot (n + 1)/2\}$.)

   (b) Use the Principle of Well-Ordering for the Natural Numbers instead of the Strong Form of Mathematical Induction to prove that every integer n such that $n > 1$ can be factored into a product of one or more primes.

   (c) Using the Principle of Well-Ordering for the Natural Numbers, prove one of the forms of the Principle of Mathematical Induction.

   (d) Using one of the forms of the Principle of Mathematical Induction, prove the Principle of Well-Ordering for the Natural Numbers.

23. The Binary Search of Phone Directory algorithm in Section 1.10.4 looks for any page (if any) containing a name *Name* in a telephone book *City*. The portion of the algorithm used in searching for the page is called *BinarySearch*. Prove that the algorithm works correctly.

1.12 Chapter Review

The language of sets was introduced. The basic operations of union, intersection, set difference, and complementation were studied. The properties of these operations were given as well as the properties of these operations when they are used with each other. One important way that union, intersection, and complementation interact is through DeMorgan's Laws. Finally, the power set of a set and the product of two sets are introduced. The proof techniques used with sets are highlighted as templates for an idea of how to approach similar proofs. The chapter then moves to the topic of determining the number of elements in a set of overlapping sets using the Principle of Inclusion-Exclusion. The last two sections introduce extremely important proof techniques for proving results about the natural numbers. Both the Principle of Mathematical Induction and the Strong Form of Mathematical Induction are explained and used in constructing proofs of statements about natural numbers. The basic idea of a pseudocode that is used to present algorithms is described for use throughout.

Set operations are used as examples of operations that define boolean algebras and lattices. Induction is used to study Fibonacci numbers and geometric series. Important examples regarding the use of induction in both forms in proving an algorithm is correct are given. For example, algorithms for computing powers, finding factorizations of an integer, and carrying out an efficient search are proven to be correct algorithms.

1.12.1   Terms, Theorems, Algorithms, and Templates

1.1   Summary

TERMS

algebraic identity	is a member of	real numbers
empty set	is an element of	set
equal	is contained in	set-theoretic notation
factor	is in	subset
finite set	is not an element of	universal set
if and only if	natural numbers	universe
implication	not finite sets	vacuously
infinite set	proper subset	Venn diagram
integers	rational number	

THEOREM

$A = B$ if and only if $A \subseteq B$ and $B \subseteq A$

TEMPLATES

Template 1.1	Element Membership in a Set	Template 1.5	Set Equality
Template 1.2	Set Inclusion	Template 1.6	Set Inequality
Template 1.3	Set Non-Inclusion	Template 1.7	Implications and If and Only If
Template 1.4	Proper Set Inclusion		

1.3   Summary

TERMS

absolute difference	disjoint sets	minimum element
analogous	distributive lattice	power set
bit representation	equivalent statements	product
boolean algebra	inclusive or	proof by cases
bottom	indirect proof	relative difference
complement	intersection ($\cap$)	set difference
complementation	inverse	statement
complemented lattice	join ($\vee$)	symmetric difference
contrapositive	lattice	top
converse	maximum element	union ($\cup$)
counterexample	meet ($\wedge$)	

THEOREMS

Absorption Law for Join	Commutative Law for Intersection
Absorption Law for Meet	Commutative Law for Join
An Absorption Law	Commutative Law for Meet
Associative Law for Intersection	Commutative Law for Union
Associative Law for Join	DeMorgan's Law for Intersection
Associative Law for Meet	DeMorgan's Law for Union
Associative Law for Union	DeMorgan's Laws

Distributive Law for Intersection   Distributive Law for Meet
Distributive Law for Join   Distributive Law for Union

TEMPLATES

Template 1.8   Proof by Cases   Template 1.10   Proof by Contradiction
Template 1.9   Disproof by Counterexample   Template 1.11   Indirect Proof

1.5   Summary

TERMS

cardinality   number of divisors
even intersection   odd intersection
hat check problem

THEOREMS

Basic Counting Theorem   Principle of Inclusion-Exclusion for
Principle of Inclusion-Exclusion   Three Sets
Principle of Inclusion-Exclusion for   Principle of Inclusion-Exclusion for Two
   Finitely Many Sets   Sets

1.7 and 1.8   Summary

TERMS

algorithm   inductive step
base step   infinite loop
condition   lemma
correctness   mathematical induction
Fibonacci numbers   perfect square
finite geometric series   pseudocode
first form   recursive definition
for loop   recursively defined sequence
geometric series   selection sort
inductive assumption   while loop
inductive hypothesis

THEOREMS

Principle of Mathematical Induction
Size of a Power Set

ALGORITHMS

Perfect Squares
Square Root I
Square Root II

TEMPLATES

Template 1.12 Using the Principle of
   Mathematical Induction

1.10   Summary

TERMS

base cases

base step(s)

closed form

inductive hypothesis

inductive step

prime numbers

recursion

recursive call

recursively defined sequence

THEOREMS

Fundamental Theorem of Arithmetic

Strong Form of Mathematical Induction

ALGORITHMS

Compute Powers

Largest Odd Divisor

Print a Prime Factorization of an Integer

Binary Search of Phone Directory

Compute F_n

TEMPLATE

Template 1.13 Using the Strong Form of Mathematical Induction

1.12.2   Starting to Review

1. Which of the following set descriptions gives the set {2, 8, 14, 20, 26, 32}?

    (a) $\{n \in \mathbb{N} : n = 2x + 6 \text{ for some integer } x \text{ such that } 1 \leq x \leq 6\}$

    (b) $\{n \in \mathbb{N} : n = 6x + 2 \text{ for some integer } x \text{ such that } 1 \leq x \leq 6\}$

    (c) $\{n \in \mathbb{N} : n = 6x + 2 \text{ for some integer } x \text{ such that } 0 \leq x < 6\}$

    (d) None of the above

2. Let $B = \{2, 3, 6, 9, 11\}$ and $C = \{1, 4, 6, 11, 15\}$. Which of the following sets are not any of $B \cup C$, $B \cap C$, and $B - C$?

    (a) {1, 6, 9, 15}

    (b) {6, 11}

    (c) {2, 3, 9}

    (d) None of the above

3. What is the contrapositive of the statement "If the sun is shining, then it is time to go outside."

    (a) If the sun is shining, then it is not time to go outside.

    (b) If it is time to go outside, then the sun is shining.

    (c) If it is not time to go outside, then the sun is not shining.

    (d) None of the above.

4. Of 26 students who are either females or biology majors, there are 17 females and 23 biology majors. How many females are biology majors?

   (a) 12
   (b) 17
   (c) 14
   (d) 9

5. Describe each of the following sets in the format $\{x : \text{property of } x\}$.

   (a) $A = \{0, 2, 4, 6, 8, \ldots\}$
   (b) $B = \{1, 2, 5, 10, 17, 26, 37, 50, \ldots\}$
   (c) $C = \{1, 5, 9, 13, 17, 21, \ldots\}$
   (d) $D = \{1, 1/2, 1/3, 1/4, 1/5, \ldots\}$
   (e) $E = \{\text{lemon, lime}, 1, 3, 5, 7, \ldots\}$

6. For $U = \{1, 2, 3, \ldots, 9, 10\}$, let $A = \{1, 2, 3, 4, 5\}$, $B = \{1, 2, 4, 8\}$, $C = \{1, 2, 3, 5, 7\}$, and $D = \{2, 4, 6, 8\}$. Determine the elements of each of the following sets

   (a) $(A \cup B) \cap C$
   (b) $A \cup (B \cap C)$
   (c) $\overline{C} \cup \overline{D}$
   (d) $\overline{C \cap D}$
   (e) $(A \cup B) - C$
   (f) $A \cup (B - C)$
   (g) $(B - C) - D$
   (h) $B - (C - D)$
   (i) $(A \cup B) - (C \cap D)$

7. List the subsets of each of the following sets:

   (a) $A = \{1, 2, 3\}$
   (b) $B = \{1, \{2, 3\}\}$
   (c) $C = \{\{1, 2, 3\}\}$

8. Find a counterexample to $A \subseteq B \Leftrightarrow A \cup B = A$.

9. List the first eight terms of the sequence defined as $c_0 = 1$, $c_1 = 3$, and $c_n = c_{n-1} + 2c_{n-2}$ for $n \geq 2$.

10. Let A be a subset of some universal set U. If A contains 58 elements and $\overline{A}$ contains 37 elements, how many elements are in U?

1.12.3 Review Questions

1. Let $A = \{1, 2, 4, 7, 8\}$, $B = \{1, 4, 5, 7, 9\}$, and $C = \{3, 7, 8, 9\}$. Let $U = \{1, 2, 3, 4, 5, 6, 7, 8, 9, 10\}$. Find set expressions using these sets and the operations of union, intersection, absolute difference, and relative difference to represent the following sets:

   (a) $\{2, 7, 9\}$
   (b) $\{3, 5, 6, 7, 9, 10\}$

2. A survey of reading habits was proposed for the city of Lewisburg. Let U be the sample set of adults in Lewisburg, F the set of females in the sample, B the set of readers who have finished five or more books in the past year (called regular book readers), and P the set of readers who read some of every issue of a periodical during the past year (called the regular periodical readers). Use set notation to identify the following sets of readers:

   (a) Females who regularly read books or periodicals
   (b) The men who read both books and periodicals regularly
   (c) Adults who regularly read either books or periodicals, but not both
   (d) The women who do not read either books or periodicals regularly
   (e) The men who read books but not periodicals regularly

   Now, describe in words the following sets:

   (f) $\overline{F} \cap P$
   (g) $F \cap \overline{B} \cap P$
   (h) $F \cap B \cap P$
   (i) $\overline{F} \cap \overline{B} \cap \overline{P}$
   (j) $F \cap (P \cup B) - F \cap (P \cap B)$

3. For sets A and B, prove that $A \cup (B - A) = A \cup B$.
4. For sets A and B, prove that $A \cap B = \emptyset \Leftrightarrow A \subseteq \overline{B}$.
5. Prove by induction that $3 + 11 + \cdots + (8n - 5) = 4n^2 - n$ for $n \in \mathbb{N}$ and $n \geq 1$.
6. Prove by induction that $2n + 1 < 3n - 1$ for $n \in \mathbb{N}$ and $n \geq 3$.
7. Prove that for every $n \in \mathbb{N}$ that $n^3 + n$ is even.
8. Prove by induction that $73 \mid (8^{n+2} + 9^{2n+1})$ for every $n \in \mathbb{N}$.
9. Prove that $b_n = 5 \cdot 2^n + 1$ is a closed form for the recursive relation $a_0 = 6, a_1 = 11,$ and $a_n = 3a_{n-1} - 2a_{n-2}$ for $n \geq 2$.
10. Let $S \subseteq \mathbb{N}$ and $3 \in S$. Also, assume that if $x \in S$, then $x + 3 \in S$. Prove that

$$\{3 \cdot n : n \in \mathbb{N}\} \subseteq S.$$

11. The country of Xabob uses currency consisting of coins with values of 3 zabots and 5 zabots. If you cannot combine some number of these coins to pay a bill, the item is free. For what number of zabots are items free? Prove your answer.
12. *Challenge:* The name *Strong Form of Mathematical Induction* suggests that that form really is a logically different assertion than the *Principle of Mathematical Induction.* In fact, however, this is not so. It is not too difficult to prove one form from the other.

   (a) Assuming the Strong Form of Mathematical Induction, prove the Principle of Mathematical Induction. You need to do the following: Assume the hypothesis of the first form of the Principle of Mathematical Induction, and using just the Strong Form of Mathematical Induction, prove the conclusion of the (first form of the) Principle of Mathematical Induction. So, assume $T \subseteq \mathbb{N}$, some $n_0 \in T$, and for every $n \geq n_0$, if $n \in T$, then $n + 1 \in T$. Then, using the Strong Form of Mathematical Induction but *not* the (first form of the) Principle of Mathematical Induction, prove that $T = \mathbb{N}$. (For a statement of the first form of the Principle of Mathematical Induction, see Section 1.7.1) (*Hint:* Let $T' = T \cup \{0, 1, \ldots, n_0 - 1\}$. Prove, using the Strong Form of Mathematical Induction but not the Principle of Mathematical Induction, that $T' = \mathbb{N}$. Then, use that to show that every natural number $n \geq n_0$ is in T.)

(b) Assuming the (first form of the) Principle of Mathematical Induction, prove the Strong Form of Mathematical Induction. You need to do the following: Assume $T \subseteq \mathbb{N}$ and that for all $n \in \mathbb{N}$, if all $k < n$ are in T, then $n \in T$. Prove, using the Principle of Mathematical Induction but not the Strong Form of Mathematical Induction, that $T = \mathbb{N}$. (*Hint:* Let $T' = \{n \in \mathbb{N} : \text{for all } k < n, k \in T\}$. Prove $T' = \mathbb{N}$, and then use that to prove $T = \mathbb{N}$.)

13. How many students are in Math347? From the survey of all the students, it was found that 43 had taken Econ103, 55 had taken Soci213, 30 had taken Musi111, 8 had taken both Econ103 and Soci213, 13 had taken both Econ103 and Musi111, 15 had taken Soci213 and Musi111, and 8 had taken none of the courses. No one had taken all three courses.

14. How many integers between 1 and 250, including 1 and 250, are divisible neither by 3 nor by 7 but are divisible by 5?

1.12.4 Using Discrete Mathematics in Computer Science

1. Prove that the Largest Odd Divisor algorithm outputs the largest odd divisor of N for all integers $N > 0$.
2. Prove that the *PrintFactors* algorithm factors natural numbers $N > 1$ into primes. Prove that, in fact, its output is a list of one or more primes whose product is N. So, for $N = 24$, the outputs are the numbers 2, 2, 2, and 3, in some order.
3. Consider the Binary Search of Phone Directory algorithm. This algorithm looks for the page (if any) containing a name *Name* in a telephone book *City*. The portion of the algorithm used in searching for the page is called *BinarySearch*. Prove that the algorithm works correctly.
4. The summation shown arises in determining how long it takes part of one particular method, called **heapsort,** to sort a list of numbers into increasing order. More precisely, heapsort often is written with a preprocessing step called *heapify*. (*Preprocessing* means that this step is performed once before the main step of the program.) This summation arises in determining how long it takes to "heapify" a list of 2^n numbers:

$$0 \cdot 2^n + 1 \cdot 2^{n-1} + 2 \cdot 2^{n-2} + 3 \cdot 2^{n-3} + \cdots + (n-1) \cdot 2^1 + n \cdot 2^0 = 2^{n+1} - n - 2$$

Prove by induction that the summation is correct for $n \geq 0$.

5. Show by induction on n that for $b \in \mathbb{N}, b \geq 2$,

$$(b-1) \cdot \sum_{i=0}^{n} b^i = b^{n+1} - 1$$

Interpret this identity in the context of number representation in the base b using the standard positional notation. Start by seeing what this means for $b = 10$ and $n = 4$.

6. (a) In the calculation of $base^{expont}$ using *FastPower*, how many copies of the algorithm will be invoked?
   (b) Show that if the *FastPower* algorithm is invoked n times (that is, n total invocations, including both the original invocation from the outside and the recursive invocations), somewhere between 0 and $2n$ multiplications will be performed.

(c) A simpler algorithm to calculate 1.001^{1000} is to multiply 1000 copies of 1.001 together, using 999 multiplication in all. Using parts (a) and (b), estimate how many fewer multiplications the *FastPower* algorithm performs.

7. Let X and Y be two lists sorted in nondecreasing order. Suppose that for some positive integer n, there is a combined total of n numbers in the two lists. Prove that X and Y can be merged into a single list of n numbers in nondecreasing order using at most $n - 1$ comparisons.

8. Prove that the following code to compute Fibonacci numbers is correct:

Algorithm: Compute F_n

INPUT: $n \in \mathbb{N}$
OUTPUT: F_n
recursiveFibonacci(n)
if $n = 0$ then
    recursiveFibonacci(0) $= 1$
else
    if $n = 1$ then
        recursiveFibonacci(1) $= 1$
    else
        recursiveFibonacci(n) $=$ *recursiveFibonacci*(n $-$ 1)
            $+$ *recursiveFibonacci*(n $-$ 2)

---

9. Prove that, at most, $n + 1$ comparisons are required to determine if a particular number is in a list of 2^n numbers sorted in nondecreasing order.

10. Prove that exactly $n - 1$ multiplications are needed to compute the product of n distinct real numbers in a fully parenthesized expression, regardless of how parentheses are used.

CHAPTER 2

Formal Logic

It is an old dream to write a formal, mathematical description of the laws of human thought. The goals are to identify what it is that makes certain arguments correct and to identify correct arguments only from their logical form. Work toward these goals is ancient. It began with the early Greeks and was extensively developed by Aristotle (384–322 BC). The study was again actively pursued in the Middle Ages. During the nineteenth and twentieth centuries, the field developed rapidly, with explosive growth starting around 1930. The understanding of formalized reasoning is one of the major topics of formal logic, and it has been extensively applied to studying mathematical proofs. In computer science, formal logic has many applications in areas such as database theory, artificial intelligence, program language design, and automated verification of software and hardware. In database theory, logic is used to formalize the definitions of queries. In artificial intelligence, logic is used to formalize human inference. Proving a program to be correct can use logic-based notions such as loop invariants and both pre- and postconditions. Formal logic also plays a major role during many phases in the design of electronic computers, including the design of efficient combinatorial networks or circuits.

This chapter provides an introduction to formal logic. First, we give the basic definitions of propositional logic. These cover the usual material expected of a discrete mathematics course—propositional logic and logical truth. Next, we introduce normal forms in propositional logic, particularly simple ways to write formulas, a topic that is now of special interest in computer science. One application of normal forms is in combinatorial network design. Examples of the relationship between normal forms and combinatorial networks will be explained as well. Finally, we discuss an extension of propositional logic involving predicates and quantifiers. These are key ideas in an extension of propositional logic to predicate logic. An important part of predicate or first-order logic is to express, in a single statement, how elements in a set of values can make the statement true.

2.1   Introduction to Propositional Logic

The simplest variant of formal logic is **propositional logic.** Its basic object is a simple, declarative sentence, called a **proposition.** Propositional logic is concerned with combining sentences, such as "The world is round" and "Columbus was right" to form "If the world is round, then Columbus was right."

A proposition is something that is either true or false; it is not both. "The cover of this book is pink" is a proposition. "Napoleon spent at least one day of his life in Paris" and "Either the butler did it with a bottle or the colonel did it with a lead pipe" are also propositions. On the other hand, "Justice," "The Queen's birthday," "Whoever is the stronger," and "Why is the world almost round?" are neither true nor false and, therefore, are not propositions.

In formal notation, the letters p, q, r, and s (plus those letters subscripted with natural numbers, such as p_1, q_2, and r_{127}) are used to stand for, or to denote, propositions. Such a variable is called a **proposition letter.** We consider proposition letters to be essentially the same as boolean (logical) variables in a programming language. T and F are **propositional constants**—that is, propositions with fixed truth values of *TRUE* and *FALSE,* respectively.

Propositional logic is concerned with certain ways in which simple sentences can be combined into more complex sentences. Several standard operations are used on propositions to form other propositions. Such an operation is called a **propositional connective.** The common propositional connectives are shown in Table 2.1.

Table 2.1   Propositional Connectives

Connective	Sample Use	Common Translation
$\neg$	$\neg p$	"not p"
$\wedge$	$p \wedge q$	"p and q"
$\vee$	$p \vee q$	"p or q (or both)"
$\rightarrow$	$p \rightarrow q$	"if p, then q," or "p implies q"
$\leftrightarrow$	$p \leftrightarrow q$	"p if and only if q," or "p is equivalent to q"

Example 1.   Let p denote "Henry eats halibut" and q denote "Catherine eats kippers."

(a) The proposition $\neg p$ is read "Henry does not eat halibut."
(b) The proposition $p \wedge q$ is read "Henry eats halibut, and Catherine eats kippers."
(c) The proposition $p \rightarrow q$ is read "If Henry eats halibut, then Catherine eats kippers."
(d) The proposition $p \leftrightarrow q$ is read "Henry eats halibut if and only if Catherine eats kippers."
(e) The proposition $(\neg p) \vee (\neg q)$ is read "Henry does not eat halibut, or Catherine does not eat kippers."
(f) The proposition $p \leftrightarrow (\neg q)$ is read "Henry eats halibut if and only if Catherine does not eat kippers."

Example 2.   Let p denote "Henry eats halibut," q denote "Catherine eats kippers," and r denote "I'll eat my hat."

(a) Write a proposition that reads "If Henry eats halibut but Catherine does not eat kippers, then I'll eat my hat."
(b) Write a proposition that reads "Either Henry eats halibut or Catherine eats kippers, but not both."

Solution.

(a) $(p \land \neg q) \to r$. Since *and* and *but* usually both get translated as $\land$, the difference between the two English words is usually an issue not of what is the case but, rather, of what we *would have expected* to be the case.

(b) $(p \lor q) \land \neg(p \land q)$.

This proposition is "logically equivalent to" the proposition in Example 1 (f), meaning that $p \leftrightarrow (\neg q)$ is an equally good answer. We shall discuss logical equivalence in the next section. ∎

Definition 1.   Let p, q, and r be propositions. The proposition $\neg p$ is the **negation** of p. The proposition $p \land q$ is the **conjunction** of p and q, and p and q are called its **conjuncts.** The proposition $p \lor q$ is the **disjunction** of p and q, and p and q are called its **disjuncts.** The proposition $p \to q$ is a **conditional,** or an **implication,** with **hypothesis** p and **conclusion** q. The proposition $p \leftrightarrow q$ is an **equivalence** or a **biconditional.**

Since the English language is often ambiguous, and the meanings of words can vary from context to context, the English translations of the symbols we have just introduced ($\neg$, $\land$, $\lor$, $\to$, and $\leftrightarrow$) do not define the meanings of the symbols precisely. A precise definition of each symbol is given by a **truth table,** which provides the truth value for the result of applying the operation on each possible set of truth values for the operands. As mentioned, we shall use the symbols T and F to denote the truth values *TRUE* and *FALSE* as well as to denote propositional constants. Table 2.2 shows the truth table for negation.

Table 2.2   Truth Table for Negation

Truth Table for $\neg$	
p	$\neg p$
T	F
F	T

Table 2.2 is read as follows: For any proposition p, if p is T, then $\neg p$ is F, and if p is F, then $\neg p$ is T. This assignment of truth values agrees with the common usage of the word *not*. Truth tables for the other propositional connectives are shown in Table 2.3.

Table 2.3   Truth Tables for Logical Connectives

Truth Table for $\land$			Truth Table for $\lor$		
p	q	$p \land q$	p	q	$p \lor q$
T	T	T	T	T	T
T	F	F	T	F	T
F	T	F	F	T	T
F	F	F	F	F	F

Truth Table for $\to$			Truth Table for $\leftrightarrow$		
p	q	$p \to q$	p	q	$p \leftrightarrow q$
T	T	T	T	T	T
T	F	F	T	F	F
F	T	T	F	T	F
F	F	T	F	F	T

As an example of using the truth table for $\wedge$, suppose you know that both p and q are T. Look in the truth table for $\wedge$ to find the row where both p and q have the value T. Then, look across that row to find the truth value of $p \wedge q$. In this case, $p \wedge q$ has the value T. Now, suppose in another instance you know that p is T and q is F. The second row of the table for $\wedge$ has the value T for p and F for q. In that row, the truth value given for $p \wedge q$ is F.

It is helpful to consider how the truth table for $\rightarrow$ relates to common usage of "if ... then." A simple requirement of a notion of "if ... then" is that "if ... then" statements should be usable in arguments. If it is true that "The carriage had mud on its tires" and is also true that "If the carriage had mud on its tires, then it is raining outside," then one can correctly infer that "It is raining outside." The truth table definition of $\rightarrow$ is that $p \rightarrow q$ is F just in case it would lead from a true hypothesis to a false conclusion. The truth table for $\rightarrow$ also corresponds to the template for proving an "if ... then" result that was introduced in Chapter 1.

2.1.1 Formulas

More complicated propositional expressions, called **formulas** or **well-formed formulas (wffs),** can be built from the proposition letters using the propositional connectives and parentheses. When we say $\phi = (p \wedge q) \rightarrow r$, we mean that ϕ is the string of symbols $(p \wedge q) \rightarrow r$. For the following formulas, we would like to know when the conclusion is necessarily true:

$\phi = (p \wedge q) \rightarrow r$, which can be paraphrased as "If p and q are both true, then r is also true."

$\phi_1 = (p \vee q) \rightarrow r$, which can be paraphrased as "If p or q (or both) is true, then r is also true."

$\phi_2 = (p \rightarrow r) \rightarrow ((p \wedge q) \rightarrow r)$, which can be paraphrased as "Suppose that if p is T, then r is T. Then, if p and q are both T, then r is T."

In the last formula, we translated two of the $\rightarrow$'s as *if ... then* and one as *suppose ... then*. We did that to make the reading easier. One advantage of a formal notation is that it lets us express concepts that cannot be expressed easily and unambiguously in everyday language.

Example 3. Translate the following sentences into a formula in propositional logic: "If Mr. Holmes told the truth and Mr. Watson did not hear anything, then it cannot be both that the butler did it and that the butler returned to his hotel room that night."

Solution.   Actually, there are many translations, depending on which parts of the sentence are chosen to be represented by proposition letters and on which proposition letters are chosen to represent them.

Let p denote "Mr. Holmes told the truth," q denote "Mr. Watson did not hear anything," r denote "the butler did it," and s denote "the butler returned to his room that night." The sentence can now be translated into propositional logic as

$$\phi = (p \wedge q) \rightarrow (\neg(r \wedge s)) \qquad \blacksquare$$

The reader is urged to do Exercise 1 in Section 2.2 before going on to the rest of the section.

The formal definition of a formula is an **inductive definition** of a set of strings. The **base cases** correspond to the base step of an inductive proof. The **closure rules** correspond to the inductive step.

Definition 2.   A **formula** is any string of symbols that is formed using the following rules:

1. **Base cases:**   Every proposition letter is a formula. T and F are formulas.
2. **Closure rules:**   Let ϕ be a formula. Then, $(\neg \phi)$ is a formula. For formulas ϕ and ψ, $(\phi \wedge \psi)$, $(\phi \vee \psi)$, $(\phi \rightarrow \psi)$, and $(\phi \leftrightarrow \psi)$ are formulas.

According to the base case alone, p, q, and T are formulas. From the base case and just one application of the closure rules, one can show that $(p \wedge q)$, $(p \vee p)$, $(p \rightarrow T)$, and $\neg q$ are formulas. From the base case and two applications of the closure rules, one can show that $(\neg(p \wedge q))$ and $(q \leftrightarrow (p \rightarrow T))$ are formulas.

It often seems that in elementary logic, most theorems are proved by induction on some integer related to formulas, such as the number of symbols, the number of parentheses, the number of propositional connectives, or the number of times the closure rules of Definition 2 were applied to generate the formula. (Let this be a hint for the Exercises.)

Theorem 1.   (Principle of Induction on Formulas)   Let $\mathcal{F}$ be a set of formulas such that:

Base cases   Each proposition letter is in $\mathcal{F}$, and T and F are in $\mathcal{F}$.

Closure rules   If ϕ, ψ are formulas in $\mathcal{F}$, so are

$$(\neg \phi), (\phi \wedge \psi), (\phi \vee \psi), (\phi \rightarrow \psi), \text{ and } (\phi \leftrightarrow \psi)$$

Then, $\mathcal{F}$ is the set of all formulas.

Proof.   Let $\mathcal{T} = \{n \in \mathbb{N} : \text{all formulas formed using } n \text{ elements of } \{\neg, \vee, \wedge, \rightarrow, \leftrightarrow\}\}$ are in $\mathcal{F}\}$. If we prove $\mathcal{T} = \mathbb{N}$, then all formulas are in $\mathcal{F}$. We will use the strong form of mathematical induction to complete this proof.

(Base step)   Let $n = 0$. All formulas using 0 instances of elements of $\{\neg, \vee, \wedge, \rightarrow, \leftrightarrow\}$ are just the proposition letters and the two logical constants T and F. Because these are just the elements in the base cases used to define $\mathcal{F}$, all these elements are in $\mathcal{F}$, and $0 \in \mathcal{T}$.

(Inductive step)   Let $n > 0$ and assume that $0, 1, \ldots, n - 1 \in \mathcal{T}$. To prove $n \in \mathcal{T}$ will be a proof by cases (see Template 1.8, Proof by Cases). We use a proof by cases because a formula formed using n instances of elements of $\{\neg, \vee, \wedge, \rightarrow, \leftrightarrow\}$ is of one of the following forms:

(a) $\neg \phi$, where ϕ is formed using $n - 1$ elements of $\{\neg, \vee, \wedge, \rightarrow, \leftrightarrow\}$
(b) $\phi \vee \psi$, where ϕ and ψ are each formed using fewer than n elements of $\{\neg, \vee, \wedge, \rightarrow, \leftrightarrow\}$
(c) $\phi \wedge \psi$, where ϕ and ψ are each formed using fewer than n elements of $\{\neg, \vee, \wedge, \rightarrow, \leftrightarrow\}$
(d) $\phi \rightarrow \psi$, where ϕ and ψ are each formed using fewer than n elements of $\{\neg, \vee, \wedge, \rightarrow, \leftrightarrow\}$
(e) $\phi \leftrightarrow \psi$, where ϕ and ψ are each formed using fewer than n elements of $\{\neg, \vee, \wedge, \rightarrow, \leftrightarrow\}$

The details of the proof in each of these cases are left as an exercise.   ∎

The theorem that follows is included because it is an example of an easy application of the Principle of Induction on Formulas: It may look rather uninteresting and technical: It deals only with counting the parentheses in a formula. Suppose, however, you were writing a computer program to check something about logical formulas. In this case, you would need to pay close attention to the parentheses. (Of course, you would have to worry about more sophisticated issues than just counting the parentheses.) Or, consider the job of a person writing a compiler for a computer language. The compiler code will have to pay close attention to)'s,]'s, and }'s, because having them misplaced causes difficulties for the program.

Theorem 2.   Every formula has an equal number of right and left parentheses.

Proof.   Let $\mathcal{F}$ be the set of formulas that have an equal number of right and left parentheses. Prove by induction on formulas that $\mathcal{F}$ is the set of all formulas.

(Base cases)   Each proposition letter is in $\mathcal{F}$, since it is a formula with no left parentheses and no right parentheses. Similarly, $T, F \in \mathcal{F}$.

(Closure rules)   Let $\phi, \psi \in \mathcal{F}$. Let ϕ have n left parentheses and n right parentheses and ψ have m left parentheses and m right parentheses. Then:

(a) $(\neg\phi)$ has $n + 1$ left parentheses (n in ϕ plus one more in front) and $n + 1$ right parentheses (n in ϕ plus one more following), so $(\neg\phi) \in \mathcal{F}$.
(b) $(\phi \wedge \psi)$ has $m + n + 1$ left parentheses (m in ψ, n in ϕ, and one more in front) and $m + n + 1$ right parentheses (m in ψ, n in ϕ, and one more following), so $(\phi \wedge \psi) \in \mathcal{F}$.
(c) $(\phi \vee \psi)$, $(\phi \rightarrow \psi)$, and $(\phi \leftrightarrow \psi)$ each have $m + n + 1$ left parentheses and $m + n + 1$ right parentheses, so each is in $\mathcal{F}$.

Therefore, by the Principle of Induction on Formulas, it follows that $\mathcal{F}$ is the set of all formulas.   ■

2.1.2   Expression Trees for Formulas

An **expression tree** is simply a visual representation for the way that a formula is built from propositions and logical operators. A proposition is represented by a single node, simply a filled-in circle, as shown in Figure 2.1.

p

Figure 2.1   Representation for *p*.

For an expression involving two propositions and a logical operator, the propositions are represented by nodes at the same level, and then at a higher level, a node represents the result of applying the operator to the two propositions. The nodes representing the propositions and the node representing the result of the operation are joined by lines. For example, the final picture for $p \vee q$ is shown in Figure 2.2.

Figure 2.2 Representation for $p \vee q$.

To introduce the representation structure for a more general formula, we will describe how you build an expression tree from the top down. To build an expression tree from an expression, first place the final expression at the top of the representation, and then put the expressions that are operated on to form the final expression underneath. Join by lines the nodes representing the expressions operated on and the node representing the result of the operation. The process can continue until the lowest level contains only propositions. The resulting picture or representation of an expression is an expression tree.

The expression tree structure gives exactly the same information as the parentheses in the formula about the order of execution, but the expression tree sometimes gives a better picture. Because this representation is so useful in evaluating an expression, we will give several more examples and then a formal description of how you can build an expression tree from the bottom up. The expression tree of $((p \wedge q) \wedge r)$ is shown in Figure 2.3.

Figure 2.3 Expression tree of $((p \wedge q) \wedge r)$.

The expression tree of $((\neg p) \vee q) \to (r \to p)$ is shown in Figure 2.4.

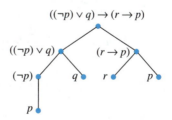

Figure 2.4 Expression tree of $(((\neg p) \vee q) \to (r \to p))$.

Definition 3. (**Expression Tree for a Formula**) The expression tree for a proposition letter p, for T, or for F consists of a single node as shown:

$$\bullet \quad \bullet \quad \bullet$$
$$p \quad T \quad F$$

If ϕ is a formula with expression tree T_ϕ, then an expression tree for $T_{(\neg\phi)}$ is

If ϕ and ψ are formulas with expression trees T_ϕ and T_ψ, respectively, then an expression tree for $T_{(\phi\wedge\psi)}$ is

Expression trees for $(\phi \vee \psi)$, $(\phi \rightarrow \psi)$, and $(\phi \leftrightarrow \psi)$ are defined analogously to the way the expression trees is defined for $(\phi \wedge \psi)$. The corresponding expression trees are

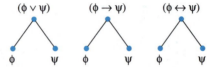

It can be proved that each formula has exactly one expression tree. This principle sometimes allows arguments that manipulate expression trees to be used as a replacement for induction on formulas. Some examples can be found in writing formal proofs for the theorems on substitution.

For any expression tree T and any node x in the expression tree, the portion T_x of the tree at or below x forms another expression tree—namely, the expression tree for x.

Definition 4. Let χ be a formula with expression tree T, and let ψ be a formula with expression tree U. Then, χ is a **subformula** of ψ if, for some node x of U, $T_\chi = U_x$.

Example 4. For the expression tree T, determine the subformulas defined by p and $(\neg(p \vee q))$.

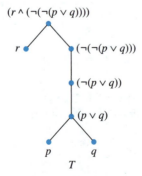

Solution.   The subtrees T_p and $T_{(\neg(p \vee q))}$ are as shown:

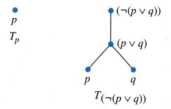

$$T_{(\neg(p \vee q))}$$

The term **syntax** refers to the rules for forming grammatically correct strings of symbols of a language. The rules specified here in the definition of the terms *formula* and *subformula* are examples of rules for forming correct strings of symbols for propositional logic. In the next section, we will discuss the **semantics** of propositional logic—that is, what the strings of symbols **mean**—though we have already begun discussing semantics by giving the truth tables.

2.1.3   Abbreviated Notation for Formulas

A formula such as

$$((((\neg(\neg p)) \wedge (\neg q)) \wedge r) \vee (((\neg(\neg q)) \wedge (\neg r)) \wedge s)) \leftrightarrow (s \rightarrow p)$$

has so many parentheses that the reader can easily get confused. Just as in ordinary arithmetic, however, formulas in informal usage are abbreviated by dropping some of the parentheses or by using different styles of parentheses, such as brackets. Some widely accepted conventions are summarized in Table 2.4.

Table 2.4   Common Abbreviations and Other Informal Usage

1. Drop the outermost set of parentheses, simplifying $(\neg p)$ to $\neg p$ and $(p \vee q)$ to $p \vee q$.
2. In a series of conjunctions nested to the left, such as $(p \wedge q) \wedge r$, drop the parentheses, writing $p \wedge q \wedge r$. Similarly, with disjunctions, abbreviate $(p \vee q) \vee r$ to $p \vee q \vee r$.
3. A $\neg$ symbol always applies to as little as possible. That is, $\neg$ is the *highest priority* operation, and $\neg a \vee b$ means $(\neg a) \vee b$.
4. The remaining operations are often given priorities as follows, from highest to lowest: $\wedge$, $\vee$, $\rightarrow$, and $\leftrightarrow$. Thus:
   (a) $\neg a \wedge b \vee c \wedge d$ abbreviates $(((\neg a) \wedge b) \vee (c \wedge d))$.
   (b) $a \rightarrow b \vee b \wedge c$ abbreviates $(a \rightarrow (b \vee (b \wedge c)))$.
   (c) $a \leftrightarrow b \rightarrow c \wedge d$ abbreviates $(a \leftrightarrow (b \rightarrow (c \wedge d)))$.
   (*Caution:* Use this rule *sparingly* to omit parentheses. Overuse of the rule creates almost-unreadable formulas. When in doubt, leave the parentheses in.)
5. In formulas with nested parentheses, it is common to replace some of the parentheses with other symbols, usually brackets that is, ([and]). So, the formula in Section 2.1.3 might be written as

$$[(\neg\neg p \wedge \neg q \wedge r) \vee (\neg\neg q \wedge \neg r \wedge s)] \leftrightarrow [s \rightarrow p]$$

2.1.4   Using Gates to Represent Formulas

At the basic hardware level, computer memory has two states, which are identified as the two **logical values** or **boolean values** of T and F. Computer operations are thought of as being composed of operations on these boolean values and, hence, as operations of propositional logic. In describing computer circuits, a specialized notation for propositional logic is used. Special physical devices, called **gates,** implement the $\wedge$, $\vee$, and $\neg$ operations. A set of gates to represent a circuit is called a **combinatorial circuit** or **combinatorial network.**

Think of a gate as representing an operation and of the wires going into the gates as representing its operands. For example, a $\wedge$ gate will let current flow out if and only if both operands (that is, both wires coming in) carry current. Notation for these gates is shown in Figure 2.5.

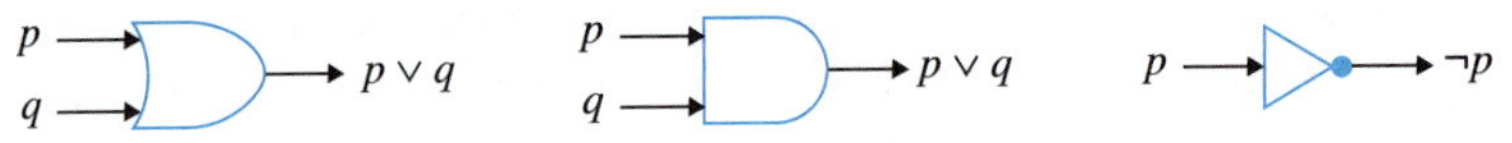

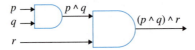

Figure 2.5   AND, OR, and NOT gates.

A combinatorial circuit is, roughly, the analogue of a formula. **Boolean circuit** notation for the formula

$$((p \wedge q) \wedge r)$$

is shown in Figure 2.6.

Figure 2.6   AND gates.

For the formula

$$((p \wedge p) \wedge p),$$

instead of having three separate p's as in an expression tree, the gate to represent it has one line that splits, as shown in Figure 2.7.

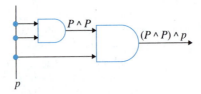

Figure 2.7   Another form of AND gates.

Since gates are used to describe computer circuits that will be implemented in a device or printed on a chip, it is common to represent more than one formula in the same diagram, as shown in Figure 2.8. The arrow in Figure 2.8 indicates that the output from gate C is an input for both gates A and B. Each of the "output wires" (A and B) corresponds to the output of a different propositional formula, as described earlier.

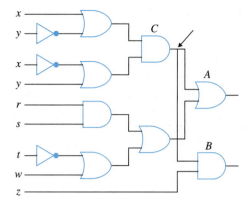

Figure 2.8 Multiple formula representation.

2.2 Exercises

1. Translate the following expressions into propositional logic. Use the following proposition letters:

   $p =$ "Jones told the truth."
   $q =$ "The butler did it."
   $r =$ "I'll eat my hat."
   $s =$ "The moon is made of green cheese."
   $t =$ "If water is heated to 100°C, it turns to vapor."

   (a) "If Jones told the truth, then if the butler did it, I'll eat my hat."
   (b) "If the butler did it, then either Jones told the truth or the moon is made of green cheese, but not both."
   (c) "It is not the case that both Jones told the truth and the moon is made of green cheese."
   (d) "Jones did not tell the truth, and the moon is not made of green cheese, and I'll not eat my hat."
   (e) "If Jones told the truth implies I'll eat my hat, then if the butler did it, the moon is made of green cheese."
   (f) "Jones told the truth, and if water is heated to 100°C, it turns to vapor."

2. Translate the following expressions of propositional logic into words using the following translation of the proposition letters:

   $p =$ "All the world is apple pie."
   $q =$ "All the seas are ink."
   $r =$ "All the trees are bread and cheese."
   $s =$ "There is nothing to drink."
   $t =$ "Socrates was a man."
   $u =$ "All men are mortal."
   $v =$ "Socrates was mortal."

(a) $(p \wedge q \wedge r) \rightarrow s$

(b) $(t \wedge u) \rightarrow v$

(c) $\neg s \rightarrow \neg v$

(d) $p \wedge (q \wedge r) \vee (t \wedge u) \vee (\neg s \vee \neg v)$

(e) $((p \vee t) \wedge (q \vee u)) \leftrightarrow (s \wedge v)$

One must sometimes be a bit creative in using language to make the results compre-hensible.

3. Let p denote the proposition "Jill plays basketball" and q denote the proposition "Jim plays soccer." Write out—in the clearest way you can—what the following proposi-tions mean:

(a) $\neg p$

(b) $p \wedge q$

(c) $p \vee q$

(d) $\neg p \wedge q$

(e) $p \rightarrow q$

(f) $p \leftrightarrow q$

(g) $\neg q \rightarrow p$

4. Let p denote the proposition "Sue is a computer science major" and q denote the proposition "Sam is a physics major." Write out what the following propositions mean:

(a) $\neg q$

(b) $q \wedge p$

(c) $p \vee q$

(d) $\neg q \wedge p$

(e) $q \rightarrow p$

(f) $p \leftrightarrow q$

(g) $\neg q \rightarrow p$

5. Jim, George, and Sue belong to an outdoor club. Every club member is either a skier or a mountain climber, but no member is both. No mountain climber likes rain, and all skiers like snow. George dislikes whatever Jim likes and likes whatever Sue dislikes. Jim and Sue both like rain and snow. Is there a member of the outdoor club who is a mountain climber?

6. Let proposition p be T and proposition q be F. Find the truth values for the following:

(a) $p \vee q$

(b) $q \wedge p$

(c) $\neg p \vee q$

(d) $p \wedge \neg q$

(e) $q \rightarrow p$

(f) $\neg p \rightarrow q$

(g) $\neg q \rightarrow p$

7. Let proposition p be T, proposition q be F, and proposition r be T. Find the truth values for the following:

(a) $p \vee q \vee r$

(b) $p \vee (\neg q \wedge \neg r)$

(c) $p \rightarrow (q \vee r)$

(d) $(q \wedge \neg p) \leftrightarrow r$

(e) $\neg r \rightarrow (p \wedge q)$

(f) $(p \rightarrow q) \rightarrow \neg r$

(g) $((p \wedge r) \rightarrow (\neg q \vee p)) \rightarrow (q \vee r)$

8. Find the expression tree for the following formulas:

(a) $(p \wedge q) \vee r$

(b) $(p \rightarrow q) \rightarrow r$

(c) $p \rightarrow (q \rightarrow r)$

9. Find the expression tree for the following formulas:

(a) $\neg p \wedge (\neg q \vee r)$

(b) $p \vee (\neg q \wedge \neg r)$

(c) $((p \vee q) \leftrightarrow r) \leftrightarrow p$

(d) $(\neg q \wedge \neg r) \leftrightarrow (p \rightarrow (q \vee r))$

10. Find the expression tree for the formula

$$(p \rightarrow ((\neg p) \rightarrow q))$$

11. Find the expression tree for the formula

$$((\neg(p \wedge q)) \vee (\neg(q \wedge r))) \wedge ((\neg(p \leftrightarrow (\neg(\neg s)))) \vee (((r \wedge s) \vee (\neg q))))$$

12. Find the expression tree for the formula

$$((((\neg(\neg p)) \wedge (\neg q)) \wedge r) \vee (((\neg(\neg q)) \wedge (\neg r)) \wedge s)) \leftrightarrow (s \rightarrow p)$$

13. Find a boolean expression to represent the following combinatorial circuits:

(a)

(b)

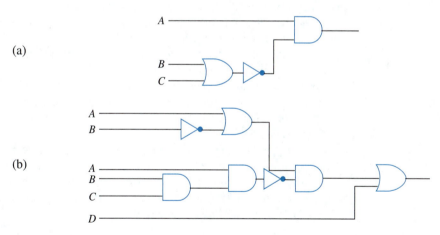

14. Draw a combinatorial circuit for each of the following boolean expressions:

(a) $(x \wedge y) \vee \neg z$

(b) $(x \wedge y) \vee (\neg x \wedge y)$

(c) $\neg(\neg x \vee y) \vee (x \wedge z)$

(d) $((x \wedge y) \vee (y \wedge z)) \vee \neg z$

(e) $(x \vee \neg(x \vee y)) \vee (\neg x \wedge \neg y)$

15. Find a boolean expression to represent each of the following combinatorial networks shown.

(a)

(b)

(c)

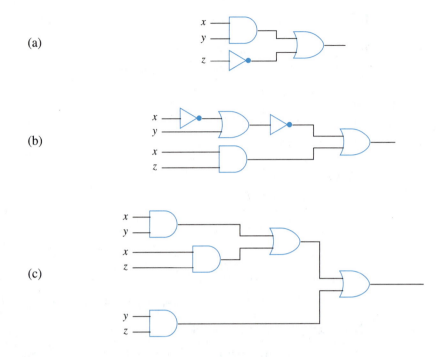

16. Prove Theorem 1, the Principle of Induction on Formulas. (*Hint:* If $\phi \vee \psi$ is a formula containing n occurrences of the logical operators, then ϕ and ψ each are formulas containing fewer than n logical operators. By the inductive hypothesis, both ϕ and ψ are in $\mathcal{F}$, so by the closure rules, $\phi \vee \psi$ is in $\mathcal{F}$.)

17. (a) What is the relationship between the number of propositional connectives in a formula and the number of parentheses? Prove your answer.
    (b) What is the relationship between the number of $\wedge$'s, $\vee$'s, $\rightarrow$'s, and $\leftrightarrow$'s in a formula and the number of proposition letters in the formula? Prove your answer.
    (c) What is the relationship between the number of $\neg$'s in a formula and the number of proposition letters in the formula? Prove your answer.
    (d) How many left parentheses may a formula contain? Prove your answer.
    (e) How many total symbols may a formula contain? Count each occurrence of each proposition letter as one symbol, so $(p_{123} \wedge p_{123})$ contains five symbols—that is, $($, p_{123}, $\wedge$, p_{123}, and $)$. For example, can a formula contain exactly two symbols? Exactly 17 symbols? Prove your answer.

2.3   Truth and Logical Truth

The **semantics** of a language is the relationship between strings of symbols in a language and their **meaning.** Consider a formula, such as

$$\phi = (\neg p \vee q) \rightarrow (r \rightarrow p)$$

How can the truth value for the formula be determined? Since this discussion is *formal* logic, one must first define what it means for ϕ to be T or F. Of course, this definition, to be useful, must match most people's intuitions.

To start, one must know what p, q, and r stand for. At first sight, one might expect to be told what sentences they stand for, such as

$p = $ "Mr. Holmes never made a mistake."
$q = $ "The professor is not a criminal."
$r = $ "Mrs. Hudson suspected the thief from the start."

For ordinary applications, that is exactly where one begins, but for the study of propositional logic, this is an unnecessary detail. In propositional logic, it matters not at all what *sentences* the proposition letters represent, only what the *truth values* of the sentences are. (This will become apparent as you see how truth values are assigned to complex formulas). Remember, F is shorthand for *FALSE* and T for *TRUE*. So, the starting point in propositional logic is an assignment of truth values to the proposition letters. For example, p may be assigned the value T, and q and r may be assigned the value F.

Definition 1. Let P be the set of proposition letters. An **interpretation** is an assignment I of a truth value (T or F) to every proposition letter in P. For $r \in P$, the assignment of a truth value to r is denoted $I(r)$.

Example 1. Let P be the set of proposition letters, and let p, q, and $r \in P$ and $X = P - \{p, q, r\}$. Let I be the following assignment of truth values to elements of P: $I(p) = F$, $I(q) = F$, $I(r) = T$, and $I(x) = F$ for every $x \in X$. Then, I is an interpretation.

An interpretation must assign a truth value to *every* proposition letter. (This is a technicality, just as it appears to be. Requiring this now simplifies the discussion a bit later.)

Once the interpretation I of the proposition letters is fixed, the interpretations of all other formulas can be computed by induction on formulas. We illustrate this here with an expression tree. The process is different from the way that arithmetic expressions are evaluated, because a value is found in a simple, bottom-up fashion.

Example 2. Determine whether the formula

$$\phi = (\neg p \vee q) \to (r \to p)$$

is T or F for an interpretation I where $I(p) = T$ and $I(q) = I(r) = F$. (*Remember:* All other proposition letters have value F.)

Solution. Mark the leaves of the expression tree of ϕ with the truth values as shown in Figure 2.9.

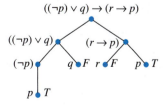

Figure 2.9 Expression tree.

Now, use these truth values to move up the tree toward the root, using the truth tables for the propositional connectives (see Tables 2.2 and 2.3 in Section 2.1). First, work one level up from the leaves. The truth table for ¬ says that if p is T, then ¬p is F. The truth table for → says that if r is F and p is T, then $r \to p$ is T.

Next, use the truth table for ∨ to compute a truth value for ¬$p \lor q$. Since the truth value of ¬p is F and the truth value of q is F, the result is F.

Finally, use the truth table for → to assign a truth value to the entire formula. The truth value of an implication for which the hypothesis is F and the conclusion is T is just T. Therefore, T is the truth value of ϕ. The steps of this evaluation are shown in Figure 2.10.

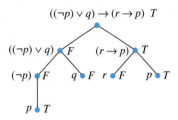

Figure 2.10   Step 3 of evaluating a formula.

The truth value of the entire formula ϕ is denoted by $I(\phi)$. In the case shown here, $I(\phi) = T$.   ■

Formally, what happened in Example 2 is an induction on formulas. The interpretation I specified the truth values for the proposition letters. The truth $I(\phi)$ for more complex formulas ϕ is defined using the truth values for simpler formulas and the truth tables for ¬, ∧, ∨, →, and ↔ as shown here:

1. $I(T) = T$, and $I(F) = F$

2. $I(\neg\phi) = \begin{cases} T & \text{if } I(\phi) = F \\ F & \text{if } I(\phi) = T \end{cases}$

3. $I(\phi \land \psi) = \begin{cases} T & \text{if } I(\phi) = I(\psi) = T \\ F & \text{otherwise} \end{cases}$

4. $I(\phi \lor \psi) = \begin{cases} F & \text{if } I(\phi) = I(\psi) = F \\ T & \text{otherwise} \end{cases}$

5. $I(\phi \to \psi) = \begin{cases} F & \text{if } I(\phi) = T \text{ and } I(\psi) = F \\ T & \text{otherwise} \end{cases}$

6. $I(\phi \leftrightarrow \psi) = \begin{cases} T & \text{if } I(\phi) = I(\psi) \\ F & \text{if } I(\phi) \neq I(\psi) \end{cases}$

Since each formula ϕ has exactly one expression tree and these rules define the truth value of each node on the tree in terms of the truth values of the nodes with edges joining them to this node, there is only one way to calculate $I(\phi)$.

Definition 2.   Let I be an interpretation of P. A formula ϕ is **true in** I if $I(\phi) = T$, and ϕ is **false in** I if $I(\phi) = F$.

In Example 2, if I is the interpretation with $I(p) = T$ and $I(q) = I(r) = F$, then $(\neg p \lor q) \to (r \to p)$ is true in I.

Example 3. Let

$$\phi = ((\neg p \lor q) \to (r \to p))$$

Find $I(\phi)$ for all interpretations I.

Solution. Three proposition letters—p, q, and r—are in the formula. Hence, the truth of the formula depends only on $I(p)$, $I(q)$, and $I(r)$. Each of $I(p)$, $I(q)$, and $I(r)$ can be one of T or F, so there are $2^3 = 8$ possible interpretations.

The calculation of the truth value for each of the eight interpretations can be shown concisely in a truth table. Start out with a truth table that has eight rows, one for each interpretation:

	p	q	r
I_0	T	T	T
I_1	T	T	F
I_2	T	F	T
I_3	T	F	F
I_4	F	T	T
I_5	F	T	F
I_6	F	F	T
I_7	F	F	F

Next, assign truth values to larger and larger subformulas until the formula itself is evaluated.

We now repeat the evaluation of the formula

$$\phi = (\neg p \lor q) \to (r \to p)$$

using this method. Evaluating $\neg p$ and $r \to p$, we get

	p	q	r	$\neg p$	$\neg p \lor q$	$r \to p$	$(\neg p \lor q) \to (r \to p)$
I_0	T	T	T	F		T	
I_1	T	T	F	F		T	
I_2	T	F	T	F		T	
I_3	T	F	F	F		T	
I_4	F	T	T	T		F	
I_5	F	T	F	T		T	
I_6	F	F	T	T		F	
I_7	F	F	F	T		T	

and in two more steps, we complete the evaluation of the formula:

	p	q	r	$\neg p$	$\neg p \vee q$	$r \to p$	$(\neg p \vee q) \to (r \to p)$
I_0	T	T	T	F	T	T	T
I_1	T	T	F	F	T	T	T
I_2	T	F	T	F	F	T	T
I_3	T	F	F	F	F	T	T
I_4	F	T	T	T	T	F	F
I_5	F	T	F	T	T	T	T
I_6	F	F	T	T	T	F	F
I_7	F	F	F	T	T	T	T

By convention, we put the truth value directly under the operation performed. The truth values in the right-most column of the table are the truth values of each of the interpretations of this formula. The truth tables show that ϕ is T in the interpretations I_0, I_1, I_2, I_3, I_5, and I_7. The truth table also shows that ϕ is F in the interpretations I_4 and I_6. ∎

2.3.1 Tautologies

Propositional logic is the study of propositions and the propositional connectives. It is the study not only of one particular interpretation of a formula but also of what can be deduced about all interpretations of a formula. Of particular interest are those formulas that are true "by virtue of pure logic." Definition 3 captures the notion of "true by virtue of pure logic," at least as closely as is possible from the standpoint of propositional logic.

Definition 3.   Let ϕ be a formula. Then, ϕ is a **tautology,** or is **logically valid,** if it is T in every interpretation. ϕ is **satisfiable** if it is T in some interpretation, and it is **unsatisfiable** if it is T in no interpretation. Unsatisfiable formulas are also called **contradictions.**

A formula is a tautology if and only if all entries under the formula in its truth table evaluation are T. For example, "John is married, or John is not married" is a logical truth. "John is married, or John is a bachelor" is not a logical truth, since it depends on the meaning of the word *bachelor.*

"John is married, and John is not married" is unsatisfiable, since the proposition "John is married" cannot be both T and F. On the other hand, "John is married, or John is a bachelor" is clearly satisfiable. Of course, every tautology is also satisfiable.

Example 4.   Construct a truth table to show that $(p \wedge q) \to p$ is a tautology.

Solution.   The truth table for $(p \wedge q) \to p$ is

p	q	$p \wedge q$	$((p \wedge q) \to p)$
T	T	T	T
T	F	F	T
F	T	F	T
F	F	F	T

Since all entries under $((p \wedge q) \to p)$ are T, the formula is a tautology. ∎

The reader should note that, intuitively, $(p \wedge q) \rightarrow p$ "asserts" that if p and q are both T, then p is T. Thus, we expect it to be a tautology.

Example 5. Construct a truth table to show that $p \rightarrow (p \vee r)$ is a tautology.

Solution. The truth table for $p \rightarrow (p \vee r)$ is

p	r	$p \vee r$	$(p \rightarrow (p \vee r))$
T	T	T	T
T	F	T	T
F	T	T	T
F	F	F	T

Again, all entries in the final column are T, so the formula is a tautology. ∎

This tautology also "asserts" an obvious truth. If p is T, then it is true that either p is T or r is T (or both).

The next two examples show how logical connectives can be expressed in terms of each other.

Example 6. Construct a truth table to show that $(p \rightarrow q) \leftrightarrow (\neg p \vee q)$ is a tautology.

Solution. This formula shows how $\rightarrow$ can be expressed using $\vee$ and $\neg$.

p	q	$p \rightarrow q$	$\neg p$	$\neg p \vee q$	$(p \rightarrow q) \leftrightarrow (\neg p \vee q)$
T	T	T	F	T	T
T	F	F	F	F	T
F	T	T	T	T	T
F	F	T	T	T	T

Since the formula involving only $\rightarrow$ is $T(F)$ if and only if the formula involving $\neg$ and $\vee$ is $T(F)$, all the entries in the final column are T, so the formula is a tautology. ∎

Example 7. Construct a truth table to show that

$$(p \leftrightarrow q) \leftrightarrow ((p \rightarrow q) \wedge (q \rightarrow p))$$

is a tautology.

Solution. This formula shows how to express $\leftrightarrow$ in terms of $\wedge$ and $\rightarrow$.

p	q	$p \leftrightarrow q$	$p \rightarrow q$	$q \rightarrow p$	$(p \rightarrow q) \wedge (q \rightarrow p)$	$(p \leftrightarrow q) \leftrightarrow$ $((p \rightarrow q) \wedge (q \rightarrow p))$
T	T	T	T	T	T	T
T	F	F	F	T	F	T
F	T	F	T	F	F	T
F	F	T	T	T	T	T

All the entries in the final column are T, so the formula is a tautology. ∎

Table 2.5 lists many commonly used tautologies. The reader should study them carefully and determine what they "assert." The names should suggest analogies to other operations. For example, $\lor$, $\land$, and $\leftrightarrow$ all obey associative laws, just as $+$ and $\cdot$ do in arithmetic.

Table 2.5   Commonly Used Tautologies

(a)	$(p \land p) \leftrightarrow p$	Idempotence
(b)	$(p \lor p) \leftrightarrow p$	Idempotence
(c)	$p \lor \neg p$	Law of the Excluded Middle
(d)	$\neg(p \land \neg p)$	
(e)	$(p \land (p \to q)) \to q$	Modus Ponens
(f)	$((p \to q) \land (q \to r)) \to (p \to r)$	The Law of Syllogism
(g)	$((p \lor q) \land \neg p) \to q$	Modus Tollendo Ponens
(h)	$((p \land q) \land r) \leftrightarrow (p \land (q \land r))$	Associative Law
(i)	$((p \lor q) \lor r) \leftrightarrow (p \lor (q \lor r))$	Associative Law
(j)	$((p \leftrightarrow q) \leftrightarrow r) \leftrightarrow (p \leftrightarrow (q \leftrightarrow r))$	Associative Law
(k)	$(p \land r) \leftrightarrow (r \land p)$	Commutative Law
(l)	$(p \lor r) \leftrightarrow (r \lor p)$	Commutative Law
(m)	$(p \leftrightarrow r) \leftrightarrow (r \leftrightarrow p)$	Commutative Law
(n)	$(p \land (r \lor q)) \leftrightarrow ((p \land r) \lor (p \land q))$	Distributive Law
(o)	$(p \lor (r \land q)) \leftrightarrow ((p \lor r) \land (p \lor q))$	Distributive Law
(p)	$\neg\neg p \leftrightarrow p$	Double negative
(q)	$\neg(p \land r) \leftrightarrow (\neg p \lor \neg r)$	DeMorgan's Law
(r)	$\neg(p \lor r) \leftrightarrow (\neg p \land \neg r)$	DeMorgan's Law
(s)	$(p \to r) \leftrightarrow (\neg r \to \neg p)$	Contrapositive
(t)	$(p \to (r \to q)) \leftrightarrow ((p \land r) \to q)$	
(u)	$((\neg p \to r) \land (\neg p \to \neg r)) \leftrightarrow p$	Contradiction
(v)	$((p \land r) \lor r) \leftrightarrow r$	Absorption
(w)	$((p \lor r) \land r) \leftrightarrow r$	Absorption
(x)	$(p \leftrightarrow q) \leftrightarrow ((p \land q) \lor (\neg p \land \neg q))$	
(y)	$\neg(p \leftrightarrow q) \leftrightarrow ((\neg p \land q) \lor (p \land \neg q))$	
(z)	$(p \to F) \leftrightarrow (\neg p)$	

The two tautologies (q) and (r) in Table 2.5, called DeMorgan's Laws, are the logical analogues of the DeMorgan's Laws of set theory (Theorem 8 in Section 1.3.2). That theorem states how the set operations of union, intersection, and complementation interact. Here, we see how conjunction, disjunction, and negation interact with propositions.

Example 8.   Let p denote "X is a bird" and r denote "X can fly." Tautology(s) from Table 2.5 states that "If X is a bird implies that X can fly" is equivalent to "If X cannot fly, then X is not a bird."

The following theorem is the basis of many proofs, notably many proofs by contradiction.

Theorem 1.   A formula ψ is a tautology if and only if $\neg\psi$ is unsatisfiable.

Proof. ($\Rightarrow$) Let ψ be a tautology, and let I be any interpretation of the proposition letters in ψ. Since ψ is a tautology, $I(\psi) = T$, so $I(\neg\psi) = F$. Hence, $\neg\psi$ is not satisfiable.

($\Leftarrow$) The converse is analogous. ∎

A proof by contradiction shows that if $I(\psi) = T$ for an interpretation, then we prove
$I(\neg\psi) = T$ in that interpretation, which is clearly a contradiction.

2.3.2 Substitutions into Tautologies

The formula $\phi = (p \wedge (p \rightarrow q)) \rightarrow q$ is a tautology. Now, replace each occurrence of p
in ϕ with another formula, say $p_1 \vee p_2$. The result is the formula

$$\phi_1 = ((p_1 \vee p_2) \wedge ((p_1 \vee p_2) \rightarrow q)) \rightarrow q$$

The reader can easily write the truth table for ϕ_1 and see that it also is a tautology. Something more general, however, is taking place here. One can think of the substitution not as
substituting the formula $p_1 \vee p_2$ into ϕ for p but, rather, as substituting the *truth value* for
$p_1 \vee p_2$ for the *truth value* of p in ϕ. Since ϕ is a tautology, any truth value for p together
with any truth value for q yields a truth value of T for ϕ. So, ϕ_1 should also be a tautology.
This intuitive argument can be formalized to prove the following theorem. (The interested
reader is invited to prove it.)

First Substitution Principle

Let ϕ be a tautology; let $p_1, p_2, \ldots, p_k$ be any proposition letters appearing in ϕ,
and let $\chi_1, \chi_2, \ldots, \chi_k$ be formulas. Form a formula ϕ_1 by *simultaneously* replacing
p_1 with χ_1, p_2 with $\chi_2, \ldots, p_k$ with χ_k wherever they occur in ϕ. Then, ϕ_1 is a
tautology.

The requirement of *simultaneous* replacement is important, since it allows, say, χ_1 to
contain a p_2 without forcing that p_2 to be replaced with χ_2. For example, again let

$$\phi = (p \wedge (p \rightarrow q)) \rightarrow q$$

and replace p with $\chi_1 = q \rightarrow r$ and q with $\chi_2 = r \rightarrow q$. The result is

$$\phi_1 = (\chi_1 \wedge ((\chi_1 \rightarrow \chi_2)) \rightarrow \chi_2$$
$$\phi_1 = (q \rightarrow r) \wedge ((q \rightarrow r) \rightarrow (r \rightarrow q)) \rightarrow (r \rightarrow q)$$

Since ϕ is a tautology, so is ϕ_1. The simultaneous replacement condition meant that the q
in χ_1 did not have to be replaced with χ_2.

2.3.3 Logically Valid Inferences

We began the study of logic to help distinguish valid from invalid arguments. We have now
covered enough material to present a formal notion of a valid argument for propositional
logic.

Definition 4.   Let S be a set of formulas. An interpretation I **satisfies** S if $I(\phi) = T$ for every $\phi \in S$. A set S of formulas is *satisfiable* if there is an interpretation I that satisfies S.

For example, $\{p, q, r\}$ is satisfiable. It is satisfied by any interpretation I where $I(p) = I(q) = I(r) = T$. However, $\{p, \neg p\}$ is not satisfiable.

One intuition is that I describes the actual state of the world and that S is a set of formulas, which can be thought of as assertions about the world. I satisfies S if each formula in S is a true statement about (the state of) the world. Another intuition is that I is a possible state of the world. Suppose it is known that all statements in S are T. Then, one can check whether a possible state I of the world matches what is known—that is, whether I satisfies the known facts S.

Theorem 2.

(a) Every interpretation satisfies $\emptyset$.
(b) If $S = \{\phi_1, \ldots, \phi_k\}$ and I is an interpretation, then I satisfies S if and only if $I(\phi_1 \wedge \ldots \wedge \phi_k) = T$.

Proof.   This Proof is left for Exercise 23 in Section 2.4.   ■

Definition 5.

(a) For formulas ψ and χ, ψ **logically implies,** χ, or ψ **tautologically implies** χ, if, for every interpretation I,

$$\text{if } I(\psi) = T, \text{ then } I(\chi) = T$$

We denote ψ logically implies χ as $\psi \models \chi$.
(b) Formulas ψ and χ are **logically equivalent,** or **tautologically equivalent,** or **equivalent,** if, for every interpretation I, we have $I(\psi) = I(\chi)$.

As a natural extension of one formula logically implying another formula, we say that for a set of formulas S, $S \models \chi$ means that inferring χ from S is logically valid.

Example 9.

(a) $p \wedge q \models p \vee q$.
(b) $p \wedge q$ is logically equivalent to $\neg(\neg p \vee \neg q)$.
(c) $p \wedge q$ and $p \vee q$ are not logically equivalent.

Solution.

(a) Suppose I is any interpretation. We need to show that if $I(p \wedge q) = T$, $I(p \vee q) = T$. So, suppose $I(p \wedge q) = T$. Then, $I(p) = T$, and $I(q) = T$. So, $I(p \vee q) = T$, as desired.

(b) We show that $I(p \wedge q)$ always equals $I(\neg(\neg p \vee \neg q))$ by building a truth table of all possibilities:

p	q	$p \wedge q$	$\neg p$	$\neg q$	$\neg p \vee \neg q$	$\neg(\neg p \vee \neg q)$
T	T	T	F	F	F	T
T	F	F	F	T	T	F
F	T	F	T	F	T	F
F	F	F	T	T	T	F

Note that all entries under $p \wedge q$ and $\neg(\neg p \vee \neg q)$ are identical—that is, that $I(p \wedge q)$ is always equal to $I(\neg(\neg p \vee \neg q))$.

(c) Let I be the interpretation where $I(p) = T$ and $I(q) = F$. Then, $I(p \wedge q) = F$, and $I(p \vee q) = T$. Because these two truth values are not the same, the two formulas are not logically equivalent.   ∎

The intuitive content of logical implication is that if $\psi \models \chi$, then it is correct to infer χ from ψ in any argument. The definition of logically implies can be extended to sets of propositions in a straightforward way. At this point, we need to understand this notion at the level of formulas only. For example, if we have two sets of formulas R and S, then R logically implies S if, for every interpretation I, I satisfies R if and only if I satisfies S.

Compare the formula $\phi \rightarrow \psi$ with the assertion "ϕ logically implies ψ." The first is just a formula. It may be T, or it may be F. We are just *discussing,* or *mentioning,* the formula. The second is an *assertion* that some logical relationship holds. Nevertheless, there is a connection between them. This connection is given in part (a) of Theorem 2.5.

Theorem 3.   Let ϕ and ψ be formulas. Then:

(a) $\phi \models \psi$ if and only if $\phi \rightarrow \psi$ is a tautology.
(b) $R \models \psi$ if and only if $R \cup \{\neg \psi\}$ is unsatisfiable.
(c) Let $R = \{\phi_1, \phi_2, \ldots \phi_k\}$. $R \models \psi$ if and only if $\phi_1 \wedge \phi_2 \wedge \cdots \wedge \phi_k \rightarrow \psi$ is a tautology.
(d) $\emptyset \models \psi$ if and only if ψ is a tautology.
(e) ϕ and ψ are logically equivalent if and only if $\phi \leftrightarrow \psi$ is a tautology.

Proof.   (a) ($\Rightarrow$)   First, suppose $\phi \models \psi$, and let I be any interpretation. It is necessary to show that $I(\phi \rightarrow \psi) = T$. The only way that $I(\phi \rightarrow \psi)$ can be F is for $I(\phi)$ to be T and $I(\psi)$ to be F. However, if $I(\phi) = T$, then, since ϕ logically implies ψ, $I(\psi)$ must also be T. Hence, $I(\phi \rightarrow \psi) = T$.

($\Leftarrow$)   Second, suppose $\phi \rightarrow \psi$ is a tautology. It is necessary to show that for any interpretation I, if $I(\phi) = T$, then $I(\psi) = T$ as well. So, suppose I is an interpretation, and suppose $I(\phi) = T$. Since $\phi \rightarrow \psi$ is a tautology, $I(\phi \rightarrow \psi) = T$. By the truth table for $\rightarrow$, if $I(\phi) = T$ and $I(\phi \rightarrow \psi) = T$, then $I(\psi) = T$.

(b)–(e) These proofs are left for Exercise 24 in Section 2.4.   ∎

The next theorem tells us that if two propositions are either always T or always F, then they are logically equivalent.

Theorem 4.

(a) Suppose ϕ and ψ are both tautologies. Then, ϕ and ψ are logically equivalent.
(b) Suppose ϕ and ψ are both unsatisfiable. Then, ϕ and ψ are logically equivalent.

Proof.   These proofs are left for Exercise 25 in Section 2.4.   ■

The result just says that since a tautology is T for every set of truth values of its propositions, its truth value will match the truth value of any other tautology for those same truth values. Similarly, the same holds for two unsatisfiable formulas.

In Table 2.6, we add a bit of notation; rather than just saying $\{p,\ p \to q\} \models q$, we replace p and q with the symbols ϕ and ψ, representing arbitrary formulas. What Table 2.6 really means is that if we replace ϕ, ψ, and χ with any formulas, the results are logical implications.

Table 2.6   Some Logically Valid Inferences and Their Traditional Names

Some Logically Valid Inferences		
a.	$\{\phi,\ \phi \to \psi\} \models \psi$	Modus Ponens
b.	$\{\phi \to \psi,\ \psi \to \chi\} \models \phi \to \chi$	Law of Syllogism
c.	$\{\phi \vee \psi,\ \neg\phi\} \models \psi$	Modus Tollendo Ponens
d.	$\neg\neg\phi \models \phi$	Double Negation
e.	$\neg\phi \to \neg\psi \models \psi \to \phi$	Contrapositive
f.	$\phi \to \psi \models \neg\psi \to \neg\phi$	Contrapositive
g.	$\{\phi \to \chi,\ \phi \to \neg\chi\} \models \neg\phi$	Proof by Contradiction
h.	$\{\neg\phi \to \chi,\ \neg\phi \to \neg\chi\} \models \phi$	Proof by Contradiction
i.	$\{\phi \vee \psi,\ \phi \to \chi,\ \psi \to \chi\} \models \chi$	Proof by Cases
j.	$\{\phi \wedge \psi\} \models \neg(\neg\phi \vee \neg\psi)$	DeMorgan's Law
k.	$\{\neg(\phi \wedge \psi)\} \models (\neg\phi \vee \neg\psi)$	DeMorgan's Law
l.	$\{\phi \vee \psi\} \models \neg(\neg\phi \wedge \neg\psi)$	DeMorgan's Law
m.	$\{\neg(\phi \vee \psi)\} \models (\neg\phi \wedge \neg\psi)$	DeMorgan's Law

Example 10.

(a) Let ϕ denote "X is a cat" and ψ denote "X is an animal." Under the assumption that ϕ is true and $\phi \to \psi$ is true, we can use Modus Ponens to conclude that X is an animal.
(b) Let ϕ denote "X is a cat" and ψ denote "X is a bird." If we are given that $\phi \vee \psi$ is true and $\neg\phi$ is true, then we can use Modus Tollendo Ponens to conclude that "X is a bird."   ■

As commented earlier, computer programs have been written to automate logical inference. Some material relevant to this are discussed in the next section and in Section 2.4.

2.3.4   Combinatorial Networks

A combinatorial network is just another representation for a formula or a set of formulas of propositional logic. Start with an assignment of truth values to the wires going into the circuit—that is, to the proposition letters. The circuit computes a group of outputs that correspond to the truth values of the corresponding formulas.

Gates and Boolean Algebra

In Section 1.3.5, the axiom system for a boolean algebra was introduced. Example 10 showed that if B is a set of elements with values from $\{0, 1\}$, and with the operations $\wedge$ and $\vee$ defined as

$\vee$	0	1
0	0	1
1	1	1

$\wedge$	0	1
0	0	0
1	0	1

then B with $\wedge$ as meet and $\vee$ as join forms a boolean algebra.

If we interpret the operation of $\vee$ as the logical operation of *OR* and $\wedge$ as the logical operation of *AND* as well as substitute T for 1 and F for 0, then these operation tables are just the tables of *AND* and *OR* introduced in Table 2.3. The logical value T satisfies the conditions for $\top$, whereas F satisfies the conditions for $\bot$. Finally, if we interpret **complementation** as $\neg x$, then the conditions for complements hold. What all this means is that there are two different but equivalent ways to represent a circuit. The first is to draw the gates, as shown in Figure 2.11.

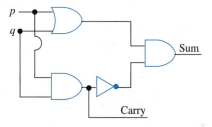

Figure 2.11   Half-adder.

The second is to represent the gates as boolean operations and the whole gate structure as a boolean expression. In Figure 2.12, we show a circuit and its equivalent boolean expression.

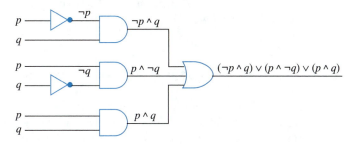

Figure 2.12   Gates for $(\neg p \wedge q) \vee (p \wedge \neg q) \vee (p \wedge q)$.

The power of this alternate representation is shown in Example 11 where we use logic—which we can also think of as boolean algebra—to find a simpler expression to represent this set of gates.

Example 11.   Use logic to show that $(\neg p \wedge q) \vee (p \wedge \neg q) \vee (p \wedge q)$ is logically equivalent to the formula $p \vee q$ thus providing us with a one-gate equivalent to the circuit in Figure 2.12.

Solution.   We use the axioms for the Commutative Law, the Distributive Law, and the basic properties of $\top$ to simplify this expression.

$$(\neg p \wedge q) \vee (p \wedge \neg q) \vee (p \wedge q)$$
$$= (p \wedge q) \vee (\neg p \wedge q) \vee (p \wedge \neg q) \vee (p \wedge q) \quad ((p \wedge q) = (p \wedge q) \vee (p \wedge q))$$
$$= ((p \vee \neg p) \wedge q) \vee (p \wedge (\neg q \vee q)) \qquad \text{(Distributive Law)}$$
$$= (\top \wedge q) \vee (p \wedge \top) \qquad\qquad\qquad \text{(property of } \top)$$
$$= q \vee p$$
$$= p \vee q \qquad\qquad\qquad\qquad\qquad\qquad \text{(Commutative Law)}$$

The simplified circuit is shown in Figure 2.13.

Figure 2.13   Equivalent, simpler circuit.   ∎

It is not always possible to have such clear reduction in the complexity of a combinatorial circuit. Example 11, however, shows how computer science can use different tools in approaching a problem.

2.3.5   Substituting Equivalent Subformulas

In many respects, logically equivalent formulas are indistinguishable from each other. The sense in which two logically equivalent formulas are indistinguishable is stated as the **Second Substitution Principle.**

Second Substitution Principle

Let ϕ_1 be logically equivalent to ϕ_2, and let ψ be any formula containing ϕ_1, possibly several times, as a subformula. Form a new formula ψ' by replacing *some* (or possibly all) of the occurrences of ϕ_1 in ψ with ϕ_2. Then, ψ is logically equivalent to ψ'.

The Second Substitution Principle is really quite useful. Consider, for example, the formula

$$\phi = (\neg p \to q) \to r$$

Since the subformula $\phi_1 = \neg p \to q$ is logically equivalent to $\phi_2 = p \vee q$ (see Example 9 in Section 2.3.1) by the Second Substitution Principle, we can transform the formula as follows:

$$\phi_1 \to r$$
$$\phi_2 \to r$$
$$(p \vee q) \to r$$

Many people would find the last formula easier to understand than the original one. We shall not prove the Second Substitution Principle; the reader is invited to prove it.

2.3.6 Simplifying Negations

When given a formula, it is often useful to find a simpler formula that is logically equivalent to the first. Here, *simpler* has no fixed meaning; it just means simpler to use in some application. For example, in programming, we write conditions saying when a loop should continue for another pass and when a loop should stop. One equivalent way of writing that condition may be easier than another for someone reading the program to understand. In the context of boolean networks, simpler can mean smaller, such as having fewer gates or taking up less area on a chip. A standard, though obviously imprecise, meaning of simpler is "easier for people to understand." This latter notion of simpler comes up often.

Consider a piece of a program:

while (not((x < 3) or (x > 5)))
{ ... }

A complex formula that is negated is usually difficult to understand. Consequently, programmers look for logically equivalent formulas where the operator *not* is "pushed inside" and applied to simpler formulas. Unfortunately, in this example, one might think the negation is logically equivalent to

while ((x ≥ 3) or (x ≤ 5))
{ ... }

which turns out to be an infinite loop. The problem, of course, is that DeMorgan's Law was not applied correctly.

Let us rewrite the condition by letting the proposition letter p stand for $x < 3$ and q for $x > 5$. Hence, $\neg p$ is true just in the case $x \geq 3$, and $\neg q$ is true just in the case $x \leq 5$. The condition at the top of the while loop can be rewritten $\neg(p \vee q)$. By DeMorgan's Law, this formula is equivalent to $\neg p \wedge \neg q$. So, the proper translation would have been

while ((x ≥ 3) and (x ≤ 5))
{ ... }

In fact, for any formula ϕ, it is possible to find an equivalent formula in which negations are applied only to proposition letters. The technique can be thought of as "moving negations inward." This technique is done in two steps.

Step 1. Find an equivalent formula containing no $\leftrightarrow$'s or $\to$'s. First, replace each subformula of the form $\phi \leftrightarrow \psi$ with the logically equivalent subformula

$$(\phi \to \psi) \wedge (\psi \to \phi)$$

This is an application of the Second Substitution Principle. Then, eliminate all $\rightarrow$'s as follows: Replace each subformula of the form $\phi \rightarrow \psi$ with the logically equivalent subformula $\neg\phi \vee \psi$.

Step 2. Apply DeMorgan's Laws,

$$\neg(p \vee q) \leftrightarrow (\neg p \wedge \neg q) \quad \text{and} \quad \neg(p \wedge q) \leftrightarrow (\neg p \vee \neg q)$$

to "push negations" in past $\wedge$ and $\vee$ and replace each double negation $\neg\neg p$ formed with the unnegated p. Ultimately, only proposition letters will be negated. By the Second Substitution Principle, the formula so formed will be equivalent to the original formula.

Example 12. For the formula

$$\neg(\neg(p \wedge \neg q) \vee (q \wedge \neg r))$$

use DeMorgan's Laws and the law of double negation to "push negations inside."

Solution. Start from the "outside" and work "inside."

1. This formula is of the form $\neg(\phi \vee \psi)$, where $\phi = \neg(p \wedge \neg q)$ and $\psi = (q \wedge \neg r)$, so we apply DeMorgan's Law to get the equivalent

$$\neg\neg(p \wedge \neg q) \wedge \neg(q \wedge \neg r)$$

2. Apply the same techniques to the "outermost" subformulas, $\neg\neg(p \wedge \neg q)$ and $\neg(q \wedge \neg r)$. By the law of double negation, the first is equivalent to $(p \wedge \neg q)$ and by DeMorgan's Laws, the second is equivalent to $\neg q \vee \neg\neg r$. So, the entire formula is equivalent to

$$(p \wedge \neg q) \wedge (\neg q \vee \neg\neg r)$$

3. Now work "inward." Again, by the law of double negation, $\neg\neg r$ is equivalent to r, so the entire formula is equivalent to

$$(p \wedge \neg q) \wedge (\neg q \vee r)$$

which is in the desired form. ∎

2.4 Exercises

1. A restaurant displays the sign "Good food is not cheap," and a competing restaurant displays the sign "Cheap food is not good." Are the two restaurants saying the same thing?

2. The country of Ost is inhabited only by people who either always tell the truth or always tell lies and who will respond to questions only with a "yes" or a "no." A tourist comes to a fork in a road, where one branch leads to the capital and the other does not. There is no sign indicating which fork to take, but Mr. Zed, who is a resident of Ost, comes along. What single question should the tourist ask Mr. Zed to determine which fork in the road to take?

3. Find the expression tree for the formula

$$p \rightarrow ((\neg p) \rightarrow q)$$

Evaluate the expression tree if proposition p is T and proposition q is F.

4. Find the expression tree for the formula

$$((p \rightarrow \neg q) \vee q) \rightarrow q$$

Evaluate the expression tree if proposition p is F and proposition q is T.

5. Find the expression tree for the formula

$$((((\neg(\neg p)) \wedge (\neg q)) \wedge r) \vee (((\neg(\neg q)) \wedge (\neg r)) \wedge s) \leftrightarrow (s \rightarrow p)$$

Evaluate the expression tree if proposition p is T, proposition q is T, proposition r is F, and proposition s is F.

6. Find the expression tree for the formula

$$((\neg(p \wedge q)) \vee (\neg(q \wedge r))) \wedge ((\neg(p \leftrightarrow (\neg(\neg s)))) \vee ((r \wedge s) \vee (\neg q))).$$

Evaluate the expression tree if proposition p is F, proposition q is T, proposition r is F, and proposition s is T.

7. Find the expression tree for the formula

$$\neg(p \wedge q) \leftrightarrow (\neg p \vee \neg q)$$

Evaluate the expression tree for all possible pairs of truth values for p and q. Use these evaluations to prove this formula is a tautology.

8. For each of the following sets of propositions, identify a logically valid inference listed in Table 2.6 that could be used to draw inferences from the formulas given. Identify the rule of inference and what the inference rule implies.

   (a) "If the sun is shining, then the courts will be open for play."
       "If the courts are open for play, then we will play at 3 PM."
   (b) "The sun is shining, or the courts are closed."
       "The sun is not shining."
   (c) "It is false that the sun is not shining."
   (d) "If the courts are not open for play, then the sun is not shining."
   (e) "If the sun is not shining, then the courts are not open for play."
       "The courts are open for play."
   (f) "If it is raining, then the courts are wet."
       "If it is raining, then the courts are closed."
       "If the courts are wet, then the courts are closed."
       "The sun is shining."

9. Let ϕ = "The home team is ahead." Let ψ = "The fans are happy." Let χ = "The visiting team is losing." For inference rules (a), (g), and (i) in Table 2.6, write out the hypothesis and the conclusion for ϕ, ψ, and χ.

10. Write the truth tables for the following formulas. Use the truth table to determine whether any of these formulas is a tautology.

   (a) $((p \rightarrow q) \wedge (q \rightarrow r)) \rightarrow (p \leftrightarrow r)$
   (b) $((p \rightarrow q) \wedge (q \rightarrow r)) \rightarrow (p \rightarrow r)$
   (c) $((p \rightarrow q) \rightarrow r) \rightarrow (p \rightarrow (q \rightarrow r))$

(d) $(p \to (r \lor q)) \to ((p \to r) \lor (p \to q))$

(e) $(p \to (r \land q)) \to ((p \to r) \lor (p \to q))$

(f) $((p \to q) \to q) \to p$

11. Construct the truth table for

$$(p \land (p \to q) \land (q \to r)) \to r$$

Simplify this expression to one using only $\land$, $\lor$, and $\neg$.

12. Show that the following formulas from Table 2.5 are tautologies:

(a) $(p \land p) \leftrightarrow p$

(b) $(p \land (p \to q)) \to q$

(c) $(p \to r) \leftrightarrow (\neg r \to \neg p)$

13. Let $\phi = (p \lor q) \to (r \land \neg s)$. For each of the following interpretations of p, q, r, and s, compute $I(\phi)$ using the truth tables for $\neg, \lor, \land, \to$, and $\leftrightarrow$:

(a) $I(p) = T, I(q) = T, I(r) = T$, and $I(s) = F$

(b) $I(p) = T, I(q) = T, I(r) = F$, and $I(s) = F$

(c) $I(p) = F, I(q) = T, I(r) = T$, and $I(s) = T$

(d) $I(p) = F, I(q) = F, I(r) = T$, and $I(s) = T$

14. Let $\phi = (p \to q) \to ((r \land \neg s) \to q)$. For each of the following interpretations of p, q, r, and s, compute $I(\phi)$ using the truth tables for $\neg, \lor, \land, \to$, and $\leftrightarrow$:

(a) $I(p) = T, I(q) = T, I(r) = F$, and $I(s) = T$

(b) $I(p) = T, I(q) = F, I(r) = T$, and $I(s) = F$

(c) $I(p) = F, I(q) = T, I(r) = T$, and $I(s) = F$

(d) $I(p) = F, I(q) = F, I(r) = T$, and $I(s) = F$

15. Let $\phi = (\neg(p \land q)) \leftrightarrow (\neg r \lor \neg s)$. For each of the following interpretations of p, q, r, and s, compute $I(\phi)$ using the truth tables for $\neg, \lor, \land, \to$, and $\leftrightarrow$:

(a) $I(p) = T, I(q) = T, I(r) = F$, and $I(s) = T$

(b) $I(p) = T, I(q) = F, I(r) = F$, and $I(s) = F$

(c) $I(p) = F, I(q) = T, I(r) = F$, and $I(s) = T$

(d) $I(p) = F, I(q) = F, I(r) = F$, and $I(s) = T$

16. Simplify the following boolean expressions:

(a) $(x \land y) \lor (x \land \neg y) \lor (\neg x \land y) \lor (\neg x \land \neg y)$

(b) $(x \land y \land z) \lor (x \land \neg y \land z) \lor (\neg x \land y \land \neg z) \lor (\neg x \land \neg y \land z)$

(c) $(x \land y \land \neg z) \lor (x \land \neg y \land z) \lor (x \land \neg y \land \neg z)$

17. Find formulas equivalent to the following formulas with all the negations "pushed inward to the proposition letters":

(a) $\neg(p \land T)$

(b) $((p \to q) \to r) \to F$

(c) $((p \to q) \to r) \to T$

(d) $(p \leftrightarrow q) \leftrightarrow r$

(e) $(p \leftrightarrow q) \leftrightarrow F$

(*Hint:* Look for a way to simplify this last one.) (*Note:* The method given to "push negations inward" does not always give the *shortest* formula that is equivalent to the given formula and has $\neg$ applied only to proposition letters.)

18. Find all truth values for which the following combinatorial circuit gives a value of T. Interpret this combinatorial circuit in terms of mechanizing majority rule for three parties. (*Hint:* If current is interpreted as a "yes" vote and no current as a "no" vote, then you should be able to see from a truth table when at least two of the three votes are in favor of the measure.)

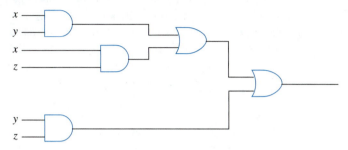

19. Prove that a combinatorial network for

$$(x \wedge y \wedge z) \vee (\neg x \wedge y \wedge z) \vee (x \wedge \neg y \wedge z) \vee (x \wedge y \wedge \neg z)$$

    can be simplified to a combinatorial network representing

$$(x \wedge y) \vee (x \wedge z) \vee (y \wedge z)$$

    (*Hint:* Replace $(x \wedge y \wedge z)$ with $(x \wedge y \wedge z) \vee (x \wedge y \wedge z)$ as often as needed.)

20. A half-adder circuit was given in the text. It adds two 1-bit numbers and produces two 1-bit outputs, a sum and a carry. To add two n-bit numbers, it is tempting to try to use n half-adders in parallel, one for each position, but this does not work. Consider the following base-2 addition:

carries	1	1	1	1	0	
		1	1	0	1	0_2
+		1	0	1	1	1_2
	1	1	0	0	0	1_2

    For example, the fourth digit of the sum, the third position from the right, is the sum of a 1 plus a 0, plus 1 carried from the position to the right of it. So, to compute that one position, one needs a circuit that computes the sum of *three* 1-digit binary numbers, the two digits and a carry. It should output the sum (the 1's position of the sum) and a carry (the 2's position of the sum). Such a circuit is called a *full-adder.*

    (a) Draw a full-adder circuit.
    (b) Draw a circuit, with one half-adder and three full-adders, for adding two 4-digit binary numbers.
    (c) Draw a circuit that implements the multiplication table (for one-digit numbers).

21. (a) The conjunction of n formulas $p_1, p_2, \ldots, p_n$ is defined to be the formula $(\ldots ((p_1 \wedge p_2) \wedge p_3) \wedge \ldots) \wedge p_n$. For $n = 0$, there is a special case: The conjunction of zero formulas is *defined* to be T. For $n = 1$, that conjunction simplifies to p_1. Let ϕ be the conjunction of $p_1, p_2, \ldots, p_n$. Prove that for any interpretation I, $I(\phi) = T$ if and only if $I(p_i) = T$ for each i such that $1 \le i \le n$. (*Hint:* Use induction.)

(b) Let ϕ be the formula

$$(\ldots((p_1 \leftrightarrow p_2) \leftrightarrow p_3) \leftrightarrow \ldots) \leftrightarrow p_n$$

for $n \geq 1$. For what interpretations I is $I(\phi) = T$? (*Hint:* The answer involves counting how many of the p_i's are true in I. Prove the result by induction on n.)

22. Two other commonly used propositional connectives are **exclusive or** (either one or the other but not both are T), denoted $\veebar$, and the **Sheffer stroke** (not both T), denoted $|$. Their truth tables are as follows:

p	q	$p \veebar q$
T	T	F
T	F	T
F	T	T
F	F	F

p	q	$p \downarrow q$
T	T	F
T	F	T
F	T	T
F	F	T

(a) Do commutative laws hold for $\veebar$ and $|$?

(b) Do associative laws hold for $\veebar$ and $|$?

(c) For what interpretations I is $I((\ldots((p_1 \veebar p_2) \veebar p_3) \veebar \ldots) \veebar p_n) = T$?

(d) Find formulas ϕ_1 and ϕ_2, (containing only proposition letters; the propositional constants T and F; the propositional connectives $\neg$, $\vee$, and $\wedge$; and parentheses) that are logically equivalent to $p \veebar q$ and $p|q$. (Compare formula x in Table 2.5 in Section 2.3.1, where such a formula is given for $p \leftrightarrow q$.)

(e) Repeat part (d) for $p \veebar p$ and $p|p$, but find the *shortest* formulas you can.

(f) Find formulas logically equivalent to $p \wedge q$, $p \vee q$, and $\neg p$ built from p and q using only $|$ and parentheses.

23. Prove both parts of Theorem 2.

24. Prove parts (b) through (e) of Theorem 3.

25. (a) Prove both parts of Theorem 4.

(b) Show that the converses to both parts of Theorem 4 need not be true.

(c) Does Theorem 4(a) remain true if the word *tautology* is replaced with *satisfiable*?

Definition   A formula ψ is an **alphabetic substitution** of a formula ϕ if ψ is formed from ϕ by replacing every occurrence of some proposition letter p in ϕ with some proposition letter q where q does not occur in ϕ. (*Note:* The relation of being an alphabetic substitution is symmetric, but it is not reflexive or transitive.) Define ψ to be an **alphabetic variant** of ϕ if there is a finite sequence of formulas $\phi_0, \phi_1, \ldots, \phi_n$ where $\phi_0 = \phi$, each ϕ_{i+1} is an alphabetic substitution of ϕ_i, and $\phi_n = \psi$.

26. (a) Show that $(p \vee q)$ is an alphabetic variant of $(q \vee p)$.

(b) Show that the relation of being an alphabetic variant is an equivalence relation.

(c) Show that if ψ is an alphabetic variant of ϕ, then ϕ is a tautology (respectively, is satisfiable, is unsatisfiable) if and only if ψ is a tautology (respectively, is satisfiable, is unsatisfiable).

(d) Show that ϕ being an alphabetic variant of ψ does not imply that ϕ and ψ are tautologically equivalent.

27. The first stage of the method described to "push negations inward" was a method to eliminate $\rightarrow$'s and $\leftrightarrow$'s. Prove that in the method to eliminate them, the process of

substituting always stops. Consider, for example, the substitution in the formula

$$(p \leftrightarrow q) \leftrightarrow (r \leftrightarrow s)$$

If the substitution is first performed on the second $\leftrightarrow$, the resultant formula is

$$((p \leftrightarrow q) \rightarrow (r \leftrightarrow s)) \wedge ((r \leftrightarrow s) \rightarrow (p \leftrightarrow q))$$

which has more $\leftrightarrow$'s to replace than in the original formula! At first sight, one might expect that if the substitutions are made in the wrong order, the process might continue generating more $\leftrightarrow$'s at each stage, and the process might continue forever. (*Hint:* One method is to, instead of just counting the number of $\leftrightarrow$ symbols, put a weight on each $\leftrightarrow$ symbol, with the weight of the $\leftrightarrow$ symbol in $\psi \leftrightarrow \chi$ being dependent on the number of $\leftrightarrow$'s in ψ and χ. If the correct method of calculating weights is used, it can be shown that the total weight of the $\leftrightarrow$'s decreases with each substitution.

28. The second stage of the procedure to "push negations inward" started with a formula whose only logical connectives are $\neg$, $\vee$, and $\wedge$ and constructed a tautologically equivalent formula with negations applied only to proposition letters.

  (a) Write an algorithm describing exactly what is done. The algorithm should work on formulas as strings of symbols. To avoid what in this case is irrelevant detail, the program should assume that all proposition letters are one character long and that any symbol encountered, except for (,), $\wedge$, $\vee$, and $\neg$, is a proposition letter. Assume that the formula contains no blanks. (It is perhaps easiest to consider the program as a function that is passed the original formula—a string—as a parameter, and then returns the equivalent formula with all the negations pushed inward. It is easiest to use recursion to handle *many* subformulas.)

  (b) Prove that your program from part (a) works. (*Hint:* if your program in part (a) uses recursion to handle subformulas, it is natural to do this proof by induction on formulas. However, the induction may not be straightforward.)

2.5   Normal Forms

Although two formulas may be logically equivalent, one may be "easier" for someone to understand or to manipulate. For example, in one formula, it may be easy to determine that the formula is satisfiable. It may be fairly obvious that one formula is a tautology but quite difficult to conclude that from the other form of the same formula. In this section, we discuss two special forms or representations for formulas logically equivalent to a given formula. These forms are called disjunctive normal forms and conjunctive normal forms. Formulas in conjunctive normal form make it easy to determine when a formula is satisfiable. Formulas in disjunctive normal form are easy to use when asking whether a formula is a tautology. These special forms have assumed prominence in computer science, in both theoretical and applied areas. The famous $\mathcal{P} \neq \mathcal{NP}$ problem deals with conjunctive normal forms, and combinatorial networks use both conjunctive and disjunctive normal forms to find representations of combinatorial circuits.

2.5.1   Disjunctive Normal Form

Consider the following two formulas:

$$\phi = (p \rightarrow (q \vee r)) \leftrightarrow (q \rightarrow p)$$

and

$$\psi = (p \wedge q) \vee (p \wedge \neg q \wedge r) \vee (\neg p \wedge \neg q)$$

The truth tables for ϕ and ψ would show that these two formulas are logically equivalent. By some measures, ψ is more complicated. For example, ϕ has four propositional connectives, whereas ψ has nine. Nevertheless, many people find ψ to be far easier to understand. The formula ψ explicitly lists three cases in which the formula is true:

(1) p and q are both T.
(2) p and r are T and q is F.
(3) p and q are both F.

For all other interpretations of p, q, and r, the truth value of ψ is F. It is not nearly so obvious what ϕ "says." Although ϕ is shorter, it also seems to be more complex.

A formula like ψ that is just a list of cases that make the formula have a truth value of T is called a **disjunctive normal form (DNF).** Each of the three cases, $(p \wedge q)$, $(p \wedge \neg q \wedge r)$, and $(\neg p \wedge \neg q)$, is called a **term.** One might think of each term as describing a single case. The entire disjunctive normal form formula is just a disjunct of terms that make the formula T. (The words *term* and *disjunctive normal form* will be defined formally below.)

The difference in comprehensibility is even more extreme if the formula ϕ is negated. The formula

$$\neg((p \rightarrow (q \vee r)) \leftrightarrow (q \rightarrow p))$$

is logically equivalent to the disjunctive normal form formula

$$(\neg p \wedge q) \vee (p \wedge \neg q \wedge \neg r)$$

The disjunctive normal form is a disjunction of only two terms, which makes it particularly easy to understand.

Definition 1.   Let p be a proposition letter. Then, p is a **positive literal,** and $\neg p$ is a **negative literal.** A **literal** is a positive literal or a negative literal.

Definition 2.   Let $\lambda_1, \lambda_2, \ldots, \lambda_m$ be a set of m literals with $m \in \mathbb{N}$. A term is a conjunction

$$\lambda_1 \wedge \lambda_2 \wedge \cdots \wedge \lambda_m$$

of m literals. A formula ϕ is in DNF if it is a disjunction $\phi_1 \vee \phi_2 \vee \cdots \vee \phi_k$ of k terms where $k \in \mathbb{N}$.

The disjunction of zero formulas is F. The conjunction of zero formulas is T. This is analogous to defining the sum of zero numbers to be zero and the product of zero numbers to be 1. For example, $F \vee p \leftrightarrow p$ is analogous to $0 + x = x$.

Example 1.

(a) $a \wedge b \wedge \neg c$ is a term.

(b) The formula

$$(a \wedge b \wedge c) \vee (\neg a \wedge \neg b \wedge \neg c) \vee (a \wedge \neg c \wedge q)$$

is in disjunctive normal form.

(c) T is a term. It is a conjunction of zero literals.

(d) a is a term. It is a conjunction of one literal.

(e) $a \wedge b \wedge \neg c$ and T are in disjunctive normal form. Each is a disjunction of one term.

(f) F is in disjunctive normal form. It is a disjunction of zero terms.

Theorem 1. Every formula is logically equivalent to a formula in DNF.

The proof of Theorem 1 is just a formalization of what is done in Example 2.

Example 2. Let ψ be the formula

$$\psi = (\neg(p \rightarrow q)) \rightarrow (q \wedge \neg r)$$

Determine a DNF for ψ.

Solution. A formula may have several equivalent formulas in DNF, but we want a systematic way to find one.

The first step in finding a DNF for ψ is to find the truth table for all the interpretations of ψ, as shown in Table 2.7.

Table 2.7 Abbreviated Truth Table for ψ

Interpretation	p	q	r	$(\neg(p \rightarrow q)) \rightarrow (q \wedge \neg r)$
I_0	T	T	T	T
I_1	T	T	F	T
I_2	T	F	T	F
I_3	T	F	F	F
I_4	F	T	T	T
I_5	F	T	F	T
I_6	F	F	T	T
I_7	F	F	F	T

The next step is to construct, for each interpretation I_i, $0 \le i \le 7$, a term that is T in that interpretation and F in all other interpretations. Such terms are listed in Table 2.8.

Table 2.8 True Terms in the Interpretations

Interpretation	Matching Term
I_0	$p \wedge q \wedge r$
I_1	$p \wedge q \wedge \neg r$
I_2	$p \wedge \neg q \wedge r$
I_3	$p \wedge \neg q \wedge \neg r$
I_4	$\neg p \wedge q \wedge r$
I_5	$\neg p \wedge q \wedge \neg r$
I_6	$\neg p \wedge \neg q \wedge r$
I_7	$\neg p \wedge \neg q \wedge \neg r$

The reader should observe that these terms have the desired properties. That is, I_0 satisfies $p \wedge q \wedge r$, and all seven other interpretations do not satisfy $p \wedge q \wedge r$.

Now, breaking into the cases where ψ is T, we construct a disjunction of terms with one corresponding to each interpretation where ψ is T:

$$\phi_\psi = (p \wedge q \wedge r) \vee (p \wedge q \wedge \neg r) \vee (\neg p \wedge q \wedge r) \vee (\neg p \wedge q \wedge \neg r)$$
$$\vee (\neg p \wedge \neg q \wedge r) \vee (\neg p \wedge \neg q \wedge \neg r)$$

Clearly, ϕ_ψ is in DNF. The only question is whether ϕ_ψ is logically equivalent to ψ. As a result of the construction, however, each term of ϕ_ψ is T for exactly one of the interpretations for which ψ is T, whereas ψ is F in all other interpretations. So, ϕ_ψ is T when ϕ is T. Each term of ϕ_ψ is F in each interpretation for which ψ is F. Therefore, ϕ_ψ is F in each interpretation for which ψ is F. Thus, ϕ_ψ is logically equivalent to ψ. ■

Example 3.   Let ψ be the formula

$$\psi = (p \rightarrow (q \vee r)) \wedge (\neg q) \wedge (\neg r) \wedge p$$

Find a DNF for ψ.

Solution.   It is easy to see that ψ is unsatisfiable. We see this from the truth table for ψ shown in Table 2.9.

Table 2.9   Abbreviated Truth Table for ψ

Interpretation	p	q	r	$p \rightarrow (q \vee r)$	$\neg q$	$\neg r$	$(p \rightarrow (q \vee r)) \wedge (\neg q) \wedge (\neg r) \wedge p$
I_0	T	T	T	T	F	F	F
I_1	T	T	F	T	F	T	F
I_2	T	F	T	T	T	F	F
I_3	T	F	F	F	T	T	F
I_4	F	T	T	T	F	F	F
I_5	F	T	F	T	F	T	F
I_6	F	F	T	T	T	F	F
I_7	F	F	F	T	T	T	F

Since at least one of the formulas $p \rightarrow (q \vee r)$, $\neg q$, $\neg r$, and p is F in each interpretation, the disjunct of these terms is always F. Therefore, the construction as in Example 2 that formed terms for interpretations satisfying ψ, would construct *no* terms. Accordingly, the formula generated as in Example 2 would be a disjunction of zero terms, which, by convention, is the formula F. The formula F is in DNF and is logically equivalent to ψ. ■

2.5.2   Application: DNF and Combinatorial Networks

To interpret the DNF for a boolean expression, we view a term as a product or a join of a set of literals. The DNF for a formula is viewed as a sum or a meet of a set of terms. The DNF for a boolean expression gives us an option to use when designing combinatorial circuits.

Example 4. Let x, y be elements of a boolean algebra. Use the DNF for the boolean expression $(x \wedge y) \vee (\neg x \wedge \neg y)$ to design a combinatorial circuit.

Solution. The boolean expression is in DNF. Therefore, the combinatorial circuit is

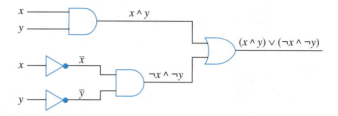

2.5.3 Conjunctive Normal Form

Consider again the formula used as a motivating example for DNFs:

$$(p \rightarrow (q \vee r)) \leftrightarrow (q \rightarrow p)$$

The formula is logically equivalent to the formula

$$(p \vee \neg q) \wedge (\neg p \vee q \vee r)$$

This logically equivalent formula is in **conjunctive normal form (CNF)**. It consists of a conjunction of two formulas that are disjunctions of literals. In this example, it is the conjunct of $(p \vee \neg q)$ and $(\neg p \vee q \vee r)$. Each disjunction of zero or more literals can be thought of as a restriction on when the formula can be T. The first restriction is that at least one of p and $\neg q$ must be T. The second is that at least one of $\neg p$, q, and r must be T. This can be thought of as a list of rules that must all be met for the formula to be satisfied. Thus, CNF formulas are often easy to understand.

Definition 3. Let $\lambda_1, \lambda_2, \ldots, \lambda_m$ be a set of m literals with $m \in \mathbb{N}$. A **clause** is a disjunction

$$\lambda_1 \vee \lambda_2 \vee \ldots \vee \lambda_m$$

of m literals. A formula ϕ is in CNF if it is a conjunction

$$\phi_1 \wedge \phi_2 \wedge \cdots \wedge \phi_k$$

of k clauses $\phi_1, \phi_2, \ldots, \phi_k$ where $k \in \mathbb{N}$.

Example 5.

(a) $a \vee b \vee \neg c$ is a clause.
(b) T is in CNF. It is a conjunction of zero clauses.
(c) F is a clause. It is a disjunction of zero literals.
(d) a is a clause. It is a disjunction of one literal.
(e) The disjunction of clauses shown is in CNF:

$$(a \vee b \vee c) \wedge (\neg a \vee \neg b \vee \neg c) \wedge (a \vee \neg c \vee q)$$

(f) $a \vee b \vee \neg c$ and F are in CNF. Each is a conjunction of one clause.

Theorem 2.   Every formula is logically equivalent to a formula in CNF.

The proof of Theorem 2 is just a formalization of what is done in Example 6.

Example 6.   Find the conjunctive normal form for the formula

$$\psi = (\neg(p \to q)) \to (q \wedge \neg r)$$

Solution.   The process starts by finding a formula in DNF that is equivalent to $\neg\psi$.

The following is an abbreviated truth table for $\neg\psi$. We will misuse the word *interpretation* exactly as we did in Example 2 in Section 2.3.

Interpretation	p	q	r	$\neg((\neg(p \to q)) \to (q \wedge \neg r))$
I_0	T	T	T	F
I_1	T	T	T	F
I_2	T	F	T	T
I_3	T	F	F	T
I_4	F	T	T	F
I_5	F	T	F	F
I_6	F	F	T	F
I_7	F	F	F	F

Now, put $\neg\psi$ into DNF:

$$\phi_{\neg\psi} = \phi_2 \vee \phi_3 = (p \wedge \neg q \wedge r) \vee (p \wedge \neg q \wedge \neg r)$$

So, ψ is logically equivalent to

$$\neg((p \wedge \neg q \wedge r) \vee (p \wedge \neg q \wedge \neg r))$$

Push the negations inside, first past the $\vee$ using DeMorgan's Law:

$$\neg(p \wedge \neg q \wedge r) \wedge \neg(p \wedge \neg q \wedge \neg r)$$

then past the internal $\wedge$'s, again using DeMorgan's Law:

$$(\neg p \vee \neg\neg q \vee \neg r) \wedge (\neg p \vee \neg\neg q \vee \neg\neg r)$$

and finally, eliminate the double negations:

$$(\neg p \vee q \vee \neg r) \wedge (\neg p \vee q \vee r)$$

Since $\phi_{\neg\psi}$ was in DNF, negating and pushing the negations inside creates a formula in CNF logically equivalent to ψ.   ∎

Example 7. Let ψ be the formula

$$\psi = \neg((p \rightarrow (q \vee r)) \wedge (\neg q) \wedge (\neg r) \wedge p)$$

Find a CNF for ψ.

Solution. The negation of ψ is equivalent to

$$(p \rightarrow (q \vee r)) \wedge (\neg q) \wedge (\neg r) \wedge p$$

which in Example 3 was shown to have F as a DNF. So, ψ is equivalent to $\neg F$, and pushing negations inward gives the CNF formula T. ∎

2.5.4 Application: CNF and Combinatorial Networks

To interpret the CNF for a boolean expression, we view a clause as a meet of a set of literals. The CNF for a formula is viewed as a join of a set of clauses. The CNF for a boolean expression gives us another option to use when designing combinatorial circuits.

Example 8. Let x, y be elements of a boolean algebra. Use the CNF for the boolean expression $(x \wedge y) \vee (\neg x \wedge \neg y)$ to design a combinatorial circuit.

Solution. The truth table for the expression is

x	y	$x \wedge y$	$(\neg x \wedge \neg y)$	$(x \wedge y) \vee (\neg x \wedge \neg y)$
T	T	T	F	T
T	F	F	F	F
F	T	F	F	F
F	F	F	T	T

The DNF for $\neg ((x \wedge y) \vee (\neg x \wedge \neg y))$ is just

$$(x \wedge \neg y) \vee (\neg x \wedge y)$$

Therefore, the CNF for $(x \wedge y) \vee (\neg x \wedge \neg y)$ is

$$\begin{aligned} \neg((x \wedge \neg y) \vee (\neg x \wedge y)) &= \neg(x \wedge \neg y) \wedge \neg(\neg x \wedge y) \\ &= (\neg x \vee \neg\neg y) \wedge (\neg\neg x \vee \neg y) \\ &= (\neg x \vee y) \wedge (x \vee \neg y) \end{aligned}$$

The combinatorial circuit for this boolean expression is

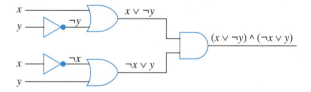

∎

2.5.5 Testing Satisfiability and Validity

It turns out to be very easy to tell when a formula in DNF is satisfiable. This is one reason why a DNF is often nice to work with.

Example 9.

(a) Show that

$$\phi = (a \wedge \neg b) \vee (\neg a \wedge \neg c \wedge b) \vee (a \wedge \neg a)$$

is satisfiable.
(b) Show that

$$\phi = (a \wedge \neg b \wedge b) \vee (\neg a \wedge \neg c \wedge b \wedge c) \vee (a \wedge \neg a)$$

is unsatisfiable.

Solution.

(a) To find an interpretation I where $I(\phi) = T$, it is enough to find an interpretation where one of the terms is true. In this case, there is an interpretation where the first term is true. If $I(a) = T$ and $I(b) = F$, then $I(a \wedge \neg b) = T$ and, hence, $I(\phi) = T$.
(b) In this case, ϕ is unsatisfiable, because every term is F in every interpretation I. For example, in the first term, we require both $I(\neg b)$ and $I(b)$ to be T in an interpretation. In the second term, we require both $I(c)$ and $I(\neg c)$ to be T in an interpretation. In the third term, we require both $I(a)$ and $I(\neg a)$ to be T in an interpretation. These conditions are clearly impossible in any interpretation. ■

Similarly, it is easy to tell when a formula in CNF is a tautology.

Example 10.
(a) Show that

$$\phi = (a \vee \neg b) \wedge (\neg a \vee \neg c \vee b) \wedge (a \vee \neg a)$$

is not a tautology.
(b) Show that

$$\phi = (a \vee \neg b \vee b) \wedge (\neg a \vee \neg c \vee b \vee c) \wedge (a \vee \neg a)$$

is a tautology.

Solution.

(a) If $I(a) = F$ and $I(b) = T$, then $I(a \vee \neg b) = F$, so $I(\phi) = F$.
(b) The first clause is a tautology, since $b \vee \neg b$ alone is a tautology. The second clause is a tautology, since $c \vee \neg c$ alone is a tautology. The third clause is also a tautology. ■

Theorem 3.

(a) Let $\phi_1, \phi_2, \ldots, \phi_k$ be clauses for $k \in \mathbb{N}$. Let

$$\phi = \phi_1 \wedge \phi_2 \wedge \ldots \wedge \phi_k$$

Then, ϕ is a tautology if and if every ϕ_i is a tautology.
(b) Let $\lambda_1, \lambda_2, \ldots, \lambda_m$ be literals for some m. Let

$$\phi_i = \lambda_1 \vee \lambda_2 \vee \cdots \vee \lambda_m$$

Then, ϕ_i is a tautology if and only if ϕ_i contains two literals, λ_a and λ_b, where $\lambda_a = \neg\lambda_b$ and $1 \le a \ne b \le m$.

Proof. The proofs of parts (a) and (b) just formalize what was done in Example 10. ∎

2.5.6 The Famous $\mathcal{P} \ne \mathcal{NP}$ Conjecture

How easy is it to test whether a formula in DNF is satisfiable or whether a formula in CNF is a tautology? One way to check whether a CNF formula ϕ is satisfiable is to write its truth table, but this can be a time-consuming process. A formula ϕ with n symbols may contain more than $n/4$ different proposition letters. The truth table has to have a row for each assignment of T's and F's to these $n/4$ proposition letters—thus, $2^{n/4}$ rows. Hence, for some formulas, the size of the truth table is exponentially larger than the size of the formula. Consequently, this does not give a practical way to check satisfiability. Just check how many rows that is for $n = 1000$, because it's common, for example, in computer hardware verification applications to have more than 1000 variables.

Another way is to find an equivalent formula ϕ' in DNF. Now, the construction given in the proof of Theorem 1 (Section 2.5.1) requires writing down the truth table for ϕ, so that method is too slow. Maybe, however, there is a faster way to find such a formula ϕ' in DNF. If so, that may provide a way to check satisfiability. Unfortunately, this approach also is not, in general, practical: The shortest such formula ϕ' may itself be far longer than ϕ.

It turns out that there is a reasonably fast algorithm for checking satisfiability for CNF formulas

$$\phi = (\lambda_0 \vee \lambda_1) \wedge (\lambda_2 \vee \lambda_3) \wedge \cdots \wedge (\lambda_{2n} \vee \lambda_{2n+1})$$

where each clause contains at most two literals—formulas in what is called **2-CNF.** However, if the clauses are allowed to contain even three literals **(3-CNF),** then the answer is unknown. This problem is called the **3-satisfiability problem.**

The 3-satisfiability problem is one of a large group of problems called $\mathcal{NP}$**–complete** problems, which will be discussed further in Section 5.3.5. Another famous $\mathcal{NP}$-complete problem is a form of the traveling salesperson problem, which will be discussed in Chapter 7. The commonly believed conjecture, called the $\mathcal{P} \ne \mathcal{NP}$ conjecture, is that no $\mathcal{NP}$-complete problems can be solved, in general, by algorithms that are even remotely practical. However, the conjecture is neither proved nor disproved (at least as of the time this book was written), and it appears to be a very hard mathematical problem. It is considered by many to be the most important unsolved problem in theoretical computer science—and one of the most important unsolved problems in all of mathematics.

2.5.7 Resolution Proofs: Automating Logic

The ancient dream of automating reasoning will require a computer program to be able to arrive at conclusions using rules of inference such as those shown in Table 2.6. One attempt to automate reasoning in a special context was made by John R. Robinson, who used a single inference rule called **resolution.** This inference rule deals exclusively with formulas in CNF (clauses). As a simple example of this inference rule, called the **resolution rule,** suppose the two clauses

$$p \vee \neg q \text{ and } r \vee q$$

are given. What conclusion is possible for the conjunction of these two clauses as a hypothesis? In the resolution system, we are interested in the implication

$$((p \vee \neg q) \wedge (r \vee q)) \rightarrow (p \vee r)$$

It is easy to prove using a truth table that this implication is always T. With this inference rule, we can then use the clause $p \vee r$ as another clause in the resolution system. The resolution rule uses only this inference rule with formulas in CNF. We often display this rule as

$$\frac{\begin{array}{c} p \vee \neg q \\ r \vee q \end{array}}{p \vee r}$$

Definition 4.   Let two clauses $c_1 = \phi \vee p$ and $c_2 = \psi \vee \neg p$ be given where p is a proposition letter and ϕ and ψ are clauses. The **resolvant** of c_1 and c_2 on p is the clause $\phi \vee \psi$.

Example 11.   Let $c_1 = p \vee \neg q \vee r$ and $c_2 = \neg p \vee r \vee s \vee t$. The resolvant of clauses c_1 and c_2 on p is

$$\neg q \vee r \vee r \vee s \vee t = \neg q \vee r \vee s \vee t$$

In a resolution proof or resolution refutation, we imagine the conjunction of a set of clauses being the hypothesis for an implication. The resolution rule can be used to see if the set of clauses is satisfiable. If the conjunction of the set of clauses is F, then the set of clauses is unsatisfiable. We formalize the idea of this proof technique in the next definition.

Definition 5.   Let S be a set of clauses. A **resolution refutation** of S is a sequence of clauses $r_0, r_1, \ldots, r_k$ such that:

(a) each r_i is either an element of S or a **resolvant** of r_j and r_k where $0 \leq j \neq k < i \leq k$, and
(b) $r_k = F$.

Example 12.   Let S be the set of clauses $\{p, \neg p \vee \neg q, \neg p \vee q \vee r, \neg r\}$. Give a resolution refutation of S.

Solution.   The right-hand column of the following table just explains why each step is valid. The left-hand column simply numbers the lines so that we can refer to them later.

Proof Step	Clause	Justification
r_0	p	Element of S
r_1	$\neg p \vee \neg q$	Element of S
r_2	$\neg q$	Resolvant of r_0 and r_1 on p
r_3	$\neg p \vee q \vee r$	Element of S
r_4	$q \vee r$	Resolvant of r_0 and r_3 on p
r_5	r	Resolvant of r_4 and r_2 on q
r_6	$\neg r$	Element of S
r_7	F	Resolvant of r_5 and r_6 on r

■

Example 13. Let $S = \{p \vee q, p \vee \neg q, \neg p \vee q, \neg p \vee \neg q\}$. Give a resolution refutation for S (plus comments, as noted in Example 12 above).

Solution.

Line	Proof Step	Justification
r_0	$p \vee q$	Element of S
r_1	$\neg p \vee q$	Element of S
r_2	q	Resolvant of r_0 and r_1 on p
r_3	$p \vee \neg q$	Element of S
r_4	$\neg p \vee \neg q$	Element of S
r_5	$\neg q$	Resolvant of r_3 and r_4 on p
r_6	F	Resolvant of r_2 and r_5 on q

A proof method is **sound** if everything that is provable is true or satisfiable. Here, that means that for any set S of clauses, if there is a resolution refutation of S, then S is unsatisfiable.

A proof method is **complete** if everything that is true is provable. If there is a resolution refutation of a set of clauses S, then S is not complete, since some things that are not true are provable—that is, any set of clauses for which there is a **resolution refutation.**

2.6 Exercises

1. Write DNFs and CNFs corresponding to each of the following truth tables:

(a)
p	q	r	Truth Value
T	T	T	T
T	T	F	T
T	F	T	T
T	F	F	F
F	T	T	F
F	T	F	F
F	F	T	T
F	F	F	F

(b)
p	q	r	Truth Value
T	T	T	F
T	T	F	T
T	F	T	F
T	F	F	T
F	T	T	F
F	T	F	F
F	F	T	T
F	F	F	T

(c)
p	q	r	Truth Value
T	T	T	T
T	T	F	T
T	F	T	F
T	F	F	T
F	T	T	F
F	T	F	F
F	F	T	T
F	F	F	F

(d)
p	q	r	s	Truth Value
T	T	T	T	F
T	T	F	T	F
T	T	F	F	T
T	F	T	F	T
F	T	T	T	F
F	T	T	F	T
F	T	F	T	T
F	F	F	F	T

(e)	p	q	r	s	Truth Value
	T	T	T	T	F
	T	T	F	T	T
	T	F	T	F	T
	F	T	F	T	F
	F	T	F	F	T
	F	F	T	F	F
	F	F	F	F	F

(f)	p	q	r	s	Truth Value
	T	T	T	F	T
	T	T	F	F	F
	T	F	F	F	T
	F	T	T	F	T
	F	T	F	T	F
	F	F	T	F	F

2. Find formulas in DNF equivalent to each of the following formulas:

   (a) $\neg(p \wedge T)$
   (b) $((p \to q) \to r) \to F$
   (c) $((p \to q) \to r) \to T$
   (d) $(p \leftrightarrow q) \leftrightarrow r$
   (e) $\neg(p \leftrightarrow q) \leftrightarrow r$
   (f) $((p \vee q) \to r) \wedge (r \to \neg(p \vee q))$
   (g) $(\neg r) \to (((p \vee q) \to r) \to \neg q)$

3. Which of the following DNF formulas are satisfiable? If the formula is satisfiable, give an interpretation that satisfies it. If it is not satisfiable, explain why not.

   (a) $(a \wedge b \wedge c) \vee (c \wedge \neg c \wedge b)$
   (b) $(a \wedge b \wedge c \wedge d \wedge \neg b) \vee (c \wedge d \wedge \neg c \wedge e \wedge f)$
   (c) $(a \wedge b \wedge c) \vee (\neg a \wedge \neg b \wedge \neg c)$

4. Find formulas in DNF equivalent to each of the following formulas, and find at least two interpretations that make each formula satisfiable:

   (a) $((p \to q) \to r) \to F$
   (b) $\neg(p \leftrightarrow q) \leftrightarrow r$
   (c) $(\neg r) \to (((p \vee q) \to r) \to \neg q)$

5. Find formulas in CNF equivalent to each of the following formulas:

   (a) $\neg(p \wedge T)$
   (b) $((p \to q) \to r) \to F$
   (c) $((p \to q) \to r) \to T$
   (d) $(p \leftrightarrow q) \leftrightarrow r$
   (e) $\neg(p \leftrightarrow q) \leftrightarrow r$
   (f) $((p \vee q) \to r) \wedge (r \to \neg(p \vee q))$
   (g) $(\neg r) \to (((p \vee q) \to r) \to \neg q)$

6. For the following formulas find equivalent formulas in CNF and DNF form. Draw combinatorial networks corresponding to the original formulas and their equivalent CNF forms.

   (a) $(p \wedge q) \leftrightarrow (p \wedge r)$
   (b) $((p \to q) \to r) \to p$

7. Which of the following formulas in CNF are tautologies? Explain, as in Example 6.

   (a) $(a \vee b \vee c) \wedge (c \vee \neg c \vee b)$
   (b) $(a \vee b \vee c \vee d \vee \neg b) \wedge (c \vee d \vee \neg c \vee e \vee f)$
   (c) $(a \vee b \vee c) \wedge (\neg a \vee \neg b \vee \neg c)$

8. Find a CNF for each of the following formulas, and prove that each formula is a tautology.

   (a) $(p \land p) \leftrightarrow p$
   (b) $(p \land (p \rightarrow q)) \rightarrow q$
   (c) $(p \rightarrow (r \rightarrow q)) \leftrightarrow ((p \land r) \rightarrow q)$
   (d) $(p \rightarrow r) \leftrightarrow (\neg r \rightarrow \neg p)$

9. (a) Show that the following formula in CNF is unsatisfiable:

$$(p \lor q) \land (p \lor \neg q) \land (\neg p \lor q) \land (\neg p \lor \neg q)$$

   (b) Show that the following formula in CNF is unsatisfiable:

$$(p \lor q \lor r) \land (p \lor \neg q \lor r) \land (\neg p \lor q \lor r) \land (\neg p \lor \neg q \lor r)$$
$$\land (p \lor q \lor \neg r) \land (p \lor \neg q \lor \neg r) \land (\neg p \lor q \lor \neg r) \land (\neg p \lor \neg q \lor \neg r)$$

   Can you find an easier argument than just writing the entire truth table?

   (c) Generalize the above to some class of CNF formulas on an arbitrary number $n \geq 1$ of proposition letters, and prove it by induction on n.

10. (a) Prove that the formula $\neg p$ is not equivalent to any formula built from the proposition letters T and F using only $\land$ and $\lor$ plus parentheses.
    (b) Prove that the formula $p \lor q$ is not equivalent to any formula built from the proposition letters using only $\leftrightarrow$.
    (c) Prove that there is a formula not equivalent to any formula built from the proposition letters using only $\underline{\lor}$. (See Exercise 22 in Section 2.4.)
    (d) Prove that there is a formula not equivalent to any formula built from the proposition letters using only $\rightarrow$ plus parentheses.

11. Write pseudocode for a program that, given a formula ϕ, finds (i) a logically equivalent formula ϕ' in CNF and (ii) a logically equivalent formula ϕ'' in DNF. The algorithm should be recursive (similar to an induction on formulas) and should not involve the construction of truth tables. Prove the algorithm works. This gives an alternate proof of the theorem that every formula is equivalent to a formula in CNF.

Definition   A *k*-**term** is a conjunction of k literals. A *k*-**DNF** formula is a disjunction of *k*-terms.

12. (a) Show that every formula containing only k (different) proposition letters is equivalent to a k-DNF formula.
    (b) Show that $p \leftrightarrow q$ is not equivalent to any 1-DNF formula.
    (c) Show that for every natural number k (including 0), there is a formula containing only $k + 1$ (different) proposition letters that is not equivalent to any k-DNF formula.

13. (a) Find the resolvant of $(p \lor q)$ and $(\neg p \lor r)$ on p.
    (b) Find the resolvant of $(p \lor q \lor r \lor s)$ and $(\neg p \lor \neg q \lor t)$ on p.
    (c) Find the resolvant of $(p \lor q)$ and $\neg p$ on p.
    (d) Find the resolvant of (p) and $(\neg p)$ on p.
    (e) Which resolvant above from parts (a) through (d) is a tautology? Which is tautologically false?

14. Write resolution refutations of the following sets of clauses. Include line numbers and justifications, as in Example 12.

(a) $\{p, \neg p \vee q, \neg p \vee \neg q \vee r, \neg r\}$

(b) $\{\neg p, p \vee q, \neg q \vee \neg r, p \vee r\}$

(c) $\{p \vee q, \neg p \vee r, \neg q \vee r, \neg p \vee s, \neg q \vee s, \neg r \vee \neg x\}$

(d) $\{p \vee q \vee r, p \vee q \vee \neg r, p \vee \neg q \vee r, p \vee \neg q \vee \neg r, \neg p \vee q \vee \neg r, \neg p \vee \neg q \vee r, \neg p \vee \neg q \vee \neg r\}.$

15. (a) Show that if r is the resolvant of two clauses c_1, c_2 on proposition letter p, then

$$\{c_1, c_2\} \models r$$

(*Hint:* For each interpretation, break into cases, depending on whether p is T or F in each interpretation.)

(b) Prove that if there is a resolution refutation of a set S of clauses, then S is unsatisfiable. (*Hint:* Use strong induction on the length of the resolution refutation.)

16. The length of a clause is the number of literals in the clause. The length of a CNF formula is the sum of the length of its clauses. The number of excess literals in a CNF formula is the length of the formula minus the number of clauses in the formula.

(a) Show that if an unsatisfiable set S of clauses contains only clauses of length 0 and 1, it has a resolution refutation. (*Hint:* Prove the following: If S contains a clause of length 0, it has [trivially] a resolution refutation. If, for some proposition letter p, S contains both p and $\neg p$, then S has a resolution refutation. Otherwise, S is satisfiable.)

(b) Show that if a set $\{\lambda_1 \vee \lambda_2 \vee \cdots \vee \lambda_k \vee \lambda_{k+1}\} \cup S$ $(k \geq 1)$ of clauses is unsatisfiable, so are $\{\lambda_1 \vee \lambda_2 \vee \lambda_k\} \cup S$ and $\{\lambda_{k+1}\} \cup S$. (*Hint:* For the first half, prove that if an interpretation I satisfies $\{\lambda_1 \vee \lambda_2 \vee \cdots \vee \lambda_k\} \cup S$, it also satisfies $\{\lambda_1 \vee \lambda_2 \vee \cdots \vee \lambda_k \vee \lambda_{k+1}\} \cup S$.)

(c) Show that for $k \geq 1$, the number of excess literals in $\{\lambda_1 \vee \lambda_2 \vee \cdots \vee \lambda_k\} \cup S$ and the number of excess literals in $\{\lambda_{k+1}\} \cup S$ are both less than the number of excess literals in $\{\lambda_1 \vee \lambda_2 \vee \cdots \vee \lambda_k \vee \lambda_{k+1}\} \cup S$.

(d) A **resolution derivation** of a clause r_k from a set S of clauses is a sequence $r_0, r_1, r_2, \ldots, r_k$ of clauses where each r_i is either an element of S or a resolvant of two previous r_j's. (Thus, resolution refutation of S is just a resolution derivation of F from S.) Show that if there is a resolution derivation of λ from S and a resolution refutation of $S \cup \{\lambda\}$, then there is a resolution refutation of S.

(e) Prove that if there is a resolution refutation ρ of $\{\lambda_1 \vee \lambda_2 \vee \cdots \vee \lambda_k\} \cup S$, then either (i) there is a resolution refutation of $\{\lambda_1 \vee \lambda_2 \vee \cdots \vee \lambda_k \vee \lambda_{k+1}\} \cup S$ or (ii) there is a resolution derivation of λ_{k+1} from $\lambda_1 \vee \lambda_2 \vee \cdots \vee \lambda_k \vee \lambda_{k+1} \cup S$. (*Hint:* Prove this by induction on the length ρ. You will have to add λ_{k+1} as a disjunct to *some* of the clauses in ρ. It is *not* true in general that if $S \models \lambda$, then there is a resolution derivation of λ from S.)

(f) Prove that resolution refutation is complete.

2.7   Predicates and Quantification

In propositional logic, our basic "objects" were entire statements, represented by proposition letters. In discussing mathematical structures, however, we want to be able go one step "lower" to assert statements, such as $x = 3$ or $x \geq y$, where the meanings of x and y are not fixed for all time. We want to allow **variables** (or **variable symbols**), such as p and q, which represent elements of some nonempty universal set. These variables are not proposition letters, because they do not evaluate to T or F. Rather, after an assignment of values to the variables, such as 2 to x and 5 to y, the predicates, such as $x = 3$ or $x \geq y$, become $2 = 3$ and $2 \geq 5$, both of which are F. We can think of similar instances in natural language when we assert "He is tall and she is of average height." The pronouns *he* and *she* can be thought of as placeholders or variables representing a range of particular men or women.

2.7.1   Predicates

A property or relationship between objects is called a **predicate.** A description of a predicate in logic is called a **formula.**

A formula such as $x < 3$ is an **atomic formula,** built with the predicate $<$. An atomic formula is a formula for which the terms do not involve any of the logical operations (and, or, implication, biconditional, negation), only proposition letters and constants from the universal set. The predicate $<$ is **binary** or (2-ary); it represents a relationship between two objects. The first object is a variable x; the second is a **constant** 3. If a specific value x_0 is substituted into the predicate for x, it becomes $x_0 < 3$. Now, if $x_0 = -37$, then $x_0 < 3$ evaluates to T. If $x_0 = 6$, then $x_0 < 3$ evaluates to F. When a predicate involves n arguments, it is said to be **n-ary;** we write an n-ary formula $P(x_1, x_2, \ldots, x_n)$.

Example 1.   The following are predicates:

(a) Let $P(x, y, z)$ denote "$x + y = z$."
(b) Let $Q(x_1, x_2)$ denote "$x_1 - x_2 \geq 0$."
(c) Let $M(x, y)$ denote "x is married to y."
(d) Let $E(x, y)$ denote "$x = y$."

Once we are given a group of predicates, we may refer to them in formulas. So, $P(x, y, z)$, $Q(x_1, x_2)$, $M(x, y)$, and $E(x, y)$ are all atomic formulas. The variable names are not important, so $P(x_1, z, x_2)$ and $Q(y, y)$ are also atomic formulas. Just as in usual notation, $Q(y, y)$ denotes "$y - y \geq 0$."

In normal uses of logic, the universal set U and the meanings of all predicates, such as $<$ or a variable P, are specified. We limit ourselves to this understanding.

2.7.2   Quantification

If we have a predicate, such as $<$, that is defined on two objects, then logic gives us two ways to specify what objects we mean. One way is to specify values from the universal set, such as $3 < 6$. Another way is called **quantification.** It comes in two forms: **universal quantification,** denoted by the symbol $\forall$; and **existential quantification,** denoted by the symbol $\exists$.

The first form is universal quantification for the predicate P, such as.

$$\forall x \, (P(x)) \quad \text{read "For all } x, \ P(x)\text{"}$$

is defined to mean "For all values x in the universal set U, the assertion $P(x)$ is true."

The second kind of quantification is existential quantification for the predicate P, such as

$\exists x \, (P(x))$ read "For some x, $P(x)$" or "There exists an x such that $P(x)$"

is defined to mean "For some value of x in the universal set U, the predicate $P(x)$ is true."

Following the $\forall x$ or $\exists x$, there is a formula in parentheses, such as $(P(x))$, although we occasionally omit the parentheses. That formula, plus the $\forall x$ or the $\exists x$ itself, is called the **scope** of the quantifier. Informally, we may leave out the parentheses, writing, for example, just $\forall x \, P(x)$, but that is informal. The definition of *scope* is defined as if we had not left the parentheses out. A similar idea of scope occurs in most computer programming languages.

In Section 2.7, we simply want to introduce predicates, formulas, and the use of quantifiers. First Order Logic deals with predicates and quantification in much the same way as propositional logic deals with propositions. We will focus on how universal and existential quantifiers interact with each other and how they interact with negation. The study of inferences in First Order Logic and other topics that we dealt with in propositional logic will not be covered.

Example 2. For the universal set $\mathbb{N}$ and the usual meanings of the symbols $-$ and $=$, determine whether

$$\exists x (x - 3 = 1)$$

is true.

Solution.   We must find some $x \in \mathbb{N}$ such that $x - 3 = 1$ is true. Choosing $x = 4$ is such a value.   ∎

Clearly, many values for x may make the predicate T in Example 2. The quantifier $\exists x$ just says that we can definitely find one, if the predicate is T. The quantifier does not exclude the possibility of finding more than one value for x that makes the predicate T.

2.7.3   Restricted Quantification

It is understood that the universe U contains every object of concern to the current discussion. In many applications, we want to discuss something more limited. Suppose $V \subseteq U$:

$\forall x \in V \, (P(x))$ read "For all x in V, $P(x)$"

is defined to mean "For all values x in V, the assertion $P(x)$ is true." Then,

$\exists x \in V(P(x))$
              read "For some x in V, $P(x)$"   or   "There exists an x in V such that $P(x)$"

is defined to mean "For some values of x in V, the assertion $P(x)$ is true."

Let $i, j \in \mathbb{N}$ such that $i < j$. A set of $j - i + 1$ consecutive storage locations that can contain the same type of values will be called an **array,** denoted as $A[i \, .. \, j]$, where A is any variable name. The contents of the individual storage locations will be denoted as $A[i], A[i+1], \ldots, A[j]$. For $N \in \mathbb{N}$, both $A[0, .. \, N - 1]$ and $A[1 .. N]$ denote an array with N elements.

Example 3. Let $V = \{1, 2, \ldots, 30\}$, and let $A[1 .. 30]$ be an array such that for each index i between 1 and 30, $A[i] = i \cdot i - 1$. For the elements $A[1], \, A[2], \ldots, A[30]$, write a predicate that says:

(a) Every entry in the array is nonnegative.
(b) The value $A[30]$ is the largest value.
(c) That every element of A is nonzero.

Solution.

(a) $\forall i \in V \, (A[i] \geq 0)$
(b) $\forall i \in V \, (A[i] \leq A[30])$
(c) $\forall i \in V \, (A[i] \neq 0)$ ■

   If $V = \emptyset$, it is understood that $\forall x \in V(S)$ is true and that $\exists x \in V(S)$ is false, no matter what S is.

2.7.4   Nested Quantifiers

The formula $\exists x \, (\exists y (P(x, y)))$ contains **nested quantifiers,** with one quantifier inside the parentheses marking the scope of the other. The obvious parentheses are usually omitted, writing "$\exists x \exists y P(x, y)$." Since the two quantifiers are both $\exists$'s, you may also see "$\exists x, y,$ $P(x, y)$."

   It is important to pay attention to the order of the quantifiers. Suppose $P(x, y)$ is "x received a higher grade on the exam than y did," and suppose U is the set of all students in a class. To show that $\exists x \, (\exists y (P(x, y)))$, we start with the quantifier on the outside: We first look for a student, $x_0 \in U$, to be represented by x. Now, we have a formula $\exists y (P(x_0, y))$. Next, look for an object $y_0 \in U$ to be represented by y. To show that the formula is true, first pick an x_0 who got a higher score, and then pick a y_0 who got a lower score.

   If we meant to choose y first, we would write $\exists y (\exists x (P(x, y)))$. In this case, it does not matter in which order we make the choices. One can see that the formula is true if and only if not all students got the same score. Just pick x_0 to be some student who got a higher score than the score achieved by y. Since not all the scores are the same, it will always be possible to make such choices.

   To show with nested universal quantification that

$$\forall x \, (\forall y \, (x \text{ and } y \text{ drive stick-shift cars and collect baseball cards}))$$

is true, you must show that for all choices of x, and then for all choices of y, both x and y drive stick-shift cars and collect baseball cards. In this case, again, the order of quantifiers does not matter: The above is true if and only if $\forall y \, (\forall x \, (x \text{ and } y \text{ drive stick-shift cars and collect baseball cards}))$ is true.

   When the quantifiers switch between $\forall$ and $\exists$, the order becomes critically important. To show that $\exists x \, (\forall y \, (P(x, y)))$, you first pick a value for x_0 for x from the universal set. Then, you must show that no matter what value y_0 is chosen for y, $P(x_0, y_0)$ is true.

Example 4.   Let $P(x, y)$ denote $x + y = 17$, and let U be the set of integers. Show that

(a) $\forall x \, (\exists y \, (P(x, y)))$ is true.
(b) $\exists y \, (\forall x \, (P(x, y)))$ is false.

Solution.

(a) First, x is specified as *any* integer. Now, you have to pick y to make $P(x, y)$ true. For example, pick $y = 17 - x$.
(b) To show this, you would have to pick a single y_0 so that for all $x \in U$, $x + y_0 = 17$. Since this must be true for *all* possible values of x, it must be true in particular for $x = 0$ and $x = 1$—that is, for $0 + y_0 = 17$ and $1 + y_0 = 17$. Those two cannot both be true.   ■

Generally, if $\exists x\,(\forall y\,(P(x, y)))$ is true, so is $\forall y\,(\exists x\,(P(x, y)))$: If the first is true, then one can pick x_0 where $\forall y\,(P(x_0, y))$. In that case, for each individual y, one can pick that same value x_0 to make $P(x_0, y)$ true, so $\forall y\,(\exists x\,(P(x, y)))$ is true. The converse, however, is false, as Example 4 in Section 2.7.4 shows.

2.7.5   Negation and Quantification

One has to be careful about how negation interacts with quantification—partly because in ordinary human conversation, people are not always very precise.

The formula $\neg(\exists x\,(P(x)))$ says that there does not exist even one x in the universal set that makes $P(x)$ true. This is the same as asserting that every x in the universal set makes P false. Thus,

$$\neg(\exists x\,(P(x))) \text{ is logically equivalent to } \forall x\,(\neg P(x))$$

Analogously, $\neg(\forall x\,(P(x)))$ says that $P(x)$ is not true for all x in the universal set. That is, there is at least one x for which $P(x)$ is false. Thus,

$$\neg(\forall x\,(P(x))) \text{ is logically equivalent to } \exists x\,(\neg P(x))$$

One important result of these rules is that we can always "push negation inward" to be an operator on a predicate rather than on a quantified formula. Often, it becomes easier to understand a formula after the negations are "pushed inward." As an illustration, consider the formulas

$\neg(\forall x\,(\exists y\,(P(x, y))))$	is logically equivalent to	$\exists x\,(\neg\,(\exists y\,((P(x, y)))))$
	which is logically equivalent to	$\exists x\,(\forall y\,(\neg\,P(x, y)))$
$\neg(\exists x\,(\forall y\,(P(x, y))))$	is logically equivalent to	$\forall x\,(\neg\,(\forall y\,(P(x, y))))$
	which is logically equivalent to	$\forall x\,(\exists y\,(\neg\,P(x, y)))$

The resulting formulas with $\neg$ applied only to atomic formulas and using only the connectives $\neg$, $\vee$, and $\wedge$ is said to be in **negation normal form.**

Example 5.   Find formulas in negation normal form equivalent to each of the following formulas. (In cases (c) and (d), the intended universal set is the set of all real numbers, but that does not affect the answers.)

(a) $\neg\forall x \in \mathbb{N}\,(x \text{ is prime} \rightarrow x^2 + 1 \text{ is even})$
(b) $\neg\exists x \in \mathbb{Q}\,(x > 0 \wedge x^3 = 2)$
(c) $\neg\exists x\,(\forall y\,(xy = y))$
(d) $\neg\forall x\,(\forall y\,(x < y \rightarrow (\exists z\,(x < z \wedge z < y))))$

Solution.   We use $\Leftrightarrow$ to mean "is logically equivalent to."

(a) $\neg(\forall x \in \mathbb{N}\,(x \text{ is prime} \rightarrow x^2 + 1 \text{ is even}))$
$\Leftrightarrow \exists x \in \mathbb{N}\,\neg\,(x \text{ is prime} \rightarrow x^2 + 1 \text{ is even})$
$\Leftrightarrow \exists x \in \mathbb{N}\,\neg\,(x \text{ is not a prime} \vee (x^2 + 1 \text{ is even}))$
$\Leftrightarrow \exists x \in \mathbb{N}\,(x \text{ is prime} \wedge \neg(x^2 + 1 \text{ is even}))$

If we go beyond pure logic and use English synonyms, we can further simplify that last expression to $\exists x \in \mathbb{N}\,(x$ is prime $\wedge\,(x^2 + 1$ is odd$))$.

(b)
$$\neg\,(\exists x \in \mathbb{Q}\,(x > 0 \wedge x^3 = 2))$$
$$\Leftrightarrow \forall x \in \mathbb{Q}\,(\neg(x > 0 \wedge x^3 = 2))$$
$$\Leftrightarrow \forall x \in \mathbb{Q}\,(\neg\,(x > 0) \vee \neg\,(x^3 = 2))$$
$$\Leftrightarrow \forall x \in \mathbb{Q}(x \leq 0 \vee x^3 \neq 2)$$

(c)
$$\neg\exists x\,(\forall y(xy = y))$$
$$\Leftrightarrow \forall x\,(\neg\exists y\,(xy = y))$$
$$\Leftrightarrow \forall x(\exists y(\neg(xy = y))$$
$$\Leftrightarrow \forall x(\exists x(xy \neq y))$$

(d)
$$\neg(\forall x\,(\forall y\,(x < y \rightarrow (\exists z\,(x < z) \wedge (z < y))))))$$
$$\Leftrightarrow \exists x(\neg(\forall y(x < y \rightarrow (\exists x((x < z) \wedge (x < y))))))$$
$$\Leftrightarrow \exists x \exists y \neg(x < y \rightarrow (\exists z((x < z) \wedge (z < y))))$$
$$\Leftrightarrow \exists x \exists y(\neg(\neg(x < y) \vee (\exists z((x < z) \wedge (z < y)))))$$
$$\Leftrightarrow \exists x \exists y((x < y) \wedge \neg(\exists z((x < z) \wedge (z < y))))$$
$$\Leftrightarrow \exists x \exists y((x < y) \wedge (\forall z \neg((x < z) \wedge (z < y))))$$
$$\Leftrightarrow \exists x \exists y((x < y) \wedge (\forall z((x \geq z) \vee (z \geq y))))$$

The last step used DeMorgan's Law.

Again, we note that putting formulas into negation normal form often—although not always—makes them more comprehensible.

2.7.6   Quantification with Conjunction and Disjunction

Predicates can be joined by the usual logical operations. Note the English translations of the following formulas:

Formula	English Translation
$\exists x\,(P(x) \wedge Q(x))$	"For some particular choice of x, both $P(x)$ and $Q(x)$ are true."
$\forall x\,(P(x) \wedge Q(x))$	"For *every* choice of x, both $P(x)$ and $Q(x)$ are true."
$\exists x\,(P(x) \vee Q(x))$	"For some particular choice of x, $P(x)$ or $Q(x)$ (or both) is true."
$\forall x\,(P(x) \vee Q(x))$	"For *every* choice of x, $P(x)$ or $Q(x)$ (or both) is true."

Example 6.   For the universal set $\mathbb{N}$, is $\exists x\,((x + 3 = 2) \wedge (x - 2 = 1))$ true?

Solution.   For any x, if $x + 3 = 2$, then x must be -1. If $x - 2 = 1$, then $x = 3$. So, no choice of x makes both true.   ∎

Example 7.   For the universal set $\mathbb{N}$, is $\exists x \, ((x - 3 = 1) \wedge (x > 3))$ true?

Solution.   Since the quantifier is $\exists x$, there need be only one such x for the formula to be true; 4 is such an x. ∎

Example 8.   For the universal set $\mathbb{N}$, which of the following formulas are true?

(a) $\exists x \, ((x + 3 = 2) \vee (x - 2 = 1))$
(b) $\exists x \, ((x \cdot x - 3 = 1) \vee (x > 3))$

Solution.

(a) True; choose $x = -1$. Because $-1 + 3 = 2$ is true, $(-1 + 3 = 2) \vee (-1 - 2 = 1)$ is also true.
(b) Also true; choose $x = 4$. Then, $4 > 3$ is true, so the disjunction is also true. How about $x = 2, 2 \cdot 2 - 3 = 1$. ∎

Example 9.   In universe $\mathbb{N}$, which of the following formulas are true?

(a) $\forall x \, ((x^2 - 2x + 1 = 0) \vee (x \geq x))$
(b) $\forall x \, ((x \leq 3) \vee (x > 3))$

Solution.   This solution is left as an exercise for the reader. ∎

In Table 2.10 we summarize the relationship between quantification and $\wedge$ and $\vee$. Since all the logical operators can be expressed in terms of $\neg$ and $\wedge$ or $\neg$ and $\vee$ (see Exercise 1 in Section 2.9.4), this table should provide a guide to answering questions about the relationships between other logical operators and quantification. Below, $\phi \Rightarrow \psi$ stands for "ϕ logically implies ψ," and $\phi \Leftrightarrow \psi$ stands for "ϕ is logically equivalent to ψ."

Table 2.10   Logical Relations for Quantified Formulas in One Variable

$\exists x \, (P(x) \wedge Q(x))$	$\Rightarrow (\exists x \, P(x)) \wedge (\exists x \, Q(x))$
$\exists x \, (P(x) \vee Q(x))$	$\Leftrightarrow (\exists x \, P(x)) \vee (\exists x \, Q(x))$
$\forall x \, (P(x) \wedge Q(x))$	$\Leftrightarrow (\forall x \, P(x)) \wedge (\forall x \, Q(x))$
$(\forall x \, P(x)) \vee (\forall x \, Q(x))$	$\Rightarrow \forall x \, (P(x) \vee Q(x))$

The formulas $\exists x \, P(x) \wedge \exists y \, P(y)$ and $\exists x \, \exists y \, (P(x) \wedge P(y))$, at first sight, both seem to say that there are (at least) two objects satisfying predicate P. This, however, is not true. There is nothing in the formula saying that $x \neq y$—that is, that x and y refer to different objects. So, both formulas say there is (at least) one object satisfying P. To say there are two different objects satisfying P, one would have to say they're different—for example,

$$\exists x \, \exists y \, (P(x) \wedge P(y) \wedge x \neq y)$$

Example 10.   For an array of 20 entries with integer entries, write a predicate that says all the elements are distinct.

Solution.   Let $V = \{1, 2, \ldots, 20\}$ represent the indexes for the entries into an array $A[1..20]$. Now,

$$\forall m \in V \, (\forall n \in V \, ((m \neq n) \rightarrow (A[m] \neq A[n])))$$

says that all the elements are distinct. In this case, another predicate is $\forall m \in V \, (\forall n \in V \, ((m < n) \rightarrow (A[m] \neq A[n])))$. We leave it for the reader to explain why. ∎

Note that in two of the lines in Table 2.10 we said "⇔" and that in two we said just "⇒." We leave it for the reader to find examples of the following:

$$(\exists x \, P(x)) \wedge (\exists x \, Q(x)) \wedge \neg \exists x \, (P(x) \wedge Q(x))$$

and
$$\forall x \, (P(x) \vee Q(x)) \wedge \neg(\forall x \, P(x)) \vee (\forall x \, Q(x))$$

2.7.7 Application: Loop Invariant Assertions

One of the most difficult aspects of computer programming is establishing whether programs produce the correct output. In principle, there is no way to establish the correctness of all correct programs. (This was proved by Alan Turing.) Tools for establishing correctness, however, do exist for many programs.

The simplest method for checking a program is to test it: Run it on some sample values, and check whether it produces the correct answers. Testing is often an effective method for showing that a program is incorrect, but unfortunately, one cannot normally check all possible inputs—nor even a significant fraction of the possible inputs. Therefore, one cannot check that a program is correct.

Another method that is often useful is to write a mathematical proof of program correctness. One of the difficulties in this case is finding tools for proving that any loops accomplish what they are supposed to.

A somewhat similar problem is encountered in making it obvious to someone else who is reading the program that the program works correctly. Many algorithms use tricks that vary from not quite obvious to totally obscure. What is an easy read and short way to present the trick and explain why the algorithm works? For example, how can one explain what a loop is accomplishing?

One method that is often useful employs **loop invariant assertions.** We will explain these in terms of a familiar algorithm, (one version of the) *BubbleSort*. We choose *BubbleSort* not because it is a good sorting algorithm—for most purposes, it definitely is not—but because it is short and easy to understand.

Algorithm: BubbleSort

INPUT: An array $A[0..N-1]$ of N integers
OUTPUT: The same array, with its elements sorted into nondecreasing order

for *limit* $= N - 2$ down to 0
    for *position* $= 0$ up to *limit*
        if $(A[position] > A[position + 1])$
            then swap the values of $A[position]$ and $A[position + 1]$

A reason this algorithm works is that after k passes through the outer loop, the largest k elements have reached the last k positions in the array—and in the correct order as well.

A formula that states this property is intuitively easy to understand, but it is not so easy to state formally. Part of the formula is that after k passes through the outer loop, the last k elements (in positions $N - k, N - k + 1, \ldots, N - 1$) are in increasing order and are larger than or equal to all elements of the array that occur in positions $0, 1, \ldots, k - 1$. We can state more, since the elements in positions $k, k + 1, \ldots, j - 1$ also have values that are less than the value of the element at position j. Thus, for any position j among the last k positions, the value in each position i where $0 \leq i \leq j$ is less than or equal to the value in position j.

Let *Ind* denote the set $\{0, 1, \ldots, N - 1\}$ of legal indices for array A:

$$\forall i \in Ind \forall j \in Ind((0 \leq i) \wedge (i \leq j) \wedge (j \geq (N - k)) \rightarrow (A[i] \leq A[j])$$

For $k = 0$, this says that

$$\forall i \in Ind \forall j \in Ind(i \leq j \wedge j \geq N \rightarrow A[i] \leq A[j])$$

which is a true predicate, because $j \geq N$ is false, making this an implication with hypothesis *FALSE*. When $k = N - 1$, we are claiming that all the elements are in increasing order. The predicate is

$$\forall i \in Ind \, \forall j \in Ind(i \leq j \wedge j \geq 0 \rightarrow A[i] \leq A[j])$$

The reader should verify that this does mean that the elements of the array are in increasing order. This predicate can be put into the code as a comment called a loop invariant assertion, as seen in the Outer Loop Invariant algorithm. (We now go back to informal usage and use both the $\leq$ and the $<$ symbol in the formula.)

Algorithm: Outer Loop Invariant

INPUT: An array $A[0 .. N - 1]$ of N integers
OUTPUT: The same array, with its elements sorted into nondecreasing order

for *limit* $= N - 2$ down to 0
   /* loop invariant for *limit* loop
      $\forall i \in Ind \forall j \in Ind((i \leq j \, \wedge \, j > limit + 1) \rightarrow (A[i] \leq A[j]))$
   */

for *position* $= 0$ up to *limit*
   if $(A[position] > A[position + 1])$ then
      swap the values

When *limit* $= N - 2$, $j > N - 1$ is false, because $0 \leq j \leq N - 1$. Therefore, the implication is *TRUE*. When *limit* $= -1$ and the loop terminates, the implication says that, for $i \leq j$ and $j > 0$, $A[i] \leq A[j]$—that is, that the elements of A are in increasing order. The **loop invariant** here is a formula that is supposed to be true at the beginning of each pass through the loop as well as true after the last pass, when control returns—(here with $limit = -1$) to test that the loop is finished. The accepted formal language for loop invariant assertions uses quantified formulas.

2.8 Exercises

Let $U = \{1, 2, 3, 4\}$ be the universal set for Exercises 1 through 4.

1. Rewrite $(\forall x \in U)P(x)$ as a conjunction that uses no quantifiers.
2. Rewrite $(\exists x \in U)P(x)$ as a disjunction that uses no quantifiers.
3. Rewrite $\neg(\forall x \in U)P(x)$ as a conjunction that uses no quantifiers.
4. Rewrite $\neg(\exists x)P(x)$ as a conjunction that uses no quantifiers.
5. For the following predicates with universal set $\mathbb{R}$, state the meaning of the predicate in a sentence. If it is false, give an example to show why. (*Example:* $\forall x(\exists y(x < y))$ says "for every real number, there is a bigger number." This is true.)

   (a) $\forall x(\exists y(x \neq 0 \rightarrow xy = a))$
   (b) $\exists y(\forall x(x \neq 0 \rightarrow xy = 1))$
   (c) $\exists x(\forall y(y \leq x))$
   (d) $\forall x(\exists y(x + y = x))$
   (e) $\exists y(\forall x(x + y = x))$
   (f) $\forall x(\forall y(\exists z(x < z \land z < y)))$
   (g) $\forall x(\forall y(x \neq y \rightarrow \exists z((x < z \land z < y) \lor (x > z \land z > y))))$
   (h) $\forall x(\forall y(\forall z((x > y \land y > z) \rightarrow x > z)))$

6. For each of the following formulas write a formula ϕ (using quantifiers) expressing the formula, find a formula in negation normal form equivalent to $\neg\phi$, and express the meaning of the negation in words.

   (a) For every x and for every y, $x + y = y + x$.
   (b) Every number x has a square root. (Do *not* use the square root symbol; use only multiplication.)
   (c) For some y, $2x^2 + 1$ is always greater than $x^2 y$. (*Hint:* In this example, "always" suggests a universal quantifier.)
   (d) For some x and y, $x < y$, and $x^3 - x > y^3 - y$.
   (e) For every x and y, there is a z where $2z = x + y$.
   (f) For every x and y, if $x^3 + x - 2 = y^3 + y - 2$, then $x = y$.

7. For each quantified formula that follows: find a universe U and predicates A and B in which the formula is true and U, A and B in which it is false.

   (a) $\forall x(((A(x) \lor B(x)) \land \neg(A(x) \land B(x)))$
   (b) $\forall x \forall y(P(x, y) \rightarrow P(y, x))$
   (c) $\forall x(P(x) \rightarrow \exists y Q(x, y))$
   (d) $\exists x(A(x) \land \forall y B(x, y))$
   (e) $\forall x A(x) \rightarrow (\forall x B(x) \rightarrow (\forall x(A(x) \rightarrow B(x))))$

8. For the following formulas, let the universe be $\mathbb{R}$. Translate each of the following sentences into a formula (using quantifiers):

   (a) There is a smallest number.
   (b) Every positive number has a square root. (Do *not* use the square root symbol; use only multiplication.)
   (c) Every positive number has a positive square root. (Again, do *not* use the square root symbol; use only multiplication.)

9. For the following formulas, let the universe be $\mathbb{R}$. Translate each of the following sentences into a formula (using quantifiers):

   (a) There is no largest number.
   (b) There is no smallest positive number.
   (c) Between any two distinct numbers, there is a third number not equal to either of them.

10. Let U be the set of all problems on a comprehensive list of problems in science. Define four predicates over U by:

    $P(x)$: x is a mathematics problem
    $Q(x)$: x is difficult (according to some well-defined criterion: it does not matter for us what the criterion is)
    $R(x)$: x is easy (according to some well-defined criterion)
    $S(x)$: x is unsolvable (if you do not know what "unsolvable" means, do not worry about it here)

    Translate into English sentences each of the following formulas:

    (a) $\forall x\, P(x)$
    (b) $\exists x\, Q(x)$
    (c) $\forall x (Q(x) \vee R(x))$
    (d) $\forall x (S(x) \rightarrow P(x))$
    (e) $\exists x (S(x) \wedge \neg P(x))$
    (f) $\neg(\forall x (\neg R(x) \vee S(x)))$
    (g) $\forall x (P(x) \rightarrow (Q(x) \leftrightarrow \neg R(x)))$
    (h) $\forall x \neg S(x)$
    (i) $\forall x (P(x) \rightarrow \neg S(x))$
    (j) $\forall x (P(x) \rightarrow (R(x) \vee S(x)))$
    (k) $\exists x (\neg Q(x) \wedge \neg R(x))$
    (l) $\exists x (R(x) \wedge S(x))$
    (m) $\forall x (Q(x) \leftrightarrow \neg R(x))$

11. Let the universe U be the set of all human beings living in the year 2001, and translate the following English sentences into quantified formulas. Let $P(x)$ stand for "x is young," $Q(x)$ for "x is female," and $R(x)$ for "x is an athlete."

    (a) "All athletes are young."
    (b) "Not all young people are athletes."
    (c) "All young people are not athletes." (*Warning:* In *informal* English, this sentence has two quite different meanings. One is "more grammatically correct" than the other, however, and that is the one we're asking for.)
    (d) "Some young people are not athletes."

    (e) "Some athletes are young females."

    (f) "All athletes are young males."

    (g) "Some athletes are female and are not young."

    (h) "Some young females are not athletes."

    (i) "All young females are athletes."

    (j) "Some athletes are not young."

    (k) "No young people are athletes."

    (l) "All athletes are either female or are young."

    (m) "If all athletes are female, then all athletes are young; otherwise, no athletes are young."

12. Give an example of a universal set U and predicates P and Q such that $(\forall x\, P(x)) \rightarrow (\forall x\, Q(x))$ is true but $\forall x (P(x) \rightarrow Q(x))$ is false.

13. Translate each of the following quantified formulas into an English sentence where the universal set is $\mathbb{R}$. Label each as true or false.

    (a) $\forall x (\exists y (xy = x))$

    (b) $\forall y (\exists x (xy = x))$

    (c) $\forall x (\exists y (xy = 1))$

    (d) $\exists y (\forall x \neq 0 (xy = 1))$

    (e) $\exists x (\forall y (xy = x))$

    (f) $(\forall x (x \neq 0 \rightarrow \exists y (xy = 1))$

14. Write a formula "saying" that at least four distinct objects satisfy predicate P.

15. For any two integers m and n, we say m divides n if there is an integer k such that $n = mk$. (Many programming languages give easy ways to say that, such as $n \% m == 0$ or n div $m = 0$.) Define $Div(m, n)$ to be m divides n. Translate each of the following propositions and quantified formulas into a clear English sentence. Label each as being true or false, with the universe as the set $\mathbb{Z}$.

    (a) $Div(5, 7)$

    (b) $Div(4, 16)$

    (c) $Div(16, 4)$

    (d) $Div(-8, 0)$

    (e) $\forall m (\forall n (Div(m, n)))$

    (f) $\forall n (Div(1, n))$

    (g) $\forall m (Div(m, 0))$

    (h) $\forall m (\forall n (Div(m, n) \rightarrow Div(n, m)))$

    (i) $\forall m (\forall n (\forall p ((Div(m, n) \wedge Div(n, p)) \rightarrow Div(m, p))))$

    (j) $\forall m (\forall n ((Div(m, n) \wedge Div(n, m)) \rightarrow m = n))$

16. Find a formula in negation normal form equivalent to the negation of

$$\exists x \forall y \forall z (P(x, y, z)).$$

17. Find formulas in negation normal form equivalent to the negations of each of the following:

    (a) $\forall x (P(x) \vee Q(x))$

    (b) $\forall x (\forall y (P(x, y) \rightarrow Q(x, y)))$

    (c) $\forall x ((\exists y\, P(x, y)) \rightarrow Q(x, y))$

    (d) $\forall x (((\exists y (P(x, y)) \rightarrow Q(x, y)) \wedge \exists z\, R(x, z)))$

(e) $\exists x((P(x) \lor Q(x)) \to R(x))$

(f) $\exists x((P(x) \to Q(x)) \land R(x))$

18. Find a formula in negation normal form equivalent to the negation of

$$\forall x \exists y((P(x, y) \land Q(x, y)) \to R(x, y))$$

19. Give a universal set U and interpretations to predicates A, B, P, and Q so that each of the following quantified formulas is false:

(a) $(\exists x A(x) \land \exists x B(x)) \to (\exists x(A(x) \land B(x)))$

(b) $(\forall x \exists y P(x, y)) \to (\exists x(\forall y P(x, y)))$

(c) $(\forall x(P(x) \to Q(x))) \to ((\exists x P(x)) \to (\forall x)Q(x))$

(d) $\forall x(\neg A(x)) \leftrightarrow \neg(\forall x A(x))$

20. To negate an expression with a single quantifier, we can replace it with the other quantifier and negate the formula inside. This generalizes to an arbitrary string of quantifiers. For instance,

$$\neg \forall x \exists y \exists z \forall t\, P(x, y, z, t)$$

is logically equivalent to

$$\exists x \forall y \forall z \exists t\, (\neg P(x, y, z, t))$$

Prove this generalization by induction.

21. Given an array *Values* with n elements

$$Values[0], Values[1], \ldots, Values[n-1]$$

each containing a real number, the following algorithm finds the sum of all the positive values in *Values*. Write an invariant for the loop.

$$rollingSum = 0$$
$$\text{for } i = \phi, 2, \ldots, n-1$$
$$\quad \text{if } Values[i] > 0$$
$$\quad\quad rollingSum = rollingSum + Values[i]$$
$$\text{Output } rollingSum$$

22. *Challenge:* A much more sophisticated sorting algorithm is the *MergeSort* algorithm. It comes in various versions; we do one here. The algorithm involves copying the list back and forth between two arrays, the input array A and an extra array B, so it takes a lot of extra space.

   Our version is not "optimized." We have attempted to keep it relatively simple to make it as easily understood as possible. Moreover, to simplify things for this exercise, we assume that the size of the input array is a power of 2; relatively easy adjustments would make it work for arrays of arbitrary sizes.

Algorithm: MergeSort

INPUT: An array $A[0..2^N - 1]$ of 2^N integers
OUTPUT: The same array, with its elements sorted into nondecreasing order

for $t = 1$ to 2^{N-1}
    $size = 2^t$
    for $position = 1$ to $2^N - 1$
        $B[position] = A[position]$
    $lo_1 = 0$
    while $lo_1 < 2^N$
        $hi_1 = lo_1 + size - 1$
        $lo_2 = hi_1 + 1$
        $hi_2 = lo_2 + size - 1$
        $position_1 = lo_1$
        $position_2 = lo_2$
        $position_3 = lo_1$
        while ($position_1 \leq hi_1$ and $position_2 \leq hi_2$)
            if $B[position_1] < B[position_2]$
            then $A[position_3] = B[position_1]$
                $position_1 = position_1 + 1$
            else $A[position_3] = B[position_2]$
                $position_2 = position_2 + 1$
            $position_3 = position_3 + 1$
        while ($position_1 \leq hi_1$)
            $A[position_3] = B[position_1]$
            $position_1 = position_1 + 1$
            $position_3 = position_3 + 1$
        while ($position_2 \leq hi_2$)
            $A[position_3] = B[position_2]$
            $position_2 = position_2 + 1$
            $position_3 = position_3 + 1$
        $lo_1 = lo_1 + 2 \cdot size$

Determine what each part of the program does (first experiment with some sample test lists), and write loop invariants for each loop that clarify why the algorithm works.

2.9   Chapter Review

Propositions are the initial focus of the chapter. After defining propositions, we introduce common operations to make formulas from propositions. The idea of a proposition being true is introduced. We can verify whether a formula is true for a particular set of truth values for its propositions using an expression tree. To find out if a formula is true for all possible truth values, we use truth tables. The notions of tautologies, contradictions, satisfiable propositions, and logically equivalent propositions give a fuller understanding of the propositional logic. Both CNFs and DNFs give a method for representing any formula using a standard format from which information is easier to determine. The last section deals with predicates and quantification. We define predicates as natural generalizations of propositions and formulas. The interaction of predicates and quantification is explored, as is the interaction of quantification with the operations that are defined on propositions.

Throughout the chapter, but independent of the core material about propositional logic, is an introduction to boolean or combinatorial circuits. First, the correspondence between logical formulas and a boolean circuit composed of gates is introduced. After showing the correlations between boolean algebras and combinatorial circuits, the results about boolean algebras become a tool for simplifying circuits. Finally, CNFs and DNFs are used to find standard representations of combinatorial circuits.

2.9.1   Terms and Theorems

2.1   Summary

TERMS

AND gate	expression tree	proposition letter
base cases	Expression Tree for a	propositional connective
biconditional	Formula	propositional constant
boolean circuit	*FALSE* (F)	(T, F)
boolean value	formula	propositional logic
closure rules	gate	semantics
combinatorial circuit	hypothesis	subformula
combinatorial network	implication	syntax
conclusion	inductive definition	*TRUE* (T)
conditional	mean	truth table
conjunct	logical value	well-formed formula
conjunction	negation	(wff)
disjunct	NOT gate	
disjunction	OR gate	
equivalence	proposition	

THEOREMS

Principle of Induction on Formulas

2.3 and 2.4   Summary

TERMS

alphabetic substitution	interpretation	semantics
alphabetic variant	logically equivalent	Sheffer stroke
complementation	logically implies	tautologically equivalent
contradiction	logically valid	tautologically implies
equivalent	meaning	tautology
exclusive or	satisfiable	true in
false in	satisfies	unsatisfiable

THEOREMS

First Substitution Principle   Second Substitution Principle

2.5 and 2.6   Summary

TERMS

2-CNF	excess literals	resolution refutation
3-CNF	k-DNF	resolution rule
3-satisfiability problem	k-term	resolvant
clause	literal	sound
complete	negative literal	term
conjunctive normal form	$\mathcal{NP}$-complete	
(CNF)	positive literal	
disjunctive normal form	resolution	
(DNF)	resolution derivation	

THEOREMS

Every Formula Is Logically Equivalent to   Every Formula Is Logically Equivalent to
a Formula in CNF   a Formula in DNF

ALGORITHMS

BubbleSort
MergeSort
Outer Loop Variant

2.7   Summary

TERMS

array	loop invariant	quantification
atomic formula	loop invariant assertion	scope
binary	n-ary	universal quantification
constant	negation normal form	variable
existential quantification	nested quantifiers	variable symbols
formula	predicate	

2.9.2   Starting to Review

1. Which of the following are propositions?

> i: "The moon is visible."
> ii: "The property tax rate will increase next year."
> iii: "No one under 18 may buy cigarettes."
> iv: "Please help me with the assignment."

(a) i and iii

(b) i and ii

(c) ii and iii

(d) All of the above

2. Write in symbolic form the statement "Claudia will sail in the regatta if the crew is ready and the weather is fair."

3. Write the converse, inverse, and contrapositive of the statement "If Sally finishes her work, she will go to the basketball game."

4. What inference rule applies to the following?

   Joe wrote a program in C, or George wrote a program in Java. If Joe wrote a program in C, then the problem was solved. If George wrote a program in Java, then the problem was solved.

   (a) Contrapositive
   (b) Proof by contradiction
   (c) Proof by cases
   (d) None of the above

5. What is the truth value that will be computed by the formula represented by the expression tree shown if $I(p) = T$, $I(q) = F$, $I(r) = T$, and $I(s) = F$ in an interpretation I?

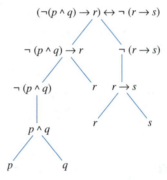

6. What is the value of the formula represented by the expression tree in Exercise 5 given the interpretation $I(p) = T$, $I(q) = F$, $I(r) = T$, and $I(s) = T$.

7. Write the following condition in an *if...then* with the negations incorporated into the conditions themselves:

$$\text{If NOT}\ \big((x < 3)\ \text{OR}\ (y > 2)\big),\ \text{then}$$

8. Construct a truth table for the proposition $\neg(p \wedge q)$.

9. Using the conjunctive normal form, identify values for which the statement

$$\neg(\neg(p \vee q) \wedge (\neg p \vee q))$$

   is true.

10. Find the DNF for the statement

$$((\neg p \wedge q) \vee r) \wedge (\neg q \vee \neg r)$$

2.9.3   Review Questions

1. Construct a truth table for the statement $\neg(p \vee q) \vee \neg(p \wedge q)$.
2. Construct a truth table for the statement $\neg(p \wedge q) \wedge (p \vee \neg q)$.
3. For which truth values does the statement $\neg(p \vee \neg q)$ have a truth value *TRUE*?

4. Form a truth table for the proposition $p \vee \neg(p \wedge q)$.

5. Use the substitution rule with $p \to q$ for p, and prove that the result is a tautology for

$$\neg q \to (q \to p)$$

6. Prove the following identities for a boolean algebra:

   (a) $(\neg p \vee q) \wedge (p \vee \neg q) = p \wedge q \vee (\neg p \wedge \neg q)$
   (b) $\neg p \vee (q \wedge r) \vee (p \vee q) \wedge (\neg p \vee r) = \neg p \vee r$
   (c) $\neg(\neg(p \vee q) \wedge \neg(q \vee r)) \vee (q \wedge r) = p \vee q$

7. Draw combinatorial circuits that realize the following formulas:

   (a) $(p \wedge q) \vee (q \wedge r) \vee (p \wedge \neg r)$
   (b) $\neg((p \wedge q) \vee p) \vee (p \wedge q)$

2.9.4   Using Discrete Mathematics in Computer Science

1. We built formulas with the logical operators $\wedge$, $\vee$, $\neg$, $\to$, and $\leftrightarrow$ and the constants T and F. In designing circuits, we described gates for only three connectives: $\wedge$, $\vee$, and $\neg$. Computer hardware designers might want to make as few kinds of gates as possible. Do they really need a $\to$ gate? (The answer turns out to be "no," but how do you know that?) Could they get along with fewer than three types of gates? A set of logical operators is called *complete* if every well-formed formula of propositional logic is equivalent to a well-formed formula using connectives from the set.

   (a) Find a formula equivalent to $a \to (b \wedge c \wedge d)$ using only the connectives $\neg$ and $\wedge$ (and not the constants T and F). Find the shortest such formula; does it have more or fewer symbols than the formula $a \to (b \wedge c \wedge d)$?
   (b) Show that the set $\{\neg, \wedge\}$ of operators is complete.
   (c) Find a formula equivalent to $a \to (b \wedge c \wedge d)$ using only the connectives $\neg$ and $\to$ (and not the constants TRUE and FALSE). Find the shortest such formula; does it have more or fewer symbols than the formula $a \to (b \wedge c \wedge d)$?
   (d) Show that the set $\{\neg, \to\}$ of operators is complete.
   (e) Find a formula equivalent to $a \to (b \wedge c \wedge d)$ using only the connective $\to$ and the constant FALSE. Find the shortest such formula; does it have more or fewer symbols than the formula $a \to (b \wedge c \wedge d)$?
   (f) Show that the set $\{FALSE, \to\}$ is complete.

2. See the definition of "complete set of operators" in Exercise 1. This problem shows that the engineers need build only one type of gate.

   (a) *NAND* has the truth table

p	q	$NAND(p, q)$
T	T	F
T	F	T
F	T	T
F	F	T

   Show that the set $\{NAND\}$ is a complete set of operators.

(b) Find a formula equivalent to $a \rightarrow (b \wedge c \wedge d)$ using only the connectives *NAND* (and not the constants TRUE and FALSE). Find the shortest such formula; does it have more or fewer symbols than the formula $a \rightarrow (b \wedge c \wedge d)$?

(c) *NOR* has the truth table

p	q	$NOR(p, q)$
T	T	F
T	F	F
F	T	F
F	F	T

Show that the set $\{NOR\}$ of operators is complete.

(d) Find a formula equivalent to $a \rightarrow (b \wedge c \wedge d)$ using only the connectives *NOR* (and not the constants TRUE and FALSE). Find the shortest such formula; does it have more or fewer symbols than the formula $a \rightarrow (b \wedge c \wedge d)$?

The *NAND* operator is often called the *Sheffer stroke* and is denoted as $p|q$. The *NOR* operator is often called the *Pierce arrow* and is denoted as $p \downarrow q$.

3. The connective *if-then-else* is defined by the following truth table:

p	q	r	if p then q else r
T	T	T	T
T	T	F	T
T	F	T	F
T	F	F	F
F	T	T	T
F	T	F	F
F	F	T	T
F	F	F	F

This connective is key in binary decision diagrams (*BDDs*), which provide one standard way for manipulation of propositional formulas in computer programs. For example, BDDs have been widely used by computer-chip designers in showing that the circuits in the chips they design match the specifications for those chips. (In BDD language, the connective is often called just *ITE*.)

(a) Find a formula equivalent to

> if a
> then
> > if b then c else d
> else
> > if e then d else c

using only the connectives $\wedge$, $\vee$, and $\neg$.

(b) Find a formula equivalent to

$$\text{if } a$$
$$\text{then}$$
$$\qquad \text{if } b \text{ then } c \text{ else } d$$
$$\text{else}$$
$$\qquad \text{if } e \text{ then } d \text{ else } c$$

in CNF.

(c) Find a formula equivalent to $\neg a$ using only the *if-then-else* connective and constants T and F.

(d) Find a formula equivalent to $(a \vee b \vee c) \wedge (\neg a \vee \neg b \vee d) \wedge (\neg c \vee \neg d)$ using only the *if-then-else* connective and constants T and F.

4. Find a DNF for the condition that there are an even number of 1's in the three binary strings p, q, and r. Draw a combinatorial network to represent the DNF. Can you simplify the combinatorial circuit using the properties of a boolean algebra?

5. Find a DNF for the condition that there are an odd number of 1's in the three binary strings p, q, and r. Draw a combinatorial network to represent the disjunctive normal form. Can you simplify the combinatorial circuit using the properties of a boolean algebra?

6. An especially simple class of CNF formulas are those built from *Horn clauses*. A Horn clause is a clause containing, at most, one positive literal. (A *pure Horn clause* is a clause containing exactly one positive literal.) Horn clauses form the basis for the computer language Prolog, which allows the programmer to input a set of requirements (specified in formal logic) and to ask the computer to find how to satisfy them all (if possible)—as opposed to the user's having to write out the case analysis.

(a) Using the atomic formulas

$$a = \text{``Tweety is a penguin.''}$$
$$b = \text{``Opus is a penguin.''}$$
$$c = \text{``Phoenix is a penguin.''}$$
$$d = \text{``Elvis lives!''}$$

express "If Tweety is a penguin and Opus is a penguin, and Phoenix is a penguin, then Elvis lives" as a Horn clause.

(b) Find a set of Horn clauses logically equivalent to $(a \wedge b \wedge c \rightarrow d \vee e) \wedge (\neg a \vee \neg e)$. Find the shortest such set of clauses.

(c) Find all satisfying truth assignments for the following set of Horn clauses:

$$\{p_1, \neg p_1 \vee p_2, \neg p_1 \vee \neg p_2 \vee p_3, \neg p_1 \vee \neg p_2 \vee p_4, \neg p_3 \vee \neg p_4 \vee p_5\}$$

Now, show that the following set of Horn clauses is not satisfiable:

$$\{p_1, \neg p_1 \vee p_2, \neg p_1 \vee \neg p_2 \vee p_3, \neg p_1 \vee \neg p_2 \vee p_4, \neg p_3 \vee \neg p_4 \vee p_5, \neg p_3 \vee \neg p_5\}$$

(d) Find all satisfying truth assignments for the following set of Horn clauses:

$$\{p_1, \neg\, p_1 \vee p_2, \neg p_1 \vee \neg p_2 \vee p_3, \neg p_1 \vee \neg p_2 \vee p_4, \neg p_3 \vee \neg p_4 \vee p_5,$$
$$\neg\, p_6 \vee p_7, \neg p_1 \vee \neg p_7 \vee p_6\}$$

Next, for each satisfying truth assignment I, let T_I be the set of truth variables assigned the value *TRUE* by I. Compare the sets T_I for the truth assignments above by $\subseteq$.

Now, show that the following set of Horn clauses is satisfiable:

$$\{p_1, \neg\, p_1 \vee p_2, \neg p_1 \vee \neg p_2 \vee p_3, \neg p_1 \vee \neg p_2 \vee p_4, \neg p_3 \vee \neg p_4 \vee p_5,$$
$$\neg\, p_6 \vee p_7, \neg p_1 \vee \neg p_7 \vee p_6, \neg p_1 \vee \neg p_6, \neg p_2 \vee \neg p_4 \vee \neg p_7\}$$

(e) *Challenge:* Prove that if ϕ is a Horn clause and if I_1 and I_2 are interpretations satisfying ϕ, then the following interpretation $I_\wedge$ also satisfies ϕ:

$$I_\wedge(x) = \begin{cases} T \text{ if } I_1(x) = T \text{ and } I_2(x) = T \\ F \text{ otherwise} \end{cases}$$

Using this, show that $p \vee q$ is not logically equivalent to (the conjunction of) any set of Horn clauses.

(f) *Challenge:* Write pseudocode for a relatively fast algorithm to determine whether a set of Horn clauses is satisfiable. Include arguments to show that (i) your algorithm returns the correct answer and (ii) your algorithm is reasonably fast (in general, *much* faster than writing the truth table for the set of Horn clauses).

7. Determine if the CNF

$$(x \vee y \vee \neg z \vee w \vee u \vee \neg v) \wedge (\neg x \vee \neg y \vee z \vee \neg w \vee u \vee v) \wedge$$
$$(x \vee \neg y \vee \neg z \vee w \vee u \vee \neg v) \wedge (x \vee \neg y)$$

or the CNF

$$(p \vee r \vee v) \wedge (\neg p \vee r \vee v) \wedge (p \vee \neg r \vee v) \wedge (\neg p \vee \neg r \vee \neg v) \wedge$$
$$(p \vee \neg r \vee \neg v) \wedge (\neg p \vee \neg r \vee \neg v) \wedge (p \vee r \vee \neg v) \wedge (\neg p \vee r \vee \neg v)$$

is satisfiable. This exercise shows that the satisfiability problem can be solved if the satisfiability problem can be solved for a CNF. The CNF satisfiability problem was the first $\mathcal{NP}$-complete problem.

8. A *first-order Horn clause* is a formula such as

$$\forall x\, \forall y\, (Loves(x, y) \vee \neg EatsGarlic(x) \vee \neg EatsGarlic(y))$$

Inside the parentheses is a disjunction of atomic formulas ($Loves(x, y)$) and negated atomic formulas ($\neg EatsGarlic(x)$). All the variables are universally quantified outside the parentheses.

Using the predicates

Trained (x, j): x is trained to do job j.
Experienced (x, j): x is experienced at job j.
Prefers (x, j_1, j_2): x prefers job j_1 to job j_2.
Hire (x, j): hire x to do job j.

State the following with sets of first-order Horn clauses:

(a) If Britney and Aaron are both trained and experienced in marketing and accounting, and if Britney prefers accounting to marketing and Aaron prefers marketing to accounting, then hire Britney to do accounting.

(b) If Harry, Hermione, and Frodo are all experienced in potions and each of them prefers potions to at least one other job, then hire them all for potions.

9. Given an array *Names* with n elements,

$$Names[0], Names[1], \ldots, Names[n-1]$$

each containing a surname (family name), the following algorithm finds the largest name (in alphabetical order). Write an invariant for the loop.

$$temp = Names[0]$$
$$\text{for } i = 1, 2, \ldots, n-1$$
$$\quad \text{if } Names[i] > temp$$
$$\quad\quad temp = Names[i]$$
$$\text{Output } temp$$

10. *Challenge:* Look up Hoare's quicksort algorithm. Write loop invariant assertions that make the logic of quicksort easy to understand. You may also want *preconditions* and *postconditions*. A precondition is a formula that the programmer assumes will be true when an algorithm is invoked (called); the programmer announces that if the precondition is not true, then the algorithm probably will not do what it is supposed to do. A postcondition is a formula expressing something that is supposed to be true after the algorithm finishes—assuming the preconditions are satisfied, of course.

CHAPTER **3**

Relations

Human language has many words and phrases to describe relationships between or among objects. It may be that for two people, A and B, that A *is a parent of* B, that A *is an ancestor of* B, that A *is taller than* B, or that A *is in front of* B. In algebra, it may be that the value of variable x *is less than* the value of variable y. In geometry, it may be that one point *lies between* two other points on a line. In set theory, it may be that a set X *is a subset of* a set Y or that X *is disjoint from* Y. At a particular moment while a computer program is running, it may be that the value of x *is less than* the value of y. All these notions are special instances of a relation.

This chapter introduces the concept of a relation to formalize the familiar notion of a relationship between or among objects. Relations provide a way of representing relationships like the ones just described, so that they can be stored, studied, and reasoned about. In this chapter, we first provide an introduction to relations, the important properties of relations, and the fundamental operations on relations. We next deal with equivalence relations, a generalization of the notion of equality, and then move on to ordering relations. These relations generalize the ordering relations on $\mathbb{R}$ ($<, >, \leq, \geq, =$). Searching and sorting operations are based on these relations. Finally, we show how the ideas in this chapter are applied in a relational database.

3.1   Binary Relations

Most card games are played with a standard **deck of 52 cards.** The deck is divided into four groups, or *suits,* called Clubs (♣), Diamonds (◇), Hearts (♡), and Spades (♠). Each suit has 13 cards, ordered in increasing order of value: 2, 3, 4, 5, 6, 7, 8, 9, 10, Jack, Queen, King, and Ace. The 2 card has the lowest value, and the Ace card has the highest value. In this ordering, we say that one card has a higher value than, or is higher than, another card if, ignoring their suits, the value of the first card occurs after the value of the second in this ordering. To focus on the ideas presented in this chapter while keeping matters simple, many of the examples that follow will use a deck with only six cards. This set of cards, called *SpecialDeck,* consists of the 10, Jack, and Queen of Hearts as well as the 10, Jack, and King of Clubs. The values of these cards are ordered as described. The *SpecialDeck* has its elements shown in Table 3.1.

Table 3.1
SpecialDeck of
Cards

SpecialDeck
10 of Clubs
Jack of Clubs
King of Clubs
10 of Hearts
Jack of Hearts
Queen of Hearts

Much of the information about suits and card values is irrelevant to many card games. Often, only two properties are important: whether two cards are in the same suit, and whether one card is higher than another. Everything else, such as the names of the suits, the names of the cards, and perhaps even the number of cards per suit, is unimportant. For example, a person would immediately be able to translate the rules of many games to a deck with five suits called *red, yellow, blue, green,* and *black,* with each suit consisting of 16 cards numbered $1, 2, 3, \ldots, 16$.

How can one abstract these important properties? Notice that both properties involve a comparison between two objects. For example, given two cards a and b, does a have a higher value than b? As a matter of convention, two elements that are related in some special way are often represented by an ordered pair. If a is higher than b, then this could be represented by the ordered pair (a, b). Formally, the relation *HigherValue* defined on *SpecialDeck* is the set of ordered pairs (a, b) where the value of card a is higher than the value of card b. This relation for *SpecialDeck* is shown in Table 3.2. Each ordered pair (a, b) in the relation contributes one row to the table, with the higher-valued card appearing first in the row and the lower-valued card second.

Table 3.2 *HigherValue* Relation

HigherValue	
Jack of Hearts	10 of Hearts
Queen of Hearts	10 of Hearts
Jack of Clubs	10 of Hearts
King of Clubs	10 of Hearts
Queen of Hearts	Jack of Hearts
King of Clubs	Jack of Hearts
King of Clubs	Queen of Hearts
Jack of Hearts	10 of Clubs
Queen of Hearts	10 of Clubs
Jack of Clubs	10 of Clubs
King of Clubs	10 of Clubs
Queen of Hearts	Jack of Clubs
King of Clubs	Jack of Clubs

Definition 1.   A **binary relation** is a set of ordered pairs. A binary relation on a set X is a set of ordered pairs of elements of X.

Example 1.   The relation *HigherValue* defined in Table 3.2 can be represented as the following set of ordered pairs:

{(Jack of Hearts, 10 of Hearts), (Queen of Hearts, 10 of Hearts), (Jack of Clubs, 10 of Hearts), (King of Clubs, 10 of Hearts), (Queen of Hearts, Jack of Hearts), (King of Clubs, Jack of Hearts), (King of Clubs, Queen of Hearts), (Jack of Hearts, 10 of Clubs), (Queen of Hearts, 10 of Clubs), (Jack of Clubs, 10 of Clubs), (King of Clubs, 10 of Clubs), (Queen of Hearts, Jack of Clubs), (King of Clubs, Jack of Clubs)}

A second important binary relation on *SpecialDeck* is *SameSuit,* which is defined as $(a, b) \in$ *SameSuit* if a and b are cards in *SpecialDeck* that belong to the same suit. For example, both the ordered pair (10 of Hearts, Jack of Hearts) and the ordered pair (Jack of Hearts, 10 of Hearts) are in *SameSuit,* but the ordered pair (10 of Hearts, Jack of Clubs) is not. The pairs in the relation *SameSuit* for *SpecialDeck* are listed in Table 3.3.

Table 3.3   *SameSuit* Relation

SameSuit			
10 of Hearts	10 of Hearts	10 of Clubs	10 of Clubs
10 of Hearts	Jack of Hearts	10 of Clubs	Jack of Clubs
10 of Hearts	Queen of Hearts	10 of Clubs	King of Clubs
Jack of Hearts	10 of Hearts	Jack of Clubs	10 of Clubs
Jack of Hearts	Jack of Hearts	Jack of Clubs	Jack of Clubs
Jack of Hearts	Queen of Hearts	Jack of Clubs	King of Clubs
Queen of Hearts	10 of Hearts	King of Clubs	10 of Clubs
Queen of Hearts	Jack of Hearts	King of Clubs	Jack of Clubs
Queen of Hearts	Queen of Hearts	King of Clubs	King of Clubs

A third relation defined on *SpecialDeck* is that of having a higher value and being in the same suit. This relation is shown in Table 3.4. Here, an ordered pair (a, b) of cards belongs to the relation *HigherValueSameSuit* if cards a and b in *SpecialDeck* have the same suit and furthermore, card a has a higher value than card b.

Table 3.4   *HigherValueSameSuit* Relation

HigherValueSameSuit	
Jack of Hearts	10 of Hearts
Queen of Hearts	10 of Hearts
Queen of Hearts	Jack of Hearts
Jack of Clubs	10 of Clubs
King of Clubs	10 of Clubs
King of Clubs	Jack of Clubs

Other familiar examples of relations arise when we consider **family trees.** Tradition-ally, a special notation is used, which goes roughly as follows: Marriages are shown with = signs. The first-generation couple sits at the top of the tree. Only their direct descendents officially belong to the tree. Marriages of descendents are indicated by an = sign and the name of the partner. With the exception of the top couple, the children of a person in the tree are drawn off a horizontal line that is joined to that person by a short vertical segment. (No horizontal line is needed for an "only" child). The horizontal line for the children of the top couple is joined to the = sign at the top, since both parents belong to the tree. The children of the first-generation couple form the second generation, the children of the second-generation couples form the third generation, and so on.

In Figure 3.1, George is the only child of Peter and Elaine. Peter is in the picture only because of his marriage to Elaine. Elaine, not Peter, is a child of Mary and John. Elaine is in the second generation, and George is in the third generation.

Figure 3.1   Examples of family tree entries.

Although the marriage of a descendent is indicated by an = sign and the name of the partner, no further information is given about these partners. For example, in the family tree of Mary and John shown in Figure 3.2, even if Peter and Harold were brothers, this would not be shown. A family tree is a rich source of information about a number of relations. In Example 2 you will list the elements of three relations that can be formed from the relationships shown in Figure 3.2.

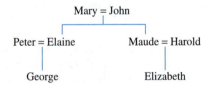

Figure 3.2   Family tree.

Example 2.   For the family tree shown in Figure 3.2, identify the elements of the relations (a) *IsMarriedTo,* (b) *IsParentOf,* and (c) *IsSameGeneration.*

Solution.

(a) *IsMarriedTo* = {(Mary, John), (John, Mary), (Peter, Elaine),
        (Elaine, Peter), (Maude, Harold), (Harold, Maude)}

A representation for the specific relation *IsMarriedTo* is shown in Table 3.5.

Table 3.5   *IsMarriedTo*
Relation

IsMarriedTo	
John	Mary
Mary	John
Peter	Elaine
Elaine	Peter
Maude	Harold
Harold	Maude

(b) *IsParentOf* = {(Mary, Elaine), (John, Elaine), (Mary, Maude),
    (John, Maude), (Peter, George), (Elaine, George),
    (Maude, Elizabeth), (Harold, Elizabeth)}

(c) Peter and Harold do not appear in the relation *IsSameGeneration,* because this relation deals with direct descendants only. In this case, the family tree has more information than is required to define this relation:

IsSameGeneration = {(Elaine, Maude), (Maude, Elaine),
    (George, Elizabeth), (Elizabeth, George), (Mary, John),
    (John, Mary), (John, John), (Mary, Mary), (Elaine, Elaine),
    (Maude, Maude), (George, George), (Elizabeth, Elizabeth)} ■

A specific computer application of relations appears in Section 3.10, which introduces the concept of relational databases. A relational database consists of a number of relations. To answer questions concerning the information contained in the relations, the user poses a question or a **query** that is processed by the database system. If, for example, a user makes a query about who is married to whom, the database system would respond with a table such as Table 3.5. In relational database systems, the answer to any query is a relation.

For example, let X be any set, then

$$Id_X = \{(x, y) : x, y \in X \text{ and } x = y\}$$

Since Id_X is a set of ordered pairs of elements in X, it defines a relation on X. This relation is called the **identity relation,** or the **equality relation,** and may be denoted as $=_X$. The **trivial relation,** or **void relation,** or **empty relation,** on any set consists of $\emptyset$. The **universal relation** on a set consists of all possible ordered pairs of elements of a set.

For any set $X \subseteq \mathbb{R}$, let

$$Lt_X = \{(x, y) : x, y \in X \text{ and } x < y\}$$
$$Le_X = \{(x, y) : x, y \in X \text{ and } x \leq y\}$$
$$Gt_X = \{(x, y) : x, y \in X \text{ and } x > y\}$$
$$Ge_X = \{(x, y) : x, y \in X \text{ and } x \geq y\}$$

Similar relations are defined on $\mathbb{N}$ and $\mathbb{R}$. When the set X is clear from the context, the subscript X will frequently be dropped. When it causes no confusion, it is convenient to use mathematical symbols for these relations and drop the subscript. Hence, we will sometimes refer to Id as $=$, to Le as $\leq$, to Lt as $<$, to Gt as $>$, and to Ge as $\geq$. Of course, to say that $(x, y) \in Id_X$, it is customary to write $x = y$, and to say that $(x, y) \in Lt_\mathbb{N}$, it is customary to write $x < y$. In a non-numeric setting, we can define a relation for any set X of words in a dictionary by saying that *word1* $<$ *word2* means that *word1* precedes *word2* in the dictionary.

Example 3.   Table 3.6 shows two subsets of the relations $=_\mathbb{N}$ and $<_\mathbb{N}$. Since both relations are infinite, the entire relation obviously cannot be displayed.

Table 3.6   Two Relations on $\mathbb{N}$

$Id_\mathbb{N}$		$Lt_\mathbb{N}$	
0	0	0	1
1	1	0	2
2	2	1	2
3	3	0	3
4	4	1	3
.	.	.	.
.	.	.	.
.	.	.	.

If R is a binary relation on a set X, then $(x, y) \in R$ may also be written as $x \, R \, y$.

3.1.1   *n*-ary Relations

There is no reason to restrict attention to relations between pairs of objects. Those are simply the most familiar examples.

Definition 2.   Let $X_1, X_2, \ldots, X_n$ be sets for some $n \in \mathbb{N}$. An ***n*-ary relation** is a set of ***n*-tuples** contained in $X_1 \times X_2 \times \cdots \times X_n$. If $X_1 = X_2 = \cdots X_n$, we say the n-ary relation is defined on X.

We have been careful to indicate how many sets are involved in a relation or how many elements are related. Often the qualifier n-ary is left off and we see references to **relations** for which the context makes clear how many sets are involved.

Example 4.   Let the set X consist of the nine positions on a tic-tac-toe board, named p_1, $p_2, \ldots, p_9$, as shown:

p_1	p_2	p_3
p_4	p_5	p_6
p_7	p_8	p_9

For three distinct positions on the board, define the relation *Between* to consist of the ordered triples (p_j, p_i, p_k) where p_j is between p_i and p_k in some row, column, or diagonal on the board. So, *Between* contains, for example, the ordered triples (p_2, p_3, p_1), (p_5, p_4, p_6), (p_5, p_1, p_9), and (p_5, p_7, p_3). Both (p_5, p_2, p_8) and (p_5, p_8, p_2) are also elements of the relation. ■

Example 5. We define a **ternary relation** R on the set $A = \{1, 2, 3, 4, 5, 6, 7\}$ as follows: such that for any $n_1, n_2, n_3 \in A$ we have $(n_1, n_2, n_3) \in R$ if and only if $n_1 \cdot n_2 = n_3$. Find all elements of R.

Solution. The triples in R are

$$\{(1, 1, 1), (1, 2, 2), (2, 1, 2), (1, 3, 3), (3, 1, 3), (1, 4, 4), (4, 1, 4), (2, 2, 4),$$
$$((1, 5, 5), (5, 1, 5), (1, 6, 6), (6, 1, 6), (2, 3, 6), (3, 2, 6), (1, 7, 7), (7, 1, 7)\}$$ ■

A 1-ary relation is also called a **unary relation** (pronounced "u"-nary relation) or a **property.** A unary relation on a set X is a set of 1-tuples of elements of X, but a 1-tuple of X is just an element of X. Hence, a unary relation on a set X is a subset of X.

Example 6. *Hearts* names a unary relation on a deck of cards. This unary relation on the standard 52-card deck is the set {2 of Hearts, 3 of Hearts, ..., Ace of Hearts}.

If R is an n-ary relation with $n \geq 2$, one can either write $R(x_1, x_2, \ldots, x_n)$ or $(x_1, x_2, \ldots, x_n) \in R$.

Theorem 1 points out that an n-ary relation on a set X is the same thing as a unary relation on the set X^n.

Theorem 1. A set R is an n-ary relation on a set X if and only if $R \subseteq X^n$.

3.2   Operations on Binary Relations

Since relations are sets, the set operations of union, intersection, and difference are well defined for relations. If only binary relations on a set X are considered, then X^2 can be considered as the universal relation, and the complement $X^2 - R$ of a relation R can also be formed. In this section, two other especially important operations on binary relations are considered, namely forming the inverse and taking the composition of two relations. The inverse operation is performed on a single binary relation; the composition operation is performed on two binary relations.

3.2.1   Inverses

For the real numbers 3 and 5, we can write $3 < 5$, but we can convey the same information by writing $5 > 3$. The two relations, $<$ and $>$, are different. More generally, for any real numbers x and y, we have $x < y$ if and only if $y > x$. This is an example of two relations being **inverses** of each other. In terms of the ordered pair notation,

$$(x, y) \in \; < \; \text{if and only if } (y, x) \in \; >$$

The formalization of this property is stated next.

Definition 3.   Let R be a binary relation. The inverse of R, denoted R^{-1}, is

$$\{(x, y) : (y, x) \in R\}$$

Producing the inverse R^{-1} of a relation R can be thought of as performing an operation on R. This operation is known as taking the inverse of R, or as inverting R.

Example 7.   Recall the relation *IsParentOf* in Example 2. Thus,

$$IsParentOf^{-1} = \{(\text{Elaine, Mary}), (\text{Elaine, John}), (\text{Maude, Mary}),$$
$$(\text{Maude, John}), (\text{George, Peter}), (\text{George, Elaine}),$$
$$(\text{Elizabeth, Maude}), (\text{Elizabeth, Harold})\}$$

The relation $IsParentOf^{-1}$ expresses the fact that one person is the child of another, so it is natural to denote this relation by a new name, such as *IsChildOf*. Hence, using the new name for $IsParentOf^{-1}$, $(a, b) \in IsParentOf$ if and only if $(b, a) \in IsChildOf$. ∎

Example 8.

$$Gt_{\mathbb{N}} = \{(1, 0), (2, 0), (2, 1), (3, 0), (3, 1), (3, 2), (4, 0), \ldots\}$$
$$Gt_{\mathbb{N}}^{-1} = \{(0, 1), (0, 2), (1, 2), (0, 3), (1, 3), (2, 3), (0, 4), \ldots\}$$

Clearly, Gt^{-1} is the relation Lt, since $a > b$ if and only if $b < a$. ∎

Theorem 2.   Let R and S be binary relations on a set X. Then,

(a) $(R^{-1})^{-1} = R$.
(b) $(R \cup S)^{-1} = R^{-1} \cup S^{-1}$.
(c) If $S \subseteq R$, then $S^{-1} \subseteq R^{-1}$.

Proof.   (a) For any $x, y \in X$,

$$(x, y) \in (R^{-1})^{-1} \Leftrightarrow (y, x) \in R^{-1}$$
$$\Leftrightarrow (x, y) \in R$$

Hence, $(R^{-1})^{-1} = R$
(b) For any $x, y \in X$,

$$(x, y) \in (R \cup S)^{-1} \Leftrightarrow (y, x) \in R \cup S$$
$$\Leftrightarrow (y, x) \in R \quad \text{or} \quad (y, x) \in S$$
$$\Leftrightarrow (x, y) \in R^{-1} \quad \text{or} \quad (x, y) \in S^{-1}$$
$$\Leftrightarrow (x, y) \in R^{-1} \cup S^{-1}$$

Hence, $(R \cup S)^{-1} = R^{-1} \cup S^{-1}$

(c) This proof is left as an exercise for the reader. ■

3.2.2 Composition

The composition of two relations produces a new relation. Some very familiar examples of relations arise in just this way. For example, we shall soon see that the relation *IsGrandparentOf* is the composition of *IsParentOf* with itself.

Definition 4. Let R and S be binary relations on the set X. The **composition** of R and S, denoted $R \circ S$, is defined as follows:

$$R \circ S = \{(x, y) \in X^2 : \text{for some } z \in X, (x, z) \in S \text{ and } (z, y) \in R\}$$

The reader may consider the notation $R \circ S$ to be backward and think that $S \circ R$ would be more natural. The motivation for writing $R \circ S$ will become obvious in the next chapter, however, when we discuss the composition of functions. Note that the composition of S and R, denoted as $S \circ R$, generally creates a different set of ordered pairs than the composition $R \circ S$ of R and S.

Example 9. The family tree diagram shown in Figure 3.2 can be used to define *IsParentOf*. Since (Mary, Elaine) $\in$ *IsParentOf* and (Elaine, George) $\in$ *IsParentOf*, (Mary, George) $\in$ *IsParentOf* $\circ$ *IsParentOf*. Working out all the possibilities for this composition gives

$$\textit{IsParentOf} \circ \textit{IsParentOf} = \{(\text{Mary, George}), (\text{John, George}),$$
$$(\text{Mary, Elizabeth}), (\text{John, Elizabeth})\}$$ ■

Clearly, $(a, b) \in$ *IsParentOf* $\circ$ *IsParentOf* means that a is the grandparent of b. As another example of composition, convince yourself that

$$\textit{IsCousinOf} = \textit{IsParentOf} \circ \textit{IsSiblingOf} \circ \textit{IsChildOf}$$

You should be able to show that George and Elizabeth in Figure 3.2 are cousins.

The composition of a relation R on a set X with the equality relation on X should always gives the relation R. We prove this in Theorem 3.

Theorem 3. Let X be any set and R be any binary relation on X. Then,

$$R = Id_X \circ R = R \circ Id_X$$

Proof. The proofs of these equalities are similar, so only the proof that $R = Id_X \circ R$ will be given. To do this proof, we show that $Id_X \circ R \subseteq R$ and $R \subseteq Id_X \circ R$. The proof follows Template 1.5 (Set Equality).

First, suppose that $(x, y) \in Id_X \circ R$. Then, by the definition of composition, there is a $z \in X$ with $(x, z) \in R$ and $(z, y) \in Id_X$. Since $(z, y) \in Id_X$, we have $z = y$. Therefore, $(x, z) = (x, y)$. Hence, $(x, y) \in R$.

Second, suppose that $(x, y) \in R$. Now, $(y, y) \in Id_X$, so $(x, y) \in Id_X \circ R$. ■

3.3   Exercises

1. For the people in the family tree (see Figure 3.2), build tables for the following relations:

    (a) *IsAncestorOf*
    (b) *IsDescendentOf*
    (c) *IsSiblingOf*
    (d) *IsCousinOf*

2. Let $M = \{1, 2, \ldots, 10\}$. Define a relation R on elements $x, y \in M$ such that $(x, y) \in R$ if and only if there is a positive integer k such that $x = ky$. Find the elements of R.

3. Find the elements in each of the following relations defined on $\mathbb{R}$:

    (a) $(x, y) \in R$ if and only if $x + 1 < y$
    (b) $(x, y) \in R$ if and only if $y < 0$ or $2x \leq 3$
    (c) $(x, y, z) \in R$ if and only if $x^2 + y = z$

4. List the 16 elements of the relation *Between* as defined in Example 4.

5. The table below gives the names of airlines and several cities that each flies to from Chicago. The table also gives the number of miles for each flight. List all the triples (X, Y, Z) of the ternary relation defined by those triples for which airline X flies Y miles to city Z.

TWA		Pan Am		Piedmont	
Topeka	603	Bombay	7809	Peoria	170
Kansas City	510	Seattle	2052	Albany	816
Phoenix	1742	Anaheim	2025	Atlanta	717

6. Let $U = \{0, 1\}$.

    (a) Let $SubsetOf = \{(X, Y) : X, Y \subseteq U \text{ and } X \subseteq Y\}$. List all ordered pairs in *SubsetOf*.
    (b) Let $StrictSubsetOf = \{(X, Y) : X, Y \subseteq U \text{ and } X \subset Y\}$. List all its ordered pairs in *StrictSubsetOf*.
    (c) $\{(X, Y, Z) : X, Y, Z \subseteq U \text{ and } X \cap Y = Z\}$ is a ternary relation. List all ordered triples in this relation.

    The relations *SubsetOf* and *StrictSubsetOf* can be defined on any set of sets. We will use these relations for other universal sets later in the text.

7. Using the family tree shown in Figure 3.2, list the elements in each of the following relations, and give these relations meaningful names.

    (a) *IsMarriedTo*$^{-1}$
    (b) *IsMarriedTo* $\circ$ *IsMarriedTo*
    (c) *IsParentOf* $\circ$ *IsParentOf*$^{-1}$
    (d) $=_{Family}$ where *Family* denotes the set of people appearing in the family tree
    (e) *IsMarriedTo* $\cap$ *IsMarriedTo*$^{-1}$
    (f) *IsParentOf* $\cap$ *IsParentOf*$^{-1}$

8. Describe the relations resulting from the inverse or composition operations. Describe the resulting relations in words.

(a) $Le_\mathbb{N} \circ Le_\mathbb{N}$

(b) $Le_\mathbb{N}^{-1}$

(c) $Lt_\mathbb{R} \circ Lt_\mathbb{R}$

(d) *Challenge:* $Lt_\mathbb{N} \circ Lt_\mathbb{N}$

(e) *Challenge:* $Lt_\mathbb{N} \circ Gt_\mathbb{N}$

(f) Let $Ne_\mathbb{N} = \{(x, y) : x, y \in \mathbb{N} \text{ and } x \neq y\}$. What is $Ne_\mathbb{N}^{-1}$?

9. Prove Theorem 2(c).

10. (a) Prove for any set X that $Id_X = Id_X^{-1}$.

   (b) Find two binary relations R and S on $\mathbb{N}$ where $R \neq Id_\mathbb{N}$ and $S \neq Id_\mathbb{N}$ such that $R = R^{-1}$ and $S = S^{-1}$.

   (c) Suppose that R is a binary relation on a set X and, for *every* binary relation S on X, $R \circ S = S$. Prove that $R = Id_X$.

11. Let $A = \{1, 2, 3, \ldots, 10\}$. Let $R = \{(1, 2), (1, 4), (1, 6), (1, 8), (1, 10), (3, 5), (3, 7), (4, 6), (6, 8), (7, 10)\}$ be a relation on A. Let $S = \{(2, 4), (3, 6), (5, 7), (7, 9), (8, 10), (8, 9), (8, 8), (9, 9), (3, 8), (4, 9)\}$ be a second relation on A. Find:

   (a) $R \circ S$

   (b) $S \circ R$

12. Show that composition of relations is an associative operation. That is, show that if R, S, and T are binary relations on a set X, then

$$R \circ (S \circ T) = (R \circ S) \circ T$$

13. Let R, S, and T be binary relations on a set X.

   (a) Prove that $R \subseteq S$ if and only if $R^{-1} \subseteq S^{-1}$.

   (b) Prove that if $R \subseteq S$, then $R \circ T \subseteq S \circ T$ and $T \circ R \subseteq T \circ S$.

   (c) If $R \circ T \subseteq S \circ T$ and $T \circ R \subseteq T \circ S$ for *some* relation T, does it follow that $R \subseteq S$?

14. Let $X = \{0, 1\}$. Let $B = \mathcal{P}(X \times X)$ be the set of all binary relations on X.

   (a) List all the elements of B.

   (b) Since elements of B are themselves relations, it makes sense to ask whether two of those relations are inverses of each other. Let

$$IsInverseOf = \{(R, S) : R \in B \text{ and } S \in B \text{ and } R = S^{-1}\}$$

   List all elements of *IsInverseOf*.

   (c) Since *IsInverseOf* is a binary relation, it has an inverse. What is $IsInverseOf^{-1}$?

   (d) What is *IsInverseOf* $\circ$ *IsInverseOf*?

3.4 Special Types of Relations

Some very common binary relations have important special properties. Three of these special properties, the reflexive, symmetric, and transitive properties, occur in relations such as *Id*, *Lt*, *Le*, and both the *SubsetOf* ($\subseteq$) and *StrictSubsetOf* ($\subset$) relations. Not all of these relations have all three of these properties, however. The properties that identify and differentiate these relations are introduced in this section.

3.4.1   Reflexive and Irreflexive Relations

Clearly, $3 \leq 3$ is true, but $3 < 3$ is not true. This distinction between $\leq$ and $<$ is captured in the next definition.

Definition 1.   Let R be a binary relation on a set X. R is **reflexive** if $(x, x) \in R$ for each $x \in X$.

It is obvious from the definition of reflexive that $Id_{\mathbb{R}}$, the equality relation on the real numbers $\mathbb{R}$, and $Le_{\mathbb{R}}$, the less than or equal relation on $\mathbb{R}$, are reflexive. It is also obvious that $Lt_{\mathbb{R}}$, the less than relation on $\mathbb{R}$, is not reflexive since there is *no* element $x \in \mathbb{R}$ for which $(x, x) \in Lt_{\mathbb{R}}$. That is, $x < x$ is never true since no number can be strictly less than itself.

The relation *IsSameGeneration* defined in Section 3.1 is reflexive since each person is in the same generation as themselves.

Theorem 1.   A binary relation R on a set X is reflexive if and only if $Id_X \subseteq R$.

A picture of $Id_{\mathbb{R}}$ is shown in Figure 3.3. The points (x, y) of the plane that represent elements of $Id_{\mathbb{R}}$ are darkened. The picture is just the familiar graph of the line $x = y$.

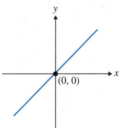

Figure 3.3   Graph of $Id_{\mathbb{R}}$.

In general, for a binary relation R defined on the real numbers $\mathbb{R}$, one can draw a picture of the relation by darkening the point (x, y) in the plane if the ordered pair of real numbers (x, y) is in R. Such a picture is called the **graph** of the relation R. Sometimes, relations have graphs that consist of a single line, but in general, graphs of relations consist of entire regions of points.

The usual convention in graphing $Le_{\mathbb{R}}$ is to draw the diagonal line $x = y$ as a darker, heavier line to show that the line is included in the graph. One can see that $Le_{\mathbb{R}}$ is reflexive from its graph since the graph of the line $x = y$ is a subset of the graph of $Le_{\mathbb{R}}$ (see Figure 3.4). Of course, making deductions from a graph is risky for essentially the same reason that making deductions from a Venn diagram is risky.

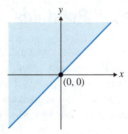

Figure 3.4   Graph of $Le_{\mathbb{R}}$.

The difference between $\leq$ and $<$ that we have discussed is formalized in Definition 2.

Definition 2.   Let R be a binary relation on a set X. R is **irreflexive** if $(x, x) \notin R$ for all $x \in X$.

Clearly, $Lt_{\mathbb{R}}$ is an irreflexive relation since $x < x$ is never true for any $x \in \mathbb{R}$. Considering relations as sets, we can characterize irreflexive relations in terms of their intersection with an identity relation.

Theorem 2.   A binary relation R on a set X is irreflexive if and only $R \cap Id_X = \emptyset$.

Example 1.   The usual convention in graphing $Lt_{\mathbb{R}}$ (see Figure 3.5) is to draw the diagonal line $x = y$ dotted to show that it is not included in the graph. Since no point on this line is in $Lt_{\mathbb{R}}$, it can be concluded that $Lt_{\mathbb{R}}$ is irreflexive.

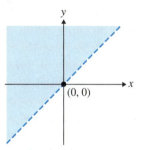

Figure 3.5 $Lt_{\mathbb{R}}$.   ■

The relation $\{(1, 1), (1, 2)\}$ on $X = \{1, 2\}$ is not reflexive, because $(2, 2) \notin R$ and it is not irreflexive because $(1, 1) \in R$.

3.4.2   Symmetric and Antisymmetric Relations

A principal distinction between the equality relation $=$ on the one hand and the relations $<$ and $\leq$ on the other is captured by the notion of symmetry.

Definition 3.   Let R be a binary relation on a set X. R is **symmetric** if $(y, x) \in R$ whenever $(x, y) \in R$.

Clearly, the relation $=$ is a symmetric relation. Neither $<$ nor $\leq$, however, is symmetric. For example, notice it is true that $3 < 5$ but not that $5 < 3$, and it is true that $3 \leq 5$ but not that $5 \leq 3$. Therefore, neither $<$ nor $\leq$ is a symmetric relation.

Example 2.   Refer to Section 3.1 for the definitions of the relations *IsMarriedTo, IsParentOf, SameSuit, HigherValue,* and *IsSameGeneration.*

(a) The relation *IsMarriedTo* is symmetric, and *IsParentOf* is not. (Mary, Elaine) $\in$ *IsParentOf* whereas (Elaine, Mary) $\notin$ *IsParentOf.*
(b) The relation *SameSuit* is symmetric, whereas *HigherValue* is not. (Jack of Hearts, 10 of Hearts) $\in$ *HigherValue,* whereas (10 of Hearts, Jack of Hearts) $\notin$ *HigherValue.*
(c) *IsSameGeneration* is symmetric.

One can see that a binary relation on $\mathbb{R}$ is symmetric if and only if its graph is symmetric about the diagonal line $x = y$. Figure 3.6 shows a symmetric relation on $\mathbb{R}$.

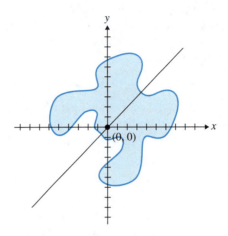

Figure 3.6   Symmetric relation on $\mathbb{R}$.

We really begin to understand the properties of relations when we understand how different concepts express the same idea. Theorem 3 relates inverses of relations to the property of a relation being symmetric.

Theorem 3.   A relation R on a set X is symmetric if and only if $R = R^{-1}$.

Proof.   Let R be a symmetric relation. Then, $(x, y) \in R$ if and only if $(y, x) \in R$, which is the case if and only if $(x, y) \in R^{-1}$. ∎

The relation shown in Figure 3.7 is not symmetric: $(0, -7)$ is an element of the relation, whereas $(-7, 0)$ is not.

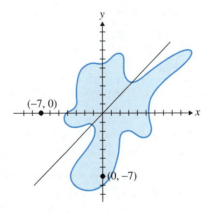

Figure 3.7   Nonsymmetric relation on $\mathbb{R}$.

Definition 4. Let R be a binary relation on a set X. R is **antisymmetric** if $(y, x) \notin R$ whenever $(x, y) \in R$ and $x \neq y$.

The relation defined on $\{1, 2, 3\}$ as $R = \{(1, 2), (2, 1), (3, 2)\}$ is neither symmetric, because $(3, 2) \in R$ but $(2, 3) \notin R$, nor antisymmetric, because both $(1, 2)$ and $(2, 1)$ are in R.

The relations $=$, $\leq$, and $<$ are all antisymmetric. A logically equivalent statement of the definition of antisymmetric is the following: If (x, y) and (y, x) are both in R, then $y = x$. To see this in terms of the logical notation introduced in Chapter 2, let p_1 be the statement "$(x, y) \in R$," p_2 the statement "$(y, x) \in R$," and p_3 the statement "$x = y$." The definition is of the form $(p_1 \wedge \neg p_3) \rightarrow \neg p_2$, which is logically equivalent to the formula $(p_1 \wedge p_2) \rightarrow p_3$.

Example 3. See Section 3.1 for the examples and the definitions of the relations *IsParentOf* and *HigherValue*.

(a) In the family tree example, the relation *IsParentOf* is antisymmetric. For example, (Mary, Elaine) $\in$ *IsParentOf*, but (Elaine, Mary) $\notin$ *IsParentOf*.
(b) In the card example, *HigherValue* is antisymmetric. We see this as (Jack of Hearts, 10 of Hearts) $\in$ *HigherValue*, but (10 of Hearts, Jack of Hearts) $\notin$ *HigherValue*.

Suppose that a binary relation R is written as a table T, as in Table 3.7(a), which repeats the information contained in Table 3.4. Now, suppose that a new table, T', is formed by interchanging the two columns of T. The resulting Table 3.7(b) corresponds to the relation R^{-1}. Theorem 3 says that R is symmetric if and only if T and T' have the same rows. The order of the rows may be different, but exactly the same rows are present. Since any n-ary relation is a set of ordered n-tuples for some $n \in \mathbb{N}$, the order in which the n-tuples are written in the table does not matter.

Table 3.7 *IsMarriedTo* (a) and *IsMarriedTo* $^{-1}$ (b) Relations

T	
John	Mary
Mary	John
Peter	Elaine
Elaine	Peter
Maude	Harold
Harold	Maude

(a)

T'	
Mary	John
John	Mary
Elaine	Peter
Peter	Elaine
Harold	Maude
Maude	Harold

(b)

An examination of the two tables shows that $T = T'$.

Example 4.

(a) For any set X, equality is a symmetric, antisymmetric, and reflexive relation on X.

(b) For any set X, the empty relation $\emptyset$ is a symmetric, antisymmetric, and irreflexive relation on X. If $X \neq \emptyset$, then the empty relation $\emptyset$ is not reflexive on X. If $X = \emptyset$, then the empty relation $\emptyset$ is (vacuously) reflexive on X.

(c) Let $R = \{(x, y) \in \mathbb{R}^2 : x < y^2\}$. R is not reflexive, irreflexive, symmetric, or antisymmetric. R is not reflexive since $(1, 1) \notin R$. R is not irreflexive since $(2, 2) \in R$. R is not symmetric since $(1, 2) \in R$ but $(2, 1) \notin R$. R is not antisymmetric since $(2, 3) \in R$ and $(3, 2) \in R$. ◼

Example 5. Define the relation *IsAncestorOf* so that x *IsAncestorOf* y means that x is a parent of y, or that x is the parent of a parent of y, or that x is the parent of a parent of a parent of y, and so on. The relation *IsAncestorOf* is an antisymmetric and irreflexive relation on the set of all people.

Example 6.

(a) The relations $<$ and $\leq$ are antisymmetric relations on $\mathbb{R}$. The relation $\leq$ is reflexive. The relation $<$ is irreflexive.

(b) The relations $\subset$ and $\subseteq$ are binary relations on the subsets of a set U. Both $\subset$ and $\subseteq$ are antisymmetric. The relation $\subseteq$ is reflexive, and $\subset$ is irreflexive.

Example 7. Let $\epsilon = 0.0005$, and let R_ϵ be the relation

$$\{(x, y) \in \mathbb{R}^2 : |x - y| < \epsilon\}$$

R_ϵ could be interpreted as the relation *approximately equal*. Prove that R_ϵ is reflexive and symmetric.

Solution. *Reflexive:* For all $x \in X$, $|x - x| = 0 < \epsilon$. *Symmetric:* For all $x, y \in \mathbb{R}$, $|x - y| = |y - x|$. So, if $|x - y| < \epsilon$, then $|y - x| = |x - y| < \epsilon$. ◼

3.4.3 Transitive Relations

To introduce the next property of relations, suppose that Sue is a parent of Joe and that Tom is a parent of Sue. We can conclude that Tom is an ancestor of Joe, but we cannot conclude that Tom is a parent of Joe. The next property, called transitivity, is a formal way of thinking about how the two relations, *IsParentOf* and *IsAncestorOf*, are different. The relation *IsParentOf* does not satisfy this next property whereas the relation *IsAncestorOf* does.

Definition 5. Let R be a binary relation on a set X. R is **transitive** if $(x, z) \in R$ whenever $(x, y) \in R$ and $(y, z) \in R$.

Example 8. Consider the relations in Examples 4 through 7.

(a) Equality is transitive.

(b) The relation $\emptyset$ is (vacuously) transitive.

(c) Over the set $\mathbb{R}$, the relations $\leq$ and $<$ are transitive.

(d) The relation *IsAncestorOf* is transitive.

(e) The relations $\subseteq$ and $\subset$ are transitive.

(f) R_ϵ is not transitive.

(g) $\{(x, y) \in \mathbb{R}^2 : x < y^2\}$ is not transitive. To see this, just note that $(9, 5) \in R$ and $(5, 3) \in R$, but that $(9, 3) \notin R$. ◼

Theorem 4. A binary relation R is transitive if and only if $R \circ R \subseteq R$.

Proof. This just restates the definition. If there is a y such that $(x, y) \in R$ and $(y, z) \in R$, then $(x, z) \in R$. ∎ ■

In Table 3.8, we summarize the properties, their characterizations, and how we prove a property holds for a relation R defined on a set X.

Table 3.8 Properties of Relations

Property	Characterization	Method of Proof
Reflexive	$Id_X \subseteq R$	Let $x \in X$. Prove $(x, x) \in R$.
Antireflexive	$Id_X \cap R = \emptyset$	Let $x \in X$. Prove $(x, x) \notin R$.
Symmetric	$R = R^{-1}$	Let $(x, y) \in R$. Prove $(y, x) \in R$.
Antisymmetric	$R \cap R^{-1} \subseteq Id_X$	Suppose that $(x, y) \in R$ and $(y, x) \in R$. Prove $x = y$.
Transitive	$R \circ R \subseteq R$	Let $(x, y), (y, z) \in R$. Prove $(x, z) \in R$.

3.4.4 Reflexive, Symmetric, and Transitive Closures

A question that arises for relations that do not possess a particular property, such as being reflexive, symmetric, or transitive, is whether more elements can be added to a relation R to produce a relation R' that does have some desired property. One obvious way is to take R' to be the universal relation (check this). What we really want to know is how to find a smallest relation R' that contains R and has some desired property, such as transitivity.

For example, how is the relation $Ge_{\mathbb{R}}$ ($\geq$) related to the relation Gt ($>$)? Clearly, $Ge_{\mathbb{R}} = Gt_{\mathbb{R}} \cup Id_{\mathbb{R}}$. The relation $Ge_{\mathbb{R}}$ turns out to be the smallest reflexive relation on $\mathbb{R}$ containing $Gt_{\mathbb{R}}$. $Ge_{\mathbb{R}}$ is called the **reflexive closure** of $Gt_{\mathbb{R}}$. (The term *reflexive closure* will be defined formally below.) More generally, Ge_X is the reflexive closure of Gt_X over any set X such that $X \subseteq \mathbb{R}$.

Suppose people are waiting in a ticket line. We say that person x is the person *InFrontOf* person y, expressed as x *InFrontOf* y, if x is the person standing immediately in front of person y. How is the relation *IsAdjacentTo* related to the relation *InFrontOf*? A person x is adjacent to a person y if x is the person in front of y or y is the person after x. Said another way, x *IsAdjacentTo* y means that x is just in front of or just behind y. It can be shown that *IsAdjacentTo* is the smallest symmetric relation containing *InFrontOf*. The relation *IsAdjacentTo* is called the **symmetric closure** of *InFrontOf*.

Finally, in the case of transitivity, we ask how the relation *IsAncestorOf* is related to the relation *IsParentOf*. A person x is the ancestor of a person y if x is a parent of y, or a parent of a parent of y, or a parent of a parent of a parent of y, and so on. The relation *IsAncestorOf* is the smallest transitive relation containing *IsParentOf*. The relation *IsAncestorOf* is called the **transitive closure** of *IsParentOf*.

To characterize the reflexive, symmetric, and transitive closures of a relation, we first define a new operation on relations.

Definition 6.   Let R be a binary relation on a set X. For $n \in \mathbb{N}$, the **nth power of R,** denoted R^n, is defined as follows:

(a) $R^0 = \{(x, x) : x \in X\} = Id_X$.
(b) $R^{n+1} = R \circ R^n$.

Let $\mathbf{R^+} = \cup_{i=1}^{\infty} R^i$ and $\mathbf{R^*} = \cup_{i=0}^{\infty} R^i$.

Example 9.   Let $A = \{a, b, c, d\}$ and let R be the relation on A consisting of the pairs $(a, b), (b, a), (b, c),$ and (c, d). Find R^+ and R^*.

Solution.
$$R^0 = \{(a, a), (b, b), (c, c), (d, d)\}$$
$$R^2 = \{(a, a), (a, c), (b, b), (b, d)\}$$
$$R^3 = \{(a, b), (b, a), (b, c), (a, d)\}$$
$$R^4 = \{(a, a), (a, c), (b, b), (b, d)\}$$

Observe that $R^2 = R^4$ and, consequently, $R^5 = R^3$. In general, $R^{2n+1} = R^3$ and $R^{2n} = R^2$ for $n \geq 1$. Therefore,

$$R^+ = \{(a, b), (b, a), (b, c), (c, d), (a, a), (a, c), (b, b), (b, d), (a, d)\}$$
$$R^* = \{(c, c), (d, d), (a, b), (b, a), (b, c), (c, d), (a, a), (a, c), (b, b), (b, d), (a, d)\}$$ ■

Example 10.   Let R be the relation *IsChildOf*.

(a) The expression $x R^2 y$ means that x is a child of a child of y, so R^2 is the same as *IsGrandchildOf*.
(b) The expression $x R^3 y$ means that x is a child of a grandchild of y or, said another way, that x is a great-grandchild of y. Hence, the relation R^3 could just as well be called *IsGreatGrandchildOf*.
(c) R^4 could just as well be called *IsGreatGreatGrandchildOf*.
(d) Relation R^+ is the same as *IsAncestorOf*.

Example 11.   Let S be the relation on $\mathbb{Z}$ that is defined by $a S b$ if and only if $b = a + 1$.

(a) It is true that $a S^0 b$ if and only if $b = a$.
(b) $a S^2 b$ if and only if, for some integer c, it is true that $c = a + 1$ and $b = c + 1$—that is, if and only if $b = a + 2$.
(c) $a S^3 b$ if and only if $b = a + 3$.
(d) $a S^n b$ if and only if $b = a + n$. (Formally, this is proved by induction on n.)
(e) $a S^+ b$ if and only if $a < b$. For if $a S^+ b$, then $a S^n b$ for some positive integer n.
(f) $a S^* b$ if and only if $a \leq b$.

Solution.   (f) ($\Leftarrow$)   By part (d), $b = a + n$, so $a < b$.
($\Rightarrow$)   Suppose, conversely, that $a < b$. Since $a, b \in \mathbb{Z}$, their difference $b - a \in \mathbb{Z}$. Since $a < b$, it follows that $b - a > 0$. Let $n = b - a$. By part (d), $a S^n b$, so $a S^+ b$.   ■

In Theorem 5 we give a characterization of the smallest reflexive, symmetric, and transitive relations containing a given relation.

Theorem 5.   Let R be a binary relation on a set X. Then:

(a) $R \cup Id_X$ is the smallest reflexive relation containing R.
(b) $R \cup R^{-1}$ is the smallest symmetric relation containing R.
(c) R^+ is the smallest transitive relation containing R.
(d) R^* is the smallest reflexive and transitive relation containing R.

Proof.   (a) By Theorem 1, a relation S on X is reflexive if and only if $Id_X \subseteq S$. So, S is reflexive and contains R if and only if $R \cup Id_X \subseteq S$. The smallest such S is $R \cup Id_X$ itself.
(b) We must prove (i) that $R \cup R^{-1}$ is symmetric and (ii) that if S is a symmetric relation on X and $R \subseteq S$, then $R \cup R^{-1} \subseteq S$.

   (i) It is enough to show that $(R \cup R^{-1})^{-1} = R \cup R^{-1}$ since the result then follows from Theorem 3 in Section 3.4.2.

$$(R \cup R^{-1})^{-1} = R^{-1} \cup (R^{-1})^{-1}$$
$$= R^{-1} \cup R$$
$$= R \cup R^{-1}$$

   (ii) Suppose S is a symmetric relation on X and $R \subseteq S$. We must show that $R^{-1} \subseteq S$. By Theorem 2 (c) in Section 3.2.1, $R^{-1} \subseteq S^{-1}$, and by Theorem 3 in Section 3.4.2, $S^{-1} = S$. So, $R^{-1} \subseteq S$.

(c) and (d) These proofs are left as Exercises for the reader.   ■

Definition 7.   Let R be any binary relation on a set X. $R \cup Id_X$ is called the *reflexive closure* of R. $R \cup R^{-1}$ is called the *symmetric closure* of R. R^+ is called the *transitive closure* of R. R^* is called the **reflexive and transitive closure** of R.

Example 12.   Let $X = \{a, b, c\}$. Define the relation R on X as $\{(a, b), (b, c)\}$. Find the reflexive, symmetric, and transitive closure of R. Also, find the reflexive and transitive closure of R.

Solution.   We must first find the following relations:

(a) $Id_X = \{(a, a), (b, b), (c, c)\}$
(b) $R^{-1} = \{(b, a), (c, b)\}$
(c) $R^0 = \{(a, a), (b, b), (c, c)\}$, $R^1 = \{(a, b), (b, c)\}$, $R^2 = \{(a, c)\}$, and $R^n = \emptyset$ for $n \geq 3$.
(d) $R^+ = \{(a, b), (b, c), (a, c)\}$ and $R^* = \{(a, a), (b, b), (c, c), (a, b), (b, c), (a, c)\}$

   So, the reflexive closure of R is

$$R \cup Id_X = \{(a, b), (b, c), (a, a), (b, b), (c, c)\}$$

The symmetric closure of R is

$$R \cup R^{-1} = \{(a, b), (b, c), (b, a), (c, b)\}$$

The transitive closure of R is

$$R^+ = \{(a, b), (b, c), (a, c)\}$$

Finally, the reflexive and transitive closure of R is

$$R^* = \{(a, a), (b, b), (c, c), (a, b), (b, c), (a, c)\}$$ ■

Example 13.   Consider the relation *Supervises* in some business. The relation is usually irreflexive, that is, people do not supervise themselves. It is also antisymmetric. Finally, it is generally not transitive. If x supervises y and y supervises z, normally x does not (directly) supervise z. The reflexive closure of *Supervises* is *SupervisesOrEquals*. The symmetric closure is *SupervisesOrIsSupervisedBy*, which is clearly an important relation in business.

Example 14.   Let U be any nonempty set. Then, $\subset$ is a relation on the subsets of U. The relation $\subset$ is transitive, but it is not reflexive and is not symmetric. The reflexive closure of $\subset$ is $\subseteq$. The symmetric closure S of this relation has no commonly used name, but for two subsets A and B of U, $(A, B) \in S$ if and only if $A \subset B$ or $B \subset A$.

As an example of the relation described in Example 14, let $U = \{0, 1\}$. The reflexive and symmetric closure of $\subset$ on U consists of the 14 ordered pairs shown in Table 3.9.

Table 3.9   Reflexive and Symmetric Closure of $\subset$ for $U = \{0, 1\}$

$(\emptyset,$	$\{0, 1\})$	$(\{0, 1\},$	$\emptyset)$	$(\emptyset,$	$\emptyset)$
$(\emptyset,$	$\{0\})$	$(\{0\},$	$\emptyset)$	$(\{0\},$	$\{0\})$
$(\emptyset,$	$\{1\})$	$(\{1\},$	$\emptyset)$	$(\{1\},$	$\{1\})$
$(\{0\},$	$\{0, 1\})$	$(\{0, 1\},$	$\{0\})$	$(\{0, 1\},$	$\{0, 1\})$
$(\{1\},$	$\{0, 1\})$	$(\{0, 1\},$	$\{1\})$		

3.4.5   Application: Transitive Closures in Medicine and Engineering

Transitive and reflexive closures are especially important in computer science. For example, suppose computers are connected to each other in a network, with each computer connected directly to a small number of other computers. Information can be passed directly from one computer to another over a connection between them. The transitive and reflexive closure of *IsConnectedTo* is *CanAccess*. This relation gives the limit of how far information from one machine may be passed along to others. The examples that follow show how the transitive closure idea leads to better understanding in fields as diverse as medicine and chip testing.

Artificial Intelligence

Many artificial intelligence applications can be phrased in terms of some (simulated) person making inferences based on some initial data. One kind of application is the expert system, in which designers try to encode the knowledge that an expert would use in approaching a problem. Suppose, for example, an expert system is used to suggest to a physician certain tests that should be run. The system might say, for example, that if the patient's weight is more than 25 percent over the recommended level to check for high cholesterol. (Drs. X, Y, and Z all told the designers of the expert system that is what they do, so it must be a reasonable rule.) And if the patient eats a high-fat diet, there should be a check for cholesterol. (Drs. X, W, and Q all said they do that.) And if there is a check of the cholesterol level, there should also be a check for high triglycerides (suggested by several other doctors.) If there is a test for triglycerides, there should also be a test for something else, and so on. This series of "rules" is stored in the program called the *expert system*. The doctor enters that the patient has a body weight 30 percent over his recommended weight and this fact triggers a series of inferences: check cholesterol level; check triglycerides level, and so on. This is a transitive closure operation: including one test triggered including another, which triggered including another, ..., until nothing else was triggered.

Often, the rules are rather more complicated, such as "if the patient's weight is 15 percent over the recommended weight *and* the patient is diabetic, then do a cholesterol test." This is a more complicated sort of closure operation, but the idea is similar.

Testing Circuits

Here, we picture a combinational electric circuit:

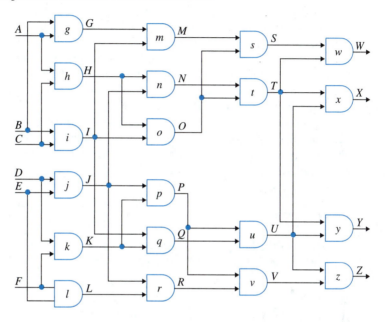

Current flows from left to right, so there are six input lines, A through F, and four output lines, W through Z. There are 20 gates, g through z. For convenience, we picture them all as and-gates, but the intention is that they might implement some AND gates, some OR gates, and some NOT gates. Define two relations between lines and gates, one "saying" that a line is an input to a gate and the other that a line is an output of a gate. The large dots indicate that a line is split, being an input for several gates, such as A is

Input		Output	
A	*g*	*g*	*G*
A	*h*	*h*	*H*
B	*g*	*i*	*I*
B	*i*	*j*	*J*
C	*h*	*k*	*K*
C	*i*	*l*	*L*
D	*j*	*m*	*M*
D	*k*	*n*	*N*
⋮	⋮	*o*	*O*
		⋮	⋮

input for both gates g and h. Otherwise, when two lines cross, such as the output line of h and the output line of i, it just means that when the circuit is fabricated, these two lines will follow this path but will not touch.

The circuit manufacturer would want to check that each gate is functioning correctly. For example, if all the lines carry 0's and 1's (designers use 1 and 0 instead of *TRUE* and *FALSE*), gate o might be "stuck at 0", that is, it might always output a 0, no matter what its input is. The manufacturer would then like to have a "test vector" for that: a set of inputs to distinguish whether gate o is stuck at 0. The first part of choosing such a test vector is to determine which output lines could be affected if gate o is malfunctioning. In this case, lines W, X, and Y could be affected. Line Z cannot be, however, since no output from gate o flows, directly or indirectly, into gate z.

The relation of one line directly influencing another is *Output ∘ Input*. The relation of directly or indirectly influencing another line—through *any number* of intermediate lines—is thus *(Output ∘ Input)**. The question above is to find all output lines where $(o, some\ output\ line) \in (Output \circ Input)^* \circ Output$.

Of course, now that designers have narrowed down which lines might be affected by a malfunction at gate o, they must go on to determine how to produce a single input that will identify the stuck-at-0 fault. However, we cannot do that without knowing what the individual gates are.

3.5   Exercises

1. Which of the following relations on the set of all people are reflexive? Symmetric? Antisymmetric? Transitive? Prove your assertions.

   (a) $R(x, y)$ if y makes more money than x.
   (b) $R(x, y)$ if x and y are about the same height.
   (c) $R(x, y)$ if x and y have an ancestor in common.
   (d) $R(x, y)$ if x and y are the same sex.
   (e) $R(x, y)$ if x and y both collect stamps.
   (f) $R(x, y)$ if x and y like some of the same music.

2. For each of the relations defined in Exercise 1, write out the condition that defines the inverse relation.

3. Which of the following relations on the set of all people are reflexive? Symmetric? Antisymmetric? Transitive? Explain why your assertions are true.

   (a) $R(x, y)$ if x and y either both like German food or both dislike German food.
   (b) $R(x, y)$ if (i) x and y either both like Italian food or both dislike it, or (ii) x and y either both like Chinese food or both dislike it.
   (c) $R(x, y)$ if y is at least two feet taller than x.

4. For each of the relations defined in Exercise 3, write out the condition that defines the inverse relation.

5. Which of the following relations on the set of people indicated are reflexive? Irreflexive? Symmetric? Antisymmetric? Transitive?

   (a) *IsSisterOf* on the set of all females
   (b) *IsBrotherOfOrEquals* on the set of all males
   (c) *IsSiblingOf* on the set of all people

(d) *IsSiblingOfOrEquals* on the set of all people

(e) *IsCousinOfOrEquals* on the set of all people

Prove your assertions.

6. Since relations are sets, it is possible to define union, intersection, relative complement, and absolute complement on pairs of relations. A natural question is which properties of the original relations still hold for the resulting new relation. Fill in the following table with Y/N, representing YES and NO, respectively. If the entry is N, find an example that shows the property is not preserved under the operation. For instance, enter a Y in the first row, second column, if the *intersection* of two *reflexive* relations is still reflexive; otherwise, enter an N.

	Union	Intersection	Relative Complement	Absolute Complement
Reflexive				
Irreflexive				
Symmetric				
Antisymmetric				
Transitive				

7. Let $A = \{a, b, c, d\}$. Define the relations R_1 and R_2 on A as

$$R_1 = \{(a, a), (a, b), (b, d)\}$$

and

$$R_2 = \{(a, d), (b, c), (b, d), (c, b)\}$$

Find

(a) $R_1 \circ R_2$

(b) $R_2 \circ R_1$

(c) R_1^2

(d) R_2^2

8. Find a set A with n elements and a relation R on A such that $R^1, R^2, \ldots, R^n$ are all distinct.

9. In the example involving the family tree (see Figure 3.2);

(a) What is the transitive and reflexive closure of *IsParentOf*?

(b) What is the transitive and reflexive closure of *IsMarriedTo*?

10. Let $X = \{a, b, c, d, e\}$. Let R_1 be the relation on X with elements $\{(a, b), (a, c), (d, e)\}$. Let R_2 be the relation on X with elements $\{(a, b), (b, c), (c, d), (d, e), (e, a)\}$. For each of these relations, find the following:

(a) The smallest relation on X that contains R and is reflexive

(b) The smallest relation on X that contains R and is symmetric

(c) The smallest relation on X that contains R and is transitive

(d) The smallest relation on X that contains R and is reflexive and transitive

11. Let $X = \{1, 2, 3, 4\}$, and define a relation R on X as

$$R = \{(1, 2), (2, 3), (3, 4)\}$$

(a) Find the reflexive closure of R.
(b) Find the symmetric closure of R.
(c) Find the transitive closure of R.
(d) Find the reflexive and transitive closure of R.

12. Let $X = \{1, 2, 3, 4, 5, 6\}$, and define a relation R on X as

$$R = \{(1, 2), (2, 1), (2, 3), (3, 4), (4, 5), (5, 6)\}$$

(a) Find the reflexive closure of R.
(b) Find the symmetric closure of R.
(c) Find the transitive closure of R.
(d) Find the reflexive and transitive closure of R.

13. Let $A = \{1, 2, 3, 4\}$. Find the transitive closure of the relation R defined on A as

$$R = \{(1, 2), (2, 1), (2, 3), (3, 4)\}$$

14. Let R be the relation on $\{a, b, c, d, e, f, g\}$ defined as

$$R = \{(a, b), (b, c), (c, a), (d, e), (e, f), (f, g)\}$$

Find the smallest integers m and n such that $0 < m < n$ and $R^m = R^n$. Identify the transitive closure of R as well as the transitive, reflexive, and symmetric closures of R.

15. Let $X = \{4, 5, 6, 7, 8\}$, and define the relation R on X as $\{(4, 5), (5, 6), (6, 7), (7, 8), (8, 4)\}$. Find the smallest integers m and n such that $R^m = R^n$, where $0 < m < n$.

16. Find the reflexive, symmetric, and transitive closures of the following relations:

(a) $=$ on $\mathbb{N}$
(b) $<$ on $\mathbb{N}$
(c) $\leq$ on $\mathbb{N}$
(d) R on $\mathbb{N}$ where $R(x, y)$ if and only if $y = x + 1$
(e) R on $\mathbb{R}$ where $R(x, y)$ if and only if $y = x + 1$
(f) R on $\mathbb{R}$ where $R(x, y)$ if and only if $|x - y| < 0.0005$

17. Show that the transitive closure of a relation R on a set X is the intersection of all transitive binary relations R' on X where $R \subseteq R'$.

18. Is there a reasonable notion of antisymmetric closure? Why, or why not?

19. Prove Theorem 5(c) as follows:

(a) Prove by induction that if R is a binary relation on a set X, then $R^m \circ R^n = R^{m+n}$ where $m, n \in \mathbb{N}$.
(b) Prove that R^+ is transitive.
(c) Prove by induction that if $R \subseteq S$ and S is a transitive binary relation, then $R^n \subseteq S$. Conclude that $R^+ \subseteq S$.

20. Prove Theorem 5(d).

Equivalence Relations

Equivalence relations generalize the familiar relation of equality (=). More specifically, equivalence relations identify elements that are the same in some respect. For instance, university students are classified by major, with two students being "related" if they have the same major. Two students are also "related" if they are in the same class, such as the sophomore class.

Definition 1.   Let R be a binary relation on a set X. R is an **equivalence relation** if R is reflexive, symmetric, and transitive.

Example 1.

(a) For any set X, the equality relation (=) is an equivalence relation on X.
(b) The relation *IsSameGeneration* (see Section 3.1) as defined using Figure 3.2 is an equivalence relation.

   The *IsSameGeneration* relation as based on Figure 3.2 is not a particularly interesting equivalence relation because there are so few elements. The reader is encouraged to construct his or her own family tree for three or four generations and see how the relation conveys information conveniently.

Example 2.   The relation *SameSuit* (see Section 3.1) shown in Table 3.3 is an equivalence relation.

Solution.   It is obvious that *SameSuit* is reflexive and symmetric, but is *SameSuit* transitive? Let cards x and y be in the same suit, and let cards y and z be in the same suit. Since y is in the same suit as z and in the same suit as x, it follows that x and z are in the same suit. Therefore, *SameSuit* is transitive.   ∎

   Recall that when we divide a natural number n by a positive number p, we obtain an integer **quotient,** which we will call q, and a **remainder,** which we will call r. That is, we get an equation

$$n = pq + r$$

where $q, r \in \mathbb{N}$ and $0 \le r \le p - 1$. For example, $7 \div 3 = 2 \cdot 3 + 1$, so the quotient is 2, the remainder is 1, and $7 = 3 \cdot 2 + 1$.
   If the remainder is zero, then $n = p \cdot q$, and we say that n is **divisible by** p.

Definition 2.   Let p be a positive integer, and let $x, y \in \mathbb{N}$. We say that x is **congruent** to y modulo p, and write $x \equiv y$ **(mod p),** if $(x - y)$ is divisible by p; that is, $(x - y) = m \cdot p$ for some integer m.

   With this terminology, we will prove that (*mod p*) is an equivalence relation for $p \ge 1$.

Example 3.   Let p be any natural number greater than zero. Then, $\equiv$ (*mod p*) is an equivalence relation on $\mathbb{N}$.

Solution.   Check that all the properties hold:

Reflexive: For any $n \in \mathbb{N}$, $(n - n) = 0 = p \cdot 0$, so $(n - n)$ is divisible by p. Therefore, $n \equiv n (mod\ p)$.

Symmetric: If $n \equiv m(mod\ p)$, then $(n - m) = pk$ for some $k \in \mathbb{Z}$. So, $(m - n) = p(-k)$, giving $m \equiv n(mod\ p)$.

Transitive: Suppose $n \equiv m(mod\ p)$ and $m \equiv k(mod\ p)$. Show that $n \equiv k(mod\ p)$. The hypothesis implies that $(n - m) = ip$ and $(m - k) = jp$ for some $i, j \in \mathbb{Z}$. Then, however,

$$(n - k) = (n - m) + (m - k) = ip + jp = (i + j)p$$

which gives $n \equiv k(mod\ p)$.

Since $\equiv (mod\ p)$ is reflexive, symmetric, and transitive, it is an equivalence relation. ∎

We will study this equivalence relation more carefully later. For now, the reader might determine the elements of this relation when $p = 8$ and the universal set is $\{0, 1, 2, \ldots, 24, 25\}$.

Example 4. Let U be any set. For $X, Y \subseteq U$, set $X \sim Y$ if $X \oplus Y = (X - Y) \cup (Y - X)$ is finite. Then, $\sim$ is an equivalence relation on the subsets of U. The relation $\sim$ is uninteresting if U is finite.

Solution.

Reflexive: $X \oplus X = \emptyset$, which is finite, so $X \sim X$.

Symmetric: If $X \sim Y$, then $X \oplus Y$ is finite. Recall that

$$X \oplus Y = (X - Y) \cup (Y - X) = (Y - X) \cup (X - Y) = Y \oplus X$$

so $Y \sim X$.

Transitive: Suppose $X \sim Y$ and $Y \sim Z$, and show $X \sim Z$. It is given that $X \oplus Y$ is finite and that $Y \oplus Z$ is finite. What must be shown is that $X \oplus Z$ is finite. Figure 3.8 shows how $(X \oplus Y) - (Y \oplus Z)$ and $(Y \oplus Z) - (X \oplus Y)$ contribute to $(X \oplus Z)$.

$$\begin{aligned} (X \oplus Y) \oplus (Y \oplus Z) &= (X \oplus (Y \oplus Y)) \oplus Z \quad (\oplus \text{ is associative}) \\ &= (X \oplus \emptyset) \oplus Z \\ &= X \oplus Z \end{aligned}$$

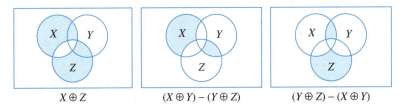

$$X \oplus Z \qquad\qquad (X \oplus Y) - (Y \oplus Z) \qquad\qquad (Y \oplus Z) - (X \oplus Y)$$

Figure 3.8 How $X \oplus Z$ is formed.

Therefore, $X \oplus Z$ is also finite, implying that $X \sim Z$. Since $\sim$ is reflexive, transitive, and symmetric, the relation $\sim$ is an equivalence relation. ∎

Some relations on $\mathbb{R}$ that were defined earlier are not equivalence relations.

Example 5.

(a) On $\mathbb{R}$, define $x \sim y$ if $|x - y| < 0.01$. Then, $\sim$ is reflexive and symmetric, but it is not transitive.

(b) On $\mathbb{R}$, the relation $Ge_{\mathbb{R}}$ is reflexive and transitive, but it is not symmetric.

Solution.

(a) Let $x = 0.0$, $y = 0.0075$, and $z = 0.015$ Then, $x \sim y$, because

$$|x - y| = 0.0075 < 0.01$$

and $y \sim z$, because

$$|y - z| = 0.0075 < 0.01$$

However, $x \not\sim z$, since

$$|x - z| = 0.015 > 0.01$$

(b) It is clear that $x \geq x$ for all $x \in \mathbb{R}$ and that for all $x, y, z \in \mathbb{R}$, if $x \geq y$ and $y \geq z$, then $x \geq z$, making the relation transitive. Since $5 \geq 3$ but $3 \not\geq 5$, the relation is not symmetric. Therefore, $Ge_{\mathbb{R}}$ is reflexive and transitive, but it is not symmetric. ■

3.6.1 Partitions

The relation *SameSuit* (see Table 3.3) is an equivalence relation on the set

$$SpecialDeck = \{10 \text{ of Hearts, Jack of Hearts, Queen of Hearts, } 10 \text{ of Clubs}$$
$$\text{Jack of Clubs, King of Clubs}\}$$

Essentially the same information can be stored in the three sets:

$$Heart = \{10 \text{ of Hearts, Jack of Hearts, Queen of Hearts}\}$$
$$Club = \{10 \text{ of Clubs, Jack of Clubs, King of Clubs}\}$$
$$Suits = \{Heart, Club\}$$

Suits consists of two sets. Each element of *SpecialDeck* is in exactly one of those sets. The cards in the first set are exactly the cards in the same suit as the 10 of Hearts. The cards in the second set are exactly the cards in the same suit as 10 of Clubs.

Definition 3. Let X be a nonempty set. A **partition** of X is a set Y of nonempty subsets of X such that every element of X is in exactly one element in Y.

A partition of *SpecialDeck* is the set *Suits*. The nonempty sets in *Suits* are *Heart* and *Club*. We can restate the definition as Theorem 1.

Theorem 1. Let X be a set, and let Y be a set of subsets of X. Then, Y is a partition of X if and only if:

(a) Each element of Y is a nonempty subset of X,
(b) For any two sets $u, v \in Y$, $u \cap v = \emptyset$ unless $u = v$, and
(c) The union of all the sets in Y is X.

Definition 4.   Let $\sim$ be an equivalence relation on a set X. For any $x \in X$, let

$$[x] = \{y \in X : x \sim y\}$$

$[x]$ is called the **equivalence class** of x.

In the example with the set *SpecialDeck* and equivalence relation *SameSuit,* we have

$$Heart = [10 \text{ of Hearts}] = [\text{Jack of Hearts}] = [\text{Queen of Hearts}]$$

and

$$Club = [10 \text{ of Clubs}] = [\text{Jack of Clubs}] = [\text{King of Clubs}]$$

The set *Suits* stores essentially the same information as the relation *SameSuit*. The next two theorems make this statement precise.

Theorem 2.   Let $\sim$ be an equivalence relation on a set X. Then:

(a) For any $x \in X, x \in [x]$.
(b) For any $x, y \in X$, either $[x] = [y]$ or $[x]$ and $[y]$ are disjoint.
(c) $\{[x] : x \in X\}$ is a partition of X.
(d) For $x, y \in X$, $x \sim y$ if and only if $y \in [x]$.

Proof.   (a) Since $\sim$ is reflexive, $x \sim x$, so $x \in [x]$.
(b) Suppose $[x]$ and $[y]$ are not disjoint, and prove $[x] = [y]$ by showing that $[x] \subseteq [y]$ and $[y] \subseteq [x]$. To prove that $[x] \subseteq [y]$, assume that $r \in [x]$, and show that $r \in [y]$—that is, that $y \sim r$.
    Since $[x] \cap [y] \neq \emptyset$, there is a $z \in [x] \cap [y]$. Thus $x \sim z$, and $y \sim z$. Since $\sim$ is symmetric, $z \sim x$. By the transitivity of $\sim$, since $y \sim z$ and $z \sim x$, $y \sim x$. Now since $y \sim x$ and $x \sim r$, $y \sim r$. So, $r \in [y]$, as required. Therefore, $[x] \subseteq [y]$.
    Analogously, $[y] \subseteq [x]$, so $[y] = [x]$.
(c) It must be shown that $\{[x] : x \in X\}$ is a set of nonempty subsets of X such that each $y \in X$ is in exactly one $[x]$. To check that the $[x]$'s are nonempty, observe that $x \in [x]$. To check that each $y \in X$ is in at least one $[x]$, observe that $y \in [y]$. To check that each $y \in X$ is in at most one $[x]$, suppose $y \in [x_1]$ and $y \in [x_2]$. Then, by part (b) $[x_1] = [x_2]$. The classes are the same, so y is in only one equivalence class.
(d) This is immediate from the definition of equivalence classes.   ■

Theorem 2 says two things. First, given an equivalence relation on a set X, its set of distinct equivalence classes form a partition of X. Second, the relation that defines two elements to be related if they are in the same element of the partition is equal to the original relation (part (d)).

Example 6.   Let *Deck* = {10 of Hearts, King of Hearts, Queen of Clubs, Ace of Clubs}. The relation *SameSuit* defined on *Deck* consists of the following ordered pairs:

(10 of Hearts, King of Hearts)	(King of Hearts, 10 of Hearts)
(10 of Hearts, 10 of Hearts)	(King of Hearts, King of Hearts)
(Queen of Clubs, Ace of Clubs)	(Ace of Clubs, Queen of Clubs)
(Queen of Clubs, Queen of Clubs)	(Ace of Clubs, Ace of Clubs)

Find the equivalence classes of this relation. Also, find the partition determined by this equivalence relation.

Solution.   The equivalence classes are

$$
\begin{aligned}
[\text{10 of Hearts}] &= \{\text{10 of Hearts, King of Hearts}\} \\
[\text{King of Hearts}] &= \{\text{10 of Hearts, King of Hearts}\} \\
[\text{Queen of Clubs}] &= \{\text{Queen of Clubs, Ace of Clubs}\} \\
[\text{Ace of Clubs}] &= \{\text{Queen of Clubs, Ace of Clubs}\}
\end{aligned}
$$

The distinct equivalence classes are the two sets

{10 of Hearts, King of Hearts}   {Queen of Clubs, Ace of Clubs}

which form a partition of *Deck*

Example 7.   Recall the equivalence relation $\equiv$ (*mod p*) of Example 3 in Section 3.6. The following are the equivalence classes of $\equiv$ (*mod* 5):

$$
\begin{aligned}
[\,0\,] &= \{0, 5, 10, 15, 20, 25, 30, \ldots\} \\
[\,1\,] &= \{1, 6, 11, 16, 21, 26, 31, \ldots\} \\
[\,2\,] &= \{2, 7, 12, 17, 22, 27, 32, \ldots\} \\
[\,3\,] &= \{3, 8, 13, 18, 23, 28, 33, \ldots\} \\
[\,4\,] &= \{4, 9, 14, 19, 24, 29, 34, \ldots\}
\end{aligned}
$$

The reader should prove that these are the equivalence classes.

In Example 8 we determine all the equivalence classes of $\equiv$ (*mod p*) for any positive integer p.

Example 8.   Let p be a positive integer. Determine the equivalence classes of $\equiv$ (*mod p*).

Solution.   We know from Example 3 in Section 3.6 that every integer is congruent to its remainder (*mod p*). Since the only possible remainders are $0, 1, \ldots, p - 1$, we have

$$
\mathbb{N} \subseteq [0] \cup [1] \cup \cdots \cup [p - 1]
$$

Thus, there are, at most, p equivalence classes—namely, $[0], [1], \ldots, [p - 1]$. We must show that these equivalence classes are all different.

Let r_1 and r_2 be two different remainders, such as $0 \leq r_1 < r_2 \leq p - 1$. We must show that $[r_1] \neq [r_2]$. Note that $r_2 - r_1$ is a positive integer less that p so that $r_2 - r_1$ is not divisible by p. Then, $r_1 \not\equiv r_2$ (*mod p*), whence $[r_1] \neq [r_2]$. Therefore, the distinct equivalence classes are $[0], [1], \ldots, [p - 1]$.

Theorem 2 says that one can go from an equivalence relation to a partition from, which one may read off the equivalence relation. Theorem 3 says that one can go from a partition to an equivalence relation, from which one may read off the partition.

Theorem 3.   Let P be a partition of a set X. For $x, y \in X$, define $x \sim y$ to mean that x and y are in the same element of the partition. Then:

(a) $\sim$ is an equivalence relation.

(b) The equivalence classes of $\sim$ are exactly the elements of P.

Proof.

(a) First, prove that $\sim$ is an equivalence relation.

Reflexive: Let $x \in X$, and show that $x \sim x$. Since P is a partition, x is in some set $Q \in P$. So, x and x are both in Q; therefore, $x \sim x$.

Symmetric: Let $x, y \in X$, and assume that $x \sim y$. That means there is a set $Q \in P$ such that $x, y \in Q$. So, y and x are in Q. Therefore, $y \sim x$.

Transitive: Suppose $x \sim y$ and $y \sim z$. Since $x \sim y$, there is a set $Q \in P$ such that $x, y \in Q$. Since $y \sim z$, z is in the same set in P as y, so $z \in Q$. Therefore, x and z are both in Q, giving $x \sim z$.

(b)

$$[x] = \{y \in X : x \sim y\}$$
$$= \{x \in X : x, y \text{ are both in the same element of } P\}$$
$$= \text{ element of } P \text{ to which } x \text{ belongs}$$

Example 9.   Let *Deck* = {10 of Hearts, King of Hearts, Queen of Clubs, Ace of Clubs}. The set

$$P = \{\{10 \text{ of Hearts, King of Hearts}\}, \{\text{Queen of Clubs, Ace of Clubs}\}\}$$

is a partition of *Deck*. Define a relation $\sim$ on *Deck* such that for $x, y \in Deck$, $x \sim y$ if and only if x and y are in the same element of P. The elements of the relation are

(10 of Hearts, King of Hearts)	(King of Hearts, 10 of Hearts)
(10 of Hearts, 10 of Hearts)	(King of Hearts, King of Hearts)
(Queen of Clubs, Ace of Clubs)	(Ace of Clubs, Queen of Clubs)
(Queen of Clubs, Queen of Clubs)	(Ace of Clubs, Ace of Clubs)

By Theorem 3 in this section, this relation is an equivalence relation for which the distinct equivalence classes are precisely the elements of P.

3.6.2   Comparing Equivalence Relations

Consider a standard deck of 52 cards, called *52Cards*. The suits are traditionally marked in two colors: Clubs and Spades are black; Diamonds and Hearts are red. The relation *Same-Suit,* consisting of all pairs of cards that are in the same suit, and the relation *SameColor,* consisting of all pairs of cards that are the same color, are both equivalence relations. The equivalence class of the 2 of Diamonds in *SameSuit* is

$$[2 \text{ of Diamonds}] = \{2 \text{ of Diamonds, 3 of Diamonds}, \dots, \text{Ace of Diamonds}\}$$

The equivalence class of the 2 of Diamonds in *SameColor* contains all the Diamonds and all the Hearts. Figure 3.9, on page 187, is a Venn diagram showing the equivalence classes of the two relations.

Each equivalence class of *SameSuit* is contained within a single equivalence class of *SameColor*.

Definition 5.   Let R_1 and R_2 be equivalence relations on a set X. R_1 **refines** R_2 if, for each $x \in X$, the equivalence class of x in R_1 is a subset of the equivalence class of x in R_2.

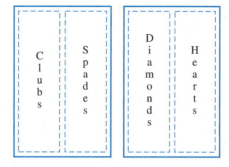

Figure 3.9   Equivalence classes of *SameSuit* (dashed lines) and *SameColor* (solid lines).

In the previous example, *SameSuit* refines *SameColor*. Also, *SameSuit* refines *Same-Suit*. Now, consider the relation *SameValue*, which is defined as consisting of all pairs of cards with the same value. The equivalence class of the 2 of Diamonds is

{2 of Diamonds, 2 of Clubs, 2 of Hearts, 2 of Spades}

This equivalence relation is shown in Figure 3.10 as a set of disjoint equivalence classes.

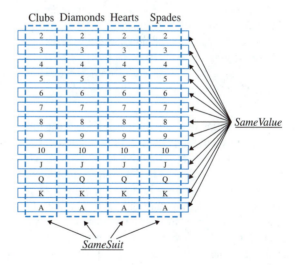

Figure 3.10   Equivalence classes *SameSuit* (vertical) and *SameValue* (horizontal).

The equivalence relation of the 2 of Diamonds under *SameValue* is not a subset of the equivalence class of the 2 of Diamonds under *SameSuit*. Hence, *SameValue* does not refine *SameSuit*. Also, *SameSuit* does not refine *SameValue*.

Theorem 4.   Let R_1 and R_2 be equivalence relations on the same set X. R_1 refines R_2 if and only if each equivalence class of R_2 is a union of equivalence classes of R_1.

Proof.   This proof is left as an exercise for the reader.

Application: *UNION-FIND*

The *UNION-FIND* algorithm has a set of elements X and a relation R defined on X as its input. The *UNION-FIND* algorithm starts with a partition of X in which each element is a set consisting of a single element of X. Each related pair of elements is processed as follows: If a related pair of elements are in different elements of the partition, those two sets of the partition are joined, forming a new partition of X with fewer elements. If the two elements are already in the same element of the partition, nothing is done.

As an example of how the algorithm operates, Table 3.10 shows a set with six elements that has a relation consisting of the pairs (0, 2), (1, 4), (2, 5), (3, 6), (0, 4), and (1, 2). The final partition has two elements, {0, 1, 2, 4, 5} and {3, 6}.

Table 3.10   *UNION-FIND* Algorithm

New Related Pair	Current Partition Defined by the Equivalence Relation
Ø	{0}, {1}, {2}, {3}, {4}, {5}, {6}
Process 0 R 2	0 and 2 are in different elements of the partition
Form new partition	{0, 2}, {1}, {3}, {4}, {5}, {6}
Process 1 R 4	1 and 4 are in different elements of the partition
Form new partition	{0, 2}, {1, 4}, {3}, {5}, {6}
Process 2 R 5	2 and 5 are in different elements of the partition
Form new partition	{0, 2, 5}, {1, 4}, {3}, {6}
Process 3 R 6	3 and 6 are in different elements of the partition
Form new partition	{0, 2, 5}, {1, 4}, {3, 6}
Process 0 R 4	0 and 4 are in different elements of the partition
Form new partition	{0, 1, 2, 4, 5}, {3, 6}
Process 1 R 2	1 and 2 are in the same element of the partition
Leave partition as is	{0, 1, 2, 4, 5}, {3, 6}

In computer science, this problem is of great interest, because it is an integral processing step in many algorithms. As an example, consider using this algorithm to find associations among a set of authors for a personal collection of journal articles about a single topic. The problem is to determine which of these authors have worked together. By starting with each author in a set by himself or herself, the articles will tell how to join pairs or sets of authors into bigger sets because they have worked together. The final outcome would be a partition of the authors such that two authors are in the same element of the partition if and only if they had worked together. The problem of determining an effective data structure for managing the information being processed is a major topic in data structures.

3.7   Exercises

1. Identify the equivalence classes of $\mathbb{N}$ for the following relations:
   (a) $\equiv (mod\ 4)$
   (b) $\equiv (mod\ 6)$

2. Determine which of the following five relations defined on $\mathbb{Z}$ are equivalence relations:

   (a) $\{(a, b) \in \mathbb{Z} \times \mathbb{Z} : (a > 0 \text{ and } b > 0) \text{ or } (a < 0 \text{ and } b < 0)\}$
   (b) $\{(a, b) \in \mathbb{Z} \times \mathbb{Z} : (a \geq 0 \text{ and } b > 0) \text{ or } (a < 0 \text{ and } b \leq 0)\}$
   (c) $\{(a, b) \in \mathbb{Z} \times \mathbb{Z} : |a - b| \leq 10\}$
   (d) $\{(a, b) \in \mathbb{Z} \times \mathbb{Z} : (a \leq 0 \text{ and } b \geq 0) \text{ or } (a \leq 0 \text{ and } b \leq 0)\}$
   (e) $\{(a, b) \in \mathbb{Z} \times \mathbb{Z} : (a \geq 0 \text{ and } b \geq 0) \text{ or } (a \leq 0 \text{ and } b \leq 0)\}$

3. Find the elements in the relation "have the same remainder when divided by 8" if the relation is defined on $\{1, 2, 3 \ldots, 24, 25\}$. Also, find the distinct equivalence classes of this equivalence relation.

4. Let *POPULATION* be the set of all people. Let R be the binary relation on *POPU-LATION* such that $(x, y) \in R$ if x is an older brother of y or $x = y$. Is R reflexive? Symmetric? Antisymmetric? Transitive? An equivalence relation?

5. Define a binary relation R on $\mathbb{R}$ as $\{(x, y) \in \mathbb{R} \times \mathbb{R} : x \text{ and } y \text{ are both positive, both negative, or both } 0\}$. Prove that R is an equivalence relation. What are its equivalence classes?

6. Define a binary relation R on $\mathbb{R}$ as $\{(x, y) \in \mathbb{R} \times \mathbb{R} : \sin(x) = \sin(y)\}$. Prove that R is an equivalence relation. What are its equivalence classes?

7. Let $A = \{a, b, c, d\}$. For each of the following partitions of A, list all the pairs of elements that form the corresponding equivalence relations:

   (a) $\{\{a, b, c\}, \{d\}\}$
   (b) $\{\{a\}, \{b\}, \{c\}, \{d\}\}$
   (c) (c) $\{\{a, b, c, d\}\}$

8. Let $A = \{a, b, c, d\}$. For each of the following partitions of A, determine the elements of the corresponding equivalence relation:

   (a) $P_1 = \{\{a, c\}, \{b, d\}\}$
   (b) $P_2 = \{\{a\}, \{b, c\}, \{d\}\}$
   (c) $P_3 = \{\{a, b\}, \{c, d\}\}$
   (d) $P_4 = \{\{a, b, c\}, \{d\}\}$

   Do any of these partitions refine any of the others?

9. Prove Theorem 1.

10. In the example *52Cards*, find a simple description for each of the following:

    (a) *SameSuit* $\cap$ *SameValue*
    (b) (*SameSuit* $\cup$ *SameValue*)*

11. (a) Draw a Venn diagram showing the equivalence classes over $\mathbb{N}$ of $\equiv$ (mod 5), $\equiv$ (mod 10), and $\equiv$ (mod 15). Which of these equivalence relations refine another one of these equivalence relations?

    (b) Let $k, m \in \mathbb{N}$. We say k is a factor of m if $m = j \cdot k$ for some j such that $j \in \mathbb{N}$ and $0 < j \leq m$. What is the relationship between whether $\equiv$ (*mod k*) refines $\equiv$ (*mod m*) and whether k is a factor of m or m is a factor of k? Prove your answer.

12. Let R and S be equivalence relations on a set X.

    (a) Show that $R \cap S$ is an equivalence relation.
    (b) Show by example that $R \cup S$ need not be an equivalence relation.
    (c) Show that $(R \cup S)^*$, the reflexive and transitive closure of $R \cup S$, is the smallest equivalence relation containing both R and S.

13. Prove Theorem 4.

14. There is an old, fallacious proof that if a relation is both symmetric and transitive, it is reflexive. We give this "proof" below. What is the error?

> Suppose R is a symmetric and transitive relation on a set X. Pick an $x \in X$. We need to show $x\, R\, x$. So, take any y where $x\, R\, y$. By symmetry, it follows that $y\, R\, x$. By transitivity, it follows that $x\, R\, x$.

15. For a relation R on a set X, let R^* denote the reflexive and transitive closure of R.

    (a) For any relation R on a set X, define a relation $\sim$ on X as follows: $x \sim y$ if and only if $x\, R^*\, y$ and $y\, R^*\, x$. Prove that $\sim$ is an equivalence relation.

    (b) Let $x_1 \sim x_2$ and $y_1 \sim y_2$. Show that $x_1\, R^*\, y_1$ if and only if $x_2\, R^*\, y_2$.

16. (a) For $k, n_1, n_2, m_1, m_2 \in \mathbb{N}$, show that if

$$n_1 \equiv n_2 (mod\, k)$$

and

$$m_1 \equiv m_2 (mod\, k)$$

then

$$n_1 + m_1 \equiv n_2 + m_2 (mod\, k)$$

and

$$n_1 \cdot m_1 \equiv n_2 \cdot m_2 (mod\ k)$$

(b) Part (a) says that if we take two equivalence classes $[\,m\,]$ and $[\,n\,]$, then we can unambiguously define $[\,m\,] + [\,n\,]$ and $[\,m\,] \cdot [\,n\,]$. Pick any $m_1 \in [\,m\,]$ and any $n_1 \in [\,n\,]$, and define

$$[\,m\,] + [\,n\,] \equiv [\,m_1 + n_1\,]$$

and

$$[\,m\,] \cdot [\,n\,] \equiv [\,m_1 \cdot n_1\,]$$

The definition is unambiguous since it doesn't matter which m_1 and n_1 we pick. Find the addition and multiplication tables for the equivalence classes of $\equiv (mod\, 4)$ and $\equiv (mod\, 5)$. (*Hint:* For both $\equiv (mod\, 4)$ and $\equiv (mod\, 5)$, your answer should include

$$[\,0\,] + [\,0\,] \equiv [\,0\,],\ [\,0\,] + [\,1\,] \equiv [\,1\,],\ [\,0\,] \cdot [\,0\,] \equiv [\,0\,]$$

and

$$[\,1\,] \cdot [\,1\,] \equiv [\,1\,]$$

but, for $\equiv (mod\, 4)$,

$$[\,2\,] + [\,2\,] \equiv [\,0\,]$$

whereas, that will be false for $\equiv (mod\, 5)$.)

Ordering Relations

In this section, we discuss two very important classes of relations, the partial orderings and the linear orderings. Partial orderings generalize the relation is a subset of ($\subseteq$), and linear orderings generalize the relation less than ($<$).

3.8.1 Partial Orderings

A typical example of a partial order, other than *IsASubsetOf*, is the relation *IsADescendantOf*. The fact that this latter relation is a partial order contributes to the difficulty in completing a person's genealogy. One of the difficulties involved in tracing a genealogy is that a line of descendants often dies out, and the search then has to find another branch of the family. The end of a line of descendants will be special elements in a partial order.

Definition 1. Let R be a binary relation on a nonempty set X. R is a **partial ordering** if R is a reflexive, transitive, antisymmetric relation.

The following are standard examples of partial orderings.

Example 1. If U is a set, then $\subseteq$ is a partial ordering on the subsets of U. This was proved in Example 6(b) in Section 3.4.2 and in Example 8(e) in Section 3.4.3.

Example 2. In Figure 3.11, there is a representation of the eight subsets of

$$U = \{0, 1, 2\}$$

Each subset is obviously a subset of itself, so the relation is reflexive. The lines going upward indicate the rest of the subset relation. Since there is a line from $\{1\}$ to $\{0, 1\}$, $\{1\}$ is shown to be a subset of $\{0, 1\}$. Since there is a line from $\emptyset$ to $\{1\}$ and another from $\{1\}$ to $\{0, 1\}$, $\emptyset$ is shown to be a subset of $\{0, 1\}$. (Thus, reflexivity, antisymmetry, and transitivity are all assumed in the way the drawing is interpreted.)

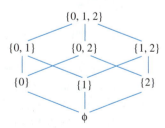

Figure 3.11 Subsets of $\{0, 1, 2\}$.

Figure 3.12 pictures the eight subsets of $\{0, 1, 2, 3\}$ having an odd number of elements. These elements also form a partial order with respect to the relation $\subseteq$. By the same argument, $\supseteq$ is also a partial ordering on any set of sets. You just need to turn the picture upside down to reverse the direction—that is, $\{0, 1, 2\} \supseteq \{0\}$.

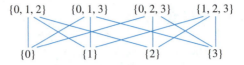

Figure 3.12   Odd subsets of $\{0, 1, 2, 3\}$.

Example 3.

(a) The relation $\leq$ is a partial ordering on $\mathbb{N}$. This follows from Example 6(a) in Section 3.4.2 and Example 8(c) in Section 3.4.3. By the same argument, $\geq$ is a partial ordering on $\mathbb{N}$.

(b) The relation $<$ is not a partial ordering, since it is transitive and antisymmetric but is not reflexive. In fact, it is irreflexive. Irreflexive relations whose reflexive closures are partial orderings are called **strict partial orderings.** So, $<$ is a strict partial ordering.

(c) On any set X, the relation $=$ is a partial ordering. This result follows from Example 4(a) in section 3.4.2 and Example 8(a) in Section 3.4.3. ∎

Example 4.   Figure 3.13 shows a subset of the family tree given in Figure 3.2.

Elaine   Maude

George   Elizabeth

Figure 3.13   Subset of family tree.

Let R be the reflexive closure of the relation "ancestor of" as defined by this subset of the family tree. Then, R is a partial ordering. The elements of the partial order are

$$\{(Elaine, George), (Maude, Elizabeth), (Elaine, Elaine), (George, George),$$
$$(Maude, Maude), (Elizabeth, Elizabeth)\}$$ ∎

Example 5.

(a) Let

$$R = \{(x, y) : x, y \in \mathbb{N} \text{ and } y \geq x \text{ and } y - x \text{ is even}\}$$

Then, R is a partial ordering on $\mathbb{N}$. (See Exercise 5 in Section 3.9.)

(b) Let $|$ denote the relation **divides** on $\mathbb{N}$. That is, $x \mid y$ if, for some $z \in \mathbb{N}$, $y = x \cdot z$. Then, the relation $|$ is a partial ordering on $\mathbb{N}$.

As another, less familiar example of a partial order, we use the relation divides on the set

$$\{0, 1, 2, 3, \ldots, 11, 12\}$$

to define a partial order by the relation $x \mid y$ if and only if $y = k \cdot x$ for some integer k. Figure 3.14 shows how these elements are related by $|$.

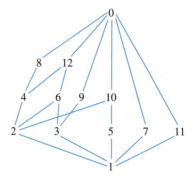

Figure 3.14   Divides for {0, 1, 2, ..., 12}.

Example 6.   Let X be a collection of finite sets taken from some universal set U. Let

$$R = \{(U, V) : U, V \in X \text{ and } |U| \le |V|\}$$

Then, R is reflexive and transitive, but it is not antisymmetric.

Solution.   Observe that if $U = \{0, 1, 2\}$, then

$$|\{0, 1\}| \, R \, |\{1, 2\}|$$

and

$$|\{1, 2\}| \, R \, |\{0, 1\}|$$

but

$$\{0, 1\} \ne \{1, 2\}$$

Therefore, R need not be antisymmetric.   ■

Example 7.   Let X be a collection of finite sets. Let

$$R = \{(U, V) : U, V \in X \text{ and } (|U| < |V| \text{ or } U = V)\}$$

The relation R with $X = \mathcal{P}\,(\{0, 1, 2\})$ is shown in Figure 3.15. The lines between levels in the figure represent the fact that the two sets are related. R is a partial ordering.

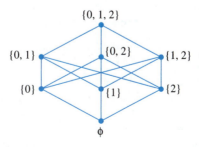

Figure 3.15 R on $\mathcal{P}(\{0, 1, 2\})$.

Solution.   We must show that R is reflexive, antisymmetric, and transitive.

Reflexive: Let $U, V \subseteq X$. If $U = V$, then $(U, V) \in R$ by definition of R.

Antisymmetric: Let $(U, V) \in R$, and suppose $U \neq V$. Then, $|U| < |V|$, so $|V| \not< |U|$. Thus, $(V, U) \notin R$.

Transitive: Let $U, V, W \subseteq X$. Let $(U, V) \in R$ and $(V, W) \in R$. Show that $(U, W) \in R$. There are four cases, depending on why $(U, V) \in R$ and why $(V, W) \in R$.

Case 1: $U = V$, and $V = W$. Then, $U = W$, so $U\ R\ W$.

Case 2: $U = V$, and $|V| < |W|$. Then, $|U| = |V| < |W|$, so $(U, W) \in R$.

Case 3: $|U| < |V|$, and $V = W$. This proof is analogous to the proof for Case 2.

Case 4: $|U| < |V|$, and $|V| < |W|$. Since $<$ is a transitive relation on $\mathbb{N}$, $|U| < |W|$. Hence, $(U, W) \in R$.

Since R is reflexive, antisymmetric, and transitive, R is a partial order.   ■

3.8.2   Linear Orderings

The relation less than $(<)$ on the integers has the property that for any n and m with $m \neq n$, either $n < m$ or $m < n$. This property is not true for the relation of set inclusion $(\subseteq)$. The set $X = \{0, 1, 2, 3\}$ has subsets $x = \{0, 2\}$ and $y = \{0, 1, 3\}$ for which neither is a subset of the other. Relations other than ones defined on a number system sometimes, however, satisfy this property, which makes it an important property of ordering relations.

Definition 2.   Let R be a binary relation on a set X. R is a **linear ordering,** or **total ordering,** on X if R is a transitive relation that satisfies the **law of trichotomy:** For every $x, y \in X$, *exactly one* of the following conditions holds: (i) $x\ R\ y$, (ii) $x = y$, or (iii) $y\ R\ x$.

Example 8.   The following are linear orderings:

(a) $<$ is a linear ordering on $\mathbb{R}$. The name *linear ordering* suggests points on a line, and $\mathbb{R}$ is the standard mathematical model of a line. Condition (ii) is never true for this relation!
(b) $<$ is a linear ordering on $\mathbb{N}$.
(c) Let M be the set of kings and queens of England since 1850. For $X, Y \in M$, set $X\ R\ Y$ if X ruled before Y. Then, R is a linear ordering on M.

The relation $\leq$ on $\mathbb{R}$ is not a linear ordering, because for any $x \in \mathbb{R}$, both $x = x$ and $x \leq x$ hold. The law of trichotomy requires that exactly one of the three properties hold.

Example 9.   (Lexicographical or Dictionary Ordering)   The alphabetical (dictionary) ordering of words is the basis for being able to sort sets of words in increasing or decreasing order. For example, let *English* be the set of words in the latest edition of the *Oxford English Dictionary,* and let $<$ be their alphabetical ordering, in which the letters of the alphabet are ordered from *a* to *z,* with *blank* being less than *a*. For this example, we will assume that all words in the dictionary begin with lowercase letters. (With computers, lowercase and uppercase letters have different representations.) Describe how two words are compared using this ordering.

Solution. Given two words, we will say that the one occurring first in the dictionary is *less than* ($<$) the other. For example,

$$elephant < tiger$$
$$aardvark < ant$$

and

$$oz < ozymandias$$

The first letters of *elephant* and *tiger* determine that *elephant* is less than *tiger*. In the second pair of words, the first two letters in the same position that are different are *a* and *n,* which occur in the second letter position. In the third pair of words, the first two letters that are different in the same position are *y* and *blank*. The rule can be thought of most easily as follows: Think of a word as an infinite string of symbols where all but the first finitely many are blank. Now, to compare two words, look for the leftmost position at which the two words contain different letters. For example,

$$
\begin{array}{ll}
a\ a\ r\ d\ v\ a\ r\ k & o\ z \\
\quad\updownarrow & \qquad\updownarrow \\
a\ n\ t & o\ z\ y\ m\ a\ n\ d\ i\ a\ s
\end{array}
$$

The smaller word of the pair is defined to be the one with the "smaller" symbol in the position where the two words first differ. What the rule says is that you should look up both words in a "dictionary" and then designate the first of the two words you come to, starting from the front of the dictionary, as the smaller word. The described ordering gives a linear ordering of all the words of a dictionary. ∎

Extended ASCII Code

The storage of uppercase and lowercase letters of the alphabet in a computer often is done by assigning an 8-bit binary code to each. A common computer code is the extended ASCII code. Special characters and numerals as well as control codes are also assigned codes, but the focus here is on the idea of what is happening. To be able to sort words, the code for "A" must be easily recognized as smaller than the code for "B," "C," and so on. The extended ASCII code for "A" is 01000001; the code for "B" is 01000010. Using the lexicographical ordering on the bit positions starting at the left, the code for "A" is clearly smaller than the code for "B":

$$
\begin{array}{ccc}
\downarrow & & \downarrow \\
01000001 < & & 01000010 \\
\text{"A"} & < & \text{"B"}
\end{array}
$$

The complete extended ASCII code assigns 8-bit binary strings to each letter of the alphabet so that

$$\text{"A"} < \text{"B"} < \text{"C"} < \cdots < \text{"X"} < \text{"Y"} < \text{"Z"}$$

3.8.3   Comparable Elements

Definition 3.   Let R be a partial or linear ordering on a set X. Elements $x, y \in X$ are said to be **comparable** under R if $x\, R\, y$ or $y\, R\, x$ (or both) holds.

Example 10.   For $X = \{0, 1, 2, 3\}$ partially ordered by the relation set inclusion $\mathcal{P}(X)$, then, $\{0, 1\}$ and $\{0, 1, 2\}$ are comparable, but $\{0, 2, 3\}$ and $\{0, 1\}$ are not.

Observe that if R is a linear ordering on a set X with $x, y \in X$ and $x \neq y$, then x and y are comparable by the law of trichotomy. Observe also that if R and S are linear or partial orders such that $R \subseteq S$, then if x and y are comparable in R, they are also comparable in S.

Theorem 1.

(a) If R is a linear ordering of a set X, then $R \cup Id_X$ is a partial ordering of X.
(b) If R is a partial ordering of X, then $R - Id_X$ is a linear ordering of X if and only if, for any $x, y \in X$, the elements x and y are comparable under R.

Proof.   (a)   This proof is left as an exercise for the reader.
(b) ($\Rightarrow$)   First, suppose $R - Id_X$ is a linear ordering. Let $x, y \in X$. It is necessary to show that x and y are comparable under R. If $x \neq y$, then x and y are comparable in $R - Id_X$ and, hence, in R by the observation before the theorem. If $x = y$, then $(x, y) = (x, x) \in R$, because $Id_X \subseteq R$.
($\Leftarrow$)   Let $x, y \in X$. Note that since R is antisymmetric,

$$(x, y) \in R - Id_X \Rightarrow (y, x) \notin R$$

Transitive: Let $(x, y), (y, z) \in R - Id_X$. Then, $(x, z) \in R$, because R is transitive. Furthermore, $x \neq z$ since $(y, z) \in R$, whereas since R is antisymmetric, $(y, x) \notin R$. Therefore, $(x, z) \in R - Id_X$.

Trichotomy: We must show exactly one of (i) $(x, y) \in R - Id_X$, (ii) $x = y$, or (iii) $(y, x) \in R - Id_X$ holds. We see that *at most* one of these can hold from the antisymmetric property of R and the obvious fact that

$$(R - Id_X) \cap Id_X = \emptyset$$

To see that *at least* one of these holds, let $x, y \in X$ with $x \neq y$. Since x and y are comparable under R, we have $(x, y) \in R$ or $(y, x) \in R$. Since $x \neq y$, either $(x, y) \in R - Id_X$ or $(y, x) \in R - Id_X$. ∎

Theorem 1 shows that there are two differences between partial and linear orderings:

1. Partial orderings are reflexive, whereas linear orderings are irreflexive.
2. Any two unequal elements of a linearly ordered set are comparable. This need not be true with partial orderings.

3.8.4   Optimal Elements in Orderings

The next property to investigate in an ordering relation is whether an ordering contains an element that is **optimal** in the sense that it is "larger" or "smaller" than any element to which it is comparable. This element may not be unique; for example, $\{\{1\}, \{1, 3\}, \{2\}\}$ under the relation $\subseteq$ has both $\{1, 3\}$ and $\{2\}$ as "larger" than any element(s) to which they are comparable. The properties of interest are more formally defined here.

Definition 4.   Let R be a partial ordering or a linear ordering on a set X. For $x, y \in X$, if $x \mathrel{R} y$ and $x \neq y$, then x is **below** y. We say x is **above** y if y is below x.

Example 11.   Let $X = \{1, 2, 3, 4\}$ be a set. $\mathcal{P}(X)$ together with $\subseteq$ is a partial order. $\{1\}$ is below $\{1, 2\}$. $\{1, 2\}$ is below $\{1, 2, 3, 4\}$. $\{2, 3\}$ is above both $\{2\}$ and $\{3\}$. $\{1, 2, 3, 4\}$ is above each element of $\mathcal{P}(X)$ distinct from itself.

Observe that the relations "above" and "below" are transitive.

Definition 5.   Let R be a partial or a linear ordering on a set X. Let $x \in X$.

(a) x is a **minimal** element of X if there is no $y \in X$ such that y is below x.
(b) x is the **minimum** element of X if x is below every other element of X.
(c) x is a **maximal** element of X if there is no $y \in X$ such that y is above x.
(d) x is the **maximum** element of X if x is above every other element of X.

In contexts where it is not clear what ordering is being discussed, write R-minimal, R-minimum, R-maximal, and R-maximum to clarify that the ordering relation is R.

Consider the ordering shown in Figure 3.16. In this ordering, A is the maximum element and the only maximal element. D, E, and F are all minimal elements. There is no minimum element.

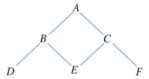

Figure 3.16   A partial ordering P.

Turning the order in Figure 3.16 upside down produces the order shown in Figure 3.17. In this ordering, A is the minimum element, and D, E, and F are maximal elements. There is no maximum element.

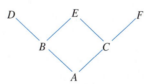

Figure 3.17   Partial ordering P upside down.

Theorem 2.   Let R be a partial ordering on X, and let $x, y \in X$.

(a) If both x and y are minimum elements, then $x = y$. This justifies speaking of *the* minimum element.
(b) If x is the minimum element of X, then x is the unique minimum element of X.
(c) If R is a linear ordering on X, then x is minimal if and only if x is the minimum element.
(d) An element $x \in X$ is R-minimal if and only if x is R^{-1}-maximal, and x is the R-minimum element if and only if x is the R^{-1}-maximum element.
(e) The analogous results to parts (a) through (d) are true, with *minimum* replaced with *maximum* and *minimal* with *maximal*.

Proof.   Proofs of (a) through (e) are left as exercises for the reader.   ■

For infinite sets like $\mathbb{Z}$, there is no minimum, maximum, minimal, or maximal element. Every finite partially ordered set has at least one minimal element and at least one maximal element. Every finite linearly ordered set has exactly one minimum element and exactly one maximum element. This result (for minimal elements and minimums) is proved in Theorem 3.

Theorem 3.   Let R be a partial ordering on a *finite* set X, and let $x \in X$.

(a) Either x is minimal or there is a minimal element $y \in X$ below x.
(b) If x is the only minimal element of X, then x is the minimum element.
(c) If R is a linear ordering, then there is a minimum element in X.

Proof.   (a) Let $z_1 \in X$. If z_1 is not minimal, there is some $z_2 \in X$ that is below z_1. If z_2 is not minimal, we can find a z_3 below z_2. Continue in this fashion. (See Figure 3.18.) Since X has only finitely many elements, the process must terminate after, at most, $|X|$ steps, finding an element z_k for which $k \le |X|$ and for which there is no element of X below z_k. Then, z_k is a minimal element below z_1.

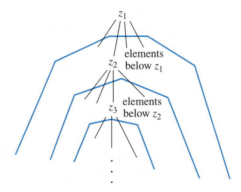

Figure 3.18   Elements below x and z.

(b) Let $z_0 \in X$ be the only minimal element, and let $z \in X$. By part (a), there is some minimal element below z. That minimal element must be z_0 itself, because z_0 is the *only* minimal element. So, z_0 is the minimum element.
(c) By part (a), there is a minimal element of X. By Theorem 2(c), that element is the minimum element.   ■

Of course, exactly analogous results hold for maximal and maximum elements in finite sets. The reader should construct examples to show that these results do not necessarily hold if the set is infinite.

3.8.5   Application: Finding a Minimal Element

The proof of Theorem 3(a) suggests an algorithm that can be used for finding a minimal element of a finite set where R is a partial or linear ordering.

Algorithm: Finding a Minimal Element

INPUT: A finite set $X = \{x_1, x_2, \ldots, x_n\}$ with an ordering relation R on X
OUTPUT: An R-minimal element of X

$y = x_1$
for $i = 2$ to n do
    if $(x_i \, R \, y$ holds) then
        $y = x_i$
print y

Example 12.   Find a minimal element in the partial order shown:

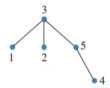

Solution.

Data Values
$x_1 = 5$
$x_2 = 3$
$x_3 = 4$
$x_4 = 1$
$x_5 = 2$

Tracing the Execution	
$y = x_1$	$y = 5$
for $i = 2$	
if $x_2 \, R \, y$	means if $3 \, R \, 5$
	R does not hold for 3 and 5
for $i = 3$	
if $x_3 \, R \, y$	means if $4 \, R \, 5$
	$y = 4$
for $i = 4$	
if $x_4 \, R \, y$	means if $1 \, R \, 4$
	R does not hold for 1 and 4
for $i = 5$	
if $x_5 \, R \, y$	means if $2 \, R \, 5$
	R does not hold for 2 and 5
Print final value: $y = 4$	

3.8.6   Application: Embedding a Partial Order

One fairly typical application of partial orderings is to schedule a set T of tasks. Usually, a set of tasks includes requirements that certain tasks be completed before others begin. If it is possible to do the tasks so that all the constraints are satisfied, these requirements may be treated as a partial ordering R on the set of tasks where, for $x, y \in T$, we have $(x, y) \in R$ if and only if x must be completed before y may be begun.

Schedules to do these tasks, on the other hand, are often linear orders, since normally, only one task can be done at a time. Hence, there is a problem of finding a linear ordering S of T so that, if $x\,R\,y$ and $x \neq y$, then $x\,S\,y$. This clearly amounts to finding a linear ordering S so that $R - Id_T \subseteq S$. For the partial ordering shown in Figure 3.17, the linear ordering S could consist of the pairs $\{(A, B), (B, D), (D, C), (C, E), (E, F)\}$ together with the pairs needed to make the relation transitive. Another linear order that would satisfy the condition consists of the pairs $\{(A, C), (C, B), (B, F), (F, E), (E, D)\}$ together with the other pairs needed to make the relation transitive. The process of finding a linear order associated with a partial order is called **embedding** a partial order in a linear order.

Example 13.   Construct a schedule for logging on to a computer and both checking email and modifying a text file. Checking email includes opening the mailer and both replying to a new message and creating a new message to another person. Modifying the text file involves opening a text editor, loading a file, editing the first paragraph of the file, inserting a separate file at the end of the file, and saving the modified file. The user is allowed to move back and forth between the mailer and the text editor for separate tasks.

Solution.   First, draw a diagram representing the dependency among various activities:

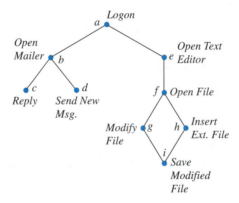

Partial order

Next, find a linear order that embeds this partial order. One result is shown here:

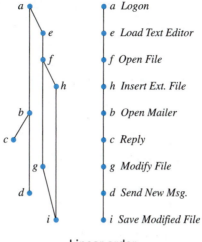

Linear order

In Chapter 6, we will examine and analyze an algorithm called *Topological Sort* that carries out the embedding of a partial order in a linear order.

3.9   Exercises

1. (a) Draw the diagram to represent the | (divides) partial order on {1, 2, 3, 4, 5, 6}.
   (b) List all the maximal, maximum, minimal, and minimum elements.
2. (a) Draw a diagram to represent the | (divides) partial order on {0, 1, 2, 3, 4, 5, 6, 7, 8, 9, 10, 11}.
   (b) Identify all minimal, minimum, maximal, and maximum elements in the diagram.
3. (a) Draw a diagram to represent the | (divides) partial order on the set {1, 2, 3, 4, 5, 6, 7, 8, 9, 10, 11}.
   (b) Identify all minimal, minimum, maximal, and maximum elements in the diagram.
4. Draw a diagram to represent the | (divides) partial order on the following:
   (a) {1, 11}
   (b) {1, 3, 7, 21}
   (c) {1, 2, 3, 4, 6, 9, 12, 18, 36}
   (d) {1, 2, 4, 8, 16, 32, 64}
5. Prove that Examples 5(a) and (b) are partial orderings.
6. Let

$$X = \{-5, -4, -3, -2, -1, 0, 1, 2, 3, 4, 5\}$$

   For $x, y \in X$, set $x \, R \, y$ if $x^2 < y^2$ or $x = y$. Show that R is a partial ordering on X. Draw a diagram of R.
7. (a) Explain why the relation "is older than or the same age" is a partial order.
   (b) Explain why the relation "is older than" is not a linear order.

8. Construct the partial order represented by the family tree shown here. The relation is "is a descendant of."

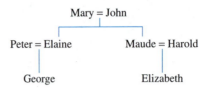

9. For the set of all people, prove that the relation "weighs no more than" is not a partial order.

10. For the set of all people, prove that the relation "weighs less than" is not a linear order.

11. (a) For $x, y \in \mathbb{N}$, define $x \mid_{pr\mathbb{N}} y$ if, for some $z \in \mathbb{N}$, $z \neq 0$, $z \neq 1$, $z \cdot x = y$. We say x is a proper divisor of y. Is $\mid_{pr\mathbb{N}}$ a linear ordering on $\mathbb{N}$?

    (b) In the real numbers $\mathbb{R}$, define $x \mid_{pr\mathbb{R}} y$ if, for some $z \in \mathbb{R}$, $z \neq 0$, $z \neq 1$, $z \cdot x = y$. Is $\mid_{pr\mathbb{R}}$ a linear ordering on $\mathbb{R}$?

12. Prove Theorem 1(a).

13. For the partial orders shown in Figures 3.11, 3.12, 3.14, and 3.15, identify all minimal, minimum, maximum, and maximal elements.

14. Suppose A, B, C, D, E, and F are tasks that must be performed with the precedence shown:

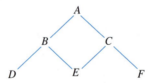

For example, E must be completed before either B or C can be performed, but D, E, and F can be completed in any order relative to one another. Let $T = \{A, B, C, D, E, F\}$, and define the partial order R on T as represented by the diagram. Find a linear order S on T where $R - Id_T \subseteq S$.

15. *Challenge:* Find a partial ordering with exactly one minimal element but where that element is not a minimum element.

16. Prove Theorem 2. (*Hint:* The proof of part (e) should be quite short.)

3.10   Relational Databases: An Introduction

A **database** is a shared collection of interrelated data designed to meet the varied information needs of an organization. To describe many interrelationships among many types of objects, there needs to be a good way to represent these interrelationships. The diagram of Figure 3.2 is a clear illustration of a family tree, but it uses certain specific facts about family relationships—for example, that each person has exactly two parents. It would be much harder to represent more complicated relationships using the same type of diagram.

A database system provides a framework for representing complex relationships. In this section, we will discuss one model for a database system called a **relational database system**. The reason we call this model a relational database system will become clear as we work through an example. To simplify the discussion, we will present simplified versions of the database operations.

3.10.1   Storing Information in Relations

To introduce some of the features of a relational database system, we consider the relational representation of a familiar problem: How can we keep track of student registrations in classes and teaching assignments of instructors. This section shows how a relational database system could be used.

The first requirement is to store the information about which students have registered for which classes at a university. In this example, John von Neumann, Emmy Noether, and Herman Hollerith are all taking English 101, section 3. George Boole, René Descartes, and Winston Churchill are taking English 101, section 4. John von Neumann and Emmy Noether are also taking English 103, section 1. George Boole and Winston Churchill are also taking Mathematics 101, section 1. Finally, René Descartes and Herman Hollerith are also taking Computer Science 103, section 3. This information is collected in Table 3.11.

Table 3.11   *Registration* Relation

Registration			
Student	*Department*	*Course*	*Section*
John von Neumann	English	101	3
Emmy Noether	English	101	3
Herman Hollerith	English	101	3
George Boole	English	101	4
René Descartes	English	101	4
Winston Churchill	English	101	4
John von Neumann	English	103	1
Emmy Noether	English	103	1
George Boole	Mathematics	101	1
Winston Churchill	Mathematics	101	1
René Descartes	Computer Science	103	3
Herman Hollerith	Computer Science	103	3

In a relational database, the n-tuples in an n-ary relation are simply called **tuples.** The relations themselves are called **tables.** Each column in a table is an **attribute,** and the values that appear in that column are referred to as **values** of that attribute.

In this example, many other 4-tuples (or **quadruples**) could be in the *Registration* relation, such as (George Boole, English 103, 4) or (Herman Hollerith, Mathematics, 103, 3). A 4-tuple is in the relation only if the student is registered for that section of that course.

Now, suppose a second relation is defined that records the professors for the various courses. It is possible to make a 5-ary relation that stores all the information in *Registration* plus the name of the professor for each course. However, the information about who is teaching a course is often used for purposes independent of determining who is registered for the course. It therefore is better to store the new information in a separate table. The information about the professors is contained in the relation *TeachingAssignments,* which is shown in Table 3.12. The value of the relational database will be seen when we explain how information from various tables can be combined to answer questions. In this case, we might want to use the two tables *Registration* and *TeachingAssignments* to list the professors of a particular student.

Table 3.12   *TeachingAssignments* Relation

TeachingAssignments			
Department	*Course*	*Section*	*Professor*
English	101	3	Geoffrey Chaucer
English	101	4	William Morris
English	103	1	Thomas Jefferson
Mathematics	101	1	David Hilbert
Mathematics	101	1	Leonardo of Pisa
Computer Science	103	3	Alan Turing

Some information from *Registration* is repeated in *TeachingAssignments*. One problem in designing the relations in a relational database systems is to manage the needed redundancy in a set of tables.

Each row in the table *TeachingAssignments* is a 4-tuple, and the relation is the set of 4-tuples that record the teaching assignments for each course. Note that David Hilbert and Leonardo of Pisa are probably team-teaching Mathematics 101, section 1.

Finally, because the total teaching program in each department is the responsibility of a department chair, a relation that gives this information is needed. This relation consists of a set of tuples of length two, or ordered pairs, as seen in Table 3.13.

Table 3.13   *DepartmentChair* Relation

DepartmentChair	
Department	*Chair*
English	Francis Bacon
Mathematics	Carl Gauss
Computer Science	Alan Turing

A set of relations, such as the three shown in this example, are the data used by a relational database system.

3.10.2   Relational Algebra

In designing the data for a database system, three things are important. First, how are the data and relationships stored? Second, how can the data be modified? Third, can information be extracted? As already noted, the data and relationships are stored in tables. Methods to modify the data will not be discussed here; a course devoted to file processing will spend much time dealing with just the problems you face in implementing a database system.

Relational databases have standard operations that act on relations. A request to extract data from the database is called a **query.** Queries use standard operations to create their output. The standard set of operations used is called the **relational algebra.** Three of the operations of the relational algebra are described in the examples that follow.

First Operation: Selection

Given a relation such as *Registration,* some users may be interested in only some of the values of an attribute. As an example, for an attribute such as *Department,* and a set of possible values for that attribute, such as {Mathematics, Computer Science}, form a new relation by selecting only the tuples with a value of *Department* that is in {Mathematics, Computer Science}.

Table 3.14 repeats the relation *Registration* so that you can easily compare this relation to the one that will be generated by this operation.

Table 3.14   *Registration* Relation

Registration			
Student	*Department*	*Course*	*Section*
John von Neumann	English	101	3
Emmy Noether	English	101	3
Herman Hollerith	English	101	3
George Boole	English	101	4
René Descartes	English	101	4
Winston Churchill	English	101	4
John von Neumann	English	103	1
Emmy Noether	English	103	1
George Boole	Mathematics	101	1
Winston Churchill	Mathematics	101	1
René Descartes	Computer Science	103	3
Herman Hollerith	Computer Science	103	3

The result of this **selection** operation is the relation R', which is shown in Table 3.15.

Table 3.15 R' Relation

R'			
Student	*Department*	*Course*	*Section*
George Boole	Mathematics	101	1
Winston Churchill	Mathematics	101	1
René Descartes	Computer Science	103	3
Herman Hollerith	Computer Science	103	3

Suppose a user wants to make a selection query of a database. A selection query returns a table with just the tuples that satisfy some condition, like students taking mathematics courses. The database contains relations $R_1, R_2, \ldots, R_n$. To specify a selection query, the user inputs three things: the name of the relation from which the selection is to be made (that is, some R_i), the (name of the) attribute on which the selection is to be made, and a finite set of possible values for that attribute. Then, the database system outputs all tuples in that relation with a value for the attribute that is in that finite set.

There is also a second form that we shall use in the exercises: The user may input the name of the relation, the names of two attributes, and $=$ or $<$. If the user inputs *Teaching-Assignments, Section, Course,* and $<$, then the user is asking for all scheduled courses (for which teachers have been assigned) where the section number is less than the course number.

What we have given here is a much more limited than the standard database definition of selection. We have adopted this definition to keep the exposition simple.

Second Operation: Projection

For any table in a relational database, it often happens that a query is only interested in one attribute. For example, in the relation R' in Table 3.14, suppose that you want to know the names of the students. The only attribute of interest is *Student.* The attributes *Department, Course,* and *Section* all may be important in other contexts, but for now, only the *Student* entries are needed. The operation that reduces a relation to a new relation consisting of some of the attributes and the entries for those attributes is called **projection.**

The second operation, or projection, is now used to find a relation that consists of some of the attributes of an existing relation. A relation, such as *Registration,* and a subset of its attributes, such as {*Student, Department,*} form the projection $R^\dagger$ of the relation onto those attributes as follows: First, delete the attributes not in {*Student, Department*} from each tuple of the relation *Registration.* The resulting relation $R^\dagger$ is shown in Table 3.16.

Table 3.16 $R^\dagger$ Relation with Duplicates

$R^\dagger$	
Student	*Department*
John von Neumann	English ←
Emmy Noether	English
Herman Hollerith	English
George Boole	English
René Descartes	English
Winston Churchill	English
John von Neumann	English ←
Emmy Noether	English
George Boole	Mathematics
Winston Churchill	Mathematics
René Descartes	Computer Science
Herman Hollerith	Computer Science

In Table 3.16, you see that the tuples (John von Neumann, English) and (Emmy Noether, English) occur twice. Since a relation is a set, it makes no sense to say twice that a tuple is an element of a set. So, the final step in forming a projection is to eliminate duplicate entries from the table $R^\dagger$ to form the relation *Registration'* shown in Table 3.17.

The projection of *Registration* tells which students are taking classes in which departments.

Table 3.17 *Registration'* Relation

Registration'	
Student	*Department*
John von Neumann	English
Emmy Noether	English
Herman Hollerith	English
George Boole	English
René Descartes	English
Winston Churchill	English
George Boole	Mathematics
Winston Churchill	Mathematics
René Descartes	Computer Science
Herman Hollerith	Computer Science

Example 1. Projection is actually a common operation in areas other than databases. Look at graphing relations on $\mathbb{R}^2$, and consider the relation

$$C = \{(x, y) \in \mathbb{R}^2 : (x - 2)^2 + (y - 2)^2 = 1\}$$

The graph of C is a circle in the plane with center $(2, 2)$ and radius 1. Figure 3.19 shows the graph of C and its projection onto the x-axis.

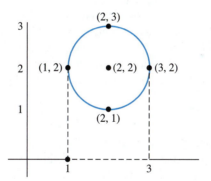

Figure 3.19 The projection of the circle $(x - 2)^2 + (y - 2)^2 = 1$ onto the x-axis.

Two values or points on the circle are projected onto each element in the open interval $(1, 3)$.

Third Operation: Join

Consider two relations, such as *Registration* and *TeachingAssignments*. Recall it was argued that since they really store different information, they need to be two separate tables. Nevertheless, some people using the system will want to know the combined information—that is, which students are taking which classes (departments, course numbers, and section numbers) taught by which professors. The **join** of the two relations puts all the information together. The relation that is needed, called *JoinedRelation,* is shown in Table 3.18. The question is how to arrive at this table starting with the tables *Registration* and *TeachingAssignments*.

Table 3.18   *JoinedRelation*

JoinedRelation				
Student	*Department*	*Course*	*Section*	*Professor*
John von Neumann	English	101	3	Geoffrey Chaucer
Emmy Noether	English	101	3	Geoffrey Chaucer
Herman Hollerith	English	101	3	Geoffrey Chaucer
George Boole	English	101	4	William Morris
René Descartes	English	101	4	William Morris
Winston Churchill	English	101	4	William Morris
John von Neumann	English	103	1	Thomas Jefferson
Emmy Noether	English	103	1	Thomas Jefferson
George Boole	Mathematics	101	1	David Hilbert
Winston Churchill	Mathematics	101	1	David Hilbert
George Boole	Mathematics	101	1	Leonardo of Pisa
Winston Churchill	Mathematics	101	1	Leonardo of Pisa
René Descartes	Computer Science	103	3	Alan Turing
Herman Hollerith	Computer Science	103	3	Alan Turing

After defining the join of two relations and giving a small example, we will present an algorithm that could be used to actually find the join of two relations.

The formation of the join of two relations is a three-step process. In the first step, we take two relations, R with attributes $A_1, A_2, A_3, \ldots, A_m$ and S with attributes $B_1, B_2, B_3, \ldots, B_n$, and form the **database Cartesian product.** The database Cartesian product $R \times S$ is a relation with attributes $A_1, A_2, \ldots, A_m, B_1, B_2, \ldots, B_n$, and its tuples are

$$\{(a_1, a_2, \ldots, a_m, b_1, b_2, \ldots, b_n) : (a_1, \ldots, a_m) \in R \text{ and } (b_1, b_2, \ldots, b_n) \in S\}$$

Note that the database definition of the term *Cartesian product* differs slightly from the set theoretic notion. The set theory definition results in 2-tuples, whereas here, all the coordinates are kept without extra parentheses.

The second step of the process involves forming the **equijoin** of R and S on attributes A_i and B_j by selecting all tuples from $R \times S$ where the values of attributes A_i and B_j are the same. To form the equijoin on several pairs of attributes, $A_{i_1}, B_{j_1}, A_{i_2}, B_{j_2}, \ldots, A_{i_k}, B_{j_k}$, perform k selections; that is, select tuples whose values on A_{i_1} and B_{j_1} are the same, whose values on A_{i_2} and B_{j_2} are the same, and so on. Often, the names of attributes A_i and B_j will in practice be the same. In that case, we may say we are taking the join on attribute A_i.

Note that in the equijoin, the attributes A_i and B_j contain the same information. The third step of the join eliminates the second of the duplicated columns: The join of R and S on attributes A_i and B_j is the projection of the equijoin on attributes $A_1, \ldots, A_m, B_1, \ldots, B_{j-1}, B_{j+1}, \ldots, B_n$. The join on several attribute pairs omits ("projects out") the second attribute from each pair.

The **natural join** of relations R and S, written $R \bowtie S$, is the join of R and S on all attribute pairs with the same name.

Example 2. Define the relations R and S as shown:

R		
Name	*Class*	*Average*
Joe	2004	3.14
Sue	2004	2.97
Mary	2005	3.76

S	
Name	*Major*
Joe	Mathematics
Sue	Computer Science
Mary	Sociology

Form the join of R and S on *Name*.

Solution. First, form the database Cartesian product $R \times S$.

Database Cartesian Product of $R \times S$				
Name	*Class*	*Average*	*Name*	*Major*
Joe	2004	3.14	Joe	Mathematics
Joe	2004	3.14	Sue	Computer Science
Joe	2004	3.14	Mary	Sociology
Sue	2004	2.97	Joe	Mathematics
Sue	2004	2.97	Sue	Computer Science
Sue	2004	2.97	Mary	Sociology
Mary	2005	3.76	Joe	Mathematics
Mary	2005	3.76	Sue	Computer Science
Mary	2005	3.76	Mary	Sociology

Now, form the equijoin: Extract the subset of $R \times S$ for which the entries for *Name* are equal, giving R'.

$R \times S$				
Name	*Class*	*Average*	*Name*	*Major*
Joe	2004	3.14	Joe	Mathematics
Sue	2004	2.97	Sue	Computer Science
Mary	2005	3.76	Mary	Sociology

Finally, project $R \times S$ on {*Name, Class, Average, Name, Major*} − {*Name*} to form the join of R and S on *Name*.

$R \times S$			
Name	*Class*	*Average*	*Major*
Joe	2004	3.14	Mathematics
Sue	2004	2.97	Computer Science
Mary	2005	3.76	Sociology

One of the problems with database queries involves the complexity of finding the join of two relations. A join on more than a single attribute can be defined. The first example had three common attributes. The algorithm for finding the join makes the complexity of this operation clearer.

Algorithm: Join Two Relations

INPUT: Relations R and S with common attributes $B_1, B_2, \ldots, B_j$
OUTPUT: Relation J that is the join of R and S on $B_1, B_2, \ldots, B_j$

$J = \emptyset$
for each tuple $x \in R$ do
    Select all tuples $y \in S$ whose values on $B_1, \ldots, B_j$
        are all the same as x's
    for each such tuple y do
        Form a tuple z by concatenating x with y
        Eliminate the duplicate entries for attributes
            $B_1, B_2, \ldots, B_j$, creating a tuple z'
        $J = J \cup \{z'\}$

Example 3. Use relational algebra, applied to the relations *Registration* and *Teaching-Assignments,* to find a list of all professors who have either René Descartes or Winston Churchill as students.

Solution. First, form the natural join of *Registration* and *TeachingAssignments*. Then, select all tuples in the joined relation with student René Descartes or Winston Churchill. The result is shown in Table 3.19.

Table 3.19 *Step1Join*

Step1Join				
Student	*Department*	*Course*	*Section*	*Professor*
René Descartes	English	101	4	William Morris
Winston Churchill	English	101	4	William Morris
Winston Churchill	Mathematics	101	1	David Hilbert
Winston Churchill	Mathematics	101	1	Leonardo of Pisa
René Descartes	Computer Science	103	3	Alan Turing

Finally, project the *Step1Join* relation onto the attribute set *{Professor}*, and remove any duplicate entries. The resulting relation is shown in Table 3.20.

The relation *Professor* answers the question of which professors have either Winston Churchill or René Descartes as a student. ∎

Table 3.20
Projection

Professor
William Morris
David Hilbert
Leonardo of Pisa
Alan Turing

3.11   Exercises

1. What operations can you apply to the sample relations in Section 3.10.1 to get the following relations?

    (a) Professors and the departments in which they teach courses.

    (b) Students and professors from whom they take courses.

    (c) Professors and the chairs of the departments in which they teach courses.

    (d) Pairs of departments that currently provide courses with the same number. So, having {English, Mathematics} in the relation would assert that both departments have courses with a number such as 101.

2. Use the operations of relational algebra and the sample relations in Section 3.10 to extract the following information:

    (a) The students taking English courses.

    (b) The students taking classes from Geoffrey Chaucer or Thomas Jefferson.

    (c) The professors teaching courses in departments chaired by Carl Gauss or Alan Turing.

    (d) The students taking classes from professors who teach some class in a department chaired by Carl Gauss or Alan Turing.

3. Add to the course scheduling database a relation showing which courses are prerequisites for which other courses. Create some sample entries to illustrate the relations.

4. (a) Rewrite the two relations *Registration* and *TeachingAssignments* as binary relations between people on the one hand and triples (*Department, Course Number, Section Number*) on the other. Does this relation make more sense?

    (b) Using this approach, why could you *not* do Exercise 1(d)? Suggest a meaningful extra relation that would allow you to do Exercise 1(d).

5. What simple operation on relations could you add to make it easy to list the *number* of students in classes taught by Alan Turing? (*Note:* This problem asks you to design a new type of query. Accordingly, it has no right or wrong answers, but some answers will be simpler than others.)

Exercises 6 through 12 ask questions about the database shown in the three relations *Students, Grades,* and *Catalog:*

Students			
SocSecNo	*Name*	*Major*	*Class Year*
247617832	Smith, John	Mathematics	2005
477677251	Brown, Mae	English	2006
149867253	Cyr, Pete	Mathematics	2005
316719842	Williams, Sue	English	2004

Grades		
SocSecNo	*CourseCode*	*Grade*
316719842	Math211	A
247617832	Engl103	B
149867253	Math214	A
149867253	Engl103	A
316719842	Math318	B
316719842	Engl224	A

Catalog		
CourseCode	*Department*	*Credits*
Math211	Mathematics	4
Engl103	English	3
Math214	Mathematics	3
Math318	Mathematics	4
Engl224	English	3

6. Find the join of *Grades* and *Catalog*.
7. Find the join of *Students* and *Grades*.
8. Find the join of *Students, Grades,* and *Catalog*.
9. Find all students who received an A in a course.
10. Find the department and number of credits for any course in which a student received an A.
11. Find all second-year students who received an A.
12. Find the departments in which a student received an A in one of that department's courses.

3.12   Chapter Review

The idea of a relation gives a format for studying mathematical and nonmathematical relationships. Forming the composition of relations and defining the inverse of a relation are fundamental operations on relations. The common properties of relations such as $=$, $<$, and $\subseteq$ are abstracted to define what it means for a relation to be reflexive, irreflexive, symmetric, antisymmetric, and transitive. Finding the reflexive, symmetric, or transitive closure of a relation identifies the smallest relation containing a given relation with a given property. Focusing on reflexive, symmetric, and transitive relations leads to equivalence relations and partitions. Focusing on antisymmetric and transitive relations leads to partial and total orders. When discussing ordering relations, it is important to understand the notion of comparable elements. Special comparable elements include minimal, minimum, maximal, and maximum elements. Finally, the chapter deals with relations in the context of the operations that are used by a relational database.

Applications in this chapter include lexicographical or dictionary ordering, finding a minimal element, and embedding a partial order in a total order. The examples dealing with relational databases point out the operations that are used for processing queries in such a database.

3.12.1 Summary

3.1 and 3.2 Summary

TERMS

binary relation

composition

deck of 52 cards

empty relation

equality relation

family tree

identity relation

inverse

irreflexiv

n-ary relation

n-tuples

property

query

relations

ternary relation

trivial relation

unary relation

universal relation

void relation

3.4 Summary

TERMS

antisymmetric

graph

irreflexive

nth power of R

R^+

R^*

reflexive

reflexive closure

reflexive and transitive closure

symmetric

symmetric closure

transitive

transitive closure

THEOREMS

A relation R on a set X is reflexive if and only if $ID_X \subset R$.

A relation R on a set X is irreflexive if and only if $R \cap Id_X = \phi$.

A relation R on a set X is symmetric if and only if $R = R^{-1}$.

A relation R on a set X is transitive if and only if $R \circ R \subseteq R$.

The reflexive closure of a relation R on a set X is $R \cup ID_X$.

The symmetric closure of a relation R on a set X is $R \cup R^{-1}$.

The transitive closure of a relation R on a set X is R^+.

The reflexive and transitive closure of a relation R on a set X is R^*.

3.6   Summary

TERMS

congruent	quotient
divisible by	refines
equivalence class	remainder
equivalence relation	$x \equiv y \ (mod\ p)$
partition	

THEOREMS

Let P be a partition of a set X. For $x \cdot y \in X$, define $x \sim y$ to mean that x and y are in the same element of the partition. Then, $\sim$ is an equivalence relation. The equivalence classes of $\sim$ are exactly the elements of P.

3.8   Summary

TERMS

above	maximal
ASCII code	maximum
below	minimal
comparable	minimum
dictionary ordering	optimal
divides	partial ordering
embedding	strict partial ordering
law of trichotomy	total ordering
lexicographical ordering	
linear ordering	

ALGORITHMS

Finding a Minimal Element

3.10   Summary

TERMS

attribute	query
cartesian product	relational algebra
database	relational database
equijoin	selection
join	table
natural join	tuples
projection	value
quadruples	

ALGORITHM

Join Two Relations

3.12.2   Starting to Review

1. Let $A = \{1, 2, 3, 4\}$. Define a relation R of A as $R = \{(1, 3), (4, 2), (2, 4), (2, 3), (3, 1)\}$. Which of the following properties does this relation not possess?

   (a) Reflexive
   (b) Symmetric
   (c) Transitive
   (d) All of the above

2. Which of the following relations defined on $X = \{1, 2, 3\}$ is an equivalence relation?

   (a) $\{(1, 2), (2, 2), (3, 3)\}$
   (b) $\{(1, 1), (2, 2), (2, 2), (2, 1), (3, 3), (1, 1)\}$
   (c) $\{(1, 1), (1, 2), (1, 3), (2, 2), (2, 1), (3, 3), (3, 1)\}$
   (d) All of the above

3. Let R be a relation on a set S. R is circular if, for $x, y, z \in S$, whenever $x\, R\, y$ and $y\, R\, z$, it follows that $z\, R\, x$. Which of the properties do a reflexive and circular relation possess?

   (a) Irreflexive
   (b) Transitive
   (c) Antisymmetric
   (d) None of the above

4. Which of the following relations defined on $X = \{1, 2, 3\}$ is a partial order?

   (a) $\{(1, 1), (2, 2), (3, 3)\}$
   (b) $\{(1, 2), (1, 2), (2, 2), (3, 3)\}$
   (c) $\{(1, 1), (2, 1), (2, 2), (1, 3), (3, 3)\}$
   (d) All of the above

5. Given the following graph of a partial order R on $X = \{1, 2, 3, 4, 5\}$, list all the ordered pairs (x, y) such that $x\, R\, y$.

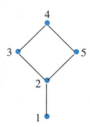

6. Let R be a partial order on a set X, and let $x \in X$. The element x is a minimal element in R if:

   (a) $x \leq y$ for all $y \in X$.
   (b) $x \leq y$ for all $y \in X$ such that $y \neq x$ and y is comparable to x.
   (c) $x \leq y$ for all y such that $y \in X$ and y is comparable to x.
   (d) None of the above.

7. Prove that $\{(1, 1), (2, 2), (3, 3), (4, 4), (5, 5), (6, 6), (1, 5), (2, 4), (2, 6), (4, 6), (6, 4), (6, 2), (4, 2), (5, 1)\}$ is an equivalence relation. Find the distinct equivalence classes for this equivalence relation.

8. If $A = \{1, 2, 3, 4, 5\}$ and R is the equivalence relation on A that induces the partition

$$A = \{1, 2\} \cup \{3, 4\} \cup \{5\}$$

what is R?

9. What are the minimal and maximal elements in the following diagram of a partial order?

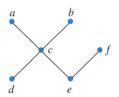

10. What is the difference between a maximal element and a maximum element in a partial order on a set X?

3.12.3   Review Questions

1. Prove or find a counterexample to the following conjectures about relations R_1 and R_2.

    (a) If R_1 and R_2 are reflexive, then $R_1 \circ R_2$ is reflexive.
    (b) If R_1 and R_2 are irreflexive, then $R_1 \circ R_2$ is irreflexive.
    (c) If R_1 and R_2 are symmetric, then $R_1 \circ R_2$ is symmetric.
    (d) If R_1 and R_2 are antisymmetric, then $R_1 \circ R_2$ is antisymmetric.
    (e) If R_1 and R_2 are transitive, then $R_1 \circ R_2$ is transitive.

2. For $x, y \in \mathbb{Z}$, define the relation R as $x\ R\ y$ if and only if $x \cdot y$ is odd. Is R Reflexive? Symmetric? Transitive? Prove, or give a counterexample.

3. Let R be a relation defined on $\{a, b, c, d\}$ such that

$$R = \{(a, a), (b, b), (c, a), (d, d), (a, b), (b, d), (a, d)\}$$

Find the symmetric closure of R.

4. Find the transitive closure of the relation $R = \{(1, 2), (2, 3), (3, 4), (4, 1)\}$. Show R^i for all values of i that give new elements of the transitive closure.

5. Define the relation R on $\mathbb{R} \times \mathbb{R}$ such that for any $(x, y), (u, v) \in \mathbb{R} \times \mathbb{R}$, we have $(x, y)\ R\ (u, v)$ if and only if $y = v$. Prove that R is an equivalence relation.

6. Let R be a binary relation on the set of all strings of 0's and 1's such that $R = \{(x, y) :$ strings x and y contain the same number of 0's$\}$. Is R Reflexive? Symmetric? Antisymmetric? Transitive? An equivalence relation?

7. The oddness or evenness of an integer is called its *parity*. Prove that the relation "have the same parity" is an equivalence relation. Find the distinct equivalence classes of this equivalence relation.

8. Four friends—Bill, Chuck, Maria, and Susie—are seated around a table. Define a relation *ARRANGE* to contain a pair $(Arr1, Arr2)$ of seating arrangements for these four people around a round table if $Arr1$ can be obtained from $Arr2$ by shifting each person

the same number of places to the right or to the left. Prove that this relation is an equivalence relation. How many equivalence classes are there, and what are the members of each equivalence class? Can you conjecture how many equivalence classes there would be if there were n friends?

9. Let R be a reflexive relation on a set A. R is an equivalence relation if and only if $(a, b), (a, c) \in R$ implies that $(b, c) \in R$.

10. Let T be a relation on A, and let R be a reflexive and transitive relation on A. Prove that T is an equivalence relation on A provided $(a, b) \in T$ if and only if $(a, b), (b, a) \in R$.

11. Let R_1 be a partial order on S and R_2 a partial order on T. For $(s_1, t_1), (s_2, t_2) \in S \times T$, define $(s_1, t_1) R_3 (s_2, t_2)$ if and only if $s_1 R_1 s_2$ and $t_1 R_2 t_2$. Prove that R_3 is a partial order.

12. Let $X = \{1, 2, 3, 4\}$, and let $\mathcal{P}(X)$ be the power set of X. Let $\mathcal{P}(X)$ be partially ordered by set inclusion. Find an embedding of this partial ordering into a total ordering.

3.12.4   Using Discrete Mathematics in Computer Science

Definition.   An **upper bound** of two elements in a partial order is an element that is greater than both of the elements. A **least upper bound** is an upper bound that is smaller than any other upper bound. A **lower bound** of two elements in a partial order is an element that is less than both of the elements. A **greatest lower bound** is a lower bound that is larger than any other lower bound.

1. Find the least upper bound and the greatest lower bound of each pair of elements in the partial order represented by the following diagram:

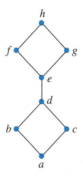

2. Find the least upper bound and the greatest lower bound of each pair of elements in the partial order represented by the following diagram:

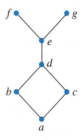

3. Define the relation D on $\mathbb{N}$ so that $n\,D\,m$ if and only if $n\,|\,m$. An upper bound of two natural numbers in D is a natural number that both divide. The smallest such natural number is called the least upper bound and is denoted as $lub(\ ,\ )$. For example, 6 is the least upper bound of 2 and 3. A lower bound of two natural numbers in D is a natural number that divides both numbers. The largest such natural number is called the greatest lower bound and is denoted as $glb(\ ,\ )$. For example, the greatest lower bound of 4 and 6 is 2. Find:

   (a) $lub(13, 29)$
   (b) $lub(12, 60)$
   (c) $glb(37, 12)$
   (d) $glb(48, 60)$

4. In drawing computer images of scenes, one must be able to tell which objects hide or partially hide, other objects from view. Imagine a scene in two dimensions consisting of a set L of line segments of various lengths drawn parallel to the x-axis. The line segments may intersect. For each of the following relations R on set L, is R antisymmetric? Transitive?

   (a) $R(\ell, m)$ if there is at least one point on segment ℓ that can look parallel to the y-axis and see a point on m (a line of sight may have zero length).
   (b) $R(\ell, m)$ if no point of ℓ can look parallel to the y-axis and see any point of m.

5. Carry out a selection sort (defined in Section 1.7.1) on the words *able, cane, bell, after, stick,* and *belt.* Explain how lexicographical ordering is used for each comparison.

6. Let $T = \{A, B, C, D, E, F\}$, and define the partial order R on T as represented by the following diagram:

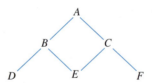

   (a) Identify all maximal, maximum, minimal, and minimum elements of the partial order represented by the diagram.
   (b) Find a linear order on T where $R - Id_T \subseteq S$.

7. (a) Prove that logical equivalence is an equivalence relation on the set of all formulas of propositional logic.
   (b) Show that as long as we have infinitely many proposition letters, there are infinitely many equivalence classes. (*Hint:* Once you see the idea, this is pretty trivial.)
   (c) Show that for logical equivalence on the set of all formulas in which the only proposition letters are $p_1, p_2, \ldots, p_n$, there are 2^{2^n} equivalence classes.

CHAPTER 4

Functions

In the study of mathematics, functions provide an important unifying concept. Functions are also familiar in computer science as components of programs that formalize the relationship between the input and the output for a computation. The problem of designing a combinatorial circuit often starts by defining a function that describes the behavior of the circuit for each possible input. Using functions to describe the behavior of a circuit, we can use techniques of Sections 2.5.2 and 2.5.4 to draw the combinatorial circuit with the same behavior. Since functions are special kinds of sets or relations, we will study them here using the ideas introduced in Chapters 1 and 3.

First, we define both functions and several fundamental properties of functions. Next, we deal with operations on functions, and basic properties of functions resulting from the operations introduced are explored. We explain special properties of functions, such as how many objects are related to a single object by a given function. Examples of functions with each property are given to help understand and differentiate among the properties that functions may possess. We discuss the Pigeon-Hole Principle and the Generalized Pigeon-Hole Principle, the applications of which include such different ideas as proving that rational numbers have a repeating decimal expansion and that two students in a small class will have a birthday on the same day of the week. Finally, we show how functions provide a way to formalize the notion of counting, and we see how to count the number of elements in both finite and infinite sets. In the context of counting rational and real numbers, Cantor's first and second diagonal arguments are introduced. These diagonal arguments come up in many computer science contexts, especially in the theory of computation and the analysis of algorithm complexity.

4.1 Basic Definitions

Intuitively, a **function** is a black box into which we put objects and out of which come other objects. A function must satisfy two rules. First, if an object is put in, then something must come out. Second, for each object input, there is only one possible output. If the same object is put in several times, then the same output must come out each time. No matter how many times one asks on what day Julius Caesar was born, the answer is always the same.

Figure 4.1 shows a picture to keep in mind when thinking about functions.

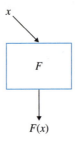

x

F

$F(x)$

Figure 4.1   Function.

Example 1.

(a) Visualize a classroom in which every student is seated at a chair. A function called *SeatOf,* outputs the chair at which a student is sitting for each student in the class.

(b) One may specify a function even though one does not have enough information whether in some or in all cases, to calculate its values. Let *BirthDate* be the function that accepts as input any person whose name appears in the current edition of the *Encyclopedia Britannica* and that outputs that person's birth date. No one knows the true birth date of Euclid, but Euclid, like every other person, did have a birth date. So, the function *BirthDate* still makes perfectly good sense.   ■

Example 2.

(a) Let $Zero_{\mathbb{R}}$ be the function that accepts as input any real number r and that always outputs 0. A function may be quite simple!

(b) Let X be any set. Let Id_X be the function that accepts as input any x in X and that outputs the same x. Id_X is called the **identity function** on X.

(c) The function **Floor** accepts any real number as input and outputs the integer formed by truncating the fractional part of the number input. For example, $Floor(3.14159) = \lfloor 3.14159 \rfloor = 3$.

(d) The function **Ceiling** accepts any real number as input and outputs the smallest integer greater than or equal to the number input. For example, $Ceiling(3.14159) = \lceil 3.14159 \rceil = 4$. This function is also referred to as the **greatest integer function.**   ■

The output of a function may be more complex to determine.

Example 3.   Let the function *ParentsOf* accept a person as input and output the *ordered pair*

(person's mother, person's father)

Example 4.

(a) By contrast with Example 3, there is no function *ParentOf* that picks out a person's parent. Such a rule is not a *function,* since there are two parents, from which one must be chosen as output.

(b) There is no function *ChildOf* that picks out a person's child. One reason this may not be a function is that some people have no children and, consequently, no object can be output. Some people also have more than one child from which to choose, and in this case, the function would not know which child to output. However, there is a function *ChildrenOf* that assigns to each person the set of that person's children. If a person has no children, the output of *ChildrenOf* is the empty set (∅).

We now define informally some basic vocabulary that will be more carefully defined later. We will illustrate these terms with the function *SeatOf* from Example 1.

The **domain** of a function is the set of all things that may be input to produce some output. The domain is usually apparent from the definition of the function. For example, the domain of *SeatOf* is the set of all students in the classroom.

The **range** of a function is the set of all things that are output. The range of *SeatOf* is the set of all occupied chairs in the classroom. Once one knows the domain of a function, one can determine the range by applying the function to each element of the domain.

The **codomain** of a function is the set of all values that are potential outputs. In informal descriptions, codomains are often not specified. For example, it is perhaps most reasonable to infer that the codomain of the function *SeatOf* is the set of all chairs in the classroom, but it is also plausible to infer that the codomain is the set of all occupied chairs. The codomain often cannot be determined from the description of the function alone; it must be inferred from the rest of the discussion. In less formal treatments, the codomain will often be implicitly defined. For example, in many mathematics courses, the codomain of most functions is implicitly $\mathbb{R}$. In other cases, as a convenience, the codomain is simply assumed to be equal to the range.

Everything so far has been intuitively expressed in terms of a black box. A formal definition of the term *function* is needed. Traditionally, there have been two ways to define this term. The first is to consider a function to be a rule. The second is to consider a function to be a specific kind of set. We will discuss the idea of a function as a rule first, since it is familiar from both computer programming and mathematics courses. After dealing with a function as a rule, we will discuss the idea of a function as a set. (We will give our formal definition in terms of sets.)

4.1.1 Functions as Rules

The notion of a function as a rule is familiar to anyone involved in computer programming. A function subprogram can be viewed as a series of instructions that tell how to calculate an output from some input.

Example 5. The following rules define functions:

(a) Let H be the function with domain and codomain equal to $\mathbb{N}$ that outputs $n/2$ for even inputs and $3n + 1$ for odd inputs.

(b) For $n \in \mathbb{N}$, compute $Fact(n) = n!$ as follows:

$$
\begin{aligned}
&\text{input } N \\
&Fact = 1 \\
&\text{while } N > 0 \\
&\qquad Fact = Fact \cdot N \\
&\qquad N = N - 1 \\
&\text{print } Fact
\end{aligned}
$$

It is important to realize that the code itself is not the function. Rather, the code is just one way to implement the rule that defines the function. The function is just the relationship between input and output. Consequently, many different rules may give rise to the same function.

Example 6. The following two algorithms compute the same function:

(a) For any $n \in \mathbb{N}$, output $\cos(n \cdot \pi)$.
(b) For any $n \in \mathbb{N}$, output $(-1)^n$.

The formal definition of equality of functions is given in Section 4.1.5. We will leave it to the reader to verify that rules (a) and (b) define the same function.

Example 7. Show that the following rule does not define a function: Let F be the rule with domain and codomain equal to $\mathbb{N}$ that outputs $n^4 - 3n$ for each n input.

Solution. $F(1)$ is not defined (since -2 is not in the codomain), so F is not a function. ∎

4.1.2 Functions as Sets

We can use the notion of a relation to define a function by allowing the elements that are related to belong to different sets. With this notion of a relation, a function is a special kind of binary relation. For sets X and Y, any subset of $X \times Y$ that "obeys" the following two rules is a function:

1. Each input corresponds to some output.
2. Each input corresponds to only one output.

The set X is the domain of the function. The set Y is the codomain of the function. The idea is that a relation consists of the set of ordered pairs for which every element of X is the first element of exactly one pair.

Definition 1. Let X and Y be sets. A **function** F with **domain** X and **codomain** Y is a subset of $X \times Y$ such that, for each $x \in X$, there is exactly one $y \in Y$ with $(x, y) \in F$. F is also called a function from X to Y. A function F from X to Y is often denoted by $F : X \rightarrow Y$.

From this point on, rather than identifying the domain and the codomain of a function as sets, we will assume that the notation $F : X \rightarrow Y$ implies this.

Example 8.

(a) Suppose a class consists of three students. Jean sits at the second chair in the first row, Michele sits at the sixth chair in the fourth row, and Paul sits at the 37th chair in the 53rd row. For this class, the function *SeatOf* is the set

$$\{(\text{Jean, Row1Seat2}), (\text{Michele, Row4Seat6}), (\text{Paul, Row53Seat37})\}$$

(b) Let

$$DayOfWeek = \{\text{Monday, Tuesday, Wednesday, Thursday,}$$
$$\text{Friday, Saturday, Sunday}\}$$

There is an obvious function:

NextDay:	DayOfWeek	→	DayOfWeek
	Monday		Tuesday
	Tuesday		Wednesday
	Wednesday		Thursday
	Thursday		Friday
	Friday		Saturday
	Saturday		Sunday
	Sunday		Monday

The binary relation defined by this function consists of the following ordered pairs:

{(Monday, Tuesday), (Tuesday, Wednesday), (Wednesday, Thursday),
(Thursday, Friday), (Friday, Saturday), (Saturday, Sunday),
(Sunday, Monday)}

Example 9. The factorial function *Fact* from Example 5(b) is the set

$$\{(0, 1), (1, 1), (2, 2), (3, 6), (4, 24), (5, 120), \ldots, (n, n!), \ldots\}$$

We now introduce a common vocabulary for functions.

Definition 2. Let $F : X \rightarrow Y$ be a function, and let $(x, y) \in F$. Then, y is the **image** of x under F, denoted by $y = F(x)$. We also say that x is **mapped to** y by F. The **range** of F is the set

$$range(F) = \{F(x) : x \in X\}$$

For $y \in Y$, the **preimage** of y under F, denoted as $F^{-1}(y)$, is the set

$$F^{-1}(y) = \{x \in X : F(x) = y\}$$

For $Y' \subseteq Y$, the preimage of Y' under F, denoted as $F^{-1}(Y')$, is the set

$$F^{-1}(Y') = \{x \in X : F(x) \in Y'\}$$

We refer to X as *domain*(F) and Y as *codomain*(F).

Example 10. For the function $F : \{1, 2, 3, 4, 5\} \rightarrow \{a, b, c, d, e\}$ defined as $F(1) = a$, $F(2) = b$, $F(3) = b$, $F(4) = d$, and $F(5) = c$, identify *domain*(F), *codomain*(F), *range*(F), $F^{-1}(a)$, $F^{-1}(\{a, b, c\})$, and $F^{-1}(e)$.

Solution. *domain*$(F) = \{1, 2, 3, 4, 5\}$; *codomain*$(F) = \{a, b, c, d, e\}$; *range*$(F) = \{a, b, c, d\}$; $F^{-1}(a) = \{1\}$; $F^{-1}(\{a, b, c\}) = \{1, 2, 3, 5\}$; $F^{-1}(e) = \emptyset$. ■

You may find the range of a function referred to as the image of the function. In Example 4.1a, the range of function *SeatOf* is the set of all chairs in the room that have someone sitting on them. For another example, the addition function $+$ on $\mathbb{Z}$ maps the ordered pair $(3, 5)$ to 8. The domain of $+$ is $\mathbb{Z} \times \mathbb{Z}$, and the codomain is $\mathbb{Z}$. If $F : X^2 \rightarrow Y$; then F is called a **binary function** from X^2 to Y. Addition as well as the other familiar arithmetic operations defined on the integers are binary functions from $\mathbb{Z}^2$ to $\mathbb{Z}$.

4.1.3   Recursively Defined Functions

When a function F is defined by a formula, we can find the value of F at any element of its domain without knowing its value at any other element of its domain. For example, consider the function $F : \mathbb{N} \to \mathbb{N}$ defined by the rule $F(n) = 3n + 2$. We can compute directly that $F(100) = 3 \cdot 100 + 2 = 302$ or that $F(3112) = 3 \cdot 3112 + 2 = 9338$.

Functions, however, are not necessarily defined in such a straightforward manner. Consider the function $G : \mathbb{N} \to \mathbb{N}$ defined as $G(0) = 2$ and, for $n > 0$, $G(n) = G(n - 1) + 3$. Then, $G(1) = G(0) + 3 = 2 + 3 = 5$. The following computation shows how $G(5)$ would be determined:

$$
\begin{aligned}
G(5) &= G(4) + 3 \\
&= G(3) + 3 + 3 \\
&= G(2) + 3 + 3 + 3 \\
&= G(1) + 3 + 3 + 3 + 3 \\
&= G(0) + 3 + 3 + 3 + 3 + 3 \\
&= 2 + 3 \cdot 5 \\
&= 17
\end{aligned}
$$

If we now wanted $G(3112)$, we would first need to compute $G(1), G(2), \ldots, G(3111)$. In this situation, we say that G is defined **recursively** or is given by a **recursive definition.**

As you might suspect from the computation of $G(5)$, the two functions F and G are actually the same; that is, $F(n) = G(n)$ for every $n \in \mathbb{N}$. In Section 1.10.1, F was described as a *closed form* for G.

Example 11.

(a) The function $F : \mathbb{N} \to \mathbb{N}$ defined as $F(n) = 3^n$ can be defined recursively as $F(0) = 1$ and $F(n) = 3 \cdot F(n - 1)$ for $n \geq 1$.

(b) The sum of the first k of n elements $a_1, a_2, \ldots, a_n$ can be defined directly as $SUM(k) = a_1 + a_2 + \cdots + a_k$ where $1 \leq k \leq n$. Recursively, the same function can be defined as $S(1) = a_1$ and $S(k) = S(k - 1) + a_k$ for $k > 1$.

(c) The sum of the first n terms of a geometric series $a + ac + ac^2 + ac^3 + \cdots + ac^{n-1}$ can be defined as $gs(0) = a$ and $gs(k) = gs(k - 1) + ac^k$ for $k \geq 1$.

(d) The harmonic sequence that consists of the terms $1, 1/2, 1/3, \ldots, 1/n, \ldots$ can have the sum of its first k terms defined as the function $H(1) = 1$ and $H(k) = H(k - 1) + 1/k$ for $k > 1$.

We introduced the Fibonacci sequence in Section 1.7.3. This sequence of values $(1, 1, 2, 3, 5, 8, 13, \ldots)$ was defined recursively; that is, no direct formula was given for finding the nth element of the Fibonacci sequence. Unlike the functions in Example 8, two terms are given as initial conditions for termination conditions in defining the nth element of the Fibonacci sequence successively in terms of smaller Fibonacci numbers. The definition of the Fibonacci sequence is $F(0) = 1$, $F(1) = 1$, and $F(n) = F(n - 1) + F(n - 2)$ for $n \geq 2$. The first five terms of the Fibonacci sequence are found as follows:

$$F(0) = 1$$
$$F(1) = 1$$
$$F(2) = F(1) + F(0) = 1 + 1 = 2$$
$$F(3) = F(2) + F(1) = 2 + 1 = 3$$
$$F(4) = F(3) + F(2) = 3 + 2 = 5$$

A recursively defined function may involve any number of initial values in determining a next value.

Example 12. Find the first six values of the function defined on $\mathbb{N}$ given by $F(0) = 2$, $F(1) = 3$, $F(2) = 5$, and $F(n) = 2F(n-1) + 3F(n-2) + F(n-3)$ for $n \geq 3$.

Solution.

$$F(3) = 2F(2) + 3F(1) + F(0) = 10 + 9 + 2 = 21$$
$$F(4) = 2F(3) + 3F(2) + F(1) = 42 + 15 + 3 = 60$$
$$F(5) = 2F(4) + 3F(3) + F(2) = 120 + 63 + 5 = 188$$

∎

4.1.4 Graphs of Functions

Since functions are relations, they have graphs. Figure 4.2 shows part of the graph of the function *Floor*.

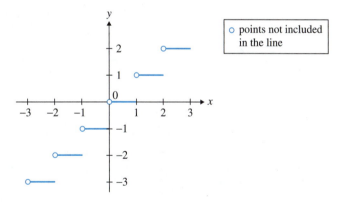

Figure 4.2 Graph of *Floor*.

Let G be the graph of a function with domain $X \subseteq \mathbb{R} \times \mathbb{R}$. G is the graph of a function if whenever $x_0 \in X$, the vertical line $x = x_0$ intersects G in exactly one point. We call this test the **vertical line test** for a function. Figure 4.3, on page 226, shows a subset of $\mathbb{R} \times \mathbb{R}$ that is not a function, since the vertical line $x = 1$ cuts the graph in two places.

When a function has a "small" set as its domain and a "small" set as its codomain, such as the function $F : \{0, 1, 2\} \rightarrow \{3, 5, 7\}$ defined as $F(0) = F(1) = 5$ and $F(2) = 7$,

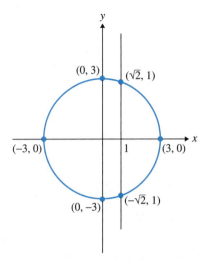

Figure 4.3 Graph of a relation that is not a function.

we often represent such functions by a diagram such as that shown on Figure 4.4. The lines joining an element on the left in Figure 4.4 with an element on the right represent the association between elements of the domain and elements of the codomain that we interpret as the rule for F. For example, we interpret the line between 0 and 5 as meaning $F(0) = 5$.

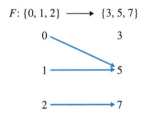

Figure 4.4 Representation of a function.

The elements of the domain and of the codomain can be listed in any order. Sometimes, a picture of this sort makes functions defined on $\mathbb{N}$ easier to understand. This representation can also be used for some "large" sets.

4.1.5 Equality of Functions

Since functions are defined as subsets of a product of two sets—that is, as sets of ordered pairs—two functions are equal when they are equal as sets.

Definition 3. Let $F, G : X \to Y$ be two functions. The functions F and G are **equal** if and only if they contain the same ordered pairs.

Example 13. Let $Sqr_{\mathbb{N}}$ be the function from $\mathbb{N}$ to $\mathbb{N}$ defined by the rule $Sqr_{\mathbb{N}}(n) = n^2$. Let $Sqr_{\mathbb{R}}$ be the function from $\mathbb{R}$ to $\mathbb{R}$ defined by the rule $Sqr_{\mathbb{R}}(r) = r^2$. Then, $Sqr_{\mathbb{N}}$ and $Sqr_{\mathbb{R}}$ are *not* the same function, since $(1.1, 1.21) \in Sqr_{\mathbb{R}}$ but $(1.1, 1.21) \notin Sqr_{\mathbb{N}}$.

Theorem 1.   Let F and G be functions such that $F = G$. Then,

$$domain(F) = domain(G)$$
$$range(F) = range(G)$$

and, for each $x \in domain(F)$, $F(x) = G(x)$.

Some authors would insist that for two functions to be equal, their codomains must also be the same. We do not insist on that condition for the equality of two functions.

Boolean Functions and Combinatorial Networks

A boolean function of n boolean variables is a function of the form

$$B : \{0, 1\} \times \{0, 1\} \times \cdots \times \{0, 1\} \to \{0, 1\}$$

The domain of B contains 2^n elements. A value of either 0 or 1 is assigned to each entry of the 2^n ordered n-tuples. An example of a boolean function on three boolean variables is shown in Table 4.1.

Table 4.1   Boolean Function of Three Variables

p	q	r	$F(p, q, r)$
1	1	1	1
1	1	0	0
1	0	1	1
1	0	0	1
0	1	1	1
0	1	0	1
0	0	1	0
0	0	0	1

This function might represent a set of switches that react in an appropriate way or a set of conditions that must be satisfied so that some action can be taken. It is useful to be able to represent functions as in Table 4.1, but the real problem is often to embed this function in a combinatorial network. We saw in Chapter 2, while discussing disjunctive normal forms and conjunctive normal forms, how we can draw the combinatorial circuit given one of these normal forms. In this case, we need to see how to represent a function in terms of one of these normal forms.

For the function in Table 4.1, a disjunctive normal form is

$$F(p, q, r) = (p \wedge q \wedge r) \vee (p \wedge \neg q \wedge r) \vee (p \wedge \neg q \wedge \neg r)$$
$$\vee (\neg p \wedge q \wedge r) \vee (\neg p \wedge q \wedge \neg r) \vee (\neg p \wedge \neg q \wedge \neg r)$$

Consequently, the combinatorial circuit is the circuit shown in Figure 4.5.

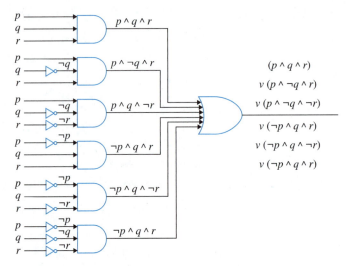

Figure 4.5   Combinatorial circuit for $F(p, q, r)$.

Notice in Figure 4.5 that there are several inputs for a gate. This is as much for convenience as for anything else, since we can obviously write a gate with three inputs as a set of gates with two inputs each, as shown in Figure 4.6.

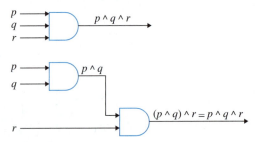

Figure 4.6   Combinatorial circuit for multiple inputs.

4.1.6   Restrictions of Functions

It is easy to write an algorithm to compute $Sqr_{\mathbb{R}}(x) = x^2$ for $x \in \mathbb{R}$. By merely asserting that only natural numbers should be used as input, one can make the same algorithm specify $Sqr_{\mathbb{N}}$, a function from $\mathbb{N}$ to $\mathbb{N}$. This is an example of restricting a function to a smaller domain. As usual, the formal definition is set theoretic.

Definition 4.   Let A, B, and C be sets such that $B \subseteq A$. Let $F : A \to C$ be a function. The **restriction** of F to B, denoted $F \mid_B$, is a function from B to C defined as the set

$$F \mid_B = \{(x, y) \in F : x \in B\}$$

Figure 4.7 shows two examples of restrictions of the function $Sqr_{\mathbb{R}}$.

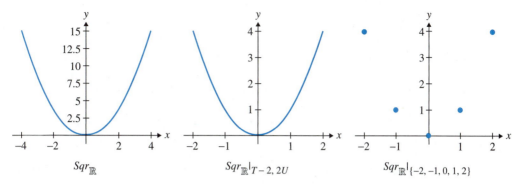

Figure 4.7   Restrictions of $Sqr_{\mathbb{R}}$.   A. $Sqr_{\mathbb{R}}$ B. $Sqr_{\mathbb{R}}|_{[-2,2]}$ C. $Sqr_{\mathbb{R}}|_{\{-2,-1,0,1,2\}}$.

4.1.7 Partial Functions

Think of a computer program as computing or specifying a function from the input of the program to its output. The input to the function is whatever string of characters is input to the program. The output is whatever string of characters has been output by the program after it has finished execution. Anyone who has programmed a computer realizes that many programs, on some input data, go into infinite loops and, by the definition above, would produce no output at all. In that case, the program is not computing a function of the input, since the definition of a function requires that there be one output for every one input. What a program computes is really what is called a partial function of its input. On each input, the program, if it produces any output at all, produces only one possible output. Thus, a partial function can be thought of as a black box into which for each input there is at most one possible output.

Another sort of partial function is the following: Suppose the amount of postage to be paid is specified in Table 4.2 (where the range (3-4] kg is understood to mean that the parcel weighs more than 3 kg but less than or equal to 4 kg). This table gives postage costs only as a partial function of the weight of the package, since the postage amount is not specified for anything weighing more than 16 kg.

Table 4.2   Postage Costs

Weight	Postage
(0–1] kg	$1.00
(1–2] kg	$1.98
(2–3] kg	$2.56
(3–4] kg	$3.11
(4–5] kg	$3.99
(5–8] kg	$5.00
(8–12] kg	$7.00
(12–16] kg	$9.00

The two examples we have given of partial functions actually reflect rather different ways in which partial functions arise. In the postage example, the function was partial because no rule was given for calculating the postage for items weighing more than 16 kg. At some later time, someone may come back and extend the rule, perhaps by specifying that items weighing (16–20] kg cost $10.89. In the computer program example, however, a rule was given for all possible input data, but that rule failed to output anything for certain values. The notion of a partial function gives a formal way to consider all programs— even ones that crash or go into infinite loops. Partial functions are particularly important in the theoretical study of computability—that is, in the study of which functions and partial functions are computable by programs, where there is assumed to be no restriction on computer memory or on computation time. There is no satisfactory way in this subject to restrict attention only to functions.

Definition 5.   A **partial function** F with domain of definition X and codomain Y is a subset of $X \times Y$ such that for each $x \in X$, there is, at most, one y with $(x, y) \in F$. Such an F is also called a partial function **F** from X to Y. When it is understood that F is partial, the notation $F : X \to Y$ is also used (but when the notation $F : X \to Y$ is used without any other comment, F is a function). The domain of a partial function $F : X \to Y$ is the set

$$\{x \in X : \text{for some } y \in Y, (x, y) \in F\}$$

If $x \in X$ is not in the domain of definition for F, then $F(x)$ is **undefined.** Other terms, such as *range* and *preimage,* are defined exactly as for functions.

When F is a partial function, the implication is not that there is necessarily any x in its domain of definition where $F(x)$ is undefined, only that there might be. Hence, every function is a partial function. When discussing both functions and partial functions that are not functions, functions are often referred to as **total functions** to emphasize the difference.

Example 14.   In Example 1(a) *SeatOf* was presented as a (total) function. It might be slightly more realistic to present it as a partial function, however, since some people in the room might not be sitting in seats. For example, they might be standing or sitting on the floor. Asking what is the seat of a standing person should get no answer.

Example 15.   The following are examples of partial functions:

(a) Subtraction $(-)$ on $\mathbb{N}$ is a partial function. Its domain of definition is $\mathbb{N}^2$, and its codomain is $\mathbb{N}$. For $i < j$, $i - j$ is not defined on $\mathbb{N}$, so the domain of subtraction is

$$\{(i, j) \in \mathbb{N}^2 : i \geq j\}$$

The range of $(-)$ is $\mathbb{N}$. To show that an arbitrary $n \in \mathbb{N}$ is in *range*$(-)$, note that $n = n - 0$.

(b) Division on $\mathbb{R}$ is a partial function. Its domain of definition is $\mathbb{R}^2$, its codomain is $\mathbb{R}$, but its domain is

$$\{(x, y) \in \mathbb{R}^2 : y \neq 0\}$$

Its range is $\mathbb{R}$. Why?

(c) For $x \in \mathbb{R}$, let *Sqrt*(x) be the non-negative square root of x. Then, *Sqrt* is a partial function, since *Sqrt*(x) is undefined for $x < 0$. The domain of definition of *Sqrt* is $\mathbb{R}$, and its codomain is $\mathbb{R}$. The range of *Sqrt* is $[0, \infty)$.

Let G be a subset of $\mathbb{R}$. G is the graph of a partial function if, whenever $x_0 \in X$, the vertical line $x = x_0$ intersects G in at most one point. We call this the vertical line test for a partial function. Figure 4.8 shows a subset of $\mathbb{R} \times \mathbb{R}$ that is not a function, because the vertical line $x = -1$ does not cross the graph. *Sqrt* is a partial function, since no vertical line defined by an element of its domain crosses the graph more than once.

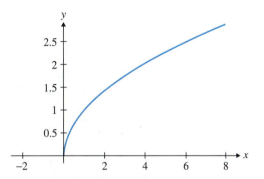

Figure 4.8 Graph of partial function *Sqrt*.

Whether a partial function is a total function depends on what the domain of definition is *defined to be*. For example, it was noted that *Sqrt* is a partial function from $\mathbb{R}$ to $\mathbb{R}$. If we declare the domain of definition to be just the set $[0, \infty)$, then *Sqrt* is a total function.

4.1.8   *1–1* and Onto Functions

Several special types of functions have turned out to be especially important. For example, the intuitive notion of counting will be formalized using the properties of functions introduced in this section.

Definition 6.   Let $F : X \to Y$ be a function. F is *1–1* if, for each $y \in Y$, there is, at most, one $x \in X$ such that $F(x) = y$.

Example 16.

(a) Let $F : \mathbb{R} \to \mathbb{R}$ be a function defined as $F(x) = 2x$. F is *1–1*.
(b) Let $G : \mathbb{N} \to \mathbb{N}$ be a function defined as $G(n) = 2n^2 + 1$. G is not *1–1*.

Solution.

(a) Since $F(x_1) = F(x_2)$ means $2x_1 = 2x_2$, it follows that $x_1 = x_2$ and F is *1–1*.
(b) Since $G(2) = G(-2)$, the function G is not *1–1*.   ∎

The function *SeatOf* (from Example 1(a) in Section 4.1) is *1–1* if and only if exactly zero or one person is sitting at each chair (and every student is seated at exactly one chair).

Figure 4.9   *1–1* Function *SeatOf*.

Figure 4.9 shows a *1–1 SeatOf* function, and Figure 4.10 shows a similar function, *SeatOf*$_1$, that is not *1–1*.

Figure 4.10   Function *SeatOf*$_1$.

The function $H(x) = x^2$ is not *1–1*. This is shown in Figure 4.11. Let G be the **graph of a function** with codomain $Y \subseteq \mathbb{R}$. G is the graph of a *1–1* function if, whenever $y_0 \in Y$, the line $y = y_0$ intersects G in, at most, one point. We call this the **horizontal line test** for *1–1* functions.

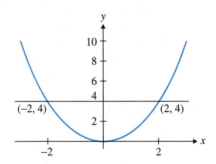

Figure 4.11 $H(x) = x^2$.

The horizontal line $y = 4$ crosses the graph in Figure 4.11 at more than one point. Therefore, G is not *1–1*. On the other hand, the function $F(x) = x^3$, as shown in Figure 4.12, on next page, is *1–1*, since each horizontal line crosses the graph in *at most* one point.

Definition 7.   Let $F : X \rightarrow Y$ be a function. F is **onto** if, for each $y \in Y$, there is at least one $x \in X$ such that $F(x) = y$.

Another way to think of the definition of onto is that a function $F : X \rightarrow Y$ is onto if and only if $range(F) = codomain(F)$. Whether a function is onto or not depends on what the codomain is *defined to be*. For example, the function $Sqrt : [0, \infty) \rightarrow \mathbb{R}$ is not onto. However, if $Sqrt$ is *defined to be* $Sqrt : [0, \infty) \rightarrow [0, \infty)$, then $Sqrt$ is onto.

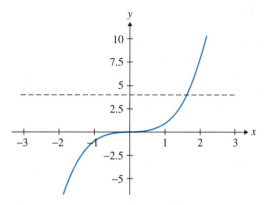

Figure 4.12 $F(x) = x^3$.

The function $SeatOf_2$, as shown in Figure 4.13, maps the set of students in the classroom *onto* the set of chairs in the classroom if every chair is occupied.

Figure 4.13 $SeatOf_2$.

The function $SeatOf_3$, as shown in Figure 4.14, is not, onto since one or more chairs remain unoccupied. In this case, two chairs are unoccupied.

Figure 4.14 $SeatOf_3$.

The function $G(x) = x^2$, as shown in Figure 4.15, is not onto.

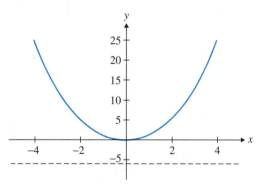

Figure 4.15 $G(x) = x^2$.

The horizontal line test for *1–1* functions can be easily modified to check whether a function is onto by simply requiring that each horizontal line defined by a member of the codomain meet the graph of the function at least once. In Figure 4.15, the horizontal line $y = -6$ does not intersect the graph of G at any point. This property of the graph of the function corresponds to the fact that there is no number x such that $G(x) = -6$. Therefore, G is not onto. On the other hand, the function $F(x) = x^3$, as shown in Figure 4.16, is onto, since each horizontal line crosses the graph in *at least* one point.

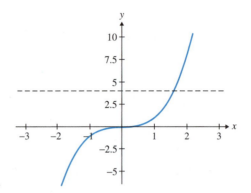

Figure 4.16 $F(x) = x^3$.

Functions that are both *1–1* and onto play a special role in counting the elements of a set. Because functions of this class have so many applications, they have been given a special name.

Definition 8.   Let $F : X \to Y$ be a function. F is a ***1–1 correspondence*** if F is both *1–1* and onto.

For example, the function *SeatOf* is a *1–1* correspondence if and only if each chair has exactly one student sitting at it. The function $F : \mathbb{R} \to \mathbb{R}$ defined by $F(x) = x^3$, as shown in Figure 4.16, is also a *1–1* correspondence. The function $G(x) = x^2$, as shown in Figure 4.15, is neither *1–1* nor onto.

The function shown in Figure 4.17 is onto but not *1–1*.

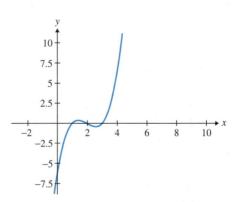

Figure 4.17   A function that is onto but not *1–1*.

The function $\exp : \mathbb{R} \to \mathbb{R}$ defined as $\exp(x) = e^x$ and shown in Figure 4.18 is *1–1* but not onto.

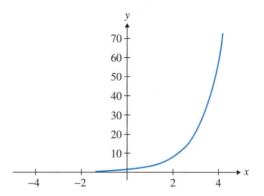

Figure 4.18   *exp(x).*

The functions defined here have been constructed to show that the two properties *1–1* and onto are *independent* of each other. Two properties of a mathematical object are independent if objects exist that can have exactly one of the properties, both of the properties, or neither of the properties. For *1–1* and onto functions, the four functions shown in Figures 4.15 through 4.18 demonstrate that the properties *1–1* and onto are independent.

Commonly used synonyms exist for the properties of functions defined in Definitions 6, 7, and 8. A *1–1* function is also called an **injective function,** or an **injection.** An onto function is called a **surjective function,** or a **surjection.** A *1–1* correspondence is called a **bijective function,** or a **bijection.** Also, a *1–1* correspondence is often referred to simply as a *1–1* **and onto function.**

Application: Hashing Functions

When you put a bank card into an ATM and enter your pin number, your bank account records must be found so that your transaction can be authorized. This is an example of information in symbolic or numeric form (the information on the magnetic stripe on the ATM card) being used to determine a location on some storage device (the physical location of your records). A function that can take information as input and find a storage address as an output is called a **hashing function.** For simplicity, at this point we will assume that a hashing function is *1–1*.

Example 17.   Define a hashing function that uses 63 storage locations as a four-stage process with surnames as input. The first step is to replace the letters of the surname with integers according to the following rule: $A \to 1, B \to 2, C \to 3, \ldots, Y \to 25, Z \to 26$. The second step is to multiply the letter value by 2^i where i is the letter's position in the word, with the leftmost character being in position 1. The third step is to add the values that represent the letters of the surname. The final step is to divide this sum by 63. The hashing value is the remainder of this division. For example, Robb has a value of 144 and a hashing value of 18. You should imagine that the information needed for Robb is in storage location 18. Carry out this hashing procedure for Smith, Jones, Brown, Zento, and Ruster.

Solution.

	Steps 1 through 3	Step 4	Hash Value
Smith $\rightarrow$	$19 \cdot 2 + 13 \cdot 2^2 + 9 \cdot 2^3 + 20 \cdot 2^4 + 8 \cdot 2^5 = 738$	$= 11 \cdot 63 + 45 \rightarrow$	45
Jones $\rightarrow$	$10 \cdot 2 + 15 \cdot 2^2 + 14 \cdot 2^3 + 5 \cdot 2^4 + 19 \cdot 2^5 = 880$	$= 13 \cdot 63 + 61 \rightarrow$	61
Brown $\rightarrow$	$2 \cdot 2 + 18 \cdot 2^2 + 15 \cdot 2^3 + 23 \cdot 2^4 + 14 \cdot 2^5 = 1012$	$= 16 \cdot 63 + 4 \rightarrow$	4
Zento $\rightarrow$	$26 \cdot 2 + 5 \cdot 2^2 + 14 \cdot 2^3 + 20 \cdot 2^4 + 15 \cdot 2^5 = 984$	$= 15 \cdot 63 + 39 \rightarrow$	39
Ruster $\rightarrow$	$18 \cdot 2 + 21 \cdot 2^2 + 19 \cdot 2^3 + 20 \cdot 2^4 + 5 \cdot 2^5 + 18 \cdot 2^6 = 1904$	$= 30 \cdot 63 + 14 \rightarrow$	14

Each of these names can be located among a set of 63 storage locations, numbered 0, 1, 2, ..., 62, by using their hash value as the location to access. ∎

If any two names give rise to the same hash value, then an auxiliary rule, called a **collision resolution strategy,** is used to make sure that each piece of information has its own storage location that can be determined from the information alone and the given collision resolution strategy.

How many students in your class can have their names hashed this way without generating a collision? (If your class has more than 63 students, simply change the function to find the remainder when you divide by some number at least as large as the size of your class.)

Application: Encryption and Decryption

In this age of electronic messaging, it is often important that only the intended receiver of an electronic message can read it. If the security of a transmission is a problem, the message can still be made secure if the original message has been **encoded** or **encrypted** so that the symbols seen make no sense unless you know how to **decrypt** the message, that is, return the encrypted message back to its original form. Here, we present an example of the process of encoding and decoding a message. The method used is very simple and not as powerful or secure as modern methods, but the example points out how an encryption scheme interacts with a message, a user, and a receiver. The difficult problem today is to find an encoding scheme that cannot be compromised through a brute force search by a computer. More complex ideas from number theory lie at the heart of the best current encryption methods. The encoding scheme presented uses a bijection from the symbol set used in writing the message to the same symbol set. The sender of the message must use the bijection to transform the message into a form that is not recognizable, and the receiver must use the inverse of the coding function to decrypt the message received to return it into plain text.

A very simple encoding scheme is to associate each letter of the alphabet (we will only deal with uppercase letters) with two digits as follows: $A \rightarrow 00$, $B \rightarrow 01$, $C \rightarrow 02, \ldots, X \rightarrow 23$, $Y \rightarrow 24$, and $Z \rightarrow 25$. Define a function $F(lettervalue) \equiv a(lettervalue) + b \,(mod\,26)$, where a and b are integers and a has no factor in common with 26 and the sum is reduced modulo 26. For example, if $a = 3$ and $b = 5$, then

$$F(X) \equiv 3(23) + 5 \,(mod\,26) \equiv 74 \,(mod\,26) \equiv 22 \,(mod\,26)$$

A message such as

$$LEAVINGTODAY = 11\,04\,00\,21\,08\,13\,06\,19\,14\,03\,00\,24$$

is transmitted as

$$F(11)\;F(4)\;F(0)\;F(21)\;F(8)\;F(13)\;F(6)\;F(19)\;F(14)\;F(3)\;F(0)\;F(24)$$

The computation is shown in Table 4.3.

Table 4.3 Encryption Computation

$F(0) = 3(0) + 5 \ (mod\ 26) = 5$	$F(3) = 3(3) + 5 \ (mod\ 26) = 14$
$F(4) = 3(4) + 5 \ (mod\ 26) = 17$	$F(6) = 3(6) + 5 \ (mod\ 26) = 23$
$F(8) = 3(8) + 5 \ (mod\ 26) = 3$	$F(11) = 3(11) + 5 \ (mod\ 26) = 12$
$F(13) = 3(13) + 5 \ (mod\ 26) = 18$	$F(14) = 3(14) + 5 \ (mod\ 26) = 21$
$F(19) = 3(19) + 5 \ (mod\ 26) = 10$	$F(21) = 3(21) + 5 \ (mod\ 26) = 16$
$F(24) = 3(24) + 5 \ (mod\ 26) = 25$	

The message that is sent is

$$12\,17\,05\,16\,03\,18\,23\,10\,21\,14\,05\,25$$

The message is transformed into the following string of symbols:

$$M\ R\ F\ Q\ D\ S\ X\ K\ V\ O\ F\ Z$$

The problem for the receiver is to know the inverse function and then apply it to each of these two digit pairs to see the original message. The inverse for $F(letter) \equiv 3(lettervalue) + 5 \ (mod\ 26)$ is a function of the same form—that is, $G(lettervalue) \equiv a(lettervalue) + b \ (mod\ 26)$ where a and b are determined as follows:

$$G \circ F(lettervalue) \equiv a(3 \cdot lettervalue + 5) + b \equiv lettervalue(mod\ 26)$$

We solve

$$3a \equiv 1(mod\ 26) \text{ and } 5a + b \equiv 0(mod\ 26)$$

to get $a = 9$ and $b = 7$. The inverse is $G(lettervalue) \equiv 9(lettervalue) + 7(mod\ 26)$. We now compose these two functions to decrypt the message as shown:

$$G \circ F(L)\,G \circ F(E)\,G \circ F(A)\,G \circ F(V)\,G \circ F(I)\,G \circ F(N)G \circ F(G)\,G \circ$$
$$F(T)\,G \circ F(O)\,G \circ F(D)\,G \circ F(A)\,G \circ F(Y)$$
$$= G(12)\,G(17)\,G(05)\,G(16)\,G(03)\,G(18)\,G(23)\,G(10)\,G(21)\,G(14)G(05)\,G(25)$$
$$= 11\,04\,00\,21\,08\,13\,06\,19\,14\,03\,00\,24$$
$$= L\,E\,A\,V\,I\,N\,G\,T\,O\,D\,A\,Y$$

4.1.9 Increasing and Decreasing Functions

The reader has probably already encountered increasing and decreasing functions in a mathematics course. It is common to speak of a function as being increasing or decreasing on an interval. The function defined on $\mathbb{R}$,

$$F(x) = x^2 - 6x + 12$$

is decreasing on $(-\infty, 3]$ and increasing on $[3, \infty)$. (You can see this from the graph of the function.) The definition of the terms increasing and decreasing uses the familiar orderings *less than* and *less than or equal* on $\mathbb{R}$.

Definition 9. Let $X, Y \subseteq \mathbb{R}$, and let $F : X \to Y$ be a function.

(a) F is **increasing** if for, all $x_1, x_2 \in X$, $x_1 < x_2$ implies $F(x_1) \leq F(x_2)$.
(b) F is **strictly increasing** if, for all $x_1, x_2 \in X$, $x_1 < x_2$ implies $F(x_1) < F(x_2)$.
(c) F is **decreasing** if, for all $x_1, x_2 \in X$, $x_1 < x_2$ implies $F(x_1) \geq F(x_2)$.
(d) F is **strictly decreasing** if, for all $x_1, x_2 \in X$, $x_1 < x_2$ implies $F(x_1) > F(x_2)$.

Example 18. The following functions are increasing:

(a) The function $F : \mathbb{R} \to \mathbb{R}$ where $F(x) = x^3$ is strictly increasing (see Figure 4.19).

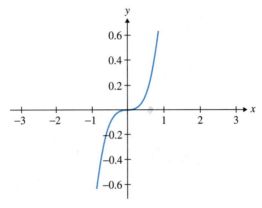

Figure 4.19 $F(x) = x^3$.

(b) The function *Floor* : $\mathbb{R} \to \mathbb{N}$ is increasing but not strictly increasing (see Figure 4.20).

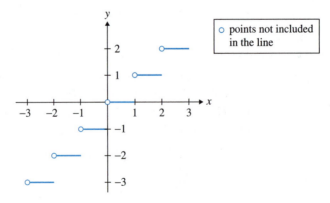

Figure 4.20 *Floor.*

Theorem 2. Suppose $X \subseteq \mathbb{R}$ and $F : X \to \mathbb{R}$ is a strictly increasing function. Then, F is *1–1*.

Proof.   This proof is left as an exercise for the reader.   ■

Of course, the definitions of the terms *strictly increasing* and *strictly decreasing* do not involve anything special about $\mathbb{R}$, just that it has the relations $<$ and $\leq$. Consequently, a similar definition could be made for any linearly ordered, or even any partially ordered, domain and codomain.

4.2   Exercises

1. Which of the following are functions? If not, why not?

   (a) X is the set of students in the discrete mathematics class. For $x \in X$, define $g(x)$ to be the youngest cousin of x.
   (b) X is the set of senators serving in 1998. For $x \in X$, define $g(x)$ to be the number of terms a senator has held.
   (c) For $x \in \mathbb{R}$, define $g(x) = |x/|x||$.

2. Let $X = \{0, 1, \ldots, 6, 7\}$ and $Y = \{8, 10, 12, \ldots, 20, 22\}$. Define $F : X \to Y$ as $F(x) = 2x + 8$. List the ordered pairs of the relation that define this function.

3. What are the domain and range of the addition function on the real numbers? On Multiplication? Subtraction? Division?

4. Find the first six terms of the sequence with the elements defined as $F(0) = 5$, $F(1) = 10$, and $F(n) = F(n-1) - 2F(n-2)$ for $n \geq 2$.

5. Find the first six terms of the sequence with the elements defined as $F(0) = 1$, $F(1) = 3$, $F(2) = 5$, and $F(n) = 3F(n-1) + 2F(n-2) - 3F(n-3)$ for $n \geq 3$.

6. Find both a function defined by a formula and a recursively defined function for the following sequences:

   (a) $1, 3, 5, 7, 9, 11, 13, \ldots$
   (b) $1, 1, 3, 3, 5, 5, 7, 7, \ldots$
   (c) $0, 2, 4, 6, 8, \ldots$
   (d) $1, 2, 4, 8, 16, \ldots$

7. Which of the following represent a partial function? A (total) function?

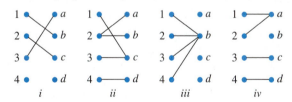

8. Let $X = \{a, b\}$.

   (a) There are nine partial functions $F : X \to X$. List them.
   (b) There are four functions $F : X \to X$. List them.
   (c) List all *1–1* functions $F : X \to X$.
   (d) List all onto functions $F : X \to X$.

9. Let $X = \{-1, 0, 1, 2\}$ and $Y = \{-4, -2, 0, 2\}$. Define the function $F : X \to Y$ as $F(x) = x^2 - x$. Prove that F is neither *1–1* nor onto.

10. List all *1–1* and onto functions from $\{1, 2, 3\}$ to itself.

11. Let A be a set with three elements and B be a set with two elements.

    (a) How many different functions are there with domain A and codomain B?
    (b) How many different functions are there with domain B and codomain A?
    (c) How many different *1–1* functions are there with domain A and codomain B?
    (d) How many different *1–1* functions are there with domain B and codomain A?

12. Determine which of the following functions are onto:

    (a) $F_1 : \mathbb{R} \to \mathbb{R}$ where $F_1(x) = x^2 - 1$.
    (b) $F_2 : \mathbb{R} \to \mathbb{Z}$ where $F_2(x) = \lceil x \rceil$ ($\lceil x \rceil$ is the "ceiling" of x).
    (c) $F_3 : \mathbb{Z} \to \mathbb{Z}$ where $F_3(x) = x^3$.
    (d) $F_4 : \mathbb{R} \to \mathbb{R}$ where $F_4(x) = x^3$.
    (e) For the linear ordering $<$ on $\mathbb{R}$, list all the increasing functions among parts (a) through (d).
    (f) For the ordering $<$ on $\mathbb{R}$, list all the strictly increasing functions among parts (a) through (d).

13. Which of the functions in Exercise 12 are *1–1*? Prove each of your answers.

14. Two months are equivalent if their 13th day must fall on the same day of the week in every (nonleap) year.

    (a) Show that the 13th day of the 12 months occur on seven different days of the week.
    (b) Conclude that there must be at least one Friday the 13th in each year.
    (c) Show that there are at most three Friday the 13th's in any year.
    (d) Show that the result is also true for leap years.

    (*Hint:* Number the days of the year from 1 (January 1) to 365 (December 31), and then show that the days representing the 13th days of these months occur on seven different days of the week.)

15. Let $A = \{1, 2, 3, 4\}$ and $B = \{a, b, c\}$. Define a function $F : A \to B$ as $F(1) = a$, $F(2) = b$, $F(3) = c$, and $F(4) = c$. List the ordered pairs of the equivalence relation R defined on A as $x\,R\,y$ if and only if $F(x) = F(y)$. List the elements of the partition of A determined by this equivalence relation.

16. Let $\mathcal{F}_{\{0,1,2\}}$ be the set of all functions with domain and codomain equal to $\{0, 1, 2\}$. For each of the following relations, prove that the relation is an equivalence relation. Also, find the distinct equivalence classes of each equivalence relation. Let $F, G \in \mathcal{F}_{\{0,1,2\}}$.

    (a) $F\,R\,G$ if and only if *range(F)* = *range(G)*.
    (b) $F\,R\,G$ if and only if *max(F)* = *max(G)*.
    (c) $F\,R\,G$ if and only if $F(0) + F(1) + F(2) = G(0) + G(1) + G(2)$. For this problem, two functions are related if the sum of their images, seen as an operation in the natural numbers and not in the function space, are equal.

17. Find two functions $F, G : \mathbb{R} \to \mathbb{R}$ where $F \neq G$ but $F\,|_{[0,1)} = G\,|_{[0,1)}$.

18. Let $F : \mathbb{R} \to \mathbb{R}$ with $F(x) = x^2$. The following is a function from $\mathbb{R}$ to $\mathbb{R}$:

$$Id_{\mathbb{R}}\,|_{[2,\infty)} \cup Zero\,|_{[0,2)} \cup F\,|_{(-\infty,0)}$$

Write an algorithm to compute this function.

19. Let A, B, and C be sets, and let $F : A \to C$ be a function. If $B \subseteq A$, prove that $F\,|_B = F \cap (B \times C)$.

20. Prove that the function $F : \mathbb{Z} \to \mathbb{Z}$ defined as $F(n) = n + 6$ is a bijection.

21. For each of the following functions, prove that the function is *1–1* or find an appropriate pair of points to show that the function is not *1–1*:

   (a) $F : \mathbb{Z} \to \mathbb{Z}$

$$F(n) = \begin{cases} n^2 & \text{for } n \geq 0 \\ -n^2 & \text{for } n \leq 0 \end{cases}$$

   (b) $F : \mathbb{R} \to \mathbb{R}$

$$F(x) = \begin{cases} x + 1 & \text{for } x \in \mathbb{Q} \\ 2x & \text{for } x \notin \mathbb{Q} \end{cases}$$

   (c) $F : \mathbb{R} \to \mathbb{R}$

$$F(x) = \begin{cases} 3x + 2 & \text{for } x \in \mathbb{Q} \\ x^3 & \text{for } x \notin \mathbb{Q} \end{cases}$$

   (d) $F : \mathbb{Z} \to \mathbb{Z}$

$$F(n) = \begin{cases} n + 1 & \text{for } n \text{ odd} \\ n^3 & \text{for } n \text{ even} \end{cases}$$

22. (a) Find functions from $\mathbb{R}$ to $\mathbb{R}$ that are:
   - i. strictly decreasing
   - ii. decreasing but not strictly decreasing
   - iii. neither increasing nor decreasing
   - iv. both increasing and decreasing
   (b) Show that no $F : \mathbb{R} \to \mathbb{R}$ is both increasing and strictly decreasing.
   (c) Find a subset $X \subseteq \mathbb{R}$ and a function $F : X \to X$ where F is both strictly increasing and strictly decreasing.

23. Construct functions with the following properties:
   (a) $F : \mathbb{N} \to \mathbb{N}$ such that $range(F) = \mathbb{N}$ and, for each $n \in \mathbb{N}$, there exist exactly two solutions for the equation $F(x) = n$.
   (b) $F : \mathbb{N} \to \mathbb{N}$ such that, for each $n \in \mathbb{N}$, there are exactly n solutions for the equation $F(x) = n$.

24. Prove Theorem 3.

25. Using the numbering scheme for the letters of the alphabet as given in Section 4.1.8, encrypt the message DISCRETE MATH IS GREAT using the function $F(letter) = 17(letter\,value) + 9(mod\,26)$. List the letters of the encrypted message. Find the inverse function, and decrypt the message. (*Hint:* $23 \cdot 17 = 1(mod\,26)$.)

26. Using the numbering scheme for the letters of the alphabet given in Section 4.1.8, encrypt the message DISCRETE MATH IS GREAT using the function $F(letter) = (11(letter\,value) + 13)\,mod\,26$. List the letters of the encrypted message. Find the inverse function, and decrypt the message. (*Hint:* $19 \cdot 11 = 1(mod\,26)$.)

27. *For the American history fan:* Consider the list of U.S. presidents up through Harry Truman. Define the following "function" on all presidents *before* Harry Truman: The *successor* of X is the person who followed X as president. Why is *successor* not a function?

28. Define a function $F : \mathbb{N} \to \mathbb{N}$ such that $F(n) = n - 10$ if $n > 100$ and $F(n) = F(F(n + 11))$ if $n \leq 100$.

    (a) Show that $F(99) = 91$.
    (b) Prove that $F(n) = 91$ for all n such that $0 \leq n \leq 100$.

29. Let A, B, and C be sets, and let $F : A \to C$ and $G : B \to C$ be functions.

    (a) What condition must F and G satisfy for $F \cup G$ to be a function from $A \cup B$ to C?
    (b) Give conditions on A and B such that $F \cup G$ is a function for *every* $F : A \to C$ and $G : B \to C$.

30. Let F be a function, and let $C, D \subseteq domain(F)$.

    (a) Prove that $range(F\,|_{C \cap D}) \subseteq range(F\,|_C) \cap range(F\,|_D)$.
    (b) Show by example that equality need not hold in part (a).

31. If looked at appropriately, the definition of a function as a set of ordered pairs and the intuitive notion that a function is something given by a rule are equivalent. Develop that equivalence here. Assume that F has a finite domain $\{0, 1, 2, \ldots, n - 1\}$ and a finite codomain $\{0, 1, 2, \ldots, m - 1\}$.

    (a) Suppose F is a function given as a set of ordered pairs. For an input x_1, give a rule for calculating $F(x_1)$. Use F (or its graph) in your rule.
    (b) Suppose the function F is given by a rule. Express F as a set of ordered pairs.

32. Find a combinatorial circuit for each of the following boolean functions:

    (a)

p	q	$F(p, q)$
1	1	1
1	0	1
0	1	0
0	0	0

(b)

p	q	r	$F(p,q,r)$
1	1	1	1
1	1	0	1
1	0	1	0
1	0	0	0
0	1	1	0
0	1	0	1
0	0	1	0
0	0	0	1

(c)

p	q	r	$F(p,q,r)$
1	1	1	0
1	1	0	1
1	0	1	1
1	0	0	0
0	1	1	1
0	1	0	1
0	0	1	1
0	0	0	0

4.3   Operations on Functions

Since functions and partial functions are special types of binary relations, all operations defined on binary relations can be applied to functions. The most interesting operations, however, are composition and inversion.

4.3.1   Composition of Functions

The definition of the composition of functions is exactly the same as that of the composition of relations. We merely restate it here using the vocabulary of functions.

Definition 1. Let both $F : X \to Y$ and $G : Y \to Z$ be partial functions. The **composition** of G and F is

$$G \circ F = \{(x, z) \in X \times Z : \text{ for some } y \in Y, \quad y = F(x) \text{ and } z = G(y)\}$$

Thus, $(G \circ F)(x) = G(F(x))$.

It turns out that the composition of two functions is always a function.

Example 1. Start with the *SeatOf* function for a class:

$$SeatOf = \{(\text{Jean, Seat2}), (\text{Michele, Seat5}), (\text{Paul, Seat3})\}$$

Assume that just before the class started, workers finished repainting the desks in the following colors:

$$ColorOfSeat = \{(\text{Seat1, red}), (\text{Seat2, red}), (\text{Seat3, green}), (\text{Seat4, green}), (\text{Seat5, red})\}$$

The definition of *ColorOfSeat* ∘ *SeatOf* is

$$\{(x, z) : \text{ for some } y, \quad y = SeatOf(x) \text{ and } z = ColorOfSeat(y)\}$$

Now, unravel that definition. Start with $x = $ Jean. Since *SeatOf* is a function, there is exactly one object $y = SeatOf$ (Jean), which is Seat2. Figure 4.21 shows this procedure.

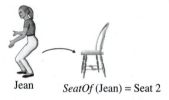

Jean   *SeatOf* (Jean) = Seat 2

Figure 4.21 Jean and *SeatOf*(Jean).

Since *ColorOfSeat* is a function, there is exactly one object $z = ColorOfSeat$(Seat2), which is red. This object can also be referred to as *ColorOfSeat*(*SeatOf*(Jean)). Figure 4.22 shows this procedure.

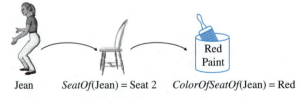

Jean   *SeatOf*(Jean) = Seat 2   *ColorOfSeatOf*(Jean) = Red

Figure 4.22 Jean, *SeatOf*(Jean) and *ColorOfSeat*(*SeatOf*(Jean)).

The same sort of analysis holds also for *ColorOfSeat* (*SeatOf* (Michele)) and *ColorOfSeat*(*SeatOf*(Paul)). The function is given as

$$ColorOf \circ SeatOf = \{(\text{Jean, red}), (\text{Michele, red}), (\text{Paul, green})\}$$

In general, composition of functions is a function. The operation can even be stated for partial functions.

Theorem 1. Let X, Y, and Z be sets. Let both $F : X \to Y$ and $G : Y \to Z$ be partial functions. Then, $G \circ F$ is a partial function from X to Z. Moreover, for every $x \in X$, the following hold:

(a) If $F(x)$ is undefined, then $(G \circ F)(x)$ is undefined.
(b) If $F(x)$ is defined but $G(F(x))$ is undefined, then $(G \circ F)(x)$ is undefined.
(c) If $F(x)$ and $G(F(x))$ are defined, then $(G \circ F)(x)$ is also defined, and $(G \circ F)(x) = G(F(x))$.

The proof of Theorem 1 is omitted, since it is just a formalization of the discussion in the example above. One important corollary to Theorem 1 is used all the time. This corollary says that the composition of functions is an associative operation, just like addition and multiplication with real numbers as well as union and intersection of sets.

Corollary 1: Let $F : X \to Y, G : Y \to Z$, and $H : Z \to W$ be functions. Then, $F \circ (G \circ H) = (F \circ G) \circ H$.

Corollary 4.1 follows from Theorem 3 by reducing both $F \circ (G \circ H)(x)$ and $(F \circ G) \circ H(x)$ to $F(G(H(x)))$.

For functions F and G, one often defines $G \circ F(x)$ to be $G(F(x))$. Since we have already defined the operation $\circ$ on relations in Section 3.2.2, we only had to show that $(G \circ F)(x)$ is the same as $G(F(x))$.

Example 2 shows that the composition of functions is not a commutative operation.

Example 2. Let $G : \mathbb{N} \to \mathbb{N}$ and $H : \mathbb{N} \to \mathbb{N}$ be given by the rules $G(n) = n^2 + 1$ and $H(n) = 2^n$. Then, $(H \circ G) : \mathbb{N} \to \mathbb{N}$, and for all $n \in \mathbb{N}$, we have $(H \circ G)(n) = 2^{n^2+1}$. By contrast, $(G \circ H)(n) = 2^{2n} + 1$.

Earlier, we studied *1–1* and onto functions. It is now natural to ask whether the composition of *1–1* functions is *1–1* or whether the composition of onto functions is onto. We answer these questions in Theorem 2.

Theorem 2. Let $F : X \to Y$ and $G : Y \to Z$ be functions.

(a) If F and G are both *1–1*, then $G \circ F$ is *1–1*.
(b) If F and G are both onto, then $G \circ F$ is onto.
(c) If F and G are *1–1* correspondences, then $G \circ F$ is a *1–1* correspondence.
(d) If $G \circ F$ is *1–1*, then F is *1–1*.
(e) If $G \circ F$ is onto, G is onto.

Proof. These proofs are left as exercises for the reader. ∎

4.3.2 Inverses of Functions

Recall the definition of the inverse of a relation given in Section 3.2.1. For any relation R defined on a set X,

$$R^{-1} = \{(y, x) \in X \times X : (x, y) \in R\}$$

Since functions are relations, they also have inverses.

Definition 2.   Let $F = \{(x, y) \in X \times Y : F(x) = y\}$ be a function. The **inverse** of F, denoted by F^{-1}, is the relation

$$F^{-1} = \{(y, x) \in Y \times X : F(x) = y\}$$

Example 3.   Consider a business where each employee has an employee number and no two employees have the same number. The function

$$EmplNoOf : Employees \rightarrow EmplNos$$

and its inverse, $EmplNoOf^{-1}$, are pictured below in Figure 4.23.

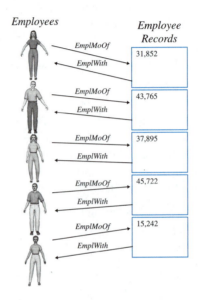

Figure 4.23   Employee functions.

The function *EmplWith*, as shown in Figure 4.23, would normally be a partial function since employee numbers are very rarely a set of consecutive integers. The gaps between employee numbers would represent values for which the function is not defined.

Example 4.   Define two functions, *Succ* and *Pred*, from $\mathbb{Z}$ to $\mathbb{Z}$. Let $Succ(z) = z + 1$ and $Pred(z) = z - 1$. We can show that $Pred^{-1} = Succ$.

Solution.

$$Pred = \{(z, z - 1) : z \in \mathbb{Z}\}$$

And

$$
\begin{aligned}
Succ &= \{(z, z + 1) : z \in \mathbb{Z}\} \\
&= \{(z_1 - 1, z_1) : z_1 \in \mathbb{Z}\} \quad \text{(substitute } z_1 = z + 1) \\
&= \{(z - 1, z) : z \in \mathbb{Z}\} \quad \text{(substitute } z \text{ for } z_1\text{—since } z \text{ is no longer in use,} \\
&\qquad\qquad\qquad\qquad\quad \text{it can be reused)} \\
&= Pred^{-1}
\end{aligned}
$$

The inverse of a function F is not always a function or a partial function. If, however, F is *1–1* or a *1–1* correspondence, then we have Theorem 3.

Theorem 3.   Let $F : X \to Y$ be a function.

(a) F^{-1} is a function from Y to X if and only if F is a *1–1* correspondence.
(b) F^{-1} is a partial function from Y to X if and only if F is *1–1*.
(c) If F is a *1–1* correspondence, then $F^{-1} : Y \to X$ is a *1–1* correspondence.

Proof.

(a) This proof is left as an exercise for the reader.
(b) F^{-1} is a partial function
$$\Leftrightarrow \text{ for each } y \in Y, \text{ there is at most one } x \in X \text{ with } (y, x) \in F^{-1}$$
$$\Leftrightarrow \text{ for each } y \in Y \text{ there is at most one } x \in X \text{ with } (x, y) \in F$$
$$\Leftrightarrow F \text{ is } 1\text{–}1.$$
(c) This proof is left as an exercise for the reader. ∎

A function whose inverse is a function is also referred to as being **invertible.**

Theorem 4.   Let X be a set, and let $F : X \to Y$ be a *1–1* and onto function.

(a) $F^{-1} \circ F = Id_X$
(b) $F \circ F^{-1} = Id_Y$

Proof.

(a) First, observe that $F^{-1} \circ F$ is a *1–1* correspondence. This follows from three facts: (i) F is given as a *1–1* correspondence; (ii) by Theorem 3(c) we have F^{-1} is a *1–1* correspondence; and (iii) by Theorem 2(c) $F^{-1} \circ F$ is a *1–1* correspondence.

Now, let $x \in X$. Since F is a total function, there is a $y \in Y$ such that $(x, y) \in F$. By the definition of an inverse, we have $(y, x) \in F^{-1}$. By the definition of composition of functions (see Section 4.3.1), it follows that $(x, x) \in F^{-1} \circ F$. That is, $Id_X \subseteq F^{-1} \circ F$.

To show that $F^{-1} \circ F \subseteq Id_X$, let $(x, x') \in F^{-1} \circ F$. Since we have just seen that $(x, x) \in F^{-1} \circ F$ and we observed that $F^{-1} \circ F$ is *1–1*, we must have $x' = x$; that is, $(x, x') \in Id_X$. Therefore, $F^{-1} \circ F \subseteq Id_X$.

(b) By Theorem 3, F^{-1} is *1–1* and onto. It follows from part (a) that $(F^{-1})^{-1} \circ (F^{-1}) = Id_Y$. By Theorem 2 in Section 3.2.1 it follows that $F \circ F^{-1} = (F^{-1})^{-1} \circ F$. Now, by part (a), $(F^{-1})^{-1} \circ F = Id_Y$. ∎

Very informally, Theorem 4 can be summarized as saying that if F^{-1} is a function at all, then F^{-1} "undoes" what F "does."

Example 5.   The function $\exp(x) = e^x$ where e, the real number $2.718281828459\ldots$, is called the exponential function base e, which is also called exp. The function $\exp : \mathbb{R} \to (0, \infty)$ is strictly increasing, *1–1*, and onto. Its inverse is called the natural logarithm function, designated ln. Hence, $y = \ln(x)$ is true if and only if $x = \exp(y)$ is true. It is also easy to show that ln is strictly increasing.

4.3.3   Other Operations on Functions

The reader is familiar with operations on polynomial functions. Consider polynomial functions $F, G : \mathbb{R} \to \mathbb{R}$ where $F(x) = x^2$ and $G(x) = 2x + 1$. Then, $(\boldsymbol{F + G})(x)$ is defined as

$$(F + G)(x) = F(x) + G(x) = x^2 + 2x + 1$$

This is a very different sort of operation on functions in that it uses the operation $+$ on $\mathbb{R}$, whereas composition and inversion operations make no reference to operations on the codomain of the function.

Definition 3.   Let $F, G : X \to \mathbb{R}$ be functions. The following are functions:

$$
\begin{aligned}
(F + G) : \quad & X \to \mathbb{R} \\
& x \to F(x) + G(x) \\
(F \cdot G) : \quad & X \to \mathbb{R} \\
& x \to F(x) \cdot G(x) \\
|F| : \quad & X \to \mathbb{R} \\
& x \to |F(x)|
\end{aligned}
$$

Define the following partial function:

$$(F/G) : X \to \mathbb{R} \text{ by the rule } (F/G)(x) = F(x)/G(x)$$

The function F/G is total if and only if $G(x) \neq 0$ for all $x \in X$.

Of course, the same definitions make sense if the codomain is $\mathbb{Q}$, $\mathbb{Z}$, or $\mathbb{N}$. In general, any operation on the codomain may be used to define an operation on functions.

Definitions such as Definition 3 create some very ambiguous notation. For x, a real number, x^{-1} denotes $1/x$. So, $F^{-1}(x)$ should denote $1/F(x)$. The symbol F^{-1}, however, also means the inverse function, which is not at all the same thing. The symbol F^{-1} usually—but not always—denotes the inverse function. In this book, we shall use F^{-1} only to denote the inverse function.

4.4   Sequences and Subsequences

This section introduces functions defined on $\mathbb{N}$ and its subsets that we commonly refer to as sequences. Subsequences are formed by using the operation of composition of functions on subsets of $\mathbb{N}$.

Intuitively, a **sequence** is a list of objects in order, such as

red, orange, yellow, green, blue, indigo, violet

where red is first, followed by orange, . . . , followed by violet. Other sequences are the prime numbers listed in increasing order:

$$2, 3, 5, 7, 11, 13, 17, 19, 23, 29, 31, 37, \ldots$$

or the natural numbers in increasing order:

$$0, 1, 2, 3, 4, 5, 6, 7, 8, 9, 10, 11, 12, \ldots.$$

Definition 3. An **infinite sequence** of elements of a set X is a function $F : \mathbb{N} \to X$. A function $F : \{0, 1, 2, \ldots, n - 1\} \to X$ for some $n \in \mathbb{N}$ is a **finite sequence** of elements of X of length n. The expression *sequence of elements of X* means a finite sequence of elements of X or an infinite sequence of elements of X.

In computer programming, finite sequences are often called **lists.** Infinite sequences are often called **streams** and sometimes also lists. Often, if F is a finite sequence of elements, then its elements are denoted not as

$$F(0), F(1), \ldots, F(n-1)$$

but, rather, as

$$x_0, x_1, \ldots, x_{n-1}$$

Similarly, an infinite sequence is usually written as

$$x_0, x_1, \ldots, x_n, \ldots.$$

An infinite sequence of real numbers is a function from $\mathbb{N}$ to $\mathbb{R}$. For example,

$$x_0 = \frac{0}{1}, \quad x_1 = \frac{1}{2}, \quad x_2 = \frac{2}{3}, \ldots, x_n = \frac{n}{n+1}, \ldots.$$

is an infinite sequence of real numbers.

For any $X \subseteq \mathbb{R}$, a sequence F of elements of X is **increasing** if, thought of as a function from $\mathbb{N}$ to X, F is increasing. Thus, the sequence

$$x_0 = \frac{0}{1}, \quad x_1 = \frac{1}{2}, \quad x_2 = \frac{2}{3}, \ldots, x_n = \frac{n}{n+1}, \ldots$$

is increasing. The terms **increasing, decreasing, strictly increasing,** and **strictly decreasing** apply to sequences in the same way.

Example 6.

(a) The elements of a sequence need not be different. For example, $0, 0, 0, 0, \ldots$ is a sequence. Formally, this sequence is given by the function $Zero : \mathbb{N} \to \mathbb{N}$ defined by the rule $Zero(n) = 0$.
(b) Let $F : \mathbb{N} \to \mathbb{Z}$ be defined by the rule $F(n) = (-1)^n$. Then, F is the sequence $1, -1, 1, -1, 1, -1, \ldots$
(c) Let $Fact(n) = n!$. Then, *Fact* defines the sequence

$$1, 1, 2, 6, 24, 120, 720, \ldots$$

An important notion associated with sequences is the notion of a subsequence. Intuitively, a subsequence is just a subset of a sequence, with the elements of the subsequence occurring in the same order as they do in the sequence.

Example 7. For the sequence of factorials

$$1, 1, 2, 6, 24, 120, 720, 5040, 40{,}320, \ldots$$

the following are subsequences:

(a) Ø; the subsequence of length 0
(b) $1, 6, 120, 5040, \ldots$; every other factorial, starting with the second one
(c) $1, 1, 2, 6, 24, 120, 720, 5040, 40{,}320, \ldots$; the entire sequence
(d) 1 the first element alone
(e) $2, 6, 40{,}320$; another finite subsequence

What, more precisely, is a subsequence? Think of an infinite sequence:

$$x_0, x_1, x_2, x_3, x_4, x_5, x_6, \ldots.$$

Pick out a subset of the subscripts, such as subscripts

$$1, 2, 4, 8, 16, 32, \ldots$$

and then list the corresponding elements of the sequence in the same order as used in the original sequence:

$$x_1, x_2, x_4, x_8, x_{16}, x_{32}, \ldots$$

The chosen subscripts themselves form a sequence:

$$i_0 = 1, i_1 = 2, i_2 = 4, i_3 = 8, i_4 = 16, i_5 = 32, \ldots$$

So, the subsequence is

$$x_{i_0}, x_{i_1}, x_{i_2}, x_{i_3}, x_{i_4}, x_{i_5}, \ldots.$$

(See Exercise 13 in Section 4.5 for missing details.) The important point is that the elements are listed in the same order as in the original sequence; that is,

$$i_0 < i_1 < i_2 < i_3 < i_4 < i_5 < \cdots$$

is itself a strictly increasing sequence.

Definition 9.   Let F be a sequence, and let S be a strictly increasing sequence of elements of the domain of F. Then, $F \circ S$ is a **subsequence** of F.

The proof that Definition 8 formalizes the previous discussion is left as an exercise.

Example 8.   In the definition of a subsequence, the sequence S was required to be strictly increasing. The sequence of elements in a sequence F are not required to be increasing; as a result, the subsequence of elements determined by $F \circ S$ need not be strictly increasing. For example, let $F : \mathbb{N} \to \mathbb{R}$ where $F(n) = (-1)^n/(n + 1)$. So, F is the sequence

$$1, -\frac{1}{2}, \frac{1}{3}, -\frac{1}{4}, \ldots.$$

(a) If S is the sequence $0, 2, 4, \ldots$ of even natural numbers, then $F \circ S$ is the subsequence consisting of every other element of the sequence F, starting with the first element:

$$1, \frac{1}{3}, \frac{1}{5}, \ldots$$

which is decreasing.

(b) If S is the sequence $1, 3, 5, \ldots$ of odd natural numbers, then $F \circ S$ is the sequence consisting of every other element of the sequence F, starting with the second element:

$$-\frac{1}{2}, -\frac{1}{4}, -\frac{1}{6}, \ldots$$

which is increasing.

4.5   Exercises

1. Let $X = \{1, 2, 3, 4\}$ and $Y = \{5, 6, 7, 8, 9\}$. Let $F = \{(1, 5), (2, 7), (4, 9), (3, 8)\}$. Show that F is a function from X to Y. Find F^{-1}, and list its elements. Is F^{-1} a function? Why, or why not?

2. Let $S = \{(0, 8), (1, 10), (2, 12), (3, 14), (4, 16), (5, 18), (6, 20), (7, 22)\}$. Is S a function? Why, or why not? Find S^{-1}, and list its elements. Is S^{-1} a function? Why, or why not? Identify the domain of S^{-1}.

3. Let $X = \{1, 2, 3, 4\}$. Let $F : X \to \mathbb{R}$ be a function defined as the set of ordered pairs $\{(1, 2), (2, 3), (3, 4), (4, 5)\}$. Let $G : \mathbb{R} \to \mathbb{R}$ be the function defined as $G(x) = x^2$. What is $G \circ F$?

4. Let $F : \mathbb{R} \to \mathbb{R}$ be defined as $F(x) = 2x + 8$. Let $G : \mathbb{R} \to \mathbb{R}$ be defined as $G(y) = (y - 8)/2$. Prove that $F \circ G = Id_{\mathbb{R}}$ and $G \circ F = Id_{\mathbb{R}}$.

5. Define the functions $F, G,$ and H as indicated in the following diagrams:

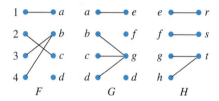

Find the following:

(a) $G \circ F$

(b) $H \circ (G \circ F)$

(c) $(H \circ G) \circ F$

6. Let $X = \{0, 1, 2\} \subseteq \mathbb{R}$. List all eight strictly increasing sequences of elements of X. The ordering is $<$ on $\mathbb{R}$. List all subsequences of the sequence x, y, z.

7. Let $A = \{1, 2, 3, 4\}$. Let the functions $F, G,$ and H be given with domain and codomain A defined as

$$F(1) = 3, \ F(2) = 2, \ F(3) = 2, \ \text{and } F(4) = 4$$
$$G(1) = 1, \ G(2) = 3, \ G(3) = 4, \ \text{and } G(4) = 2$$
$$H(1) = 2, \ H(2) = 4, \ H(3) = 1, \ \text{and } H(4) = 3$$

Find the following:

(a) $F \circ G$

(b) $H \circ F$

(c) $G \circ H$

(d) $F \circ G \circ H$

8. Let A be a rule for defining a function $F : \mathbb{N} \to \mathbb{N}$ such that F is *1–1* and onto. Show how to construct a rule for defining F^{-1}.

9. For sets $X, Y,$ and Z, let $F : X \to Y$ and $G : Y \to Z$ be *1–1* correspondences. Prove that $(G \circ F)^{-1} = F^{-1} \circ G^{-1}$.

10. Find the first six terms of the sequences defined for $n \geq 0$ as:

    (a) $H(n) = n^2 (n+1)^2/4$
    (b) $G(n) = 2^n - 1$
    (c) $F(n) = (-1)^n 2^n - 3^n$

11. Find the first six terms of the sequences defined as:

    (a) $H(0) = 0$ and $H(n) = H(n-1) + n^3$ for $n \geq 1$
    (b) $G(0) = 0$ and $G(n) = 2G(n-1) + 1$ for $n \geq 1$
    (c) $F(0) = 2$ and $F(n) = 3F(n-1) - n + 3$ for $n \geq 1$

12. Find a recursively defined function that gives the terms of the following sequences:

    (a) $2, 5, 8, 11, 14, \ldots$
    (b) $3, 6, 12, 24, 48, \ldots$

13. The formal definition of a sequence was in terms of a function F, with domain either $\mathbb{N}$ or $\{0, 1, \ldots, n-1\}$. (If $n = 0$, then $\{0, 1, \ldots, n-1\} = \emptyset$.) The formal definition of a subsequence involves a sequence F and a strictly increasing sequence S of elements of the domain of F. Since S is a sequence, S is, formally, another function as above. In parts (a) through (e) of Example 7, identify the functions S and $F \circ S$ as sets of ordered pairs.

14. Prove the following:

    (a) Theorem 2(a)
    (b) Theorem 2(b)
    (c) Theorem 2(d)
    (d) Theorem 2(e)

15. Prove the following:

    (a) Theorem 3(a)
    (b) Theorem 3(c)

16. Let A and B be nonempty sets, and let $F : A \rightarrow B$ be a function. Prove that the following are equivalent:

    (a) F is onto.
    (b) There is a function $G : B \rightarrow A$ such that $F \circ G = Id_B$.
    (c) For any set C and for functions $H_1 : B \rightarrow C$ and $H_2 : B \rightarrow C$, if $H_1 \circ F = H_2 \circ F$, then $H_1 = H_2$.

17. Let A and B be nonempty sets, and let $F : A \rightarrow B$ be a function. Prove that the following are equivalent:

    (a) F is 1–1.
    (b) There is a function $G : B \rightarrow A$ such that $G \circ F = Id_A$.
    (c) For any set C and for functions $H_1 : C \rightarrow A$ and $H_2 : C \rightarrow A$, if $F \circ H_1 = F \circ H_2$, then $H_1 = H_2$.

18. Let A and B be sets with $A_1, A_2 \subseteq A$, and let $F : A \rightarrow B$. Let $F(A_i)$ denote $\{F(x) : x \in A_i\}$ for $i = 1, 2$. Show that:

    (a) If $A_1 \subseteq A_2$, then $F(A_1) \subseteq F(A_2)$.
    (b) $F(A_1 \cup A_2) = F(A_1) \cup F(A_2)$.
    (c) $F(A_1 \cap A_2) \subseteq F(A_1) \cap F(A_2)$.

    (d) $F(A_1) - F(A_2) \subseteq F(A_1 - A_2)$.

    (e) $A_1 \subseteq F^{-1}(F(A_1))$.

    (f) Find an example in which $A_1 \subset A_2$ but $F(A_1) = F(A_2)$.

    (g) Find an example in which $A_1 \neq F^{-1}(F(A_1))$.

19. Let A be any nonempty set, and let $\mathcal{F}_A$ be the set of all functions from A to $\mathbb{R}$.

    (a) Why is $F + G \in \mathcal{F}_A$ for all $F, G \in \mathcal{F}_A$.

    (b) Prove $(F + G) + H = F + (G + H)$ for all $F, G, H \in \mathcal{F}_A$.

    (c) Let $Zero \in \mathcal{F}_A$ be defined by $Zero(a) = 0$ for all $a \in A$. Prove that $Zero + F = F$ for all $F \in \mathcal{F}_A$.

    (d) For $F \in \mathcal{F}_A$, define $\bar{F}$ by $\bar{F}(a) = -F(a)$ for each $a \in A$. Prove that $F + \bar{F} = Zero = \bar{F} + F$ for all $F \in \mathcal{F}_A$.

20. Let $\mathcal{F}_A$ be defined as in Exercise 19. For each $F, G \in \mathcal{F}_A$, define $F \cdot G(a) = F(a) \cdot G(a)$.

    (a) Why is $F \cdot G \in \mathcal{F}_A$ for all $F, G \in \mathcal{F}_A$?

    (b) Prove that $F \cdot G = G \cdot F$ for all $F, G \in \mathcal{F}_A$.

    (c) Prove that $(F \cdot G) \cdot H = F \cdot (G \cdot H)$ for all $F, G, H \in \mathcal{F}_A$.

    (d) Prove that the function $U : A \to \mathbb{R}$ defined by $U(a) = 1$ for all $a \in A$ satisfies $U \cdot F = F = F \cdot U$ for all $F \in \mathcal{F}_A$.

    (e) Prove that $(F + G) \cdot H = F \cdot H + G \cdot H$ for all $F, G, H \in \mathcal{F}_A$ with $F + G$ defined as in Exercise 19.

    (f) Prove there are $F, G \in \mathcal{F}_A$ such that $F \neq Zero$ and $G \neq Zero$ but $F \cdot G = Zero$.

21. (a) Let $F : A \to B$ be a function. Prove that F is onto if and only if $F^{-1}(B_1) \neq \emptyset$ for each nonempty subset B_1 of B.

    (b) Let $F : A \to B$ be a function. Prove that F is onto if and only if $F(F^{-1}(B_1)) = B_1$ for all $B_1 \subseteq B$.

22. Let X be a set, and let $\mathcal{F}_X$ be the set of all *1–1* functions from X onto X. We have two operations on functions in $\mathcal{F}_X$: $\circ$ and $^{-1}$. Prove the following statements called group axioms. (If the results are already proved in the book, note where to find the proofs.)

    (a) For all $F, G \in \mathcal{F}_X$, $F \circ G \in \mathcal{F}$.

    (b) For all $F, G, H \in \mathcal{F}_X$, $(F \circ G) \circ H = F \circ (G \circ H)$ (Associative Law).

    (c) For all $F \in \mathcal{F}_X$, $F \circ Id_X = Id_X \circ F = F$. (Identity Axiom).

    (d) For all $F \in \mathcal{F}_X$, there exists an F^{-1} such that $F \circ F^{-1} = F^{-1} \circ F = Id_X$ (Inverse Axiom).

23. An operation $\otimes$ on a set Y is **commutative** if for all $y, z \in Y$, $y \otimes z = z \otimes y$. For X and $\mathcal{F}_X$ as defined in Exercise 22, prove that $\circ$ need not be commutative on $\mathcal{F}_X$.

24. Let $F : A \to B$ be a function. Define $G : \mathcal{P}(B) \to \mathcal{P}(A)$ by $G(B_1) = F^{-1}(B_1)$. Prove that G is *1–1* if and only if F is onto.

4.6   The Pigeon-Hole Principle

After a February town meeting in rural New England, the 200 people attending entered the parking garage to get their cars and drive home. (Assume that no one was walking home in the typical winter weather.) An observer counted 65 cars exiting the parking garage.

What can we conclude about the function that maps people to the cars in which they are riding? Is it *1–1*? Is it onto? How far is it from being *1–1*? From being onto? A similar question could be how many of the students in your class have a birthday on the same day of the week this year. The results of this section will help you answer these and similar questions as well as see applications of the results presented in other contexts.

For the remainder of this chapter, we will discuss only total functions.

4.6.1   *k* to 1 Functions

The first step in answering the questions posed at the beginning of this section is to determine if more than one element in a function's domain must be mapped to a single element in the function's range.

Definition 1.   Let $F : X \rightarrow Y$ be a function. Let $k \in \mathbb{N}$. F is **k to 1** if, for each $y \in Y$, there are at most k different x's in X with $F(x) = y$. Alternatively, for each $y \in Y$,

$$|\{x \in X : F(x) = y\}| \leq k$$

Example 1.

(a) Let $F : \{0, 1, 2, 3\} \rightarrow \{a, b, c, d\}$ be a function defined as

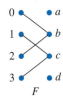

   F is *2–1*.

(b) Let $F : \{0, 1, 2, 3\} \rightarrow \{a, b, c\}$ be a function defined as

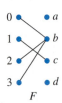

   F is *3–1*. F is neither *2–1* nor *1–1*.

It may seem strange that a *1–1* function is 58 to 1, but it certainly is by the definition. If k elements are mapped to a given element, the function is not m to 1 for any $m < k$ but, rather, is m to 1 for any $m \geq k$. Definition 1 gives a way to talk about the more important fact that is the smallest integer for which some element in the codomain has k preimages.

Theorem 1.   Let $F : X \rightarrow Y$ is k to 1, and let $|Y|$ be finite, with $|Y| = n$. Then, X is finite, and $|X| \leq k \cdot n$.

Proof.   For each of the n elements of Y, there are at most k elements of X mapped to each element of Y. So, only $k \cdot n$ elements can be mapped to all the elements of Y. However, every element of X is mapped to some element of Y, so there are at most $k \cdot n$ elements in X. ∎

4.6.2 Proofs of the Pigeon-Hole Principle

Consider again the New England town meeting example. If at most one person left in each car, then at most 65 people left the garage. This notion is formalized in the contrapositive to Theorem 1, which is used more often than the theorem itself. The contrapositive is so important that we state it separately, in two variants. The contrapositive has two names: the *Pigeon-Hole Principle* and the **Dirichlet Drawer Principle.** A more colorful description of this principle is often given in terms of pigeons and nesting holes. Suppose m pigeons are placed into n nesting holes where $m > n$. Then, (at least) one nesting hole contains (at least) two pigeons.

Theorem 2. (Generalized Pigeon-Hole Principle) Let $F : X \to Y$ be onto where $|X| = m$ and $|Y| = n$. Then, there is a $y \in Y$ that is the image of at least

$$\left\lceil \frac{m}{n} \right\rceil$$

elements of X.

Proof. Suppose no y is the image of more than $\left\lceil \frac{m}{n} \right\rceil - 1$ elements of X. Then, the total number of elements in X is at most

$$m \le n \left(\left\lceil \frac{m}{n} \right\rceil - 1 \right) < n \left(\left(\frac{m}{n} \right) + 1 - 1 \right) = m$$

This contradiction proves the result. ∎

The formulation of the Generalized Pigeon-Hole Principle involved the **ceiling** function. It can also be stated using the **floor** function since for $m > n > 0$, we have

$$\left\lceil \frac{m}{n} \right\rceil = \left\lfloor \frac{m-1}{n} \right\rfloor + 1$$

Example 2. Suppose a class has 89 students. How many students (at least) must have a birthday in the same month.

Solution. Use the Generalized Pigeon-Hole Principle, and calculate

$$\left\lceil \frac{89}{12} \right\rceil = 8$$

The same answer can be found by computing

$$\left\lfloor \frac{89-1}{12} \right\rfloor + 1 = 8$$
∎

Theorem 3. (Pigeon-Hole Principle) Let X and Y be sets, and let $F : X \to Y$ where X and Y are finite and $|X| > |Y|$. There is a $y \in Y$ that is the image of at least two elements of X.

Proof.   By Theorem 2, if $m = |X|$ and $n = |Y|$ with $m > n$, then $m \geq n + 1$, which implies

$$\left\lceil \frac{m}{n} \right\rceil \geq \left\lceil \frac{n+1}{n} \right\rceil \geq 2$$

     ■

Theorem 3 gives a condition that ensures a function is not *1–1* when X and Y are finite sets.

Example 3.   The setting for this example is a room containing 367 people.

(a) At least two of these people have the same birthday.
(b) At least 31 of these people were born in the same month (though possibly in different years).
(c) Provided no one is more then 121 years old, at least four of these people are the same age (number of years).

Proof.

(a) Let X be the set of people. Let Y be the set of the 366 possible birthdays. Let *BirthDay* : $X \rightarrow Y$ map each person to his or her birthday. Then, by the Pigeon-Hole Principle (Theorem 3), *BirthDay* is not *1–1*.
(b) Let X be the set of people, and let Y be the set of the 12 months. Let

$$\text{\textit{BirthMonth}} : X \rightarrow Y$$

be the mapping that takes a person as input and gives the month containing that person's birthday as output. By the Generalized Pigeon-Hole Principle (Theorem 4.2), at least $\lfloor (367 - 1)/12 \rfloor + 1 = 30 + 1$ people will have the same birth month.
(c) This proof is left as an exercise for the reader.   ■

Theorem 4 deals with the question of how far a function is from having a single image for each element of the domain by guaranteeing that some element will have many images. Theorem 5 is related to the idea that a function is *1–1*.

Theorem 4.   Let $F : X \rightarrow Y$ where X and Y are finite. Let $k \in \mathbb{N}$ and $|X| > k \cdot |Y|$. Then, F is not *k to 1*.

Theorem 5.   Let $F : X \rightarrow Y$ where X and Y are finite with $|X| = |Y|$. Then, F is *1–1* if and only if F is onto.

Proof.

($\Rightarrow$)   Prove the contrapositive: If F is not onto, then F is not *1–1*. If F is not onto, then there is a $y \in Y$ that is not in the range of F. So, F is also a function from X to $Y - \{y\}$. $|Y - \{y\}| < |Y| = |X|$, so by the Pigeon-Hole Principle, F is not *1–1*.

($\Leftarrow$)   Prove the contrapositive: If F is not *1–1*, then F is not onto. Suppose F is not *1–1*, and let $n = |X|$. There are at least two elements of X with the same image. Pick two, and call them x_1 and x_2. Let the remaining elements of X be $x_3, x_4, \ldots, x_n$. Now, count the elements in

$$\text{\textit{range}}(F) = \{F(x) : x \in X\} = \{F(x_1), F(x_2), F(x_3), \ldots, F(x_n)\}$$

Since $F(x_1) = F(x_2)$, $F(x_1)$ and $F(x_2)$ account for one element in $range(F)$. The elements

$$F(x_3), F(x_4), \ldots, F(x_n)$$

account for at most $n - 2$ more. That leaves at most $n - 1$ elements in $range(F)$. Since Y has n elements, $range(F)$ is not all of Y, so F is not onto. ∎

Theorem 6. Let X and Y be sets and X be finite. Let $F : X \rightarrow Y$ be *1–1* and onto. Then, Y is finite, and $|X| = |Y|$.

As an example of applying Theorem 6, suppose a professor enters a classroom that she knows contains 55 chairs. Now, suppose the professor can see that all students are seated, no chairs are empty, and no chair has two people sitting in it. If the professor wants to know how many students are seated in the room, it is not necessary to count them. Let the function *SeatOf* map each student to the chair that student occupies. The professor has observed that *SeatOf* is *1–1* and onto. Therefore, the number of students equals the number of chairs: 55.

4.6.3 Application: Decimal Expansion of Rational Numbers

We will present several examples and prove several theorems that are applications of the Pigeon-Hole Principle. These results are interesting in their own right, but they also give insight regarding possible applications of the Pigeon-Hole Principle.

The first application concerns converting fractions to decimals or, in more formal language, expressing **rational numbers as decimals.** A rational number of the form

$$0.d_1 d_2 \cdots d_n d_i d_{i+1} \cdots d_n d_i d_{i+1} \cdots d_n \cdots$$

with the digits $d_i d_{i+1} \cdots d_n$ repeating is denoted as

$$0.d_1 d_2 \cdots d_{i-1} \overline{d_i d_{i+1} \cdots d_n}$$

The following are examples of the conversion of a fraction to a decimal with the special notation for the repeated digits:

$$\frac{1}{10} = 0.1 = 0.10000 \cdots 00 \cdots = 0.1\overline{0}$$

$$\frac{102}{25} = 4.08 = 4.08000 \cdots 00 \cdots = 4.08\overline{0}$$

$$\frac{1}{3} = 0.333333 \cdots 33 \cdots = 0.\overline{3}$$

$$\frac{2}{11} = 0.181818 \cdots 1818 \cdots = 0.\overline{18}$$

The decimal expansions of $1/10$ and $102/25$ are **finite** or **terminating.** All but a finite number of their decimal digits are 0. The decimal expansions of $1/3$ and $2/11$ are **nonterminating** or **infinite.** All these decimal expansions are **repeating:** After a certain point, the decimal expansion can be generated by repeating a block of digits infinitely many times. The decimal expansions of $1/10$ and $102/25$ have repeating 0's. The expansion of $1/3$ has repeating 3s. The expansions of $2/11$ has the two digits, 18, repeating. There are some rationals that have two repeating decimal expansions, such as $1.0000\cdots$ and $0.\overline{9}$. (See Exercise 14 in Section 4.5.)

Study the example for finding the decimal expansion of the rational number 3/7, as shown in Figure 4.24, to gain an insight regarding how the formal proof that follows will proceed.

$$
\begin{array}{r}
.4\ 2\ 8\ 5\ 7\ 1\ 4 . \ . \ . \\
7\overline{)3.0\ 0\ 0\ 0\ 0\ 0\ 0 . \ . \ .} \\
\underline{2\ 8} \\
2\ 0 \\
\underline{1\ 4} \\
6\ 0 \\
\underline{5\ 6} \\
4\ 0 \\
\underline{3\ 5} \\
5\ 0 \\
\underline{4\ 9} \\
1\ 0 \\
\underline{7} \\
3\ 0 \\
.\\
.\\
.
\end{array}
$$

Figure 4.24   Long division to calculate decimal expansion of 3/7.

The decimal expansion of 3/7 can be seen to repeat. After the sixth decimal place has been calculated, the remainder is 3, and the rest of the division process corresponds to dividing 7 into 0.000003. That is the same process as dividing 7 into 3, except that it is shifted six decimal places to the right. Therefore, exactly the same sequence of quotients and remainders will be generated as before, causing the sequence to repeat.

The ideas shown in these examples will now be incorporated into the general result about the existence of repeating decimal expansions for rational numbers.

Theorem 7.   A real number is rational if and only if it has a repeating decimal expansion.

Proof.

($\Rightarrow$)   Suppose a real number r is rational. Show that it has a repeating decimal expansion. First, express r as the fraction j/k where $0 < j < k$. Now, consider the long division of k into j. (For an illustration, look again at the computation of the decimal expansion of 3/7 in Figure 4.24.) The first division produces one digit (the tenths digit) of the quotient and a remainder $r_1 < k$. The remainder is contained in $\{0, 1, \ldots, k-1\}$. To prepare for the next division, concatenate a 0 on the the end of r_1. Now, repeat the procedure to calculate another digit (the hundredths) of the quotient and another remainder $r_2 < k$, follow this by concatenating another 0 on the end of r_2, repeat again, and so on.

The only k possible remainders at each step are $0, 1, 2, \ldots,$ and $k-1$. Hence, after at most $k+1$ steps of the division, two of the remainders must be equal. Then, however, the process must start repeating, as in the previous illustrations. It is important that at the end of each step, the same digit, 0, and not any other digit, is always concatenated onto the remainder. The digit concatenated is always 0, because j and k are both integers and $0 < j < k$. This guarantees that when remainders are equal, the entire process repeats. So, r has a repeating decimal expansion.

($\Leftarrow$)   Suppose a real number r has a repeating decimal expansion. Again, for convenience, we will limit the proof to decimals in the interval $(0, 1)$. For illustration, use

$$r = 0.4579909909909909\ldots909909\ldots$$

with repeating block of digits 909. It is easier to work with expansions in which the repeating part appears beginning immediately to the right of the decimal point. To accomplish this proof, we will need to multiply the decimal by some power of 10. This is really just for our convenience and does not affect the proof.

If the repeating part has length k that begins j digits to the right of the decimal point, multiply r by 10^{j+k}. In the illustration,

$$10^7 \cdot r = 4579909.909909909\ldots909\ldots.$$

Now, multiply r by 10^j, and subtract the product from $10^{j+k} \cdot r$, giving $d = (10^{j+k} - 10^j) \cdot r$. In the illustration,

$$10^7 r = 4579909.909909909909\ldots909\ldots$$
$$- 10^4 r = -4579.909909909909\ldots909\ldots$$
$$\overline{}$$
$$(10^7 - 10^4)\ \ r = 4575330.000000000000\ldots000\ldots$$

Since all digits past the decimal point match, the subtraction results in all 0's to the right of the decimal point. Therefore, the difference d is an integer. It follows from this computation that $r = d/(10^{j+k} - 10^j)$. Therefore, r is a rational number.   ∎

4.6.4   Application: Problems with Divisors and Schedules

In scenarios as diverse as studying for exams or finding divisors of sums of numbers, the Pigeon-Hole Principle can provide answers to many questions.

Example 4.   Let $m \in \mathbb{N}$. Given m integers $a_1, a_2, \ldots, a_m$, there exist k and l with $0 \le k < l \le m$ such that

$$a_{k+1} + a_{k+2} + \cdots + a_l$$

is divisible by m.

Solution.   Consider the m sums:

$$a_1, a_1 + a_2, a_1 + a_2 + a_3, \ldots, a_1 + a_2 + \cdots + a_m$$

If any of these sums is divisible by m, then the conclusion follows. If not, then we may suppose that each sum has a nonzero remainder when divided by m. The possible remainders are

$$1, 2, 3, \ldots, m - 1$$

Since there are m sums and only $m - 1$ possible remainders, at least two of the sums must have the same remainder when divided by m (according to the Pigeon-Hole Principle). Therefore, there are integers k and l with $l > k$ such that

$$a_1 + a_2 + \cdots + a_k$$

and

$$a_1 + a_2 + \cdots + a_l$$

have the same remainder r when divided by m. That is, there are integers c, d, and r such that

$$a_1 + a_2 + \cdots + a_k = cm + r$$

and

$$a_1 + a_2 + \cdots + a_l = dm + r$$

Subtracting the k-element sum from the l-element sum gives

$$a_{k+1} + a_{k+2} + \cdots + a_l = (d - c)m$$

Therefore,

$$a_{k+1} + a_{k+2} + \cdots + a_l$$

is divisible by m.   ∎

Example 5 shows how a scheduling decision can be better understood.

Example 5.   The local softball league wants to schedule at least one game every day during the 11-week summer season. To keep the fields in good condition, it is decided to schedule no more than 12 games in any week. Show that there is a succession of days during which exactly 21 games are scheduled.

Solution.   Let a_1 be the number of games scheduled for day 1. In general let a_i where $1 \leq i \leq 77$ be the total number of games played on days 1 through i. The sequence of numbers $a_1, a_2, \ldots, a_{77}$ is strictly increasing since at least one game is played each day. Since $a_1 \geq 1$ and at most 12 games are played in a week, we have $a_{77} \leq 132$. The sequence $a_1 + 21, a_2 + 21, \ldots, a_{77} + 21$ is also an increasing sequence. Each of the 154 numbers $a_1, a_2, \ldots, a_{77}, a_1 + 21, a_2 + 21, \ldots, a_{77} + 21$ is an integer between 1 and 153. Since there are 154 numbers, then by the Pigeon-Hole Principle, two of them must be equal. No two of the numbers $a_1, a_2, \ldots, a_{77}$ are equal, however, and no two of the numbers $a_1 + 21, a_2 + 21, \ldots, a_{77} + 21$ are equal. Therefore, there are i and j such that

$$a_i = a_j + 21$$

Thus, on days $a_{j+1}, a_{j+2}, \ldots, a_i$, 21 games are scheduled.   ∎

It would be nice if we knew how many days were used for these 21 games. The only thing we can say for sure is that the number of days is no more than 21 and no less than 11. In 7 days 12 games can be played. During a second week an additional 12 games can be played. Since at least one game must be played each day, a total of 21 games cannot occur in fewer than 11 days.

4.6.5   Application: Two Combinatorial Results

The two results included here are probably surprising as far as finite sequences of natural numbers. The first proves that two elements of certain finite sequences must have the property that one divides the other. The second proves that some sequences always have an increasing or decreasing subsequence that is at least of a length given as a function of the number of elements in the sequence. Both of these results are credited to the eminent mathematician Paul Erdös (1913–1996, b. Hungary).

To appreciate these two results, it is helpful to experiment with some subsets of a set, say $\{1, 2, 3, \ldots, 17\}$, and see how large a subset you can find so that no elements of the subset divides any other element of the subset. For a second experiment, write down these 17 elements in an arbitrary order (not in increasing or decreasing order), and see how long a subsequence you can find that is either increasing or decreasing. For example, try

$$12, 6, 3, 7, 8, 1, 17, 16, 14, 15, 13, 2, 9, 10, 4, 11$$

You should be able to find an increasing subsequence of length six but no increasing subsequence of longer length. The theorems will tell us what we can always expect as answers for these two problems.

Theorem 8.   (Erdös)   Let

$$X \subseteq \{1, 2, 3, 4, \ldots, 2n - 1\}$$

and $|X| \geq n + 1$. There are two numbers $a, b \in X$ with $a < b$ such that a divides b.

Proof.   For $x \in X$, let $F(x)$ be the largest odd divisor of x. So

$$F(1) = 1, F(2) = 1, F(3) = 3, F(4) = 1, F(5) = 5, F(6) = 3, \ldots$$

and so forth. For $x \in X$, there are n possible values for $F(x)$, namely $1, 3, 5, \ldots, 2n - 1$. There are at least $n + 1$ elements of X. So, F is not *1–1* on X. Pick two elements of X whose images under F are the same, and call the smaller one a and the larger one b. Now, let

$$k = F(a) = F(b)$$

So $a = 2^i \cdot k$, and $b = 2^j \cdot k$ where $i < j$. Then, $b = a \cdot 2^{j-i}$, so a divides b.   ∎

Theorem 8 is the "best possible" result. That is, if the hypothesis instead required only that $|X| = n$, then the result would be false. To prove this for any n, choose

$$X = \{n, n + 1, n + 2, \ldots, 2n - 1\}$$

Then, $|X| = n$. Now, show that no element of X is a factor of any other. Because if a were a factor of b, then $a \cdot c = b$ for some c. Since $a < b$, we must have $c > 1$. However, $a \geq n$ and $b \leq 2n - 1$, so

$$n \cdot c \leq a \cdot c = b \leq 2n - 1$$

Hence,

$$1 < c \leq (2n - 1)/n < 2$$

However, there is no integer c between 1 and 2.

Theorem 9 tells that in a sequence of $n^2 + 1$ elements, for any $n \in \mathbb{N}$ there is always a subsequence of at least $n + 1$ elements that is either increasing or decreasing. Even in choosing a sequence of random numbers, this behavior occurs.

Theorem 9.   (Erdös and Szekerés)   Let $n \in \mathbb{N}$ and $k = n^2 + 1$. Let

$$a_1, a_2, a_3, \ldots, a_k$$

be any sequence of k distinct numbers. Then, the sequence has either an $n + 1$ element increasing subsequence or an $n + 1$ element decreasing subsequence.

Example to Motivate Proof.   Let $n = 3$, and consider the 10-element sequence

$$5\ 0\ 6\ 4\ 9\ 8\ 2\ 1\ 7\ 3$$

The goal is to find either a four-element **increasing subsequence** or a four-element **decreasing subsequence**.

For each element of the sequence, find the longest increasing subsequence starting with that element. For example, starting with 5, there are three increasing subsequences of length three (5 6 9, 5 6 8, and 5 6 7), but none of length four. Starting with 0, there are several increasing subsequences of length three but none of length four. Under each number, write the length of the longest increasing subsequence starting with that number:

$$
\begin{array}{cccccccccc}
5 & 0 & 6 & 4 & 9 & 8 & 2 & 1 & 7 & 3 \\
\downarrow & \downarrow & \downarrow & \downarrow & \downarrow & \downarrow & \downarrow & \downarrow & \downarrow & \downarrow \\
3 & 3 & 2 & 2 & 1 & 1 & 2 & 2 & 1 & 1
\end{array}
$$

with $LongestIncSeq(*)$ labeling the arrows.

If any of these subsequences had length four or greater, then that subsequence would be the example needed. In this case, there is no such subsequence, since each of the 10 elements of the sequence mapped to 1, 2, or 3. By the Generalized Pigeon-Hole Principle, we know that at least four elements of the sequence must map to the same value. In this example, each element of the subsequence (6, 4, 2, 1) maps to 2, and each element of the subsequence (9, 8, 7, 3) maps to 1. Both of these subsequences are decreasing subsequences of length four.

Proof. Let $k = n^2 + 1$ and the sequence $a_1, a_2, \ldots, a_k$ be given as in the statement of Theorem 9. For each a_i, define a function F such that $F(a_i)$ is the length of the longest increasing subsequence starting with a_i.

We first show that if $i < j$ and $a_i < a_j$, then $F(a_i) > F(a_j)$. This follows because, if, say, $F(a_j) = l$, then there is a length-l increasing subsequence beginning with a_j : $a_j = b_1 < b_2 < \cdots < b_l$. Then, $a_i < b_1 < b_2 < \cdots < b_l$ is a subsequence of length $l + 1$, which implies that $F(a_i) \geq l + 1 > F(a_j)$. In particular,

$$\text{if } i < j \text{ and } F(a_i) = F(a_j), \text{ then } a_i > a_j$$

Case 1:   For some i such that $1 \leq i \leq k$, we have $F(a_i) > n$. Then, there is an increasing subsequence of length $n + 1$ starting with a_i.

Case 2:   There is no i such that $1 \leq i \leq k$ with $F(a_i) > n$. Consequently, the range of F is a subset of the n-element set $\{1, 2, 3, \ldots, n\}$. By the Generalized Pigeon-Hole Principle, at least

$$\left\lfloor \frac{(k-1)}{n} \right\rfloor + 1 = \left\lfloor \frac{(n^2 + 1) - 1}{n} \right\rfloor + 1 = n + 1$$

elements of the sequence will be mapped to the same element of $\{1, 2, \ldots, n\}$. By the remark before Case 1, these $n + 1$ elements form a decreasing sequence.   ■

4.7   Exercises

1. Prove that in any set of 27 words, at least two must begin with the same letter assuming at most a 26-letter alphabet.
2. Prove that in any group of five integers, at least two have the same value under the (*mod* 4) operation.

3. Prove that in any class of more than 101 students, at least two must receive the same grade for an exam with grading scale of 0 to 100.

4. Prove that for any 44 people, at least four must be born in the same month.

5. Prove that in any class of 35 students, at least seven receive the same final grade, where the scale is A-B-C-D-F.

6. Area codes are used to distinguish phone numbers for which the last seven digits are the same. If you have 35,000,000 phone numbers in a state and an area code can distinguish approximately 900,000 phone numbers, how many area codes are needed to distinguish the phone numbers of this state?

7. There are 35,000 students at State University. Each student takes four different courses each term. State University offers 999 courses each term. The largest classroom on campus holds 135 students. Is this a problem? If so, what is the problem?

8. At Bridgetown University, there are 45 time periods during the week for scheduling classes. Use the Generalized Pigeon-Hole Principle to determine how many rooms (at least) are needed if 780 different classes are to be scheduled in the 45 time slots.

9. Suppose someone (say, Aesop) is marking days in some leap year (say, 2948). You do not know which days he marks, only how many. Use this to answer the following questions. (*Warning:* Some, but not all, of these questions use the Pigeon-Hole Principle.)

   (a) How many days would Aesop have to mark before you can conclude that he marked two days in January?

   (b) How many days would Aesop have to mark before you can conclude that he marked two days in February?

   (c) How many days would Aesop have to mark before you can conclude that he marked two days in the same month?

   (d) How many days would Aesop have to mark before you can conclude that he marked three days in the same month?

   (e) How many days would Aesop have to mark before you can conclude that he marked three days with the same date (for example, the third of three different months, or the 31st of three different months)?

   (f) How many days would Aesop have to mark before you can conclude that he marked two consecutive days (for example, January 31 and February 1)?

   (g) How many days would Aesop have to mark before you can conclude that he marked three consecutive days?

10. Prove that for any collection of n people, two persons have the exact same number of acquaintances in the group provided that each person has at least one acquaintance.

11. There are five suburbs in the city of Melbourn. How many all-stars must be picked from each suburb to guarantee that at least five players come from the same suburb?

12. A bowl contains raspberry and orange lollipops, with 15 of each. How many must be drawn one at a time to ensure that you have at least three orange lollipops?

13. A man has 10 black socks and 11 blue socks scrambled in a drawer. Still half-asleep, the man reaches in the drawer to get a pair of matching socks. How many socks should he select, one at a time, before he will be sure that he has a matching pair. How many selections are needed to be sure he has a blue pair?

14. Prove that:

   (a) $0.999999\ldots99\ldots = 1$

   (b) $0.346270\overline{0} = 0.346269\overline{9}$

15. Construct a sequence of 16 integers that has no increasing or decreasing subsequence of five elements.

16. During a month with 30 days, a team will play at least one game a day but no more than 45 games in all 30 days. Show that there is a stretch of consecutive days during which the team plays exactly 14 games. (*Hint:* Let a_i be the number of games played on or before the ith day for $1 \leq i \leq 30$.)

17. A widget-maker makes at least one widget every day but not more than 730 widgets in a year. Given any n, show that the widget-maker makes exactly n widgets in some set of consecutive days. For some n, it may take more than a single year.

18. A student has 37 days to prepare for an exam. From past experience, he knows that he will need no more than 60 hours of study. To keep from forgetting the material, he wants to study for at least one hour each day. Show that there is a sequence of successive days during which he will have studied exactly 13 hours.

19. For any four integers, none of which is even and none of which is a multiple of 5, prove that some consecutive product of these ends in the digit 1. A consecutive product is one term, two terms in a row, three terms in a row, or all four terms. For example, for the four integers $a, b, c,$ and d a consecutive product would be $a \cdot b$ but not $a \cdot c$. (*Hint:* Prove that if $b \cdot c, b \cdot c \cdot d$ do not end in a 1, and if there is no integer ending in 1 among $a, b, c,$ and d, then $a, a \cdot b, a \cdot b \cdot c,$ and $a \cdot b \cdot c \cdot d$ are all distinct. Use Theorem 3 in Section 4.6.2).

20. Select 100 integers from the integers $1, 2, \ldots, 200$ such that no one of the chosen values is divisible by any other chosen value. Show that if one of the 100 integers chosen from $1, 2, \ldots, 200$ is less than 16, then one of those 100 numbers is divisible by another.

21. Prove the assertion in Example 4(c).

22. (a) Find two functions $F, G : \mathbb{N} \to \mathbb{N}$ that are *1–1* but not onto.
    (b) Find a function $G : \mathbb{N} \to \mathbb{N}$ that is onto but not *1–1*.
    (c) *Challenge:* Suppose $G : \mathbb{N} \to \mathbb{N}$ is onto but not *1–1*, and suppose G is specified by an algorithm A. Show that there is an algorithm A' that computes a function $F : \mathbb{N} \to \mathbb{N}$, where $G \circ F = Id_{\mathbb{N}}$. Also, show that F must be *1–1* but not onto. We have not been precise about what an "algorithm" is; you might choose to interpret an "algorithm" as being a function written in some programming language. (*Hint:* A' can use A as a subprogram.)

23. *Infinite Pigeon-Hole Principle:* Suppose X is an infinite set and Y is a finite set. Now, suppose $F : X \to Y$. Show there is a $y \in Y$ such that for infinitely many $x \in X$ such that $F(x) = y$.

4.8   Countable and Uncountable Sets

In this section, we develop the notion of counting the elements of a set or cardinality more carefully. The modern notion of cardinality is credited to Georg Cantor (1845–1918, b. Russia), who found an abstract notion of counting that enabled mathematicians to speak of the cardinality of an infinite set. The notion also enabled mathematicians to discuss the cardinality of finite sets more precisely. In Section 1.5.1, we gave a provisional definition for counting the number of elements in a set that can now be made more precise.

Consider checking to see whether the sets {red, blue, green} and {Jean, Michele, Paul} have the same number of elements. Of course, each set has three elements, and one merely counts the elements:

$$\{(0, \text{red}), (1, \text{blue}), (2, \text{green})\}$$

and

$$\{(0, \text{Jean}), (1, \text{Michele}), (2, \text{Paul})\}$$

Each of the two sets of ordered pairs are *1–1* functions, the first from {0, 1, 2} onto {red, blue, green}—call it *Color*—and the second from {0, 1, 2} onto {Jean, Michele, Paul}— call it *Person*. Consider the set {0, 1, 2} only as an intermediary. The function $Color^{-1} \circ Person$ is a *1–1* function (Theorem 3(c) in Section 4.3.2 and Theorem 2(a) in Section 4.3.1) (from {red, blue, green} onto {Jean, Michele, Paul}. This function shows, without explicitly counting, that the two sets have the same number of elements in the sense that we now make more precise.

Definition 1. **(Cantor)** Let X and Y be sets. Then, the cardinality of X is less than or equal to the cardinality of Y, written $|X| \le |Y|$, if there is a *1–1* function $F : X \to Y$. The cardinality of X is equal to the cardinality of Y, written $|X| = |Y|$, if there is a *1–1* correspondence $F : X \to Y$. The cardinality of X is less than the cardinality of Y, written $|X| < |Y|$, if $|X| \le |Y|$ and $|Y| \nleq |X|$.

The definition of $|X| = |Y|$ generalizes Theorem 5 in Section 4.6. Notice that we have not defined the term *cardinality* of X here, only some relationship between X and Y.

Using these notions, one can define the usual notion of cardinality for a finite set.

Definition 2. Let X be a set and $n \in \mathbb{N}$. If X has the same cardinality as the set {0, 1, 2, ..., n − 1}, then the **cardinality** of X is n. We say X is **finite** if X has cardinality equal to some natural number. We say X is **infinite** if X is not finite.

At this point, the careful reader should note that, since we have redefined (or perhaps, at last, defined) a term we have used throughout this book, some of our earlier proofs may have been false, relying on unprovable intuitions. In fact, as the reader surely suspects, the earlier results are not false by this definition; however, the proofs may have important parts missing. This is not a book about the foundations of mathematics, so we shall not go back to recheck any proofs. We shall make one further remark, however: The entire discussion of the Pigeon-Hole Principle depended critically on the result that if m and n are natural numbers and $m > n$, then the sets {0, 1, 2, ..., m − 1} and {0, 1, 2, ..., n − 1} do not have the same cardinality. After the previous discussion, the student likely has no idea what one is allowed to use in proving such a result. In the development of the foundations of mathematics, this theorem can be proved by induction on m. The interested reader is invited to look for a simple proof.

The following properties of cardinalities are easy to prove. They are also suggested by the notations for equal (=) and less than or equal ($\le$). but it is, of course, very dangerous to assume such results by analogy based on notation.

Theorem 1.   (Properties of Cardinalities)   Let X, Y, and Z be sets.

(a) $|X| \le |X|$.
(b) If $|X| \le |Y|$ and $|Y| \le |Z|$, then $|X| \le |Z|$.
(c) $|X| = |X|$.
(d) If $|X| = |Y|$, then $|Y| = |X|$.
(e) If $|X| = |Y|$ and $|Y| = |Z|$, then $|X| = |Z|$.

Proof.   This proof is left as an exercise for the reader.   ■

One is tempted to restate parts (c) through (e) of Theorem 1 by stating that the relation *has the same cardinality as* is an equivalence relation. This is not done, however, because if it is an equivalence relation, then it is an equivalence relation on the set of all sets—but the set of all sets does not exist! Even so, it is correct to say that *has the same cardinality as* is an equivalence relation on any set of sets.

A fundamental result regarding cardinality uses the work of Georg Cantor, Friedrich Schröder (1841–1902, b. Germany), and Sergi Bernstein (1880–1968, b. Ukraine).

Theorem 2.   (Cantor-Schröder-Bernstein)   Let X and Y be sets, and let $|X| \le |Y|$ and $|Y| \le |X|$. Then, $|X| = |Y|$.

The proof of the Cantor-Schröder-Bernstein Theorem is fairly complicated, and we refer the reader to a text about set theory. A related question is whether there exist two sets X and Y where $|X| \not\le |Y|$ and $|Y| \not\le |X|$. This question turns out to be related to whether one accepts a famous and, sometimes, controversial axiom called the *Axiom of Choice*. The reader is also referred to books on set theory and the foundations of mathematics for discussion of this issue.

4.8.1   Countably Infinite Sets

Cantor's definition allows a more careful development of the study of finite sets, but the study of infinite sets under Cantor's definition is sometimes surprising. The simplest infinite sets are $\mathbb{N}$ and those other sets with the same cardinality as $\mathbb{N}$.

Definition 3.   Let X be a set. X is **countably infinite** if $|X| = |\mathbb{N}|$. If X is countably infinite, the cardinality of X is $\aleph_0$ (pronounced **aleph nought**), written $|X| = \aleph_0$. X is **countable** if it is either finite or countably infinite. If a set is not countable, then it is **uncountable.**

In Definition 3 the object $\aleph_0$ was left undefined. For set theorists, $\aleph_0$ is another name for $\mathbb{N}$, but the symbol $\aleph_0$ is used almost exclusively to denote the cardinality of $\mathbb{N}$.

Theorem 3.   Any countably infinite set is infinite.

Proof.   As noted earlier, $\mathbb{N}$ is infinite. Now, suppose a set X is both countably infinite and finite. Then, there are *1–1* correspondences $F : X \to \mathbb{N}$ and $G : X \to \{0, 1, 2, \ldots, n\}$ for some natural number n. However, then $F^{-1} \circ G : \mathbb{N} \to \{0, 1, 2, \ldots, n\}$ is *1–1* and onto by Theorem 3(c) in Section 4.3.2, contradicting the result, noted above, that $\mathbb{N}$ is infinite.   ■

Theorem 5 in Section 4.6.2 says that for finite X and $F : X \to X$, F is *1–1* if and only if F is onto. This result fails for infinite X. Earlier (see Section 4.1.8), a *1–1* function from

$\mathbb{R}$ to $\mathbb{R}$ was given that is not onto, and an onto function from $\mathbb{R}$ to $\mathbb{R}$ was given that is not *1–1*. We recommend the reader find a *1–1* function from $\mathbb{N}$ to $\mathbb{N}$ that is not onto and an onto function from $\mathbb{N}$ to $\mathbb{N}$ that is not *1–1*.

The next three results give proofs that some of the common sets we use are indeed infinite.

Theorem 4. *Evens* $= \{n \in \mathbb{N} : n = 2k$ for some $k \in \mathbb{N}\}$ is countably infinite.

Proof. Let

$$F : \mathbb{N} \rightarrow Evens$$

be defined by $F(n) = 2n$. The graph of this function is shown in Figure 4.25.

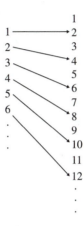

Figure 4.25 Bijection $F(n) = 2n$ maps $\mathbb{N}$ to *Evens*.

Show that F is both *1–1* and onto. First, show that F is *1–1*. Suppose $F(m) = F(n)$. That is, suppose $2m = 2n$. Dividing both sides by 2 gives $m = n$. So, F is *1–1*. Now show that F is onto. Suppose k is even, and show that $k = F(n)$ for some $n \in \mathbb{N}$. Since k is even, $k = 2n_0$ for some $n_0 \in \mathbb{N}$. Now, $F(n_0) = 2n_0 = k$ as required. ■

Theorem 5 tells us that there were infinitely many base cases in our proof of the Fundamental Theorem of Arithmetic.

Theorem 5. The set of prime positive integers is countably infinite.

Proof. First, show that the set of primes is not finite. Suppose it were, and let the set of primes be

$$\{p_0, p_1, \ldots, p_{n-1}\}$$

The set is nonempty since 2 is a prime. Now, let $k = p_0 \cdot p_1 \cdots p_{n-1} + 1$. None of the given primes divides k, because each divides $k - 1$. Thus, there must be some other prime that divides k—perhaps even k itself. In any case, the existence of at least one more prime contradicts our assumption that $p_0, p_1, \ldots, p_{n-1}$ are the only primes. Therefore, there must be infinitely many primes.

Now, show that the set of primes is countably infinite. List the primes in increasing order:

$$2, 3, 5, 7, 11, 13, 17, 19, 23, 29, 31, 37, 41, 43, 47, 53, 59, \ldots$$

Then, define a function

$$P : \mathbb{N} \rightarrow \{2, 3, 5, 7, 11, 13, 17, 19, 23, 29, 31, 37, 41, 43, 47, 53, 59, \ldots\}$$

by the rule that $P(n)$ is the nth prime on the list for $n = 0, 1, 2, 3, \ldots$. We claim that P is a *1–1* and onto function. Function P is *1–1*, because it is strictly increasing. Function P is also onto, because the nth prime is always larger than n, so if k is prime, then $k = P(n)$ for some $n \in \{0, 1, 2, \ldots, k - 1\}$. ∎

The first somewhat surprising result that follows from Theorem 5 is that there are no more integers than there are prime numbers, even though there certainly are gaps between consecutive primes.

Theorem 6. $\mathbb{Z}$ is countably infinite.

Proof.   This part depends on listing the integers in a special order—in order of their absolute values. First, list the integer with absolute value 0:

$$0$$

Then, add to the list the integers with absolute value 1:

$$0, -1, 1$$

and so forth:

$$0, -1, 1, -2, 2, -3, 3, -4, 4, -5, 5, \ldots, -n, n, \ldots$$

We must now formalize this idea of a function. For any $n \in \mathbb{N}$, define $G(n)$ to be the nth number on this list. It is apparent that every integer is listed exactly once on the list. In fact, it is easy to see that $G(0) = 0$, and for $n \geq 1$, we have $G(2n - 1) = -n$ and $G(2n) = n$. It follows that the function $G : \mathbb{N} \rightarrow \mathbb{Z}$ is *1–1* and onto. ∎

4.8.2   Cantor's First Diagonal Argument

Cantor found two important proof techniques for showing that two sets had the same cardinality. The first of these arguments, called **Cantor's first diagonal argument,** is used in proving that $|\mathbb{Z}| = |\mathbb{Q}|$.

Theorem 7. $\mathbb{Q}$ is countably infinite.

Proof.   The proof depends on listing all the rational numbers in a special order. For each rational number r, pick out its expression p/q in lowest terms where $q > 0$. **Lowest terms** means p and q have no common divisor. For every rational number p/q written in lowest terms, compute the number $|p| + q$. Since p and q are integers and $q > 0$, $|p| + q$ is a positive integer. For $|p| + q = 1$ and 2, we have

$$\frac{0}{1} \quad |p| + q = 1$$

$$\frac{-1}{1}, \frac{1}{1} \quad |p| + q = 2$$

Then, for $|p| + q = 3$, we have

$$\frac{-2}{1}, \frac{-1}{2}, \frac{1}{2}, \frac{2}{1}$$

For $|p| + q = 4$, skipping $-2/2$ and $2/2$ (since they are not in lowest terms), we have

$$\frac{-3}{1}, \frac{3}{1}$$

and so forth.

The order in which the distinct rationals are listed is shown in Figure 4.26, where the rationals p/q are represented as points on the plane with x coordinate p and y coordinate q. As the indicated path is followed, just rationals not already occurring on the list are added to the list.

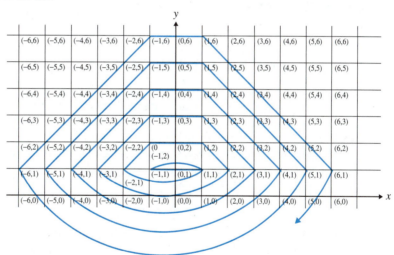

Figure 4.26   Order for listing elements of $\mathbb{Q}$.

For any positive integer n, there are only finitely many different rational numbers p/q with $|p| + q = n$. In fact, since q must be greater than zero, q must be one of $1, 2, 3, \ldots, n$ for a total of n choices. There are two choices for p, $n - q$ and $-(n - q)$, giving a total of $2n$ choices for p/q. Among these $2n$ choices, some, such as $0/2$ and $2/2$, will not be in lowest terms and so will be ignored.

Let p/q be a rational number such that $|p| + q = n$. Then, there are fewer than

$$2 \cdot 1 + 2 \cdot 2 + 2 \cdot 3 + \ldots + 2 \cdot n = n \cdot (n + 1)$$

rationals that could be listed in front of p/q. Hence, every rational number ultimately appears on the list. Furthermore, since each rational number is listed only in lowest terms, each rational number is listed only once.

Set $G(n)$ to be the nth rational number on the list above. Then, by the discussion above, $G : \mathbb{N} \to \mathbb{Q}$ is *1–1* and onto.   ∎

The proof of Theorem 7 is especially important. This method of proof is called *Cantor's first diagonal argument*. Another view of how the positive elements of $\mathbb{Q}$ are arranged and counted shows why this is called a diagonal argument (see Figure 4.27). Only elements of $\mathbb{Q}$ with no common factors in the numerator and the denominator are shown so that no element is counted more than once.

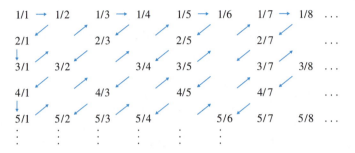

Figure 4.27   Counting positive rational numbers.

4.8.3   Uncountable Sets and Cantor's Second Diagonal Argument

After one sees Cantor's proof that $\mathbb{Q}$ is countably infinite, it is natural to conjecture that all infinite sets are countably infinite. Hence, it may be a bit surprising that sets that are not countable (or uncountable) exist. Cantor proved that $\mathbb{R}$ is not countable. That proof involves using the decimal expansions of real numbers.

Several familiar facts about the decimal expansion of rational numbers were presented in Section 4.6.3. The fundamental idea needed here is that for any real number, there is a decimal expansion of the form

$$c = c_0.c_1\, c_2\, c_3\, \ldots\, c_i \ldots$$

where c_0 is in $\mathbb{Z}$ and $c_i \in \{0, 1, \ldots, 9\}$ where $i \in \mathbb{N} - \{0\}$. A thorough study of decimal expansions requires careful development of the real numbers and will not be discussed in this book. One property of some real numbers is that they have two decimal expansions—for example,

$$0.233999\ldots = \frac{117}{500} = 0.2340000\ldots.$$

Important properties of the decimal representations of real numbers needed here are listed in Theorem 8.

Theorem 8.

(a) Every real number has at least one decimal expansion, and no real number has more than two decimal expansions.

(b) Every decimal expansion is the decimal expansion of some real number.

(c) If real numbers x and y have the same decimal expansion, they are equal.
(d) If a real number has two decimal expansions, then one of them terminates in an infinite string of 0's and the other in an infinite string of 9's.

Of course, not all finite sets have the same cardinality, but by now, the reader is surely wondering how two infinite sets could have different cardinalities. We are about to prove that the cardinality of the open interval $(0, 1)$ of real numbers is strictly greater than the cardinality of $\mathbb{N}$ or, in other words, that $(0, 1)$ is uncountable. We will do this following Template 1.10, based on the fact that if $(0, 1)$ were countable, we would be able to list all its elements in a (countable) sequence without omitting any or duplicating any. Let us assume that some function F defined on $\mathbb{N}$ lists all the decimal expansions of the numbers in $(0, 1)$. For the sake of developing the example to illustrate the idea of the proof, let us suppose the function F, with $F(0) = 0.254257\cdots$, $F(1) = 0.751999\cdots$, $F(2) = 0.485259\cdots$, $F(3) = 0.254157\cdots$, and continuing until our list contained every element of $(0, 1)$.

We need to show that such a list cannot contain all real numbers. We first display our countably infinite sequence in a table whose *diagonal* elements—the first decimal digit of $F(0)$, the second decimal digit of $F(1)$, ..., the nth decimal digit of $F(n-1)$—appear in boxes:

$$
\begin{aligned}
F(0) &= 0.\boxed{2}54157\cdots \\
F(1) &= 0.7\boxed{5}1999\cdots \\
F(2) &= 0.48\boxed{5}259\cdots \\
F(3) &= 0.000\boxed{0}08\cdots
\end{aligned}
$$

We now show how to construct a number $d = 0.d_1 d_2 d_3 \cdots$ that is not on the above countably infinite list. Since we want to avoid having $d = F(0)$, we decide to make the first digit of d different from the first digit of $F(0)$—that is, we choose $d_1 \neq 2$. To be definite, let us take $d_1 = 5$. Now, the second digit of $F(1)$ is 5, so to avoid having $d = F(1)$, we take the second digit of d to be, say, 4. We continue in this fashion, taking care that d_n— the nth digit of d—is always different from the nth digit of $F(n-1)$. To be systematic, we always choose $d_n = 5$ *unless* the nth digit of $F(n-1)$ is 5, in which case we choose $d_n = 4$. (So, in our example, we have $d = 0.5445\cdots$.)

Thus, we have created a number $d = 0.d_1 d_2 d_3 \cdots \in (0, 1)$, the decimal expansion of which is different from all the decimal expansions on this countably infinite list. Furthermore, since we avoided using 0's and 9's in d, we know by Theorem 8(d) that even when some real number has a second decimal expansion, that decimal expansion cannot be d. So, we have achieved a contradiction: We have created a real number d that could not possibly be on the countable list $F(0)$, $F(1)$, $F(2)$, ..., which supposedly included *all* real numbers. This example shows that the function chosen, F, did not work. The proof that the reals are uncountable must show that no such function exists.

The proof of the next theorem formalizes this intuitive argument using an arbitrary function. The argument is called **Cantor's second diagonal argument.**

Theorem 9. (Cantor) $\mathbb{R}$ is uncountable.

Proof.   It is enough to show that $(0, 1)$ is uncountable, since clearly, $|(0, 1)| \leq |\mathbb{R}|$. (See Exercise 7 in Section 4.9 for a function that is a bijection from $(0, 1)$ to $\mathbb{R}$.)

We assume that F is any *1–1* correspondence from $\mathbb{N}$ onto $(0, 1)$. Let $F(n) = \emptyset.f_1 f_2 f_3 \ldots$ be a decimal representation of $F(n)$. Construct a new decimal $d = 0.d_1 d_2 d_3 \cdots$ by putting $d_n = 4$ if the f_n is 5 and $d_n = 5$ otherwise. Then, d differs from $F(n)$ in the nth digit. Since no digit of d is 0 or 9, by Theorem 8(d), d is the only decimal expansion of the real number that it represents—call it D. So, if $F(n)$ has two decimal representations, then D is certainly not equal to $F(n)$, and otherwise, $D \neq F(n)$ because $d_n \neq f_n$. This contradicts our assumption that F is onto. Since there is no *1–1* correspondence from $\mathbb{N}$ to $\mathbb{R}$, we conclude the reals are not countable.   ■

Corollary 1:   Not every real number is rational.

Proof.   Since $\mathbb{Q} \subseteq \mathbb{R}$, either $\mathbb{Q} = \mathbb{R}$ or there is a real number $r \in \mathbb{R} - \mathbb{Q}$. Since $\mathbb{Q}$ is countable and $\mathbb{R}$ is uncountable, $\mathbb{Q} \neq \mathbb{R}$. Therefore, there is a real number that is not rational.   ■

Corollary 2:   The set $\mathbb{R} - \mathbb{Q}$ is uncountable.

Proof.   This proof is left as an exercise for the reader.   ■

Working with infinite cardinalities gives us the impression that uncountable sets are much larger than countable sets. So Corollary 2 states that almost all real numbers are irrational. But it does not demonstrate any particular number to be irrational! It has been known since the days of Pythagoras, an early Greek mathematician, that $\sqrt{2}$ is irrational. Perhaps more stunning is the next theorem.

Definition 4.   A real number r is ***algebraic*** if there is a polynomial $P(x)$ of the form

$$P(x) = a_n \cdot x^n + a_{n-1} \cdot x^{n-1} + \cdots + a_2 \cdot x^2 + a_1 \cdot x + a_0$$

where each a_i is an integer and r is a root of the equation—that is,

$$P(r) = a_n \cdot r^n + a_{n-1} \cdot r^{n-1} + \cdots + a_2 \cdot r^2 + a_1 \cdot r + a_0 = 0$$

For example, every rational number is algebraic: p/q where p and q are integers is a root of the polynomial equation $qx - p = 0$. Also, $\sqrt{2}$ and $\sqrt[5]{\sqrt[17]{93} + \sqrt[193]{217} - (333/41)}$ are algebraic, as are almost all numbers humans normally write down.

Theorem 10.   The set A of algebraic numbers is countably infinite. Furthermore, $\mathbb{R} - A$ is uncountable.

Proof.   This proof is left as an exercise for the reader.   ■

A number that is not algebraic is **transcendental.** Theorem 10 states that almost all real numbers are transcendental, yet it is very difficult to show that any particular real number is transcendental! The two standard examples are π and e. Proving that π is transcendental is very complicated.

A more difficult question to answer is whether any $X \subseteq \mathbb{R}$ exists where $|\mathbb{N}| < |X| < |\mathbb{R}|$. The conjecture that there is no such set X is called the **continuum hypothesis.** The conjecture is a famous and much-studied question. Finally, work of the famous mathematicians Kurt Gödel (1906–1978, b. Austria–Hungary) and Paul Cohen (1934–, b. United

States) showed that the continuum hypothesis is neither provable nor disprovable from the standard axioms for set theory (unless it is possible to prove a contradiction from Zermelo-Fränkel set theory).

4.8.4 Cardinalities of Power Sets

A variant of Cantor's second diagonal method can be used to prove that for any set X, $|X| < |\mathcal{P}(X)|$. Of course, that result is already known for finite sets X (see Theorem 2 in Section 1.7.4), but it is interesting that a single proof works for all sets, both finite and infinite. This proof looks even more like the proof of Russell's paradox (see Section 1.1.1). It follows that if $F : X \to \mathcal{P}(X)$, then for each $x \in X$, $F(x)$ is a subset of X. So it makes sense to ask whether $x \in F(x)$.

Theorem 11. Let X be any set. Then, $|X| < |\mathcal{P}(X)|$.

Proof.   To show that $|X| \leq |\mathcal{P}(X)|$, find a *1–1* function $F : X \to \mathcal{P}(X)$. The function F mapping each $x \in X$ to $\{x\}$ is easily seen to be such a function.

To show that $|\mathcal{P}(X)| \not\leq |X|$, show that there is no *1–1* correspondence $F : X \to \mathcal{P}(X)$. This can be accomplished by showing that no function $F : X \to \mathcal{P}(X)$ is onto. So, suppose $F : X \to \mathcal{P}(X)$ is an onto function. Let $Y = \{x \in X : x \notin F(x)\}$. Show that $Y \notin range(F)$. Now, suppose it were, say, $Y = F(y)$, and ask whether $y \in F(y)$:

$$y \in F(y) \Leftrightarrow y \notin Y \text{ by definition of } Y$$
$$\Leftrightarrow y \notin F(y) \text{ by assumption that } Y = F(y)$$

which is a contradiction. So, the assumption that $Y \in range(F)$ is false, and we conclude that F is not onto.   ∎

The cardinality of the set of all subsets of $\mathbb{N}$ is called $2^{\aleph_0}$. It is equal to the cardinality of $\mathbb{R}$, and it is denoted by c, for continuum.

4.9 Exercises

1. Show that if $X \subseteq Y$, $|X| \leq |Y|$.
2. Prove that the sets $\mathcal{X} = \{2n + 1 : n \in \mathbb{Z}\}$, $\mathcal{Y} = \{10j : j \in \mathbb{Z}\}$, and $\mathcal{Z} = \{3n : n \in \mathbb{Z}\}$ have the same cardinality.
3. In the first quadrant of the x–y plane, draw a path that passes exactly once through each point with both coordinates being integers. Each stopping place on the path should only be one unit right, one unit up, one unit left, or one unit down from the previous stopping place. Start the path at $(0, 0)$. Use the path to construct a bijection from $\mathbb{N}$ to $\mathbb{N} \times \mathbb{N}$.
4. Show that the following sets are countably infinite:
    (a) $\{q \in \mathbb{Q} : q > 10\}$
    (b) $\{q \in \mathbb{Q} : q^2 < q\}$
    (c) $\{q \in \mathbb{Q} : q = i/j \text{ where } i \text{ is odd and } j \text{ is even}\}$

5. (a) Prove that if X and Y are countable sets, so are $X \cup Y$, $X \cap Y$, $X - Y$, and $X \times Y$, (*Caution: Countable* means either finite or countably infinite, so there may be separate cases to consider.)

   (b) If X and Y are countably infinite, which of the following sets must be countably infinite: $X \cup Y$, $X \cap Y$, $X - Y$, and $X \times Y$?

6. Prove that every subset of $\mathbb{N}$ is countable.

7. Prove that the function $F : (0, 1) \to \mathbb{R}$ defined as $F(x) = (1/2 - x)/(x (1 - x))$ is a bijection.

8. Find *1–1* and onto functions from $\mathbb{R}$ to the following sets:

   (a) $(0, 1)$
   (b) $[0, 1]$
   (c) $(0, 1) - \{1 / 2\}$
   (d) $\mathbb{R} - \mathbb{Q}$, the irrationals

9. Prove Theorem 1.

10. Prove Corollary 2 to Theorem 9.

11. A chain-letter scheme is a famous (and usually illegal) get-rich-quick scheme. A person X receives a letter with, say, five names on it. X sends \$10 to the person whose name is at the top of the list. X then deletes that name from the top of the list, adds his or her own name to the bottom of the list, and sends the letter to five "friends," all within one day. In around two weeks, X is supposed to receive \$31,250.

    Suppose every person who receives the letter follows the instructions (including sending \$10 to the person listed first!). Show that if there are only finitely many people, the scheme cannot work (in some sense of "cannot work" that you should make precise). Show that if there are countably infinitely many people, the scheme can work.

12. Show for the natural numbers $\mathbb{N}$ that

$$|\mathcal{P}(\mathbb{N})| > |\mathbb{N}|$$

13. *Challenge:* Show that $|\mathbb{R}| = |\mathcal{P}(\mathbb{N})|$. (*Hint:* Use the Cantor-Schröder-Bernstein Theorem. To show that $|\mathbb{R}| \le |\mathcal{P}(\mathbb{N})|$, you might use function $D : \mathbb{R} \to \mathcal{P}(\mathbb{Q})$ where for $r \in \mathbb{R}$, $D(r) = \{q \in \mathbb{Q} : q < r\}$.)

14. Show that $\sqrt{2}$ is algebraic.

15. Show that there are infinite sets

$$X_0, X_1, X_2, \ldots, X_k, X_{k+1}, \ldots$$

where for each $k \in \mathbb{N}$, $|X_k| < |X_{k+1}|$.

16. *Challenge:* Show how to modify Cantor's second diagonal argument so that the real number produced is always irrational. (*Hint:* There is more than one way to do this.)

17. (a) Show that the set of all finite sequences of elements of the one-element set $\{0\}$ is countably infinite.

    (b) Show that the set of all finite sequences of elements of the two-element set $\{0, 1\}$ is countably infinite.

    (c) *Challenge:* Show that the set of all finite sequences of natural numbers is countably infinite. (*Hint:* Use a diagonal argument.)

18. (a) Show that the set of all infinite sequences of elements of the one element set $\{0\}$ is finite.

(b) Show that the set of all infinite sequences of elements of the two element set $\{0, 1\}$ has the same cardinality as $\mathcal{P}(\mathbb{N})$.

(c) *Challenge:* Show that the set of all infinite sequences of elements of $\mathbb{N}$ has the same cardinality as $\mathcal{P}(\mathbb{N})$.

4.10 # Chapter Review

Functions describe the transition from input information to output information. Functions are a special class of relations, and they can be defined and studied in that context. Functions can also be defined either as sets or as rules of correspondence. In addition, the rules of correspondence can either give the value of the function directly from the input value or recursively, in terms of other values of the function. There are even rules of correspondence that do not always give an output value for each possible input value. These correspondences are called partial function. Partial functions are especially important in describing the behavior of programs.

Two key properties that a function may possess are being *1–1* and/or being onto. The notion of functions being *1–1* and onto leads to two major topics. The first focuses on functions that are *1–1* or measures how far a function is from being *1–1*. The principles introduced are the Pigeon-Hole Principle and the Generalized Pigeon-Hole Principle. Operations such as composition and taking inverses are also defined on functions, and the class of function called sequences is described formally. The second topic deals with counting the elements of a set. The notion of cardinality makes precise what it means for two sets to have the same number of elements. Infinite sets, such as $\mathbb{Q}$ and $\mathbb{R}$, are shown to have different number of elements. Various relationships between other infinite sets are explored. Two important tools for studying the cardinality of a set are Cantor's first diagonal argument and Cantor's second diagonal argument. The class of function called sequences are described formally.

Among the applications discussed with this material is a characterization of the properties of inverse and of composition of functions when the properties of the functions are given. It is also shown how boolean functions can be realized by combinatorial networks. The Pigeon-Hole Principle and the Generalized Pigeon-Hole Principle can be applied in such diverse settings as determining what competitive schedules must look like and guaranteeing that a set of integers contains at least two with a common divisor. Rational numbers are shown to have repeating decimal representations. Both $\mathbb{N}$, $\mathbb{Z}$, and $\mathbb{Q}$ are shown to have the same number of elements.

4.10.1 Terms, Theorems, and Algorithms

4.1 Summary

TERMS

1–1 and onto function	bijective function
1–1 correspondence	binary function
bijection	ceiling

codomain	injection
collision resolution strategy	injective function
decreasing	mapped to
decrypt	one-to-one (1–1)
domain	onto
encoded	partial function
encrypted	preimage
equal	range
floor	recursive definition
function	recursively
graph of a function	restriction
graph of a partial function	strictly decreasing
greatest integer function	strictly increasing
hashing function	surjection
horizontal line test	surjective function
identify function	total function
image	undefined
increasing	vertical line test
independent	

4.3–4.5 Summary

TERMS

commutative	infinite sequence
composition	invertible
decreasing	inverse
F	list
$(F + G)$	sequence
$(F \cdot G)$	stream
(F/G)	strictly decreasing
finite sequence	strictly increasing
increasing	subsequence

4.6 Summary

TERMS

ceiling	k to 1
decreasing subsequence	nonterminating
Dirichlet Drawer Principle	rationals numbers as decimals
finite	repeating
floor	terminating
increasing subsequence	

THEOREMS

Erdös	Generalized Pigeon-Hole Principle
Erdös and Szekerés	Pigeon-Hole Principle

4.8 Countable and Uncountable Sets

TERMS

aleph nought ($\aleph_0$)
algebraic
Cantor's first diagonal argument
Cantor's second diagonal argument
cardinality
continuum hypothesis
countable

countably infinite
finite
infinite
lowest terms
transcendental
uncountable

THEOREMS

Cantor-Schröder-Bernstein
Properties of Cardinalities
$\mathbb{Q}$ is countably infinite

$\mathbb{R}$ is uncountable
$\mathbb{Z}$ is countably infinite

4.10.2 Starting to Review

1. Which of the following are functions?

   i. X is the set of students at Purdue University. For $x \in X$, define $g(x)$ to be the oldest brother of x.

   ii. X is the set of governors of Oregon. For $x \in X$, define $g(x)$ is the year that x was first sworn into office as governor.

   iii. For $x \in \mathbb{R}$, define $g(x) = x/|x|$.

  (a) i
  (b) ii
  (c) iii
  (d) None of the above

2. Let $X = \{1, 2, 3, 4\}$ and $Y = \{a, b, c, d\}$ be sets. Define the following subsets of $X \times Y$:

$$F_1 = \{(1, b), (4, d), (2, c), (3, a)\}$$
$$F_2 = \{(3, b), (1, d), (4, c), (2, b)\}$$
$$F_3 = \{(4, c), (2, a), (3, b), (1, d)\}$$

Which of the sets F_1, F_2, and F_3 are *1–1* functions?

  (a) F_1, F_2, F_3
  (b) F_1, F_3
  (c) F_2, F_3
  (d) F_1, F_2

3. Let $X = \{11, 12, 23, 44\}$ and $Y = \{r, s, t, v\}$ be sets. Define the following subsets of $X \times Y$:

$$F_1 = \{(11, r), (44, v), (12, s), (23, t)\}$$
$$F_2 = \{(23, s), (11, v), (44, t), (12, s)\}$$
$$F_3 = \{(44, t), (12, r), (23, s), (11, v)\}$$

Which of the sets F_1, F_2, and F_3 are onto functions?

(a) F_1, F_2, F_3
(b) F_2, F_3
(c) F_1, F_3
(d) F_1, F_2

4. Which of the following is the correct definition of a decreasing function?

(a) If for all $x_1, x_2 \in X$, $x_1 > x_2$ implies $F(x_1) \leq F(x_2)$.
(b) If for all $x_1, x_2 \in X$, $x_1 < x_2$ implies $F(x_1) < F(x_2)$.
(c) If for all $x_1, x_2 \in X$, $x_1 < x_2$ implies $F(x_1) \leq F(x_2)$.
(d) None of the above.

5. For $n = 0, 1, \ldots, 10$, list the values of the following functions defined recursively:

(a) function $s(n)$:

> if $n = 0$ then
>> return (0)
> else
>> return $(2n + s(n - 1))$

(b) function $p(n)$:

> if $n = 0$ then
>> return (1)
> else
>> return $(n \cdot p(n - 1))$

(c) function $d(n)$:

> if $n = 10$ then
>> return (100)
> else
>> return $(d(n + 1) - 10)$

6. Express the function $F(x) = \sqrt{x^2 + 3x + 2}$ as the composition of two simpler functions, neither of which is the identity function.

7. Let F and G be the functions defined on $\{1, 2, 3, 4, 5\}$ as follows:

$F:$			$G:$		
1	$\rightarrow$	2	1	$\rightarrow$	3
2	$\rightarrow$	3	2	$\rightarrow$	1
3	$\rightarrow$	5	3	$\rightarrow$	2
4	$\rightarrow$	4	4	$\rightarrow$	5
5	$\rightarrow$	1	5	$\rightarrow$	4

Prove that F and G are 1–1. Also, prove that $F \circ G$ and $G \circ F$ are both 1–1 and onto.

8. Let $X = \{-1, 0, 1, 2\}$ and $Y = \{-4, -2, 0, 2\}$. Define the function $F : X \rightarrow Y$ as $F(x) = x^2 - x$. Prove F is neither 1–1 nor onto.

9. If a class has 89 students, how many (at least) must have a birthday on the same day of the week?

10. Of 15 A's, 20 B's, and 25 C's, how many letters must be chosen so that 12 identical letters will always be included in the selection?

11. From the standard deck of 52 cards, how many cards must be chosen so that three cards from the same suit will always be included in the selection?

4.10.3   Review Questions

1. We define three functions, with definitions involving unspecified constants a and b. In each case, whether the function defined as onto depends on the values of a and b. For what values of the constants a and b are the following functions onto?

   (a) $F_1 : \mathbb{Q} \rightarrow \mathbb{Q}$ where $F_1(x) = ax + b$ with $a, b \in \mathbb{Q}$
   (b) $F_2 : \mathbb{Z} \rightarrow \mathbb{Z}$ where $F_2(x) = ax + b$ with $a, b \in \mathbb{Z}$
   (c) $F_3 : \mathbb{N} \rightarrow \mathbb{N}$ where $F_3(x) = ax + b$ with $a, b \in \mathbb{N}$

2. Let A and B be sets with $B_1, B_2 \subseteq B$, and let $F : A \rightarrow B$ be a function. Show that:

   (a) If $B_1 \subseteq B_2$, then $F^{-1}(B_1) \subseteq F^{-1}(B_2)$.
   (b) $F^{-1}(B_1 \cup B_2) = F^{-1}(B_1) \cup F^{-1}(B_2)$.
   (c) $F^{-1}(B_1 \cap B_2) = F^{-1}(B_1) \cap F^{-1}(B_2)$.
   (d) $F^{-1}(B_1 - B_2) = F^{-1}(B_1) - F^{-1}(B_2)$.
   (e) $F(F^{-1}(B_1)) \subseteq B_1$.
   (f) Find an example where $B_1 \subset B_2$ but $F^{-1}(B_1) = F^{-1}(B_2)$.

3. Show that the set $\{3, 6, 9, 12, \ldots\}$ is countable.

4. Prove that the function $F : [0, 1] \rightarrow (0, 1)$ defined as $F(0) = 1/2$, $F(1/n) = 1/(n + 2)$ for $n \in \mathbb{N}$ and $n \geq 1$, and $F(x) = x$ for all other $x \in [0, 1]$ is a bijection.

5. (a) The lattice points in the plane $\mathbb{R}^2$ are the points (x, y) where x and y are both integers. Prove that there are only countably many lattice points in $\mathbb{R}^2$.
   (b) The lattice points in three-dimensional space $\mathbb{R}^3$ are the points (x, y, z) where x, y, and z are all integers. Prove that there are only countably many lattice points in $\mathbb{R}^3$.

6. Let p be any natural number and $\mathbb{Z}_p = \{0, 1, \ldots, p - 1\}$ be the integers modulo p. For each integer r where $1 \leq r < p$, the function $F_r : \mathbb{Z}_p \rightarrow \mathbb{Z}_p$ is the function $F_r(n) = r \cdot n(mod\ p)$. Show that for each p, every F_r is a bijection if and only if p is a prime. (*Hint:* Examine cases for which $r \cdot n = r \cdot 0(mod\ p)$, and then use the Pigeon-Hole Principle.)

7. Let S be a set of six positive integers with a maximum that is at most 14. Show that the sums of the elements in all the nonempty subsets of S cannot all be distinct.

8. If 11 integers are selected from $\{1, 2, 3, \ldots, 100\}$, prove that there are at least two, say, u and v, such that

$$0 < |\sqrt{u} - \sqrt{v}| < 1$$

9. *Uses First-Order Logic:* This exercise concerns a result needed for the definition of *subsequence*. We took it as obvious there, but it does need proof. So, let $X \subseteq \mathbb{N}$.

   Part of the job here is to use as simple set theory as possible to define the functions. You may assume that (i) sets $\emptyset$, X, $\mathbb{N}$, and $\mathbb{N} \times X$ exist; (ii) for every $i, j \in \mathbb{N}$, (i, j)

exists; (iii) for any x, $\{x\}$ exists; (iv) for any sets x, y, $x \cup y$ and $x - y$ exist; and (v) the collection of elements of any set satisfying some formula of First-Order Logic (with quantifiers ranging over numbers and sets) exists. You are to use just this much set theory to show that the function defined by recursion exists. The problem is that induction may naturally be used to prove that arbitrarily large, finite sets of ordered pairs exist, but it does not allow one to conclude that a particular infinite set of ordered pairs exists.

Prove the following:

(a) Suppose X is finite. Let $n = |X|$. There is at most one increasing function from $\{0, 1, \ldots, n - 1\}$ onto X. (*Hint:* Suppose there are two, say, F and G. Prove by induction on $i < n$ that $F(i) = G(i)$.)

(b) If X is infinite, there is at most one increasing function from $\mathbb{N}$ onto X.

4.10.4   Using Discrete Mathematics in Computer Science

1. Let F be the function that maps strings of characters and blank spaces onto strings of characters by removing all blank spaces and all vowels. For example, $F(\text{"dog cat"}) = \text{"dgct."}$ Let G be the function that maps strings of characters onto integers such that the value of a string is simply the number of characters in the string. What is $F(\text{"george washington"})$? What is $G(\text{"george washington"})$? What is $G \circ F(\text{"george washington"})$? The function F is a simple example of a compression technique.

2. Let $F : \mathbb{R} \to \mathbb{R}$ be a function with $F(x) = x^2$. Define a relation R on $\mathbb{R}$ such that $x\,R\,y$ for any $x, y \in \mathbb{R}$ if $F(x) = F(y)$. Prove R is an equivalence relation, and find its equivalence classes.

3. A computer is used at least one hour a day and at most 102 hours in 12 days. Prove that in at least one pair of the six pairs of nonoverlapping consecutive days, the computer was used for 17 hours.

4. Triangle ACE is equilateral, with $AC = 1$. If five points are selected from the interior of the triangle, there are at least two whose distance apart is less than one-half. (*Note:* This is a simple version of a problem in the area called computational geometry that is actively studied in computer science.)

5. Prove there is no surjection from $\mathbb{N}$ to the set of functions with domain and codomain $\mathbb{N}$.

6. Construct a bijection from the set of all polynomials of one variable with coefficients in $\mathbb{N}$ to $\mathbb{N}$.

7. The circumference of each of two concentric disks is divided into 200 sections. For the outer disk, 100 of the sections are painted red, and 100 of the sections are painted blue. For the inner disk, the sections are painted red and blue in an arbitrary manner. Show that it is possible to align the two disks so that 100 or more of the sections on the inner disk have their color matched with the corresponding section on the outer disk.

8. Define a hashing function as a two-stage process that uses Social Security numbers as input. The first step is to form a new number consisting of every other digit of the Social Security number, starting with the leftmost digit. The hashing value is the remainder when this number is divided by 31. For example, 357-75-8564 hashes to $37554 = 13(mod\ 31)$. Carry out this hashing procedure for the Social Security num-

bers 534-27-3175, 289-13-3754, 413-39-4431, 978-65-4891, 534-75-9614. How many students in your class can have their Social Security numbers hashed this way without generating a collision?

9. Define a hashing function on a Social Security number as follows: Let the digits of the Social Security number be $x_1x_2x_3x_4x_5x_6x_7x_8x_9$ and form the three digit number $y_1y_2y_3$ as follows:

$y_1 = (x_1 + x_4 + x_7) \ (mod \ 5)$ (use the remainder when $x_1 + x_4 + x_7$ is divided by 5)
$y_2 = (x_2 + x_5 + x_8) \ (mod \ 10)$
$y_3 = (x_3 + x_6 + x_9) \ (mod \ 10)$

Calculate the hash value for the following social security numbers:

(a) 234-54-7654
(b) 534-37-9021
(c) 435-54-6782
(d) 537-98-9092
(e) 239-67-4397

Do not deal with collisions, if any occur.

10. A half-adder takes two boolean inputs and produces a sum digit and a carry digit. The output for a half-adder can be represented by a function as

Input and Output For a Half-Adder			
Input		Output	
p	q	s	c
1	1	0	1
1	0	1	0
0	1	1	0
0	0	0	0

Draw a combinatorial network that represents this network with two outputs.

CHAPTER 5

Analysis of Algorithms

We have frequently used the word *algorithm* to mean the idea, or plan of attack, behind a computer program. More precisely, an **algorithm** can be thought of as a sequence of steps, in which each step is unambiguously defined and can be executed mechanically in some finite amount of time. An algorithm may have some input data, and it must have some output. Many definitions used in explaining algorithms also require that an algorithm **halts** (or stops or terminates) after some finite number of steps. We will not require this. Therefore, we can say that the principle behind a computer program is an algorithm even when the computer program goes into an infinite loop on some inputs and does not output anything. However, when we say that *A* is an algorithm to solve a particular problem, we will mean that *A* halts on all inputs—or at least on all input data that make sense for the problem. A more detailed theoretical study of algorithms requires a more formal definition such as the notion of a *Turing machine*, which is beyond the scope of our needs here.

Although experimental timing of programs is obviously important, it is quite inadequate for dealing with questions such as these. In this chapter, we introduce a mathematical study, called *computational complexity,* that deals with these issues. This study is highly important wherever complicated computer programs are needed, including areas such as the design of layouts for computer chips and numerical analysis.

First, we study a general mathematical measure of (approximately) how fast functions grow. For this, we will introduce some very standard vocabulary: asymptotic domination, $\mathbf{O}(*)$, and $\Theta(*)$. The motivation for looking at how fast a function grows is as follows: Define the amount of time taken by an algorithm to be a function of the input data. Say that (i) A_1 and A_2 are two algorithms to solve the same problem, (ii) they are encoded as programs P_1 and P_2, and (iii) for input data D, $T_1(D)$ and $T_2(D)$ are the amounts of time taken by P_1 and P_2 on data set D. Now, let the number of elements in D increase: If $T_1(D)$ grows much faster than $T_2(D)$ as D increases in size, then, except for small inputs D, algorithm A_2 is more efficient than algorithm A_1. (The word *small* here is relative, as we will see.)

Next, we will use these ideas of growth rates in analyzing algorithms. From looking at the algorithm alone, one can deduce a great deal of information about the time functions above. This also means that as you write an algorithm, you can, before you ever start coding it, make some educated choices about better ways to organize it. We will also briefly indicate some variants of this notion regarding computational complexity to show other measures for algorithms that are of interest to computer science.

Finally, we briefly introduce another topic: Are there any functions that are, in principle, not computable by any computer program—on any computer, no matter how big, and no matter how long we wait for a program to finish? We will prove the fundamental result of Turing's about the "unsolvability of the halting problem."

5.1   Comparing Growth Rates of Functions

There are three ideas to keep in mind as we develop a formal setting for determining the complexity of an algorithm:

1. Since you can time the program on only a finite (usually small) number of input data sets, how can you (reasonably) estimate the time on other input sets? For example, can you (reasonably) extrapolate from the time you have taken to run your program on test data sets to the expected time on real (and, probably, much larger) data sets?
2. Computer A may do one operation (say, swapping two integers) twice as fast as computer B. Computer B may do some other operation (say, comparing two real numbers) twice as fast as computer A. Can you (reasonably) compare the speeds of two algorithms without knowing on which machine each program will be run?
3. Can you design your algorithm so that your program will "scale well"—in other words, so that as the size of the data set increases, the time needed by the program will not increase too rapidly? You may have seen algorithms to search for a number in a sorted array (a special kind of list). Two standard methods for this purpose are called *linear search* and *binary search.* When these two algorithms are presented, people justify *BinarySearch* as being obviously faster. How can they (reasonably) say this?

We would like to approximate how the time taken by a program increases as the size of the input data set grows. Moreover, we would like to make an approximation that we can compute by looking at a (detailed) algorithm alone, so that the program designer can make reasonable choices about which algorithms to use without having to code an actual algorithm.

Before we are ready to approximate the growth rate of algorithms, however, we need to understand how the growth rate of functions can be compared. Once we understand how functions can be compared, we will show how a function can be associated with the running time of an algorithm. We then compare algorithms by comparing the functions associated with describing their running times.

5.1.1   A Measure for Comparing Growth Rates

There has been much discussion among computer scientists about how to measure the growth rate of algorithms. In this area, two major simplifications are useful:

1. If we have two algorithms, A_1 and A_2, where A_1 always runs faster on "small" inputs and A_2 always runs faster on "large" inputs, then we—at least under this measure—prefer A_2: It is "almost always" faster. This measure ignores differences between the two functions on "small" inputs, and here, "small" is taken to mean "smaller than any fixed finite size." (This is a huge issue to ignore, but the approximation turns out to be highly useful.)

2. Sometimes, two different computers execute various instructions at different relative speeds. C_1 may be twice as fast as C_2 at doing floating point operations, and C_2 may be twice as fast as C_1 at doing logical operations. The measure ignores constant factors, however, such as the two times factors.

We are interested in timing functions that input the size of the input data and output the amount of time taken by the program. The input to such a function is a natural number, and the output is a non-negative real number. Keep this idea in mind as we explore the next notion.

Definition 1. Let $F, G : \mathbb{N} \to [0, \infty)$ be functions. F **asymptotically dominates** G if, for some integer $N_0 \in \mathbb{N}$ and some real number $c \in (0, \infty)$ such that for all $n \geq N_0$,

$$G(n) \leq c \cdot F(n)$$

The condition "for all $n \geq N_0$" in Definition 1 reflects the fact that finitely many integers can be ignored, since we are looking for a starting place for all following integers to be instances where the condition holds: We do not worry about whether the inequality may fail for some elements of the finite set $\{0, 1, 2, \ldots, N_0 - 1\}$. The multiplicative constant c in Definition 1 reflects the fact that integer multiples of a function are considered to be the "same" in complexity.

Example 1. Let $F, G : \mathbb{N} \to [0, \infty)$ be functions defined as $G(n) = 3n$ and $F(n) = n^2$. F asymptotically dominates G.

Solution.   Let $N_0 = 0$ and $c = 3$. It follows that for $n \geq N_0$,

$$
\begin{aligned}
G(n) &= 3n \\
&\leq 3n^2 \\
&\leq 3F(n) \\
&\leq cF(n)
\end{aligned}
$$
∎

Some authors allow F and G to take on negative values. Typically, the convention is to study $|F(n)|$ and $|G(n)|$ in those cases.

It remains for us to develop the notion of asymptotic domination so that we can show how a function represents the complexity of an algorithm. This notion is captured in the "big **O**" notation.

Definition 2. Let $F, G : \mathbb{N} \to [0, \infty)$ be functions. Then,

$$\mathbf{O}(F) = \{G : F \text{ asymptotically dominates } G\}$$

The expression "F asymptotically dominates G" is usually not written out in full. It is far more common to write "$G \in \mathbf{O}(F)$." The expression "$\mathbf{O}(F)$" is pronounced "big-Oh of F," or **order of F**. (With abuse of notation, people sometimes write "G is $\mathbf{O}(F)$," or even "$G = \mathbf{O}(F)$." However, "$\mathbf{O}(G) = F$" is never considered to be acceptable.)

If $G \in \mathbf{O}(F)$ and $F \in \mathbf{O}(G)$, the intuition is that F and G grow at approximately the same rate. To understand this notion, we present some examples. Although the definition is in terms of functions from $\mathbb{N}$ into $[0, \infty)$, we actually graph the functions on the usual coordinate system for the (x, y)-plane.

Example 2.   Let $F_1, G_1 : \mathbb{N} \to [0, \infty)$ where $F_1(n) = n$ and

$$G_1(n) = \begin{cases} n - 2 & \text{for } n \geq 2 \\ 0 & \text{otherwise} \end{cases}$$

as shown in Figure 5.1.

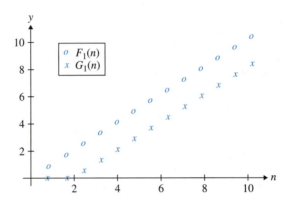

Figure 5.1 $F_1(n)$ and $G_1(n)$.

Then, $F_1 \in \mathbf{O}(G_1)$ and $G_1 \in \mathbf{O}(F_1)$.

Solution.   Obviously, $F_1(n) \geq G_1(n)$ for all n. Intuitively, however, the functions seem to grow at the same rate. Indeed, for $n \geq 2$, their graphs have the same slope.

Showing that $G_1 \in \mathbf{O}(F_1)$ is easy. Clearly, $G_1(x)$ is always less than or equal to $F_1(x)$. So, pick $c = 1$ and $N_0 = 0$, and the definition of asymptotic domination is satisfied.

To show $F_1 \in \mathbf{O}(G_1)$ is only slightly harder: Let $c = 2$, and let $N_0 = 4$. For all n such that $n \geq 4$, $n - 4 \geq 0$; hence,

$$
\begin{aligned}
F_1(n) &= n = n + 0 \\
&\leq n + (n - 4) \\
&= 2n - 4 \\
&= 2(n - 2) \\
&= 2 \cdot G_1(x) \\
&= c \cdot G_1(x)
\end{aligned}
$$
∎

In Example 2, $G_1(n)$ could have been zero for all $n \leq N_0$ where $N_0 > 2$, and a similar proof would still work.

Example 3 shows that behavior at a single point will make no difference to the **O**-relation.

Example 3.   Let $F_2, G_2 : \mathbb{N} \to [0, \infty)$ where $F_2(n) = n$ and

$$G_2(n) = \begin{cases} 100{,}000{,}000{,}000 & \text{for } n = 0 \\ n & \text{otherwise} \end{cases}$$

Then, $F_2 \in \mathbf{O}(G_2)$ and $G_2 \in \mathbf{O}(F_2)$.

Solution.   Here, $G_2(0)$ is much larger than $F_2(0)$. For all other values of n, we have $G_2(n) = F_2(n)$. The intuition is that one value, or any finite number of values, makes no difference to the growth rate of a function.

To show that $F_2 \in \mathbf{O}(G_2)$ and $G_2 \in \mathbf{O}(F_2)$, choose $c = 1$ and $N_0 = 1$ for both implications. Then, for all n such that $n \geq 1$, it follows that

$$F_2(n) \leq 1 \cdot G_2(n)$$

and

$$G_2(n) \leq 1 \cdot F_2(n)$$

This finishes the proof.   ■

The values of F_2 and G_2 in Example 3 could have differed at any point—say, n_0. At n_0, let $G_2(n_0) = F_2(n_0) + 100{,}000$. The only adjustment to the proof would be to choose N_0 such that $N_0 > n_0$.

We will see in Example 4 what is involved in proving that one function is not asymptotically dominated by another function. The functions in Example 4 show that the relation asymptotically dominates is not a symmetric relation.

Example 4.   Let $F_4, G_4 : \mathbb{N} \to [0, \infty)$ where $F_4(n) = n$ and $G_4(n) = n^2$, as shown in Figure 5.2.

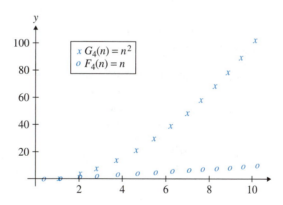

Figure 5.2 $F_4(n) = n$ and $G_4(n) = n^2$.

Then, $F_4 \in \mathbf{O}(G_4)$, and $G_4 \notin \mathbf{O}(F_4)$.

Solution.   From the graphs of the functions, G_4 seems to grow faster than F_4. More importantly, G_4 seems to grow faster than $2 \cdot F_4$, faster than $3 \cdot F_4$, and for each real number c such that $c > 0$, faster than $c \cdot F_4$, as shown in Figure 5.3 on page 288.

The intuition is that G_4 grows much faster than F_4. We verify this intuition by proving that (a) $F_4 \in \mathbf{O}(G_4)$ but (b) $G_4 \notin \mathbf{O}(F_4)$:

(a) Pick $c = 1$ and $N_0 = 1$. For $n > 1$, it is obvious that $n^2 > n$.

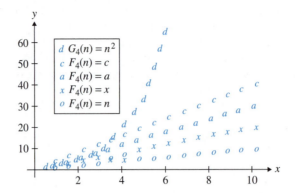

Figure 5.3 Comparison of multiples of a function: $G_4(n) = n^2$, $F_4(n) = n$, $2F_4(n)$, $3 \cdot F_4(n)$ (shown a), and $4 \cdot F_4(n)$.

(b) We will show that for any N_0 and any c, there is at least one $n \geq N_0$ where

$$G_4(n) \leq c \cdot F_4(n)$$

does not hold. Let N_0 and c be given. If $N_0 > c$, then pick $n = N_0$ so that the inequality fails. Let $n \geq N_0$, and since $n > c$, $n^2 > cn$. That is, the inequality fails.

If $N_0 \leq c$, pick any integer $n > c$. (We cannot just pick $n = c$, because we did not require c to be an integer.) For this n, we have $n \geq N_0$ and, again, $n^2 > cn$, so the inequality fails. In both cases, we have shown integers n such that $n \geq N_0$ that are counterexamples to the required inequality. ∎

Although Example 4 deals with n and n^2, the proof can be used as a model for n^k and n^{k+1} for $k = 3, 4, 5, \ldots$. In Theorem 2 in Section 5.1.2, we will prove that asymptotic domination is a transitive relation. This result will imply $n^i \in \mathbf{O}(n^j)$ for $i, j \in \mathbb{N}$ and $i \leq j$.

A function F asymptotically dominates a function G if $G(n) \leq c \cdot F(n)$ for all but finitely many values of n. It is unimportant what happens on those finitely many values. In fact, it turns out to be convenient not even to require that F and G be defined on those values. So, we extend the previous definitions to partial functions that are defined on all but finitely many values.

Definition 3. Let $F, G : \mathbb{N} \to [0, \infty)$ be partial functions. Then, F **asymptotically dominates** G if, for some $N_0 \in \mathbb{N}$, and some $c \in (0, \infty)$, for all $n \geq N_0$, both $F(n)$ and $G(n)$ are defined and $G(n) \leq c \cdot F(n)$.

In the rest of this section, we assume that $F : \mathbb{N} \to [0, \infty)$ will mean that F is a partial function and that $F(n)$ is defined for all but finitely many n in $\mathbb{N}$.

There are many partial functions from $\mathbb{N}$ to $[0, \infty)$ with growth rates that are commonly discussed but are undefined for finitely many arguments. The following are typical of such partial functions:

(1) $\ln(n)$ is undefined for $n = 0$ but is defined for all $n \geq 1$.
(2) $\ln \circ \ln(n)$ is undefined for $n = 0, 1$ but is defined for all $n \geq 2$.

(3) $F : \mathbb{N} \to [0, \infty)$, given by the rule $F(n) = n - 2$, is undefined for $n = 0, 1$ (since $0 - 2$ and $1 - 2$ are less than 0) but is defined for all $n \geq 2$.

5.1.2   Properties of Asymptotic Domination

Most of the theorems in this section are chosen so that we could prove some fundamental results about how polynomial functions are related with respect to complexity. Theorems 1–4 that follow are useful for that purpose but apply much more generally to arbitrary functions.

Theorem 1.   Let $F : \mathbb{N} \to [0, \infty)$, and let $a \in (0, \infty)$. Then, F asymptotically dominates $a \cdot F$.

Proof.   By the general assumption, there is a natural number N_0 such that $F(n)$ is defined for all $n \geq N_0$. Therefore, $a \cdot F(n) \in [0, \infty)$ for all n such that $n \geq N_0$. Now, choose $c = a$. Then, for all $n \geq N_0$,

$$a \cdot F(n) \leq c \cdot F(n)$$

Therefore, $a \cdot F \in \mathbf{O}(F)$.   ■

Corollary 1:   For any *positive* real number a and any $F : \mathbb{N} \to [0, \infty)$, both $a \cdot F \in \mathbf{O}(F)$ and $F \in \mathbf{O}(a \cdot F)$.

Proof.   By Theorem 1, $a \cdot F \in \mathbf{O}(F)$. However, since $a > 0$, $(1/a)$ also exists and is positive. So, since $F = (1/a)(a \cdot F)$, $F \in \mathbf{O}(a \cdot F)$.   ■

In the case $a = 1$ in Theorem 1, the result says that the relation asymptotically dominates is reflexive.

Theorem 2 tells us that the asymptotically dominated relationship is a transitive relation.

Theorem 2.   Let $F, G, H : \mathbb{N} \to [0, \infty)$, with $H \in \mathbf{O}(G)$ and $G \in \mathbf{O}(F)$. Then, $H \in \mathbf{O}(F)$.

Proof.   Since $H \in \mathbf{O}(G)$, there are constants $c_H \in [0, \infty)$ and $N_H \in \mathbb{N}$ such that for all $n \geq N_H$, $G(n)$ and $H(n)$ are both defined and $H(n) \leq c_H G(n)$. Since $G \in \mathbf{O}(F)$, there are constants $c_G \in [0, \infty)$ and $N_G \in \mathbb{N}$ such that for all $n \geq N_G$, $F(n)$ and $G(n)$ are both defined and $G(n) \leq c_G F(n)$. Let N_0 be the maximum of N_G and N_H, and let $c = c_H \cdot c_G$. Then, for any $n \geq N_0$,

$$
\begin{aligned}
H(n) &\leq c_H \cdot G(n) \\
&\leq c_H \cdot (c_G \cdot F(n)) \\
&= (c_H \cdot c_G) \cdot F(n) \\
&= c \cdot F(n)
\end{aligned}
$$

Therefore, $H \in \mathbf{O}(F)$.   ■

One feature in the proof of Theorem 2 deserves comment. We picked N_0 to be the maximum of N_G and N_H and c to be $c_H \cdot c_G$. We do not know whether smaller values for c or N_0 might have "worked": That depends on further information about G, H, c_G, c_H,

N_G, and N_H that we do not have. However, to verify that $H \in \mathbf{O}(F)$, we were not required to find the smallest possible constants c and N_0; we were only required to find some such constants.

In Theorem 3 we see that if two functions G and H are both asymptotically dominated by a third function F, then both $G + H$ and $|G - H|$ are functions that are asymptotically dominated by F. For example, if $G(n) = n^2$, $H(n) = n$, and $F(n) = n^3$, then $n^2 + n \in \mathbf{O}(n^3)$.

Theorem 3.   Let $F, G, H : \mathbb{N} \to [0, \infty)$, with $G \in \mathbf{O}(F)$ and $H \in \mathbf{O}(F)$. Then, $G + H \in \mathbf{O}(F)$ and $|G - H| \in \mathbf{O}(F)$.

Proof.   Since $G \in \mathbf{O}(F)$, there exist constants $c_G \in [0, \infty)$ and $N_G \in \mathbb{N}$ such that for all $n \geq N_G$, $F(n)$ and $G(n)$ are both defined and $G(n) \leq c_G \cdot F(n)$. Similarly, there are constants $c_H \in [0, \infty)$ and $N_H \in \mathbb{N}$ such that for all $n \geq N_H$, $F(n)$ and $H(n)$ are both defined and $H(n) \leq c_H \cdot F(n)$.

Let $c = c_G + c_H$, and let N_0 be the maximum of N_G and N_H. Then, for all $n \geq N_0$, both $F(n)$ and $G(n) + H(n)$ are defined, and

$$G(n) + H(n) \leq c_G \cdot F(n) + c_H \cdot F(n) = c \cdot F(n)$$

Furthermore, for all $n \geq N_0$, $|G(n) - H(n)|$ is defined, and

$$|G(n) - H(n)| \ \leq \ |G(n)| + |H(n)| \ = \ G(n) + H(n) \ \leq \ c \cdot F(n)$$

Hence, $G + H \in \mathbf{O}(F)$ and $|G - H| \in \mathbf{O}(F)$.   ∎

Corollary 2:   Let $F, G_1, G_2, \ldots, G_k : \mathbb{N} \to [0, \infty)$ for some $k \in \mathbb{N}$ such that for $1 \leq i \leq k$, each $G_i \in \mathbf{O}(F)$. Then, for any real numbers $a_1, a_2, \ldots, a_k$,

$$|a_1 G_1 + a_2 G_2 + \cdots + a_k G_k| \in \mathbf{O}(F)$$

Proof.   Use induction on k. Separate arguments are needed depending on whether a_k is positive (or zero) or negative. The results in both Theorems 1 and 2 are used. The details are left as an exercise for the reader.   ∎

We can use our results to define an equivalence relation that is closely related to asymptotic domination.

Theorem 4.   Let $F, G : \mathbb{N} \to [0, \infty)$.

(a) $F \in \mathbf{O}(G)$ if and only if $\mathbf{O}(F) \subseteq \mathbf{O}(G)$.
(b) $\mathbf{O}(F) = \mathbf{O}(G)$ if and only if $F \in \mathbf{O}(G)$ and $G \in \mathbf{O}(F)$.
(c) The relation Θ, defined as $\{(F, G) : \mathbf{O}(F) = \mathbf{O}(G)\}$, is an equivalence relation.

Proof.   This proof is left as an exercise for the reader.   ∎

With the proof of Theorem 4(c) we see what condition is needed to define a symmetric relation based on asymptotic domination. We showed that in general, asymptotic domination is reflexive and transitive, but the functions in Example 4 tell us there is no hope of making the relation symmetric. The relation Θ tells us there is an equivalence relation that depends on a slightly stronger relation than asymptotic domination. An ongoing problem in computer science is to determine what functions are in each equivalence class of the equivalence relation Θ.

The relation Θ considers seemingly different functions as being equivalent, at least in the sense that they determine the same set $\mathbf{O}(*)$. For example, there are many unexpected functions F where $\mathbf{O}(F) = \mathbf{O}(x)$. Two such functions are shown in Figure 5.4.

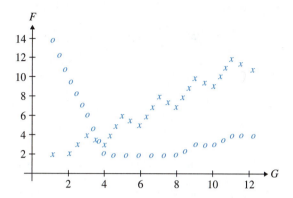

Figure 5.4 $F, G \in \mathbf{O}(x)$.

5.1.3   Polynomial Functions

A polynomial such as

$$17n^{137} + 197n^{45} - 20{,}143n^{14} + 3$$

is a rule for computing a function. In a calculus class, for example, a polynomial function is usually defined as having domain and range $\mathbb{R}$. For the purposes of this chapter, a polynomial is a rule for defining a partial function with domain of definition $\mathbb{N}$ and codomain $[0, \infty)$.

Definition 4.   A **polynomial function** (of n) of **degree** k is a function

$$P(n) = a_k\, n^k + a_{k-1}\, n^{k-1} + \cdots + a_2\, n^2 + a_1\, n + a_0$$

where $a_k \neq 0$. The **zero polynomial** of n is the function $F(n) = 0$. The zero polynomial has degree -1. In general, a polynomial function is a polynomial function of any degree. (You must read this definition carefully in other contexts, because some authors say the zero polynomial has degree 0.)

According to the conventions of this section, for $n \in \mathbb{N}$, the polynomial function is defined at n if the polynomial of degree k, evaluated at n, gives a non-negative value. Also, by convention, $P(n)$ should be defined for all but finitely many n's. For the polynomial

$$P(n) = a_k\, n^k + a_{k-1}\, n^{k-1} + \cdots + a_2\, n^2 + a_1\, n + a_0$$

(where $a_k \neq 0$) to satisfy this assumption, it is necessary that $a_k > 0$.

The first result about polynomials shows that polynomials of different degree have different growth rates. The essential step in this proof is true in the more general case of the real numbers. We state the general result first.

Theorem 5.   Let $k_1, k_2 \in \mathbb{R}$ with $0 < k_1 < k_2$. Let x^{k_1} and x^{k_2} be functions of x, where $x \in \mathbb{R}$. Then:

(a) $x^{k_1} \in \mathbf{O}(x^{k_2})$
(b) $x^{k_2} \notin \mathbf{O}(x^{k_1})$

Proof.   Similar to that for Example 4. The details are left as an exercise for the reader.

■

Corollary 3:   Let $i, j \in \mathbb{N}$ with $i < j$. Let n^i and n^j be functions of n where $n \in \mathbb{N}$. Then:

(a) $n^i \in \mathbf{O}(n^j)$
(b) $n^j \notin \mathbf{O}(n^i)$

Theorem 5 is a first step in determining how polynomials of different degrees are related by the relation asymptotically dominates. For example, $n^5 \in \mathbf{O}(n^8)$, but $n^8 \notin \mathbf{O}(n^5)$. Both Theorem 1 in Section 5.1.2 and Theorem 5 tell us how monomials are related. Theorem 3 in Section 5.1.2 could be used to build polynomials from the individual terms (monomials). In Theorem 6, we show how you can prove directly that a polynomial and its terms are related to the monomial that is its highest power.

Theorem 6.   For a polynomial function of degree k:

$$P(n) = a_k \, n^k + a_{k-1} \, n^{k-1} + \ldots + a_2 \, n^2 + a_1 \, n + a_0$$

Therefore:

(a) $P(n) \in \mathbf{O}(n^k)$
(b) $n^k \in \mathbf{O}(P(n))$

Motivation for Proof.   Consider the polynomial $x^3 - 3x^2 - 1$. For $x = 10$, the value of 10^3 contributes approximately 70 percent of the value of this polynomial at $x = 10$. For $x = 100$, the value of 10^3 contributes approximately 97 percent of the value of the polynomial at $x = 100$. So, the intuition is that for large enough values of n, or all n greater that some N_0, $P(n)$ is almost the same as $a_k \cdot n^k$ and, hence, grows approximately as fast. The major task of the proof is to establish that intuition. We will make no effort to choose N_0 as small as possible. Rather, we will choose an N_0 that makes the proof easy. (Part (a) can also be proved using Theorem 5 and Corollary 2.)

Proof.   Set

$$N_0 = 2 \cdot \left\lceil \frac{|a_{k-1}| + |a_{k-2}| + \cdots + |a_1| + |a_0|}{a_k} \right\rceil + 2$$

We leave it as an exercise for the reader to show that for all $n \geq N_0$,

$$\frac{1}{2} a_k \, n^k \leq P(n) \leq \frac{3}{2} a_k \, n^k$$

Choose N_0 as above and $c = (3/2)a_k$. Then, $P(n) \leq (3/2)a_k \, n^k$ for all $n \geq N_0$. This proves part (a). Now, choose N_0 as above and $c = 2/a_k$. Then, $n^k < (2/a_k)P(n)$ for all $n \geq N_0$. This proves part (b).   ■

Corollary 4:   For any integer $k \geq -1$, the polynomial functions in $\mathbf{O}(n^k)$ are the polynomial functions of degree less than or equal to k.

We now know that the complexity of any polynomial is completely determined by its degree. For example, $P(n) = n^3 - 3n^2 + n$ is $\mathbf{O}(n^3)$, and $R(n) = n^3 + 3n$ is also $\mathbf{O}(n^3)$.

5.1.4   Exponential and Logarithmic Functions

Before proving some needed facts about the growth rate of the exponential and logarithmic functions, we need to review some properties of these functions.

Familiar Facts About Exponential and Logarithmic Functions

(a) For $m, n, r \in \mathbb{R}$ where $m, n, r > 0$, $m < n$ if and only if $m^x < x^x$.
(b) The exponentiation function to a base greater than 1 is an increasing function: For $n, r, s \in \mathbb{R}$ where $n > 1$ and $r, s > 0$, $0 < r < s$ if and only if $n^r < n^s$.
(c) A **logarithmic function** to a base greater than 1 is an increasing function: For $n, r, s \in \mathbb{R}$ where $n > 1$ and $r, s > 0$, $r < s$ if and only if $\log_n(r) < \log_n(s)$.
(d) For $b, x \in \mathbb{R}$ where $x > 0$, $b > 0$, and $b \neq 1$, $\log_b(x) = \ln(x)/\ln(b)$.
(e) For all $m, r, s \in \mathbb{R}$ where $m > 0$, $m^{r+s} = m^r \cdot m^s$.
(f) For all $m, r, s \in \mathbb{R}$ where $m > 0$, $(m^r)^s = m^{rs}$.
(g) For all $m, n, r \in \mathbb{R}$ where $m, n > 0$, $\log_m(n^r) = r \log_m(n)$.

We will not prove these facts in this text. Part (c) is equivalent to part (b) since the functions $\log_n(x)$ and n^x, as functions of x, are inverse functions.

The most commonly seen functions that grow faster than the polynomial functions are the **exponential functions.** Figure 5.5 shows an example for each of these kinds of functions.

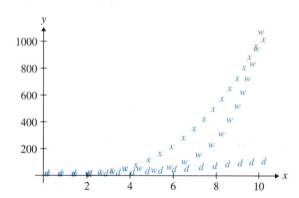

Figure 5.5 n^2 (d), n^3 (x), and 2^n (w).

Theorem 7.   Let $k \in \mathbb{R}$ such that $k \geq 2$. Then:

(a) $n^k \in \mathbf{O}(2^n)$
(b) $2^n \notin \mathbf{O}(n^k)$

Proof.   (a) We first establish an inequality involving n^k and then define N_0 and c:

$$n^k \leq n^c \quad c = \lceil k \rceil \text{ and}$$
$$n \leq 2^l \quad l = \lceil \log_2(n) \rceil \text{ then}$$
$$n^k \leq 2^{cl}$$

Choose $N_0 = cl$. If $n > N_0$, then

$$n^k \leq 2^{N_0} < 2^n$$

(b) Suppose $2^n \in \mathbf{O}(n^k)$ for some natural number k. By part (a), $n^{k+1} \in \mathbf{O}(2^n)$. It follows by the transitivity of asymptotic domination (Theorem 3) that $n^{k+1} \in \mathbf{O}(n^k)$, contradicting Theorem 5.   ■

Hand-checking values in the proof of Theorem 7(a) will make the reader suspect (correctly) that this constant is much larger than needed. It does, however, make the proof easier to establish.

A more complete description of the general relation between exponential functions is proven in Theorem 8. Examples of this result include $\pi^n \in \mathbf{O}(4^n)$, but $4^n \notin \mathbf{O}(\pi^n)$.

Theorem 8.   Let $n \in \mathbb{N}$.

(a) Let a and b be real numbers such that $1 < a < b$. Then, $a^n \in \mathbf{O}(b^n)$ and $b^n \notin \mathbf{O}(a^n)$.
(b) Let a, b, and c be real numbers, with $0 < a < b$ and $c > 1$. Then, $c^{an} \in \mathbf{O}(c^{bn})$ and $c^{bn} \notin \mathbf{O}(c^{an})$.

Proof.   This proof is left as an exercise for the reader.   ■

As another example that will be of use in analyzing sorting algorithms, we can show that the function $F(n) = n$ grows faster than the function $G(n) = \ln(n)$.

Example 5.   Let $F(n) = n$ and $G(n) = \ln(n)$ be functions from $\mathbb{N}$ to $\mathbb{R}$. Then, $G \in \mathbf{O}(F)$, but $F \notin \mathbf{O}(G)$, as shown in Figure 5.6.

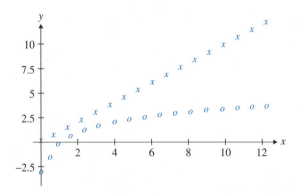

Figure 5.6 $F(n) = n$ (shown x) and $G(n) = \ln(n)$ (shown o).

Solution.   First, show that $\ln(n) \in \mathbf{O}(n)$. For all $n \in N$, we have $n < 2^n$. Since $e > 1$, by facts (c) and (g) cited previously about logarithmic functions, $\ln(n) < n \ln(2)$. This establishes the result with $N_0 = 1$ and $c = \ln(2)$. (We set $N_0 = 1$, because $\ln(0)$ is undefined.)

Now, show that $n \notin \mathbf{O}(\ln(n))$. Working toward a contradiction, suppose $n \in \mathbf{O}(\ln(n))$. That is, for some integer N_0 and some real number c, $n \leq c \cdot \ln(n)$ for all $n \geq N_0$. Then, however, for any integer n such that $n \geq max\{N_0, 1\}$,

$$2^n < e^n \qquad \text{(fact (a))}$$
$$e^n \leq e^{c(\ln(n))} = n \cdot e^c \quad \text{(fact (b) and assumption of proof)}$$
$$2^n < n \cdot e^c$$

This implies that $2^n \in \mathbf{O}(n)$, contradicting Theorem 7. ∎

Complexity Comparisons for Various Functions

To see the difference in the time requirement for processing data sets of arbitrary size, we will assume a single machine cycle will require 10^{-6} seconds to be completed. Table 5.1 gives the time required to process a data set of size n, for six different values of n, if it takes $\log_2(n)$ (n, n^2, n^5, and 2^n, respectively) machine cycles to make the computation. For example, in the column labeled n^2 for the row labeled $n = 100$, 100^2 operations are needed to complete execution. The time is $(10^2)^2 10^{-6}$ seconds $= 10^{-2}$ seconds.

Table 5.1 Complexity Table for Several Functions

$F(n)$	$\log_2(n)$	n	n^2	n^5	2^n
$n = 10$	3×10^{-6} sec	10^{-5} sec	10^{-4} sec	0.1 sec	10^{-3} sec
$n = 20$	4×10^{-6} sec	2×10^{-5} sec	4×10^{-4} sec	3 sec	1 sec
$n = 50$	6×10^{-6} sec	5×10^{-5} sec	3×10^{-3} sec	5 min	36 yrs
$n = 100$	7×10^{-6} sec	10^{-4} sec	10^{-2} sec	3 hrs	4×10^{16} yrs
$n = 1000$	1×10^{-5} sec	10^{-3} sec	1 sec	32 yrs	3.9×10^{287} yrs
$n = 100,000$	2×10^{-5} sec	0.1 sec	2.7 hrs	3×10^{11} yrs	$> 10^{30,089}$ yrs

To better appreciate the figures in the last column, note that the current conjectured age of the universe is around 2×10^{10} years. Now, suppose we could speed up the computation by a factor of 10^{10} (which is not conceivable at the time of this writing). Also, suppose we could somehow replace every elementary particle in the universe with one of those computers (a physical impossibility, of course)—less than 10^{90} computers by current estimates—and *somehow* simply split the computation equally among all those computers (and such a simple, optimal split also is not currently conceivable). That reduces the largest running time to over $10^{29,989}$ years—or over $10^{29,979}$ times the current age of the universe. (Computer users do not like to wait that long for answers.)

The discussion in this section has shown that a function F is more complex than a function G when $G \in \mathbf{O}(F)$ but $F \notin \mathbf{O}(G)$. To show how important this idea is in computer science, we used some relatively familiar functions for a table to show how they differed when acting on inputs of various sizes. It is instructive to see how important the function that describes the complexity of a computation really is by comparing various entries in Table 5.1.

5.2   Exercises

1. Find a real number c and an $N_0 \in \mathbb{N}$ such that $n^2 + 5n < cn^2$ for all $n \in \mathbb{N}$ with $n \geq N_0$.

2. Find a real number c and an $N_0 \in \mathbb{N}$ such that $n^3 + 5n^2 + 2n < cn^3$ for all $n \in \mathbb{N}$ with $n \geq N_0$.

3. Find a real number c and an $N_0 \in \mathbb{N}$ such that $n^2 + 3n < cn^3$ for all $n \in \mathbb{N}$ with $n \geq N_0$.

4. Find a real number c and an $N_0 \in \mathbb{N}$ such that $n^3 - 3n^2 + 4n < cn^3$ for all $n \in \mathbb{N}$ with $n \geq N_0$.

5. (a) Find a real number c and an $N_0 \in \mathbb{N}$ such that $n^2 < cn^3$ for all $n \in \mathbb{N}$ with $n \geq N_0$.
   (b) Find a real number c and an $N_0 \in \mathbb{N}$ such that $5n < cn^3$ for all $n \in \mathbb{N}$ with $n \geq N_0$.

6. Using the proof of the Corollary 1 to Theorem 1 as a model, find a real number c and an $N_0 \in \mathbb{N}$ such that $5n^3 \in \mathbf{O}(n^3)$ for all $n \in \mathbb{N}$ with $n \geq N_0$.

7. Using the proof of the Corollary 1 to Theorem 1 as a model, find a real number c and an $N_0 \in \mathbb{N}$ such that $7n^4 \in \mathbf{O}(n^4)$ for all $n \in \mathbb{N}$ with $n \geq N_0$.

8. (a) Find a real number c and an $N_0 \in \mathbb{N}$ such that $n^2 - 3n \in \mathbf{O}(n^3)$.
   (b) Find a real number c and an $N_0 \in \mathbb{N}$ such that $2n^2 + 7n \in \mathbf{O}(n^3)$.
   (c) Using part (a), part (b), and Theorem 3, prove that $3n^2 + 4n \in \mathbf{O}(n^3)$.
   (d) Using part (a), part (b), and Theorem 3, prove that $|n^2 - 10n| \in \mathbf{O}(n^3)$.
   (e) Using part (a), part (b), and Theorem 3, prove that $5(n^2 - 3n) + 6(2n^2 + 7n) \in \mathbf{O}(n^3)$.

9. (a) Find a real number c and an $N_0 \in \mathbb{N}$ such that $2n^2 + 7n \in \mathbf{O}(n^3)$.
   (b) Find a real number c and an $N_0 \in \mathbb{N}$ such that $3n^2 - 7n \in \mathbf{O}(n^3)$.
   (c) Using part (a), part (b), and Theorem 3, find c and N_0 to prove that $5n^2 \in \mathbf{O}(n^3)$.
   (d) Using part (a), part (b), and Theorem 3, find a real number c and an $N_0 \in \mathbb{N}$ to prove that $|n^2 - 14n| \in \mathbf{O}(n^3)$.
   (e) Using part (a), part (b), and Theorem 3, find a real number c and an $N_0 \in \mathbb{N}$ to prove that $5(2n^2 + 7n) + 3(3n^2 - 7n) \in \mathbf{O}(n^3)$.

10. (a) Find a real number c and an $N_0 \in \mathbb{N}$ such that $n^2 + 3n \in \mathbf{O}(n^3)$.
    (b) Find a real number c and an $N_0 \in \mathbb{N}$ such that $3n^3 \in \mathbf{O}(n^4)$.
    (c) Using part (a), part (b), and Theorem 2, find c and N_0 to prove that $n^2 + 3n \in \mathbf{O}(n^4)$.

11. (a) Find a real number c and an $N_0 \in \mathbb{N}$ such that $2n^2 - 5n \in \mathbf{O}(n^3 + 5n^2)$.
    (b) Find a real number c and an $N_0 \in \mathbb{N}$ such that $6n^3 + 5n^2 \in \mathbf{O}(n^4)$.
    (c) Using part (a), part (b), and Theorem 2, find c and N_0 to prove that $2n^2 - 5n \in \mathbf{O}(n^4)$.

12. Using the proof of Theorem 6 as a guide:
    (a) Find a real number c and an $N_0 \in \mathbb{N}$ such that $n^2 - 3n \in \mathbf{O}(n^3)$.
    (b) Show that $n^3 \notin \mathbf{O}(n^2 - 3n)$.

13. Using the proof of Theorem 6 as a guide:
    (a) Find a real number c and an $N_0 \in \mathbb{N}$ such that $n^2 + 5n + 3 \in \mathbf{O}(n^3)$.
    (b) Show that $n^3 \notin \mathbf{O}(n^2 + 5n + 3)$.

14. Show that:
    (a) $\sin(n) \in \mathbf{O}(1)$—that is, the sine function is asymptotically dominated by the constant function 1.
    (b) $n\sin(n) \in \mathbf{O}(n)$.
15. Prove that $\sqrt{n} \in \mathbf{O}(n)$ and that $n \notin \mathbf{O}(\sqrt{n})$.
16. Prove or find a counterexample. If $\mathbf{O}(G) = \mathbf{O}(F)$ and $\mathbf{O}(H) \subset \mathbf{O}(F)$, then for all but finitely many $n \in \mathbb{N}$, $H(n) < G(n)$.
17. Prove that $\ln(n) \in \mathbf{O}(n)$ but that $n \notin \mathbf{O}(\ln(n))$ for $n \in \mathbb{N}$.
18. Prove that $n^3 \in \mathbf{O}(n^{\pi})$ and that $n^{\pi} \notin \mathbf{O}(n^3)$.
19. Prove that for all $b, c > 1$, $\mathbf{O}(\log_b) = \mathbf{O}(\log_c)$.
20. Complete the proof of Corollary 2 to Theorem 3.
21. Prove Theorem 4.
22. Prove Theorem 5.
23. Prove Theorem 8.
24. Finish the proof of Theorem 6. For a polynomial function

$$P(n) = a_k\, n^k + a_{k-1}\, n^{k-1} + \cdots + a_2\, n^2 + a_1\, n + a_0$$

where $a_k > 0$ and for

$$N_0 = 2 \left\lceil \frac{|a_{k-1}| + |a_{k-2}| + \cdots + |a_1| + |a_0|}{a_k} \right\rceil + 2$$

prove that for all $n \geq N_0$,

$$a_k\, n^k/2 \leq P(n) \leq 3a_k\, n^k/2$$

25. Consider a polynomial

$$P(n) = a_k\, n^k + a_{k-1}\, n^{k-1} + \cdots + a_2\, n^2 + a_1\, n + a_0$$

as a function from $\mathbb{R}$ to $\mathbb{R}$. Show how to compute a number N so that all roots of $P(n) = 0$ occur between $-N$ and N. Prove that your answer is correct.
26. For functions $F, G : \mathbb{N} \to (0, \infty)$, F is said to be in $o(G)$ if

$$\lim_{x \to \infty} \frac{F(x)}{G(x)} = 0$$

    (a) Show that if $F \in o(G)$, then $F \in \mathbf{O}(G)$ and $G \notin \mathbf{O}(F)$.
    (b) Use this result to give an alternative proof of Theorem 5.
27. For functions $F, G : \mathbb{N} \to (0, \infty)$, F is said to be in $\theta(G)$ if

$$\lim_{x \to \infty} \frac{F(x)}{G(x)} = c$$

for some $c \in (0, \infty)$.

    (a) Show that if $F \in \theta(G)$, then $\mathbf{O}(G) = \mathbf{O}(F)$.
    (b) Use this result to give an alternative proof of Theorem 6.

(c) Find functions $F, G : \mathbb{N} \to [0, \infty)$ where $\mathbf{O}(G) = \mathbf{O}(F)$ but

$$\lim_{x \to \infty} \frac{F(x)}{G(x)}$$

does not exist.

28. There are many other natural ways to prove Theorem 7.

   (a) Prove Theorem 7 using the result of Exercise 12(a).
   (b) Prove Theorem 7 using the Taylor series for e^n.
   (c) Prove directly that for all large enough sets X, there are more subsets of X than there are k-tuples of elements of X. Use this to give a proof of Theorem 7.

29. Find functions $F, G : \mathbb{N} \to [0, \infty)$ where $F \in \mathbf{O}(G)$ and $G \notin \mathbf{O}(F)$ but

$$\lim_{x \to \infty} \frac{F(x)}{G(x)}$$

does not exist.

30. (a) Find two functions F and G such that $F \notin \mathbf{O}(G)$ and $G \notin \mathbf{O}(F)$. (*Hint:* For $F \notin \mathbf{O}(G)$, we need $F(n) > 1 \cdot G(n)$ infinitely often, $F(n) > 2 \cdot G(n)$ infinitely often, and so on. Similarly, for $G \notin \mathbf{O}(F)$, we need $G(n) > 1 \cdot F(n)$ infinitely often, $G(n) > 2 \cdot F(n)$ infinitely often, and so on.)
   (b) Find two functions, F and G, as in part (a), where in addition F and G are both strictly increasing.

31. Is it true that for all functions F, $\mathbf{O}(F) = \mathbf{O}(F + 1)$? Is it true for all increasing functions F?

5.3   Complexity of Programs

Recall the questions from the beginning of this chapter:

1. Since you can time a program on only a finite (usually small) number of input data sets, how can you (reasonably) estimate the time on other sets of input data?
2. Computer A may do one operation (say, swapping two integers) twice as fast as computer B. Computer B may do some other operation (say, comparing two real numbers) twice as fast as computer A. Can you (reasonably) compare the speeds of two algorithms without knowing on which machine each program will be run?
3. Can you design your algorithm so that your program will "scale well"—in other words, so that as the size of the data set increases, the time needed by the program will not increase too rapidly.

All these questions were used to motivate a comparison of the times taken by algorithms and programs "up to $\mathbf{O}$." In this section, we will show some basic techniques for computing that approximate time—without any actual timing.

Before continuing, we will expand on the first question so that you will understand some of the problems we need to resolve. Suppose you record how much time program P takes on machine M with data sets of size 10 ($T(10)$) and a data set of size 100

($T(100)$), and suppose you need to use this information to estimate how much time P will take on a data set of size 10,000. (That is, suppose you need to estimate $T(10,000)$). Can you do so?

One method is to find a function that passes through the given points and then evaluate this function at other points. We call such a function an **approximating curve.** A simple solution is the following: Plot the two points, $(10, T(10))$ and $(100, T(100))$ on a graph, draw a straight line through them, and extend the line over larger values of x. See where the straight line crosses the line $x = 10,000$, as shown in Figure 5.7. Use this value for $T(10,000)$.

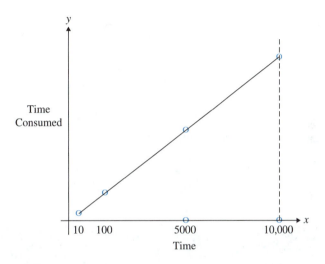

Figure 5.7   Linear extrapolation of time.

This approximation makes a totally unjustified assumption. For it to be reasonable to approximate the function T with a straight line, that time function must be a linear function (or be fairly close to one at all points), and there is no reason to assume that T is linear. You could improve the approximation a bit by measuring time for three sizes of data sets (say, 10, 100, and 1000) and choosing the parabola that passes through those three points as the approximating curve. However, that approximation would only be reasonable if you had reason to believe the function was either quadratic or linear. For example, for some computer programs, T will be an exponential function of the size of the data set. If that is the case, then no amount of curve fitting with polynomial functions would give reasonable grounds for extrapolation such as this.

The ideas presented in this section will provide tools for one to determine that the time consumption of an algorithm is approximately linear, approximately quadratic, approximately exponential, or whatever. After coming to such a general conclusion, you can (if you are very careful) legitimately use tools such as the curve-fitting procedure just described to approximate times.

We are not saying that actual timing statistics are unimportant. We know this can be a very important way to compare algorithms. All we are saying is that timing statistics are usually far from enough.

The purely theoretical approach introduced here makes an approximation of the complexity to an algorithm that ignores almost all differences among machines, languages, and details of how the algorithm is coded. If functions $T_1(n)$ and $T_2(n)$ measure the times taken by programs 1 and 2 on a data set of size n, then this approach asks only whether $T_1 \in \mathbf{O}(T_2)$ and whether $T_2 \in \mathbf{O}(T_1)$. This approach is said to measure only **time complexity,** not time in all its detail. Despite being highly theoretical, this approach has very frequently proved to be an excellent guide for practical decisions. For example, it allows one to conclude that a sorting algorithm that is in $\mathbf{O}(n^2)$ is clearly inferior to a sorting algorithm that is in $\mathbf{O}(n \ln(n))$. For the standard sorting algorithms, unless you are sorting a very short list (of fewer than eight elements or so) or a list that is already almost in order, an algorithm in $\mathbf{O}(n \ln(n))$ is decidedly preferable.

There are many variations on the way to measure complexity. Different approaches may work with different idealized models of what a computer is or of what is to be measured. What we do here is typical of most current discussions of complexity. At the end of this section, we will discuss briefly some common variants of the notion of algorithm complexity.

5.3.1   Counting Statements

To illustrate how to measure an algorithm's complexity, we start with some very simple algorithms. The following algorithm tests whether a list of three numbers is in increasing order:

Algorithm: Detecting Order for Three Values

INPUT: Values for three variables A, B, and C
OUTPUT: Message indicating whether $A \leq B \leq C$ is *TRUE*

if ($A \leq B$ and $B \leq C$) then
    print "List is in order."
else
    print "List is out of order."

Typically, you pick a single operation, or a small set of operations, called **key operations,** to analyze algorithms. You then count, for inputs of various size, how many times these key operations are performed when an algorithm is executed. In the algorithm just cited, we would count the number of comparisons made between elements of the list. In this case, two comparisons obviously are made, no matter what the input list is. However, counting statements executed just by looking at the program and its input often is not this easy. Therefore, we illustrate here some simple techniques for doing this.

One of the basic tenets of **structured programming** is that all programs should be built from just a few, simple, well-understood control structures. Although object-oriented programming has become a powerful new paradigm for program design, it remains necessary to use the fundamental ideas of structured programming in implementing algorithms. The three control structures used in structured programming are **sequence, selection,** and **repetition.**

Control Structures

Sequence: A list of steps $s_1 : s_2 : \ldots : s_k$ to be performed in the order given.
Selection: A choice of steps. In many programming languages, constructs such as *if-then*, *if-then-else*, *select*, and *case* provide the methods to make choices.
Repetition (also called **iteration** and **looping**): A block of code is executed repeatedly, either for a certain number of times or until some condition becomes true. In many programming languages, constructs such as *for*, *while*, *do-while*, and *repeat-until* provide for repetition. (Repetition can also be accomplished with recursion. The analysis is often similar, but we will not discuss recursion in this chapter.)

The remaining algorithms of this section, which demonstrate how to determine the complexity of code using these control structures, will be simple sorting algorithms. The routine *Swap(A, B)* will simply interchange the values of these two variables A and B so that the original value of A ends in B's location and the original value of B ends in A's location.

Algorithm: Ordering Three Values I

INPUT: Distinct alphabetic values for three variables A, B, and C
OUTPUT: The list of alphabetic values ordered by increasing value

```
if (A > B) then
    Swap(A, B)
if (B > C) then
    Swap(B, C)
if (A > B) then
    Swap(A, B)
print A, B, C
```

Assume that when the Ordering Three Values I algorithm starts, the three alphabetic values a, b, and c (ordered as $a < b < c$) are assigned to the variables A, B, and C *in some order*. At the end of the algorithm, the values should satisfy $A < B < C$. There are six possible input orders for the three letters a, b, and c. The reader should check that the algorithm correctly sorts all six possible initial conditions.

Each *if-then* statement makes a single comparison. Three *if-then* statements are executed in a sequence, so three comparisons are made.

Complexity of Sequence

If, for some $n \in \mathbb{N}$, an algorithm has blocks $B_1, B_2, \ldots, B_n$ in sequence and, on input I to the algorithm, block B_i executes the key operations at most $F_i(B_i)$ times for $1 \leq i \leq n$, then the entire algorithm executes the key operations at most

$$F_1(B_1) + F_2(B_2) + \cdots + F_n(B_n)$$

times.

If we choose comparison as the key operation in the Ordering Three Values I algorithm and the blocks are just the three *if* statements, then

$$F_1(B_1) = F_2(B_2) = F_3(B_3) = 1$$

since the process of *Swap* will not include any comparisons. The complexity of this algorithm with this key operation is just the constant value 3, and the algorithm is said to be **O**(1).

5.3.2   Two Algorithms Illustrating Selection

The next algorithm is like the previous one, except that the writer observed that if one first tests whether $A > B$ or $B > C$, then one can use only two tests, not three, for that case.

Algorithm: Ordering Three Values II

INPUT: Distinct values for three variables A, B, and C
OUTPUT: The list of letters ordered by increasing value

```
if (A > B or B > C) then   /* list is out of order, so */
     if (A > B) then Swap(A, B)
     if (B > C) then Swap(B, C)
     if (A > B) then Swap(A, B)
print A, B, C
```

In the Ordering Three Values II algorithm, only two comparisons are made if the list is in order. On the other hand, if the list is out of order, the algorithm makes two comparisons to identify that fact plus a sequence of three more comparisons to sort the values, for a total of five comparisons at worst on any input data. The complexity of this algorithm with comparison as the key operation is the constant value 5. This algorithm also is a member of **O**(1).

Complexity of Selection

Let $n \in \mathbb{N}$. Let a step of an algorithm be to make a selection for execution of one of n blocks, $B_1, B_2, \ldots, B_n$, and, on input I, where each B_i executes the key operations at most $F_i(B_i)$ times for $1 \leq i \leq n$ and it then takes $G(I)$ executions of the key operations to make the selection of the block to execute. Then, the key operations are executed at most

$$G(I) + \max\{F_1(B_1), F_2(B_2), \ldots, F_n(B_n)\}$$

times.

The next algorithm also sorts a three-element list. To some people, it is more obvious than the preceding algorithm, since it breaks the algorithm down into the six possible cases, asking first which element is least. For easy reference to individual lines of the algorithm, we put line numbers on the left ends of the lines. We assume the procedure *Swap* interchanges the values of the two elements that are in the storage locations named by its arguments.

Algorithm: Ordering Three Values III

INPUT: Distinct values for three variables A, B, and C
OUTPUT: The list of values in increasing order

```
 1. if (C < A and C < B) then
 2.      if (A < B) then              /* C < A < B, so */
 3.           Swap(A, B)
 4.           Swap(A, C)
 5.      else                         /* C < B ≤ A, so */
 6.           Swap (C, A)
 7. else
 8.      if (B < A and B < C) then
 9.           if (A < C) then         /* B < A < C, so */
10.                Swap (A, B)
11.           else                    /* B < C ≤ A, so */
12.                Swap (A, B)
13.                Swap (B, C)
14.      else
15.           if (C < B) then         /* A ≤ C < B, so */
16.                Swap (B, C)
17. print A, B, C
```

This algorithm nests selections within selections within selections. The outermost selection is in line 1, with its corresponding false range beginning at line 6. We will concentrate on the selection at line 1. The condition at line 1 requires two comparisons. The test in line 1 has two "selection blocks," lines 2 through 6 and lines 8 through 16, one of which is then executed.

1. If the test in line 1 ($C < A$ and $C < B$) returns the value true, one more comparison (line 2) is made.
2. Otherwise, the algorithm next tests ($B < A$ and $B < C$) given in line 8, which requires two more comparisons. Then, whether this condition is true or false, the code makes exactly one more comparison (line 9 or line 15). This gives a total of three comparisons if the test of line 1 fails.

Now, translate this information into the notation of complexity. The notation says that $F_1(B_1)$ is the number of comparisons needed to execute lines 2 through 6, the "true range" of the test in line 1. Let $F_2(B_2)$ be the number of comparisons needed to execute lines 8 through 16, the "false range" of that test. $G(I)$ is the number of comparisons needed to execute the test in line 1 itself, which is 2 for all I. As we noted above, $F_1(B_1) = 1$, and $F_2(B_2) = 3$. So,

$$G(I) + max\{F_1(B_1),\ F_2(B_2)\} = 2 + 3 = 5$$

The entire algorithm makes at most five comparisons on any input data. (This is the same upper bound as the worst case for the Ordering Three Values II algorithm.) Again, this algorithm is $O(1)$.

Suppose, however, the procedure *Swap* were chosen as the key operation instead of the comparison operation. The third algorithm makes at worst two calls to *Swap*, whereas the other two algorithms will actually make three calls to *Swap* if the list is exactly out of order. This illustrates a potential pitfall of the method described here. If one chooses the key operations badly, then the results can be misleading. The examples above illustrate a potential trade-off. One sorting algorithm may require fewer comparisons but more calls to *Swap* than another may require. Which algorithm is better may depend on the data being sorted and the computers being used.

In the case of sorting algorithms, the key operations chosen are usually either the operation of interchanging (swapping) two elements in memory or determining which of two elements is larger (comparing), since those operations tend to determine the time consumption of most sorting algorithms. If both comparisons and interchanges are chosen as key operations, then the analysis is quite realistic for most sorting algorithms.

5.3.3   An Algorithm Illustrating Repetition

One would never expect to write an algorithm to sort a three-element list. If you want to sort an n-element list where n is at least 5, it is impractical to write an algorithm like the ones previously discussed, since the algorithm gets too long. Nevertheless, the second algorithm we present is easily generalized into an algorithm called *BubbleSort*, probably the easiest of all sorting algorithms. Ease of coding is its primary virtue. It is also one of the slowest sorting algorithms, although if "optimized," it can run quickly on certain very special data sets.

The command

$$\text{for } i = 1 \text{ to } N$$
$$S$$

is a command to repeat operation S over and over, once with $i = 1$, once with $i = 2, \ldots$, and once with $i = N$. If $N < 1$, then operation S is not executed at all. (See Section 1.8.1 for further discussion of this command.)

Algorithm: BubbleSort I

INPUT: A list of N distinct integers $List[1], \ldots, List[N]$
OUTPUT: The same integers rearranged so that the values of
$List[1], \ldots, List[N]$ are in increasing order

BubbleSort-I(List, N)
1. for *Pass* = 1 to $N - 1$ do
2.   for *Position* = 1 to $N - 1$ do
3.   if ($List[Position] > List[Position + 1]$) then
4.   Swap ($List[Position]$, $List[Position + 1]$)

The BubbleSort I algorithm contains two nested loops. We will first consider the inner loop, or lines 2 through 4. This loop compares adjacent elements in the list, and whenever it finds a pair that is out of order, it interchanges them. In the following example of a four-element list, there are three adjacent pairs, so the algorithm interchanges at most three times. We trace the results of lines 2 through 4 the first time they are executed—that is, when *Pass* = 1. We assume that $a < b < c < d$ is the final order.

                Pass = 1

c	b	d	a	Initial list
<u>c</u>	<u>b</u>	d	a	Compare first pair
<u>b</u>	<u>c</u>	d	a	Out of order, so swap
b	<u>c</u>	<u>d</u>	a	Compare second pair
b	c	<u>d</u>	a	In order; don't swap
b	c	<u>d</u>	<u>a</u>	Compare last pair
b	c	<u>a</u>	<u>d</u>	Out of order, so swap
b	c	a	d	The final list for *Pass* = 1

The *for* statement causes lines 3 and 4 of the program, called the *body of the loop*, to be executed $N - 1$ times, the first time with the value of *Position* equal to 1, the second time with the value of *Position* equal to 2, and so on. Each time the loop is executed, the algorithm executes the key operation of comparison one time. Hence, the algorithm makes $N - 1$ comparisons each time the inner loop is executed. The outer loop is executed $N - 1$ times.

It is important to note that testing whether to end the loop may itself involve a key operation. In this case (depending on the computer language and what looping command is used), if the loop is executed N times, the test is usually performed $N + 1$ times, once before each time the loop is executed and once again at the end. The result of that final test shows that it is not necessary to execute the loop another time.

Complexity of Repetition

If an algorithm contains a loop B, where:

- On input I, the loop is executed at most n times for some $n \in \mathbb{N}$, and
- On input I, for $1 \leq i \leq n$, the ith execution of the body of the loop executes the key operations at most $F_i(B_i)$ times, and
- The test of whether to stop or to execute the loop another time executes the key operations at most c times,

then the number of times the entire algorithm executes the key operations during the loop is at most

$$\sum_{i=1}^{n} (c + F_i(B_i)) + c$$

Normally, when the loop controls the number of times the body of the loop is executed, the value of c is zero, because incrementing and testing a counter is usually not a key operation. We can now compute the complexity of the BubbleSort I algorithm, with comparison of the elements of the list as the key operation. Note that the termination condition here does not involve a key operation—it compares indices, not array elements—so $c = 0$.

Analysis of the Inner Loop.   The block B_i consists of lines 3 and 4. Hence, the number of comparisons is always 1; that is, $F_i(B_i) = 1$ for $i = 1, \ldots, N - 1$. So, by the previous formula, the total number of comparisons is

$$\# \text{ comparisons} \leq \sum_{Position=1}^{N-1} 1 = N - 1$$

Analysis of the Outer Loop.   Here each block B_i of the outer loop is just the (whole) inner loop. So, $F_i(B_i) = N - 1$. The outer loop is executed $N - 1$ times. Hence, the total number of comparisons is

$$\# \text{ comparisons} \leq \sum_{Pass=1}^{N-1} (N - 1) = (N - 1)^2$$

Thus, the number of comparisons made by the BubbleSort I algorithm on an input list of size N is

$$\mathbf{O}((N - 1)^2) = \mathbf{O}(N^2 - 2N + 1) = \mathbf{O}(N^2)$$

by Theorem 6 of Section 5.1.3.

5.3.4 An Algorithm Illustrating Nested Repetition

Now look at the entire BubbleSort I algorithm. It is easy to see that after the first pass, the largest element has reached the end of the list. This effect appears to be the origin of the name "bubble sort"—the larger elements seem to bubble down through the list. After the second pass, the second largest element is in place, and so on. After pass $N - 1$, the second smallest element has bubbled down to position 2, and the smallest element is left in position 1. So, in particular, at most $N - 1$ passes are needed.

BubbleSort I can be improved. Since the largest element bubbles to $List[N]$ in the first pass, there is no reason to look at the last position again. On the second pass, the second largest element bubbles to $List[N - 1]$, and that position need not be examined again, and so on.

Algorithm: BubbleSort II (slightly optimized)

INPUT: A list of N distinct integers $List[1], \ldots, List[N]$
OUTPUT: $List[1], \ldots, List[N]$, with values in increasing order

$BubbleSort\text{-}II(List, N)$
1. for $Pass = 1$ to $N - 1$ do
2. $Limit = N - Pass$
3.   for $Position = 1$ to $Limit$ do
4.   if $(List[Position] > List[Position + 1])$
5. $Swap(List[Position], List[Position + 1])$

In BubbleSort II, the first pass through the loop in lines 3 through 5 still performs $N - 1$ comparisons. In the second pass, however, the size of the list is reduced by 1, so the second pass performs $N - 2$ comparisons. The third pass reduces the number of elements considered again by one and performs $N - 3$ comparisons. Here, the function $F_i(B_i) = i - 1$. So, the total number of comparisons is equal to

$$\sum_{i=1}^{N-1} i = 1 + 2 + 3 + \cdots + (N - 2) + (N - 1) = N(N - 1)/2$$

Example 1. For the slightly optimized BubbleSort II algorithm, count the number of $Swap$'s.

(a) If the input list is out of order in the way shown here:

$$List[1] \geq List[2] \geq List[3] \geq \cdots \geq List[N - 1] \geq List[N]$$

the algorithm makes exactly $N \cdot (N - 1)/2$ calls to $Swap$.

(b) If the input list is arranged so that

$$List[1] \leq List[2] \leq List[3] \leq \cdots \leq List[N - 1] \leq List[N],$$

the algorithm makes no calls to $Swap$.

Proof.   This proof is left as an exercise for the reader.   ■

One might ask how much more efficient *BubbleSort* can be made. Further improvements are possible. For example, if for an entire pass no calls to *Swap* have been made, then the list is in order, and *BubbleSort* may stop. This modification makes some varieties of *BubbleSort* very efficient for lists that are nearly in order before the sorting starts. However, *BubbleSort* is inherently slow for arbitrary lists. More complicated analyses show that:

- If a list of N elements is ordered as in Example 1(a), *any BubbleSort* must make at least $N \cdot (N - 1)/2$ calls to *Swap*. So, although such algorithms run faster on "nice" input, they run no faster on the "worst" input.
- If one considers all possible original orderings of a list of N distinct objects, then *on average, any BubbleSort* must make $N \cdot (N - 1)/4$ calls to *Swap*, so the average is no better than half the worst-case time.

We discussed earlier one standard way to count how many key operations are performed. For the moment, the reader just needs to be aware that this is not the only way. Note that in our discussions of the complexity of selection (see Section 5.3.2) and of repetition (see Section 5.3.3), we concluded only that the number of steps taken by the entire algorithm was at most something. Some other argument might tell us that the number is, indeed, far less. Return to the discussion of repetition: Sometimes, it turns out that although in the ith time through a loop we might execute a large number $F_i(B_i)$ of key operations, most of the times through the loop, on any given input data set, we execute far fewer than $F_i(B_i)$ key operations. In this case, the total time taken by the repetition may be far lower than the bound that we calculated. We will not discuss this issue further at this point.

5.3.5   Time Complexity of an Algorithm

After the preliminaries that show us how functions correspond to a way of measuring the number of key operations executed in an algorithm, we return to the original goal of measuring the time complexity of an algorithm. We will again use *BubbleSort* as an example— both because it's easy to analyze and because the reader will be able to understand the analysis without having to understand more complex code. The goal is to measure time consumption not as a function of the particular input but, rather, as a function of the size of the input.

Time complexity is measured by the number of times that the key operations are performed, but how is the size of the input measured? For sorting algorithms, and for many other algorithms, size is usually taken to be the number of input data—that is, the number of data to be sorted.

For a fixed input size n and a fixed set of key operations, we need to consider all possible input sets of that size. We must count the number of times that key operations are executed by the algorithm on each possible set of size n. If there is a maximum possible (largest) number of steps, then that number is called the **worst-case behavior** of the algorithm on input size N. If for each input size N there is a maximum possible number *worst(n)* of steps, then the function *worst* is called the **worst-case complexity** of the algorithm. For example, the partially optimized *BubbleSort* presented earlier makes $N \cdot (N - 1)/2$ comparisons and at worst $N \cdot (N - 1)/2$ calls to *Swap*.

It is most common to look only at the growth rate—up to $\mathbf{O}(*)$—of the worst-case complexity function. For example, choose as key operations comparisons of list elements, swaps of list elements, or both. By any of these measures, *BubbleSort* is worst-case $\mathbf{O}(N^2)$, and no matter how *BubbleSort* is improved, it still is worst-case $\mathbf{O}(N^2)$. Other sorting algorithms are worst-case $\mathbf{O}(N \cdot \ln(N))$—that is, algorithms that are, usually, much more efficient. The discussion of decision trees in Section 6.12.4 of Chapter 6 deals with a lower bound for a sorting algorithm of the same sort as *BubbleSort*. (We will clarify then what "of the same sort" means.)

Polynomial Time Algorithms

Any form of bubble sort algorithm, though intrinsically slow, can still be used for reasonably large data sets. A general question is to characterize what calculations are feasible. Of course, that is really just a matter of how the word *feasible* is defined, but there is a significant underlying intuition. Originally asked by theoreticians, the question also has practical significance. For example, a basic goal of much modern research in cryptology, the study of encrypting and decrypting messages, regards finding methods to encrypt messages where breaking the code is, in principle, infeasible.

Definition 1. An algorithm A is **polynomial time** if it is worst-case $\mathbf{O}(n^k)$ for some integer k. That is, the function measuring its worst-case time complexity is less than or equal to some polynomial function.

A **decision problem** is a problem whose answer is simply *TRUE* or *FALSE*.[1] The set of all decision problems answered by polynomial-time algorithms is often referred to as $\mathcal{P}$. Sometimes, the notation $\mathcal{P}$ is also used to describe the set of polynomial-time algorithms.

The set of polynomial-time algorithms has been exceedingly important in studies of computer algorithms. The interpretation given is that if an algorithm is polynomial time, then it is conceivable that people might someday build a computer on which it can be run in a reasonable amount of time for a relatively large input set. If an algorithm has a worst-case time consumption greater than any polynomial, it is not.[2]

Nondeterministic Polynomial Problems

The rest of this subsection develops material presented in Section 2.5.6, where we first mentioned polynomial-time algorithms and the $\mathcal{P} \neq \mathcal{NP}$ conjecture. Readers who have skipped that material may want to read that section.

Example 2. There is an $\mathbf{O}(n^2)$ algorithm such that:

Given: A propositional formula ϕ, such as,

$$\phi = (a \vee b \vee c \vee d \vee \neg b) \wedge (c \vee d \vee \neg c \vee e \vee f)$$

---

[1] Many computer programs solve decision problems. Programmers write functions that return boolean values to aid in similar programming circumstances. For example, *while(NOT* IsValidInput(*InputData))* is true when *IsValidInput* returns a value of *FALSE*. Sometimes, the purpose of an entire program is to compute a single boolean result, such as "Can I buy this house with house payments of less than $1000 per month?" or "Did any shop in my department have a below-average productivity during the month of December?" or "Can a salesman visit all these cities and have to drive less than 1000 miles?"

[2] If quantum computers become practically feasible, this view might change.

and a truth assignment I to its variables—typically, here, one lists the true literals, such as $a, \neg b, c, \neg d, e,$ and $\neg f$.

Returns: Truth value $I(\phi)$.

Solution.   See Exercise 5 in Section 5.4, where a proof is sketched for ϕ in conjunctive normal form (CNF). (A proof for more general formulas ϕ would involve a digression into "parsing": Given a formula, how can we construct its parse tree?)   ■

When people formally measure the complexity of such algorithms, they are likely to use variant definitions of complexity, as in variants 1 and 2 in the next section.

In Section 2.5.6, we noted that it is generally assumed—but not proved—that no fast algorithm exists that, given a propositional formula ϕ, can decide whether ϕ is **satisfiable.** This latter problem is called the **propositional satisfiability** problem, which is sometimes referred to simply as SAT.

What we meant in Section 2.5.6 by "fast" was "polynomial time." It is commonly assumed that there is no polynomial-time algorithm to settle propositional satisfiability. However, nobody can prove it! This is often considered to be the most significant open problem in theoretical computer science—and one of the most significant open problems in all of mathematics.

Propositional satisfiability is the problem of the paradigm **nondeterministic polynomial time ($\mathcal{NP}$).** We use satisfiability to motivate an almost-precise definition of $\mathcal{NP}$. We do not give the most common definition, which is in terms of something called "nondeterministic algorithms." The definition we give, however, would be equivalent if that one small bit of precision were added.

A problem is said to be in $\mathcal{NP}$, or nondeterministic polynomial time, if there is a polynomial time **verifier algorithm** V for the problem. In other words, V is an algorithm that takes any input σ and a purported proof that the answer for σ should be *TRUE* and, in polynomial time, verifies that the proof is correct. (Such an algorithm is sometimes called a "guess-and-check algorithm": First, it guesses C, and then it checks the result. Here, C stands for **certificate.**) The set of all $\mathcal{NP}$ problems is called, simply, $\mathcal{NP}$.

Note that in the definition of $\mathcal{NP}$, it makes no difference whether there are any, or many, k-ary relations C where $A(\sigma, C) = \mathit{FALSE}$. It matters only whether there exists one C for which $A(\sigma, C) = \mathit{TRUE}$.

An example of a nondeterministic polynomial time problem is to determine if a CNF for which each term consists of exactly three literals is satisfiable. The problem can be solved for a given interpretation for each of the literals by examining each of the n clauses in some fixed order and determining whether each clause is *TRUE* or *FALSE* in this interpretation. The CNF is satisfiable if each of its clauses is *TRUE* in the given interpretation. The reader can devise a polynomial time algorithm for determining whether a clause with three literals is *TRUE* or *FALSE* in a given interpretation. After repeating this process at most n times, once for each clause, the answer will be found in polynomial time.

The definition above is not the traditional definition of the term *nondeterministic polynomial problem,* but it is equivalent.

Every decision problem in $\mathcal{P}$ is also, trivially, in $\mathcal{NP}$. (Why?) The famous $\mathcal{P} \neq \mathcal{NP}$ conjecture is the statement that this is true—that is, that not every problem in $\mathcal{NP}$ is also in $\mathcal{P}$.

Stephen Cook proved that if propositional satisfiability is in $\mathcal{P}$, then $\mathcal{NP} \subseteq \mathcal{P}$. In fact, Cook proved a stronger result—that satisfiability is $\mathcal{NP}$-*complete*—but we shall not even define that notion here.

5.3.6   Variants on the Definition of Complexity

This material is included primarily to make you aware that there are many variants on the complexity analysis we have done. We will list several variants here. Some give finer measures for circumstances when the simplifying assumptions of our previous model are too extreme. Others measure other important properties of algorithms.

Variant 1: Ways to Measure Input Data Size

We measured the size of the input by counting the number of input values. So, a very long number counts the same as a very short number. For sorting algorithms, and for many other algorithms, this is a reasonable simplifying assumption. Even so, for many algorithms, it is highly unrealistic to count a one-digit number the same as a 1000-digit number. A common measure is the number of characters it takes to represent the data. So, for numbers, one can count digits—base 2 or base 10. For character strings, one can count the number of characters.

For example, code-breaking programs may take only a single integer as data, but the code breaking takes longer and longer as the numbers get larger. There may be no theoretical maximum at all on the number of key operations performed. There is only a finite number of n-digit numbers, however, so as long as the program always terminates, there is a maximum on the number of key operations that can be performed on any input from that finite set.

With this measurement, counting just the number of input data, as we did for sorting algorithms, would be viewed only as a useful approximation of the number of input characters.

Yet another measure of the size of the input is simply the numbers themselves.

Variant 2: Choices of Key Operations

We have already noted the importance of choosing realistic key operations. We assumed that each basic operation, such as comparison, addition, subtraction, multiplication, or division, takes $\mathbf{O}(1)$ time. For most programs, this is a reasonable assumption. However, suppose we had to multiply arbitrarily large integers accurately. Ordinary long multiplication of two six-digit numbers takes 36 multiplications of one-digit numbers, as shown in Figure 5.8.

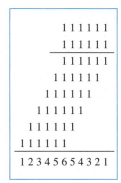

Figure 5.8   Multiplying six-digit numbers.

Consider the problem of performing multiplication with arbitrarily large integers. Ordinary long multiplication of two 100-digit numbers takes 10,000 single-digit multiplications. Sometimes, it is far more reasonable to count operations on single-digit numbers as the key operations. Of course, when operations on single-digit numbers are taken as key operations, the size of the input is generally measured in terms of the number of digits.

Some very important issues in modern cryptology hinge on the problem of factoring numbers into primes. Breaking some modern codes involves factoring very large numbers (say, 400-digit numbers into primes), and that is presumed to take too long to be feasible. (If computers get substantially faster, one just increases the number of digits a bit.) The standard translation of "too long to be feasible" is that there is no polynomial-time algorithm for which the size of the input is measured as the number of digits and the key operations are operations on individual digits.

Variant 3: Average Time Complexity

Often, worst-case complexity is not particularly important; what is more important is how long it takes to run the algorithm on average. (For example, codes that are worst-case difficult to break but, on average, are very easy to break might not be of much use.) We will not deal with average time complexity in this book. Average time complexity is harder to compute than worst-case complexity. Also, problems exist in deciding how to define the idea. For example, many analyses assume that all data sets of size n are equally likely to be input, which may be an unrealistic assumption.

Variant 4: Space Complexity

Another measure of the resources that a program requires is how much computer memory the program takes. In this case, one can study analogues to all the variants discussed previously. One usually measures the space needed by *excluding the space needed to represent input and output*.

Variant 5: Parallel and Distributed Algorithms

Another area of algorithm research concerns parallel and networked computation. A computer may have many separate processors linked together or several computers scattered across a network with huge network-traffic capacity, and work can be divided among the various processors so that each execute their part simultaneously. If there are two processors, then the speed of doing a job can usually be decreased, though by no more than a factor of approximately two—and usually noticeably less. Some current research centers use a very large number of processors, as many as the programmers can figure out how to use for the individual data. So, for sorting 1000 data, the model might allow 1000 separate processors.

All the calculations presented in this book use a sequential model for computation. The time needed on machines using parallel models for computation will normally be substantially less. Analysis of such algorithms, though similar to what we have presented, must account for simultaneous computations. Currently, there are many architectures for parallel computers and many ways of measuring complexity on such machines, depending on such things as whether all processors share a common memory or each has its own memory. If we have only a fixed number k of processors, then previous complexity bounds are changed by at most $1/k$. If we allow the number of processors to grow as the size of the data grows, however, then problem complexity is sometimes reduced significantly.

At the time of this writing, development of quantum computers, which would depend on quantum-mechanical principles to implement massive amounts of parallelism, look to be conceivable. For such computers, complexity analysis might also change.

5.4   Exercises

1. Calculate how many times statement S is executed in each block of code. Simplify all your answers.

(a) $I = 1$
   while $I \leq N$
      S
      $I = 2 \cdot I$

(b) $Counter1 = 1$
   while $Counter1 \leq N - 1$
      $Counter2 = 1$
      while $Counter2 \leq N - 1$
         S
         $Counter2 = Counter2 + 1$
      $Counter1 = Counter1 + 1$

(c) $Counter1 = 1$
   while $Counter1 < X$
      $Counter2 = 1$
      while $Counter2 \leq X$
         S
         $Counter2 = Counter2 + 2$
      $Counter1 = Counter1 + 1$

(d) $Counter1 = 1$
   while $Counter1 \leq N$
      $Counter2 = 1$
      while $Counter2 \leq N$
         S
         $Counter2 = Counter2 + 1$
      $Counter3 = 1$
      while $Counter3 \leq N$
         $Counter4 = 1$
         while $Counter4 \leq N$
            S
            $Counter4 = Counter4 + 1$
         $Counter3 = Counter3 + 1$
      $Counter1 = Counter1 + 1$

(e) *Counter1* = 1
    while *Counter1* ≤ *X*
        Counter2 = 1
        while *Counter2* ≤ *X*
            S
            Counter2 = 2 ∗ *Counter2*
        Counter1 = *Counter1* + 1

(f) *Counter1* = 1
    while *Counter1* ≤ *N* − 1
        Counter2 = 1
        while *Counter2* ≤ *Counter1*
            S
            Counter2 = *Counter2* + 1
        Counter1 = *Counter1* + 1

2. Confirm the claims of parts (a) and (b) of Example 1.
3. Show that the following SelectionSort algorithm, on any input list of size n, makes exactly $n(n-1)/2$ comparisons and exactly $n-1$ calls to *Swap*.

Algorithm: SelectionSort

INPUT: A list of distinct values $List[1], \ldots, List[N]$
OUTPUT: $List[1], \ldots, List[N]$ with values in increasing order

SelectionSort(List, N)
    for *Position* = 1 to *N* − 1 do
        Small = *List*[*Position*]
        Place = *Position*
        for *I* = *Position* + 1 to *N* do
            if (*Small* > *List*[*I*]) then
                Small = *List*[*I*]
                Place = *I*
        Swap(*List*[*Position*], *List*[*Place*])

---

4. Suppose you record the run times of program P. You find out that on inputs of length 1, its run time is always 1/8 second, and on inputs of length 10, its run time is always 1/4 second. Under each of the following assumptions, calculate how long it will take to run the program on inputs of length 20, 100, 1000, and 10,000:

   (a) For some constants a and b (the values of which you must determine), running the program on a data set of size d always takes $ad + b$ seconds.
   (b) For some constants a and b, running the program on a data set of size d always takes $ad^2 + b$ seconds.

(c) For some constants a and b, running the program on a data set of size d always takes $a \cdot 2^{bd}$ seconds.

5. Write an algorithm in pseudocode that follows the description in Example 2 for the special case where the formula ϕ is in CNF. For the sake of simplicity, assume that each proposition letter is a lowercase letter $(a\text{–}z)$, but try to ignore the fact that there are only 26 such letters. Also, assume that each of the symbols $(\wedge, \vee, \neg)$ is also just one character, some ASCII character other than a through z. (*Hint:* Suppose you are given the formula from that example,

$$\phi = (a \vee b \vee c \vee d \vee \neg b) \wedge (c \vee d \vee \neg c \vee e \vee f)$$

and the example truth assignment of *TRUE* to a, $\neg b$, c, $\neg d$, e, and $\neg f$. Your program should scan through ϕ from left to right, keeping track of (i) the value *so far* of the current clause—the disjunction of all the literals seen so far in that clause—and (ii) the value *so far* of the entire formula—the conjunction of the values of all the clauses completed so far.)

   Since this can be done with just one scan through the formula, you may be tempted to think the algorithm is linear time, but it is not. Why? For formulas not in CNF, the algorithm is also $\mathbf{O}(n^2)$ but more complicated: One must first *parse* the formula to get an expression tree and then keep track of truth values on the tree. The student may have seen such problems in a data structures or algorithms course.

6. For each of the problems (a)–(d) below;

   (i) Write an algorithm in pseudocode to solve the problem (be sure your algorithm works correctly if $m = 0$ or $n = 0$; it should not make any assignments to elements of the array), and

   (ii) Calculate how many assignment statements and how many comparisons the algorithm causes to be executed as a function of m and n. In this case, count assignments to and comparisons of *index variables,* as well as assignments to and comparisons of positions in the array. Simplify your answers.

   (a) Initialize all the elements of an $m \times n$ array to 0.
   (b) Initialize all the elements of an $m \times n$ array that lie on or above the diagonal to 0. (Here, by "diagonal" we mean positions $[r, c]$ where $r = c$.)
   (c) Initialize all the elements of an $m \times n$ array that lie above the diagonal to 0.
   (d) Initialize all the elements on the diagonal of an $m \times n$ array on the diagonal to 0.

7. It is tempting to compute the complexity of an algorithm by counting statements, as we did with the *BubbleSort* example, but only keeping track of the number of steps along the way up to $\mathbf{O}(*)$. This turns out not to work with loops. For example, it is possible that each time through the loop, the number of statements executed is in $\mathbf{O}(1)$, but that the number of statements executed by a loop of length n is not in $\mathbf{O}(n)$. Find an example. (*Hint:* Each time through the loop, the number of statements executed may be in $\mathbf{O}(1)$, but the constants c and N_0 may change).

8. Compare the graphs of $F_1(x) = 3 \ln(x + 1)$, $F_2(x) = 2x$, $F_3(x) = x^2$, $F_4(x) = x^3$, $F_5(x) = 2^{x-1}$, and $F_6(x) = 3^{x-1}$. What does this suggest about the usefulness of nonpolynomial-time algorithms?

9. Is it reasonable to consider all polynomial-time algorithms to be practical? Why, or why not?

10. Suppose somebody manages to prove that the time taken by some frequently used algorithm is in $\mathbf{O}(n^{n^n})$. Why is this probably uninteresting information?

11. What is the output of each of the following blocks of code?

    (a)

    ```
    cntr = 0
    for i = 0 to N
        cntr = cntr + 1
    print cntr
    ```

    (b)

    ```
    cntr = 0
    for i = 0 to N
        for j = 0 to N
            cntr = cntr + 1
    print cntr
    ```

    (c)

    ```
    cntr = 0
    for i = 0 to N
        for j = i to N
            for k = j to N
                cntr = cntr + 1
    print cntr
    ```

5.5 Uncomputability

In the first half of the twentieth century, a group of mathematicians set out to formalize the notion of a function being computable by an algorithm. Of particular concern was identifying which functions, in principle, are computable and which are not. For this research, the idea of an algorithm was not what could be done on a computer. In fact, much of this work was done before the first modern computer was built! Rather, the idea of an algorithm was what a person could do with paper and pencil by following rules in a mechanical fashion. The research was motivated in part by mathematicians who conjectured that maybe no algorithms exist to solve certain problems. If one wants to argue that no algorithm exists, then one must have some formal way to talk about the set of all algorithms. Standard examples of algorithms included operations such as multiplication and long division.

Several formalisms were proposed. Each arguably captures the intuition of the term **algorithm.** Perhaps the most important point is that although, on the surface, they present very different notions of what an algorithm *is*, it has been proven that all the major formalisms define exactly the same class of functions that are computable by algorithms. The widely accepted philosophical statement that these models do, indeed, exactly capture the intuition of the expression "computable by an algorithm" is called *Church's thesis.* We shall not discuss these models here. Instead, we shall present an informal definition that we expect to be more accessible. (Of course, it is certainly also very revealing to examine the classical models, and that is normally expected in any formal treatment of computability theory.)

One basic assumption of "in principle computable" is that there should be no limitations on resources. There are no time limits: The computer just keeps calculating until it is done. There are no memory restrictions: Whenever the computer needs more memory, it asks for it and gets it. In most current languages, by adding the properties of unlimited time and memory resources, one can implement arithmetic procedures on arbitrarily large integers and do string manipulation functions on arbitrarily long (but finite) character strings. To keep things simple, however, we shall merely assume that our language allows arbitrarily long character strings. Other than that, we use our algorithm notation; the same constructions could be carried out in almost any common modern programming language.

The idea of a function subprogram, although we have used it informally before, needs to be made explicit for this discussion. In particular, we need to formalize the idea of a **return value**—a value returned by the function and then used by the rest of the program. A function subprogram is a string of characters of the form *functionName*($para_1$, $para_2$, ..., $para_N$) where N can be any integer. The terms $para_1$, $para_2$, ..., $para_N$ are valid variable or data types in the program that are made available to the code of the function subprogram. We call the statement *functionName*($para_1$, $para_2$, ..., $para_N$) a **function call.** The result of executing the function call is the "return" of a single value to the program that is then substituted for the function call. The returned value may be a numeric value, a boolean value, or a character string. At some point of the program, it should be made clear what kind of value is being returned by a function call. How that is done, however, is not necessary for us to consider during this discussion.

At some level, humans ordinarily interact with computers through strings of characters. We shall simplify the context by assuming that each program takes just one character string as input. That is, all its input is concatenated together into one (usually very long) string. Similarly, assume that the program either has no output (if it never halts) or outputs a single (perhaps very long) character string. Furthermore, consider a program itself to be a single, long character string—again simply a concatenation of all the lines of code of the algorithm.

Definition 1. (Rather arbitrary, and adopted for convenience.) An **algorithm** A is a procedure written in some programming language that is passed a single character string *In,* which is referred to as the algorithm's input, and returns a character string *Out,* which is referred to as the algorithm's return value (and which we may also think of as an output). The string *Out* is also called $A(In)$. On some (or all) inputs, an algorithm may never halt; thus, it may return nothing. If it halts but has no apparent output, as when the algorithm stops because of a run-time error, we shall think of it as returning the empty string.

Example 1.
```
Foo(word1)
    if word1 < "D"
        return word1
    else
        return "Wrong starting letter"
```

is an algorithm that takes any string as input and determines whether the word begins with one of the letters "A", "B", or "C". (For simplicity, we assume the only other possibilities are D–Z.) If the word satisfies the condition, the word itself is returned. If the word does not satisfy the condition, the string "Wrong starting letter" is returned.

We stipulate that the machine on which algorithm A is run can store arbitrarily long strings. Presumably, then, it can also store arbitrarily large integers, arbitrarily large arrays, and arbitrarily long lists, but allowing just arbitrarily long strings suffices for this discussion.

If A is not a legal algorithm, or if running A on input In causes a run-time error (such as treating the letter s as a digit), our convention is that A halts on input In and returns the empty string.

5.5.1   The Halting Problem

Recall the term *decision problem* from Section 5.3.5.

Definition 2.   A set D of strings is **decidable** if there is a **decision algorithm** A that, for an arbitrary string d, returns $A(d) = TRUE$ if $d \in D$ and $A(d) = FALSE$ if $d \notin D$. A set D that is not decidable is called **undecidable.**

By definition, a decision algorithm cannot have run-time errors. A decision algorithm must always halt. It cannot run on forever (for example, in an infinite loop).

Given the expectations many people have for the power of the computer, it is perhaps surprising to find basic problems for which no algorithms exist. The standard example, credited to Alan Turing (1912–1954, b. England), is:

The Halting Problem

Given an algorithm A and an input In, will A ultimately halt on In, or will it run on forever?

Here, A need not be a decision algorithm. It may stop and return the empty string. Thus, it is assumed that A is an algorithm for a partial function, and the problem is to determine whether $A(In)$ is defined or undefined.

Now, we simply recast our discussion in terms of making a decision about a set of strings. To do this, we use a string—say, %%—that we assume does not occur elsewhere in the programming language. We just use it as a separator, so that a program can easily extract, from a concatenated string $s_1\%\%s_2$, the two separate strings s_1 and s_2.

Definition 3.   The **halting problem** is the set *Halt* of all strings $A\%\%In$ where A is an algorithm and A halts on input In.

The reader may think of it as being odd to input an algorithm as data to a program. However, since an algorithm is a string, it is perfectly legal to do so. Even so, using an algorithm as data is, indeed, common. For example, a compiler is a program that takes in one program (in a "high-level" language) as input and returns a translated program (in a "low-level" language). Also, a computer itself may take two strings of data (a machine-language program and a data file), and run the program on the data file. Here, it is possible that A halts on input In and returns the empty string. In this case, the string $A\%\%In \in Halt$.

An algorithm H solves the halting problem if it is a decision algorithm for the set $D = Halt$. In other words, H is an algorithm with the following properties:

1. H is a legal algorithm with no run-time errors, since it cannot return the empty string.
2. H halts on all inputs, since it must return $TRUE$ or $FALSE$ on every input.
3. If an arbitrary string d belongs to $Halt$, then H returns $TRUE$ on input d.

4. If an arbitrary string *d* does not belong to *Halt,* either because *d* does not have the form *A%%In* or because *d* does have this form but *A* does not halt on *In,* then *H* returns *FALSE*.

Consider a simple-minded attempt at solving the halting problem. Given the string *d = A%%In,* run the given program *A* on the input *In* and see what happens. Suppose algorithm *A* is executed for 1,000,000 steps on input *In*. If, in 1,000,000 or fewer steps, *A* comes to a halt, then *A* halts on input *In*. However, what if it does not halt within 1,000,000 steps? It might halt in exactly 1,000,001 steps. It might halt in exactly 9,876,543,210 steps. It might never halt. The fact that the algorithm has not halted in any specific amount of time does not, by itself, say anything about whether the algorithm will ultimately halt.

On the other hand, as anyone who has ever checked someone else's program knows, there are many cases where a person can just look at a program and tell whether it will go into an infinite loop. So, there seems to be some hope for testing whether a program will halt by examining a program. Unfortunately, it is easy to write a program that is so complicated that other people are totally baffled by it. Therefore, perhaps Turing's theorem (see Theorem 1) now seems to be very plausible.

The proof of the theorem is reminiscent of the ancient *liar's paradox:* "This sentence is a lie." Work through it: If it is a lie, then it is not a lie, and if it is not a lie, then it is a lie. This paradox tells us, in part, why logicians are so careful in defining the syntax and semantics of their logics (see Chapter 2): They want to make sure they do not allow "This sentence is a lie" as a legal sentence in the formal logic.

Theorem 1.   (Unsolvability of the Halting Problem, Turing)   The halting problem, *Halt,* is undecidable.

Proof.   Suppose such an algorithm *H* existed. Let *NegSelfRef* be the following algorithm, with the input of a character string *A* representing an algorithm.   ■

Algorithm: NegSelfRef

INPUT: Character string *A*
OUTPUT: That is the question

```
if (H(A%%A) = FALSE)
    return TRUE
else
    while (0 = 0) do /* repeat forever:                    */
        x = 0        /* something trivial executed         */
                     /*      ...infinitely many times ...*/
    return FALSE     /* so this statement will             */
                     /*            never be reached        */
```

Note that if *H* works as we supposed above, *NegSelfRef(A)* halts if and only if *H(A, A) = FALSE,* if and only if the calculation of *A* does not halt on input *A*.

Now ask: Does *NegSelfRef* halt on input *NegSelfRef?* Substituting *NegSelfRef* for *A*, we derive that *NegSelfRef* (*NegSelfRef*) halts if and only *NegSelfRef* (*NegSelfRef*) does not halt. This is (blatantly!) a contradiction, so no such *H* exists.

Once one problem was proved to be undecidable, researchers used a technique called **reduction** to show that many other problems are also undecidable. The technique used is to *assume* that problem *P* is decidable by an algorithm *T* and then use *T* (technically, writing another algorithm calling *T* as a subfunction) to solve the halting problem, contradicting a known result. (Similar techniques are used to show that if CNF satisfiability is testable in polynomial time, so is every other $\mathcal{NP}$ problem. See Section 2.5.6.

Another basic theorem of Turing was the existence of a **universal Turing machine.** In the terminology used here, that translates into a program *Interpret,* which accepts as input a string *A%%In* and, essentially, runs algorithm *A* on input *In*. This is now a familiar idea. BASIC and LISP interpreters do this, and it is similar to programs that compile and then run other programs. Turing's viewpoint looks slightly different: There is a single *machine* that can take a set of instructions as input and simulate any other *machine*. Such a machine is, of course, just a general-purpose, programmable computer. Turing's idea foreshadowed the modern computer, for which the program is input as just another set of data. Proving the theorem requires extensive work (essentially, writing the interpreter and showing that it works), and we shall not do it here.

Theorem 2.   (Existence of Universal Algorithm, Turing)   There is an algorithm *Interpret* where, for each pair of strings *A* and *In*, *Interpret*(*A%%In*) = *A*(*In*).

In Chapter 4, we discussed two ways in which functions can be defined to be partial:

1. The person specifying the function did not specify its value in all cases.
2. The rule defining the function somehow specified that the function be partial.
3. One could question how different these two ways are. In particular, if a function is specified by a rule that makes it partial, is that merely because the wrong rule was chosen to specify the function? Before we answer this question, however, we need to formalize the ideas.

Definition 4.   An algorithm A_2 *extends* an algorithm A_1 if, for all input strings *In*, if A_1 halts on input *In*, then A_2 also halts on input *In* and $A_2(In) = A_1(In)$. A_2 is a **total algorithm** for A_1 if $A_2(In) = A_1(In)$ whenever A_1 returns *TRUE, FALSE,* or the empty string and A_2 returns *TRUE* or *FALSE* for every other input.

Given any algorithm *A*, can one always find an algorithm *B* extending *A* where *B* defines a total algorithm?

Theorem 3.   There is an algorithm *A* that cannot be extended to a total algorithm.

Proof.   Let *A* be the following algorithm, taking one string *In* as input:

> *otherOutput* = *Interpret*(*In%%In*)
> > /* remember that the line above might never finish */
> If *otherOutput* = the empty string then
> > output "0"
> else
> > output the empty string.

Note what *A* does: It treats its input *In* as if it is an algorithm and tries to run *In* on input *In*. If running *In* on input *In* halts—that is, if *In*(*In*) is defined—then *A* returns something *other than In(In)*. Otherwise, A does not halt; it just keeps on interpreting the computation of *In(In)* forever.

Now suppose an algorithm B computes a total algorithm extending A. Then, $B(B)$ is defined. So, $A(B)$ is defined and does not equal $B(B)$. Therefore, B does not extend A—a contradiction. ∎

The reader should note how this argument resembles both the proof of the undecidability of the halting problem and Cantor's second diagonal argument (in his proof of the uncountability of the reals).

5.6   Chapter Review

This chapter presents a formal way of deciding the level of difficulty of a program and then uses these ideas to compare programs. The definition of what it means for one function to asymptotically dominate another function is the start of this analysis. First, we deal with proving the basic properties of this relation and with showing that polynomials of different degrees are actually functions of different complexity. Next, we show how basic programming structures, such as sequence, selection, and repetition, can have functions defined directly from the code that describes the complexity of these programming structures. Finally, we describe the halting problem of Turing that proves there are programs for which it is impossible even to determine if they terminate.

5.6.1   Terms, Theorems, and Algorithms

5.1   Summary

TERMS

$\Theta(F)$	exponential function	order of F
algorithm	halts	polynomial function
asymptotic domination	logarithmic function	zero polynomial
degree	$\mathbf{O}(F)$	

THEOREMS

$F \in \mathbf{O}(aF)$ and $aF \in \mathbf{O}(F)$

$H \in \mathbf{O}(G)$ and $G \in \mathbf{O}(F)$, then
  $H \in \mathbf{O}(F)$

$G \in \mathbf{O}(F)$ and $H \in \mathbf{O}(F)$, then
  $G + H \in \mathbf{O}(F)$

$G \in \mathbf{O}(F)$ and $H \in \mathbf{O}(F)$, then
  $|G - H| \in \mathbf{O}(F)$

$F \in \mathbf{O}(G)$ if and only if $\mathbf{O}(F) \subseteq \mathbf{O}(G)$

$\mathbf{O}(F) = \mathbf{O}(G)$ if and only if $F \in \mathbf{O}(G)$
  and $G \in \mathbf{O}(F)$

$a^n \in \mathbf{O}(b^n)$ and $b^n \notin \mathbf{O}(a^n)$ where $1 <$
  $a < b$

5.3   Summary

TERMS

approximating curve	Complexity of Sequence	looping
average time complexity	Control Structures	nondeterministic polyno-
certificate	decision problem	mial time $(\mathcal{NP})$
Complexity of Repetition	iteration	parallel and distributed
Complexity of Selection	key operations	algorithms

polynomial time ($\mathcal{P}$)	selection	time complexity
propositional satisfiability	sequence	verifier algorithm V
repetition	space complexity	worst-case behavior
satisfiable	structured programming	worst-case complexity

ALGORITHMS

BubbleSort I	Ordering Three Values II
BubbleSort II	Ordering Three Values III
Detecting Order for Three Values	Selection Sort
Ordering Three Values I	

5.5   Summary

TERMS

algorithm	function call	total algorithm
decidable	halting problem	undecidable
decision algorithm	reduction	universal Turing mechine
decision problem	return value	

ALGORITHMS

Existence of Universal Algorithm, Turing	Unsolvability of the Halting Problem,
NegSelfRef	Turing

5.6.2   Starting to Review

1. Express in words what $G \in \mathbf{O}(F)$ means. Express in words what $G \notin \mathbf{O}(F)$ means.
2. Let $F(n) = n$ and $G(n) = 2n - 3$ be functions defined on $[0, \infty)$. Find a real number c and an $N_0 \in \mathbb{N}$ such that $F \in \mathbf{O}(G)$.
3. Let $F(n) = n$ and $G(n) = 2n + 3$ be functions defined on $[0, \infty)$. Find a real number c and an $N_0 \in \mathbb{N}$ such that $F \in \mathbf{O}(G)$.
4. Let $F(n) = n$ and $G(n) = n^2 + 3$ be functions defined on $[0, \infty)$. Show that $F \in \mathbf{O}(G)$ but $G \notin \mathbf{O}(F)$.
5. Show that $\mathbf{O}(1) \subseteq \mathbf{O}(\log (n))$.
6. Let $F(x) = 7x$ be a function defined on $[0, \infty)$. Find a real number c and an $N_0 \in \mathbb{N}$ such that $F \in \mathbf{O}(a^2)$.
7. Let $F(n) = n^2 - n + 550$ and $G(n) = 59n + 50$ be function defined on $[0, \infty)$. Determine n such that F takes less time at n than G.
8. Show that $\sum_{i=0}^{n} i^2 \in \mathbf{O}(n^3)$. Conjecture about the complexity of $\sum_{i=0}^{n} i^m$ for $m \in \mathbb{N}$.
9. Explain in words how the complexity of a sequence of statements is computed.
10. Explain in words how the complexity of a selection statement of the form *if-then-else* is computed.
11. Explain in words how the complexity of a loop of the form *for i = 0 to n* is computed.

5.6.3   Review Questions

Prove the following set of inclusions:

$$\mathbf{O}(1) \subset \mathbf{O}(\log(n)) \subset \mathbf{O}(\sqrt{n}) \subset \mathbf{O}(n) \subset \mathbf{O}(n\log(n))$$
$$\subset \mathbf{O}(n^2) \subset \mathbf{O}(2^n) \subset \mathbf{O}(n \cdot 2^n) \subset \mathbf{O}(3^n) \subset \mathbf{O}(n!) \subset \mathbf{O}(n^n)$$

by means of solving Exercises 1 through 10. (Remember to show inequality.)

1. Prove that $\mathbf{O}(1) \subset \mathbf{O}(\log(n))$.
2. Prove that $\mathbf{O}(\log(n)) \subset \mathbf{O}(\sqrt{n})$.
3. Prove that $\mathbf{O}(\sqrt{n}) \subset \mathbf{O}(n)$.
4. Prove that $\mathbf{O}(n) \subset \mathbf{O}(n \log(n))$.
5. Prove that $\mathbf{O}(n \log(n)) \subset \mathbf{O}(n^2)$.
6. Prove that $\mathbf{O}(n^2) \subset \mathbf{O}(2^n)$.
7. Prove that $\mathbf{O}(2^n) \subset \mathbf{O}(n \cdot 2^n)$.
8. Prove that $\mathbf{O}(n \cdot 2^n) \subset \mathbf{O}(3^n)$.
9. Prove that $\mathbf{O}(3^n) \subset \mathbf{O}(n!)$.
10. Prove that $\mathbf{O}(n!) \subset \mathbf{O}(n^n)$.

5.6.4   Using Discrete Mathematics in Computer Science

1. Let $F_1, F_2, G_1, G_2 : \mathbb{N} \to [0, \infty)$. Prove the following, or find counterexamples:

   (a) If $F_1 \in \mathbf{O}(G_1)$ and $F_2 \in \mathbf{O}(G_2)$, then $F_1 + F_2 \in \mathbf{O}(G_1 + G_2)$.
   (b) If $F_1 \in \mathbf{O}(G_1)$ and $F_2 \in \mathbf{O}(G_2)$, then $|F_1 - F_2| \in \mathbf{O}(|G_1 - G_2|)$.
   (c) If $F_1 \in \mathbf{O}(G_1)$ and $F_2 \in \mathbf{O}(G_2)$, then $F_1 \cdot F_2 \in \mathbf{O}(G_1 \cdot G_2)$ where $F_1 \cdot F_2(x) = F_1(x) F_2(x)$.
   (d) If $F_1 \in \mathbf{O}(G_1)$ and $F_2 \in \mathbf{O}(G_2)$, then $F_1/F_2 \in \mathbf{O}(G_1/G_2)$ where $F_1/F_2(x) = F_1(x)/F_2(x)$ provided $F_2(x), G_2(x) \neq 0$.
   (e) If $F_1 \in \mathbf{O}(G_1)$ and $F_2 \in \mathbf{O}(G_2)$, then $F_1 \circ F_2 \in \mathbf{O}(G_1 \circ G_2)$.

2. Use Exercise 1 and the chain of set inclusions proven in Section 5.6.3 to determine a simpler bound for the complexity of the following functions:

   (a) $2n^2 + n - n\log(n)$
   (b) $n + n2^n + e^{\frac{n}{2}}$
   (c) $n^{\frac{1}{3}}(n^2 + 3n)$
   (d) $2^n(n^2 - 2n)$
   (e) $3n^2 - 2n + 5 + n^{\frac{1}{2}}\log(n)$

3. (a) Find the complexity of the following algorithm for computing factorials:

Algorithm: Compute $n!$

INPUT: $n \in \mathbb{N}$
OUTPUT: $n!$
$3 = 1$
for $i = 1$ to n
    $x = x \cdot i$
print x

(b) Find the complexity of the following algorithm for computing the mean of n values:

Algorithm: Mean

INPUT: The number of values n and an array $a[1 .. n]$ containing them
OUTPUT: The mean of the values $a[1], a[2], \ldots, a[n]$

$sumOfValues = 0$
for $i = 1$ to n
    $sumOfValues = sumOfValues + a[i]$
print $sumOfValues/n$

---

4. Find the complexity of the following algorithm for computing Fibonacci numbers:

Algorithm: Compute F_n

INPUT: $n \in \mathbb{N}$
OUTPUT: F_n

if $n = 0$ then
    $nthFib = 1$
else
    if $n = 1$ then
        $nthFib = 1$
    else
        $fibnMinusTwo = 1$
        $fibnMinusOne = 1$
        for $i = 2$ to n
            $nthFib = fibnMinusOne + fibnMinusTwo$
            $fibnMinusTwo = fibnMinusOne$
            $fibnMinusOne = nthFib$
print $nthFib$

---

5. Find the complexity for each of the following algorithms that evaluate a polynomial of degree n at a value x_0. Assume multiplication is the key operation.

a.

Algorithm: Evaluate $P(x) = a_0 + a_1 x + a_2 x^2 + \cdots + a_n x^n$ at x_0

INPUT: $n \in \mathbb{N}$, the coefficients of P stored in $a[0 .. n]$, and a real number x_0
OUTPUT: $P(x_0)$
$Poly = a[0]$
$x = 1$
for $i = 1$ to n
    $x = x_0 \cdot x$
    $Poly = a[i] \cdot x + Poly$
print $Poly$

b.

Algorithm: Evaluate $P(x) = a_0 + a_1 x + a_2 x^2 + \cdots + a_n x^n$ at x_0

INPUT: $n \in \mathbb{N}$ and a real number x_0 and the coefficients of a polynomial of degree n stored in $a[0 .. n]$
OUTPUT: $P(x_0)$
$Poly = a[0]$
for $i = 0$ to n
    $x = 1$
    for $j = 1$ to i
        $x = x \cdot x_0$
    $Poly = Poly + a[i] \cdot x$
print $Poly$

6. The code that follows implements Horner's algorithm for evaluating polynomials. Find its complexity, assuming multiplication is the key operation. Here, the phrase "for i ... down to 0" down to means to subtract 1 each time through the loop until $i < 0$.

Algorithm: Horner

Evaluate $P(x) = a_0 + a_1 x + a_2 x^2 + \cdots + a_n x^n$ at x_0

INPUT: $n \in \mathbb{N}$ and a real number x_0 and the coefficients of a polynomial of degree n stored in $a[0 .. n]$
OUTPUT: $P(x_0)$

$Poly = x_0 \cdot a[n] + a[n - 1]$
for $i = n - 2$ down to 0
    $Poly = Poly \cdot x_0 + a[i]$
print $Poly$

7. The Insertion Sort algorithm that follows is a good sorting algorithm to use when (i) the size of the list is very small (say, <10) or (ii) the list is already almost in order. In this respect, it is like bubblesort, but it is normally a bit better.

 Calculate the complexity of the Insertion Sort algorithm on an input *List* of size N. Let comparison of list elements be the key operation. (*Hint:* For this version, which sorts the list into increasing order, the worst-case behavior occurs when the list is in decreasing order.)

Algorithm: InsertionSort

INPUT: A list *List* of values, $List[1], \ldots, List[N]$
OUTPUT: The same elements, but in increasing order

for *positionToFix* = 2 to N do
    valueToPlace = List[*positionToFix*]
    candidatePos = *positionToFix*
                /* locate place to put *valueToPlace*, */
                /* relative to previous elements,   */
                /* and open up room to insert it there.*/
    while *candidatePos* > 1 and List[*candidatePos* − 1] > *valueToPlace*
        List[*candidatePos*] = List[*candidatePos* − 1]
        candidatePos = *candidatePos* − 1
    List[*candidatePos*] = *valueToPlace*

8. One variant of the Merge Sort algorithm was given in Exercise 22 of Section 2.8. Analyze the complexity of *MergeSort*. Count copying an element from array A to array B, or from array B to array A, as the key operation. Note how much faster *MergeSort* is, at least asymptotically, than *SelectionSort*, *BubbleSort*, and *InsertionSort*.

9. *Challenge:* This returns to an issue raised in Section 2.5.6. Show that for formulas ϕ in CNF, the shortest equivalent disjunctive normal formula (DNF) ϕ' may be exponentially longer than ϕ. Stated more precisely:

   i. Count the length of a formula as the total number of symbols in the formula where a proposition letter p_i is counted as one symbol and, if a single symbol occurs more than once in the formula, each occurrence is counted. So, the length of

   $$(((p_{9,876,543,210} \vee p_{9,876,543,210}) \vee p_{9,876,543,210}) \wedge (p_1 \vee \neg p_{9,876,543,210}))$$

   is 18: There are five occurrences of proposition letters, four open parentheses, four closed parentheses, one $\neg$ sign, three $\vee$'s, and one $\wedge$.

   ii. For each CNF formula ϕ, compute the length $dnfl(\phi)$ of the shortest DNF equivalent to ϕ. Now, for an integer n, define $DNFL(n)$ to be the maximum value of $dnfl(\phi)$ for all formulas ϕ of length less than or equal to n.

   iii. Show that for some real number r, $r^n \in \mathbf{O}(DNFL)$.

10. The following is a simple-minded "algorithm" to check whether a CNF formula in variables $p_1, \ldots, p_n$ is satisfiable. The "algorithm" has n nested loops, so we just use three dots to indicate that an obvious block of code has been omitted. (By our rules, this really is not an algorithm, because we had no formal mechanism to cover a variable number of nested loops in an algorithm. However, the reader should be able to understand exactly what it does.)

    The notation "for $p_1^{val} = F, T$" below says to execute the following code twice, once with $p_1^{val} = F$ and once with $p_1^{val} = T$.

Algorithm: GreedySAT

INPUT: a set C of clauses
OUTPUT: "satisfiable" or "unsatisfiable"

1. N for $p_1^{val} = F, T$
2.   set $C_1 = C$
3.   if $p_1^{val} = T$
4.   remove from C_1 all clauses containing p_1.
5.   remove $\neg p_1$ from all clauses in C_1 it occurs in
6.   else
7.   remove from C_1 all clauses containing $\neg p_1$
8.   remove p_1 from all clauses in C_1 it occurs in
9. if C_1 is empty,
          output "satisfiable" and stop
10. else
11.   for $p_2^{val} = F, T$
12.   set $C_2 = C_1$
13.   if $p_2^{val} = T$
14.   remove from C_2 all clauses containing p_2
15.   remove $\neg p_2$ from all clauses in C_2 it occurs in
16.   else
17.   remove from C_2 all clauses containing $\neg p_2$
18.   remove p_2 from all clauses in C_2 it occurs in
19.   if C_2 is empty, output "satisfiable"
20.   else

21. $\vdots$

22.   for $p_n^{val} = F, T$
23.   set $C_n = C_{n-1}$
24.   if $p_n^{val} = T$
25.   remove from C_n all clauses containing p_n
26.   remove $\neg p_n$ from all clauses in C_n it occurs in
27.   else
28.   remove from C_n all clauses containing $\neg p_n$
29.   remove p_n from all clauses in C_n it occurs in
30.   if C_n is empty, output "satisfiable" and stop
31. output "unsatisfiable"

(a) Trace through the execution of the code on the set $C = \{p_3, \neg p_3 \lor p_2, \neg p_2 \lor p_1\}$ (so, for $n = 3$). Show what each C_i is after lines 8, 18, and 28. If the algorithm stops with a "stop" command, say which step caused it to stop.

(b) Trace through the execution of the code on the set $C = \{p_1 \vee p_2, p_1 \vee \neg p_2, \neg p_1 \vee p_2, \neg p_1 \vee \neg p_2\}$ (so, for $n = 2$). Show what each C_i is after lines 8, 18, and 28. If the algorithm stops with a "stop" command, say what step caused it to stop.

(c) Suppose C contains all possible clauses on the n proposition letters p_1, $p_2, \ldots, p_n$. Prove that the test in step 30 will be executed exponentially (in n) many times.

(d) *Challenge:* Prove that *GreedySAT* correctly determines whether C is satisfiable.

11. Consider the following two algorithms to check whether a number n is prime. Let computation of mod be the key operation.

   The first algorithm directly reflects the definition of "prime." Primality testing has been used in cryptography (secret coding), particularly in building "public key cryptograms."

Algorithm: SimplestPrimalityTest

INPUT: An integer $n > 1$
OUTPUT: "prime" or "composite"

for $d = 2$ to $n - 1$
    if $mod(n, d) = 0$
output "composite" and stop
output prime

Algorithm: ShortenedPrimalityTest

INPUT: An integer $n > 1$
OUTPUT: "prime" or "composite"

$d = 2$
while $(d^2 \leq n)$
    if $mod(n, d) = 0$
output "composite" and stop
    $d = d + 1$
output prime

(a) Show that if n is a k digit prime number, *SimplestPrimalityTest* executes the key operation approximately 10^k times.

(b) Suppose that cracking a public key cryptogram depends on finding the smallest 50-digit prime. Suppose that a single execution mod of takes 0.01 nanoseconds (10^{-11} seconds)[3] and that the rest of the above program takes no time at all. Approximately how long would it take, in years, to verify that a 50-digit number is prime?

(c) If n is a k digit prime number, approximately how many times does *ShortenedPrimalityTest* execute the key operation? How much does this speed up our time from part (b)?

(d) Show that *ShortenedPrimalityTest* correctly determines whether input integer n is prime.

In fact, Manindra Agrawal, Neeraj Kayal, and Nitin Saxena of the Indian Institute of Technology showed that primality can be tested in polynomial time in the number of digits in n.

More recent work has been based not on testing primality but, rather, on factoring large integers, which seems to be harder.

---

[3] This is sure to be an unrealistic assumption no matter how fast microprocessors get: With numbers of this size, the amount of time to do one mod operation will almost certainly depend on the length of the number.

CHAPTER **6**

Graph Theory

Computer scientists use graphs to model problems as diverse as how to detect a deadlock condition in an operating system and how to plan efficient routings for transportation networks. Some problems in artificial intelligence use efficient searching procedures on the class of graphs called *trees* as effective heuristics. The scheduling of a sequence of subassemblies so that no assembly process starts without all its required subassemblies having been completed is also a problem that can be modeled and solved using directed graphs. Some of the most difficult $\mathcal{NP}$-complete problems involve finding special subgraphs within large graphs. This chapter introduces graphs as a structure that is used to represent and solve many problems in a variety of areas in computer science.

The chapter is organized into four main parts. The first introduces the terms that are used to describe this structure. The second focuses on ideas dealing with graphs that are connected. As an application of this idea, we will examine Euler's theorem, the first theorem of graph theory. This theorem explains when it is possible, for example, to stroll across a series of bridges and return to the starting point without crossing any bridge twice. The theorem can even tell how to plot a graph without taking the pen off the paper too many times. The third part of the chapter covers the special graphs called trees. Trees are a common data structure in computer science that are used to solve searching and sorting problems. Finally, the last part of the chapter deals with directed graphs, which extend the idea of a graph to include the notion of direction. Directed graphs are used to represent and solve problems such as scheduling a production of subassemblies with no unnecessary delays or designing a one-way street grid.

We begin the study of this important mathematical theory by examining a problem for which graph theory may be used to build a model of the problem.

6.1 Introduction to Graph Theory

The problem we use to introduce graph theory involves filling positions from a pool of qualified applicants.

Job Assignment Problem

GIVEN: A list of jobs to be filled and a list of job applicants who may be qualified for more than one job.

FIND: An assignment of each job to a qualified applicant with no applicant assigned to more than one job.

For this problem, every job needs a qualified applicant, but not every applicant needs to be assigned to a job. Obviously, a necessary condition for the assignment problem to have a positive solution is that the number of applicants must be at least as large as the number of jobs. Clearly, however, this is not a sufficient condition, because two of the job applicants could be qualified for exactly one and the same job as well as not qualified for any other job.

The first step toward the problem's solution will be to find an appropriate model. In this case, we will use a graph. To begin, we present an intuitive notion of graphs; a more formal introduction to this structure will follow.

The data for the Job Assignment Problem is shown in Table 6.1.

Table 6.1   Qualified Applicants and Jobs

Qualified Applicant	Job
Jack	technical writer
Jack	market analyst
James	sales representative
Jane	photographer
Jane	copy editor
Jane	production manager
John	photographer
John	technical writer
June	copy editor
June	production manager
June	sales representative

Table 6.1 suggests that the information can be represented as a relation between pairs of objects, with one object being a qualified applicant and the other being a job. A graphical representation of this information for Jane is shown in Figure 6.1.

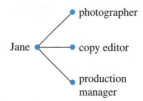

Figure 6.1   Applicant and jobs.

A line or edge from a person to a job represents the fact that the person is qualified for that job. The picture that represents all the data in the problem is shown in Figure 6.2.

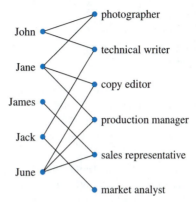

Figure 6.2   Set of applicants and set of jobs.

The structure in Figure 6.2 is called a *graph*. The assignment problem is to choose a set of the edges (lines joining pairs of points) such that each job is at the end of at most one edge and no applicant is at the end of more than one edge. In this example, there are more jobs than applicants, so it will be impossible to fill all the jobs. One solution that finds qualified applicants for five of the jobs is shown in Figure 6.3.

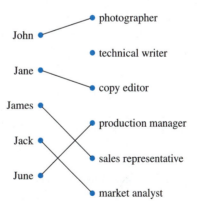

Figure 6.3   One set of job assignments.

An efficient algorithm for finding a solution to the assignment problem is an application of the theory of flows in networks. A solution of the problem using flows has complexity $\mathbf{O}(|V|^3)$ where V is the set of points in the graph. A discussion of flows in networks can be found in most books dealing with algorithms.

6.1.1   Definitions

The Job Assignment Problem gives an insight regarding how graph theory can be used to model a problem. Before exploring more uses of this theory, however, a more careful definition of graphs is needed.

Definition 1.   A **graph** $G = (V, E)$ consists of a finite nonempty set V, the elements of which are the **vertices** of G, and a finite set E of unordered pairs of distinct elements of V called the **edges** of G.

We can think of vertices as points and edges as lines joining pairs of points. Even though edges are defined as unordered pairs of distinct vertices, it is common to denote an edge consisting of the vertices a and b as (a, b). In graph theory usage, the notation (b, a) denotes the same edge as (a, b). For any edge $e = (a, b)$ in a graph, the vertices a and b are the **ends** of e, e is **incident** to both a and b, and the vertices a and b are **adjacent.** We often call adjacent vertices **neighbors.** If two distinct edges have a common end, then the edges are adjacent. For any vertex v, the **degree** of v, denoted *deg(v),* is the number of edges incident to it. A vertex is **even** (or **odd**) if its degree is even (or odd). A vertex of degree zero is called an **isolated vertex.**

These terms are illustrated in Figure 6.4.

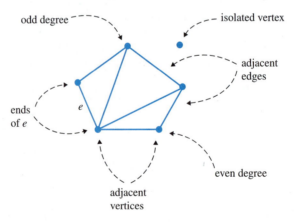

Figure 6.4   Names of parts of a graph.

A sequence of n nonnegative numbers, $d_1, d_2, \ldots, d_n$, is said to be **graphical** if there exists a graph having n vertices with the degrees of the vertices being $d_1, d_2, \ldots, d_n$. The degrees of the vertices of a graph are called a **degree sequence.** Without loss of generality, one can assume that $d_1 \leq d_2 \leq \cdots \leq d_n$. The graph in Figure 6.4 has degree sequence 0, 2, 2, 3, 3, 4.

Definition 2.   A graph on n vertices having each pair of distinct vertices joined by an edge is called a **complete graph** and is denoted by K_n. Complete graphs are often called **cliques.** A graph in which each vertex has the same degree is called a **regular graph.** A regular graph is ***n*-regular** if each **vertex** has degree n. Graphs that are 3-regular are often called **cubic graphs.**

Some examples of complete graphs and regular graphs are shown in Figure 6.5.

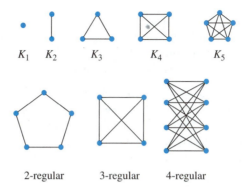

K_1 K_2 K_3 K_4 K_5

2-regular   3-regular   4-regular

Figure 6.5 Complete and regular graphs.

Another family of graphs that are used in matching problems, resource allocation problems, and computer architecture modeling is the family of bipartite graphs.

Definition 3. Let $G = (V, E)$ be a graph. G is a **bipartite graph** if its vertex set V can be partitioned into two nonempty disjoint subsets V_1 and V_2, called a **bipartition,** so that each edge has one end in V_1 and one end in V_2. A **complete bipartite graph** is a bipartite graph with bipartition V_1 and V_2 in which each vertex of V_1 is joined by an edge to each vertex of V_2. A complete bipartite graph with $|V_1| = m$ and $|V_2| = n$ is denoted as $K_{m,n}$.

Examples of bipartite graphs and the bipartitions they determine are shown in Figure 6.6.

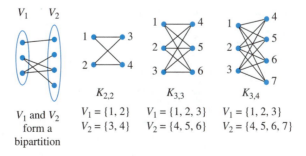

V_1 V_2

V_1 and V_2 form a bipartition

$K_{2,2}$

$V_1 = \{1, 2\}$
$V_2 = \{3, 4\}$

$K_{3,3}$

$V_1 = \{1, 2, 3\}$
$V_2 = \{4, 5, 6\}$

$K_{3,4}$

$V_1 = \{1, 2, 3\}$
$V_2 = \{4, 5, 6, 7\}$

Figure 6.6 Bipartition and complete bipartite graphs.

6.1.2 Subgraphs

In the Job Assignment Problem, the graph theory answer consists in identifying a graph that is formed by some of the vertices and some of the edges of the graph that models the problem. This idea of looking at part of a graph is made precise by the notion of a subgraph.

Definition 4.   A graph $H = (V_1, E_1)$ is a **subgraph** of $G = (V, E)$ provided that $V_1 \subseteq V$, $E_1 \subseteq E$, and for each $e \in E_1$, both ends of e are in V_1. H is a **spanning subgraph** of G if H is a subgraph of G and $V_1 = V$. H is an **induced subgraph** of G if H is a subgraph of G such that E_1 consists of all the edges of G with both ends in V_1.

An induced subgraph is found by choosing a subset of the vertex set of a graph and then defining the edge set to be all the edges of the original graph with both ends in the chosen subset of vertices. An induced subgraph with vertex set V is often denoted as $<V>$. Examples of the different kinds of subgraphs are shown in Figure 6.7.

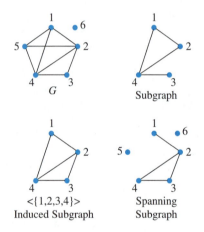

Figure 6.7   Subgraph, induced subgraph, and spanning subgraph.

The notation $E(G)$ (respectively, $V(G)$) denotes the edges (respectively, vertices) of the graph G. When more than a single graph is being discussed, this notation makes it clear which edges and vertices are being considered. Normally, we use the letters alone, as in $G = (V, E)$.

Two useful operations for combining graphs include the union and the intersection of two graphs. Let $G_1 = (V_1, E_1)$ and $G_2 = (V_2, E_2)$ be graphs.

The **union** of G_1 and G_2, denoted by $G_1 \cup G_2$, is the graph G_3 defined as $G_3 = (V_1 \cup V_2, E_1 \cup E_2)$. The **intersection** of G_1 and G_2, denoted by $G_1 \cap G_2$, is the graph G_4 defined as $G_4 = (V_1 \cap V_2, E_1 \cap E_2)$.

Another operation that is used with a single graph is complementation. For this definition, we need an analogue of a universal set. In this case, we use the complete graph on the vertex set of the graph for which we would like to find the complement. Let $G = (V, E)$ be a subgraph of $K_{|V|}$, the complete graph on $|V|$ vertices. The **complement** of G in $K_{|V|}$, denoted as $\overline{G} = (V_1, E_1)$, is the subgraph of $K_{|V|}$ with $V_1 = V$ and $E_1 = K_{|V|}(E) - E$.

Example 1. Find the union and intersection of G and H as shown:

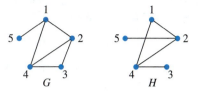

Solution.

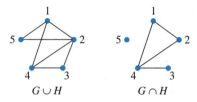

Example 2. Let G and H be graphs as shown:

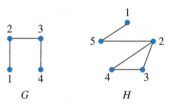

Find $\overline{G}$ and $\overline{H}$.

Solution.

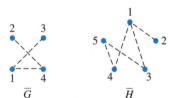

Distributed Network Architecture

A computer architecture for a set of processors (often called a *topology*) can be designed so that each processor has its own memory and only certain pairs of processors are directly linked to each other. For parallel processing of sorting algorithms, the topology chosen can affect the efficiency of the algorithm. One widely studied topology is the **hypercube.** A hypercube of size n, or an ***n*-cube,** denoted as Q_n, consists of 2^n processors indexed by the integers $\{0, 1, 2, \ldots, 2^n - 1\}$. Processors A and B are directly connected if and only if the binary representations for A and B differ in exactly one position. For example, if $n = 3$ and the two processors are 110 and 101, then the processors are not directly connected, because they differ in both the second (1 and 0) and the third (0 and 1) positions. If the two processors are 110 and 100, then the two processors are directly connected, since the only difference is in the second position (1 and 0). An integer's binary representation can be found by writing the number as a sum of powers of 2 and then using the coefficients of this expansion as the **binary representation.** For example,

$$61 = 1 \cdot 2^5 + 1 \cdot 2^4 + 1 \cdot 2^3 + 1 \cdot 2^2 + 0 \cdot 2^1 + 1 \cdot 2^0 = 111101_2$$

Example 3.   Draw Q_2 with the four vertices labeled 00, 01, 10, and 11.

Solution.

Obviously an architecture with only four nodes is pretty simple. Hypercubes can also be used in much larger examples. To make sure you understand how the graphs are formed, the next example shows Q_3. Other properties of hypercubes will be explored in the exercises.

Example 4.   Draw a graph with eight vertices, labeled with the elements of {000, 001, 010, 011, 100, 101, 110, 111}, the edges of which correspond to the edges of Q_3.

Solution.

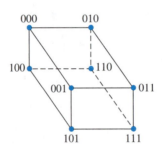

The picture shown in Example 4 should appear to be a "cube." In Exercise 11 of Section 6.6, you can see how well you can represent Q_4 as a "cube."

6.2   The Handshaking Problem

A university holds a reception for graduates of the past five years. During the course of the reception, each attendee shakes hands numerous times. What, if anything, can be said about the number of times that hands are shaken? For example, can each person shake a different number of hands?

To find a model, we put this problem in the context of graph theory. We model this problem with the graph $G = (V, E)$, where

$$V = \{\text{people at the reunion}\}$$

$$E = \{(u, v) : u, v \in V \text{ and } u \text{ and } v \text{ shake hands during the reception}\}$$

A typical example of such a graph is shown in Figure 6.8. In this graph, there is a pair of vertices that have the same degree. This means that this pair of people—Sue and Phil—shook the same number of hands. Is it an accident that a pair of vertices of the graph have the same degree? Theorem 1 says that in any graph with at least two vertices, there are always at least two vertices of the same degree.

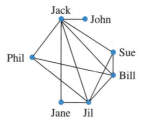

Figure 6.8   Handshaking graph.

Theorem 1.   (Handshaking Theorem)   Let G be a graph with at least two vertices. At least two vertices of G have the same degree.

Proof.   The proof is by induction on the number of vertices n in a graph. Let $n_0 = 2$ and $\mathcal{T} = \{n \in \mathbb{N} :$ any graph with n vertices has at least two vertices of the same degree$\}$.

(Base step)   For n_0, the only graphs to consider are the graph consisting of two isolated vertices and the graph having a single edge. Clearly, the result holds for each of these graphs. Therefore, the base case $n_0 = 2$ is true and $n_0 \in \mathcal{T}$.

(Inductive step)   Let $n \geq n_0$. Show that if $n \in \mathcal{T}$, then $n + 1 \in \mathcal{T}$. Assuming that any graph on n vertices with $n \geq 2$ has two vertices of the same degree, we must prove that any graph on $n + 1$ vertices has two vertices of the same degree. Let $G = (V, E)$ be a graph with $n + 1$ vertices where $n + 1 \geq 3$. Clearly, $0 \leq deg(v) \leq n$ for any $v \in V$. If there is an isolated vertex in G, then by the induction hypothesis, the subgraph of G consisting of all the vertices but one isolated vertex must have two vertices with the same degree. Adding an isolated vertex to the subgraph with at least two vertices having the same degree gives the result for G. If there is no isolated vertex in G, then all the degrees of vertices $v \in V$ satisfy $1 \leq deg(v) \leq n$. In this case, we have at most n different values for the degrees of vertices in G. Since G has $n + 1$ vertices, then by the Pigeon-Hole Principle, at least two vertices of G have the same degree. Therefore, $n + 1 \in \mathcal{T}$.

By the Principle of Mathematical Induction, $\mathcal{T} = \{n \in \mathbb{N} : n \geq 2\}$.   ■

It was indeed no accident that two people, at least, shook the same number of hands.

The graph shown in Figure 6.8 also has four odd vertices. Is it an accident that there are an even number of odd vertices? Before proving the theorem that answers this question, we need to prove a result that will be used in its proof.

Theorem 2.   Let $G = (V, E)$ be a graph. Then, $\sum_{v \in V} deg(v) = 2 \cdot |E|$.

Proof.   The sum

$$\sum_{v \in V} deg(v)$$

represents the sum of the degrees of each of the vertices. This sum counts the number of edges twice, since each edge is incident to two vertices. Therefore,

$$\sum_{v \in V} deg(v) = 2 \cdot |E|$$

■

We will now use this result to prove another property of graphs that can be interpreted for the handshaking graph to mean that an even number of people shook an odd number of hands.

Theorem 3.   In any graph, the number of odd vertices is even.

Proof.   Let $G = (V, E)$ be a graph. By Theorem 2, we have

$$\sum_{v \in V} deg(v) = 2 \cdot | E |$$

The next step is to rewrite the left-hand side of the equation as

$$\sum_{v \in V} deg(v) = \sum_{v \ odd} deg(v) + \sum_{v \ even} deg(v)$$

We now have

$$\sum_{v \ odd} deg(v) + \sum_{v \ even} deg(v) = 2 \cdot | E |$$

or

$$\sum_{v \ odd} deg(v) = 2 \cdot | E | - \sum_{v \ even} deg(v)$$

Since the right-hand side of the equation is an even number, the left-hand side of the equation is an even number. Therefore, an even number of vertices are included in the sum on the left-hand side, since the sum of an odd number of odd numbers would be odd.   ∎

Theorem 3 has proved it was no accident that an even number of people shook an odd number of hands. Although this result was motivated by the handshaking problem, you will find other important applications for these two theorems in many instances when a graph models a problem.

6.3   Paths and Cycles

The graph theory notions defined here are important not only because of their application to describing problems but also because they give important tools to use in problem solving. For example, the design for the layout of circuits in computer chips can be described in terms of these notions.

Definition 5.   Let $G = (V, E)$ be a graph. A **trail** in a graph G is a sequence of not necessarily distinct vertices $v_1, v_2, \ldots, v_k$ of G such that $(v_i, v_{i+1}) \in E$ for $1 \leq i \leq k - 1$ and the edges are distinct. A trail with k vertices is denoted by $Tr(k)$. If all the vertices in a trail are distinct, the trail is called a **path.** The **length** of a path or a trail is the number of edges it contains. A path or trail of length zero consists of a single vertex. A path of length n is denoted by P_n. The **distance** between two vertices a and b is the length of a shortest path joining a to b. If no path between a and b exists, then the distance is commonly defined to be infinity or, in some instances, left undefined.

Although a trail is defined as a sequence of vertices and certain associated edges, we often display a **trail** or a path simply as if it were a subgraph with the vertices and edges joined. In some cases, this will not be a subgraph, because vertices are repeated to indicate that the trail returns to a vertex that has already been included. Figure 6.9 illustrates how to find a path joining the ends of any trail with repeated occurrences of a vertex.

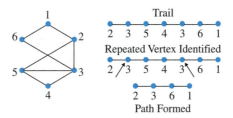

Figure 6.9   Refining a trail to find a path.

In addition to trails and paths in a graph we are often interested in trails that have the same starting and ending vertex.

Definition 6.   Let $Tr(k)$ be a trail in a graph $G = (V, E)$ for some $k \in \mathbb{N}$. Let the vertices of $Tr(k)$ be $v_1, v_2, \ldots, v_k$. A trail for which $v_1 = v_k$ is a **circuit.** If $Tr(k)$ has its first $k - 1$ vertices distinct and $v_1 = v_k$, then $Tr(k)$ is a **cycle.** A cycle with three edges is a **triangle.** A cycle with k edges where $k \geq 3$ is a **k-cycle,** denoted as C_k. A graph without cycles is **acyclic.** A cycle that contains all the vertices of a graph is a **Hamiltonian cycle.**

Examples of all these notions are found in Figure 6.10.

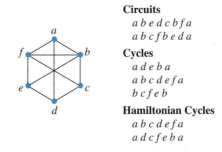

Circuits
   $a\,b\,e\,d\,c\,b\,f\,a$
   $a\,b\,c\,f\,b\,e\,d\,a$

Cycles
   $a\,d\,e\,b\,a$
   $a\,b\,c\,d\,e\,f\,a$
   $b\,c\,f\,e\,b$

Hamiltonian Cycles
   $a\,b\,c\,d\,e\,f\,a$
   $a\,d\,c\,f\,e\,b\,a$

Figure 6.10   Circuits and cycles.

6.3.1   Hamiltonian Cycles

A graph that contains a Hamiltonian cycle is called a **Hamiltonian graph** after William Hamilton (1788–1856, b. Scotland), who even invented a game based on this notion. The question of whether a graph is Hamiltonian is not an easy one to answer. A graph $G = (V, E)$ with as few as $|V|$ edges could be a Hamiltonian graph, but it need not be. Many of the known sufficient conditions for a graph to be Hamiltonian involve restrictions on the degrees of the vertices. For example, if every vertex has degree at least $|V|/2$, then the graph must be Hamiltonian. (See Exercises 40 and 41 in Section 6.6.) Showing that a graph is non-Hamiltonian is often established by observations about conditions Hamiltonian graphs must satisfy.

Conditions Satisfied by Hamiltonian Graphs

1. A Hamiltonian graph cannot contain a vertex of degree zero or one.
2. If a vertex has degree two, then both edges incident to that vertex must be in every Hamiltonian cycle in the graph.
3. Every vertex in a Hamiltonian cycle must be at the end of two edges.
4. If a vertex has degree greater than two and two of its edges are chosen to be in a Hamiltonian cycle, then the other edges incident to the vertex can be removed from further consideration.
5. In constructing a Hamiltonian cycle edge by edge, an edge cannot be chosen that completes a cycle that does not include all the vertices of the graph.

The next two examples show how such observations can be used.

Example 5.   Prove that neither of the following graphs is Hamiltonian:

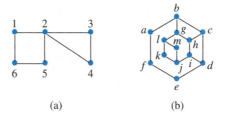

(a)   (b)

Solution.

(a) By condition 3, vertex 2 must be an end to one of the following six pairs of edges:

i. $(1, 2)$ and $(2, 3)$
ii. $(1, 2)$ and $(2, 4)$
iii. $(1, 2)$ and $(2, 5)$
iv. $(2, 5)$ and $(2, 3)$
v. $(2, 5)$ and $(2, 4)$
vi. $(2, 3)$ and $(2, 4)$

The six cases are very similar, so we will just show the first case. Suppose a Hamiltonian cycle in the graph contains the edges $(1, 2)$ and $(2, 3)$. Edges $(2, 5)$ and $(2, 4)$ can be removed from further consideration, since we have chosen $(1, 2)$ and $(2, 3)$ to be the two edges incident to vertex 2 in a Hamiltonian cycle. We must then complete the Hamiltonian cycle in the graph shown in Figure 6.11.

Figure 6.11   Remaining graph for Hamiltonian cycle construction.

Clearly, it is impossible to find any cycle in the graph shown. The remaining cases are left as an exercise for the reader.

(b) First, observe that the edges (l, k), (k, j), (l, m), and (m, j) must be included in any Hamiltonian cycle by condition 2. Clearly, this is impossible, because these four edges form a cycle with a smaller total number of vertices than the number of vertices in the graph, which contradicts condition 5.   ■

The next example involves a graph that is very famous, because it is a counterexample to a number of important conjectures about the structure of cubic graphs.

Example 6.

(a) Prove that the Petersen graph shown here is non-Hamiltonian.
(b) Prove that by removing any single vertex and its incident edges, the resulting graph is Hamiltonian.

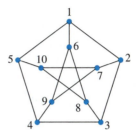

Petersen graph.

Solution.

(a) Suppose the Petersen graph has a Hamiltonian cycle. Two of the edges incident to vertex 1 must be in the Hamiltonian cycle by condition 3. There are three possibilities:

   i.   (1, 2) and (1, 5)
  ii.   (1, 2) and (1, 6)
 iii.   (1, 6) and (1, 5)

Cases ii and iii can be dealt with using a single argument because of the symmetry of the graph. We will show that Case i is impossible and leave the proof that Case ii is impossible as Exercise 26 in Section 6.6.

Case i: Suppose (1, 2) and (1, 5) are contained in the Hamiltonian cycle. It follows by condition 4 that (1, 6) can be deleted from further consideration. We must finish a Hamiltonian cycle in the graph in Figure 6.12.

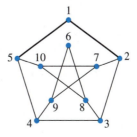

Figure 6.12   Start of Hamiltonian cycle.

Now, the second edges incident to vertices 2 and 5 must be one of the following four pairs of edges:

(i)   (2, 7) and (5, 10)
(ii)   (2, 7) and (5, 4) or, symmetrically, (5, 10) and (2, 3)
(iii) (2, 3) and (5, 4)

If edges (2, 7) and (5, 10) are chosen, the edges (5, 4) and (2, 3) can be deleted by condition 4. The edges (4, 9), (4, 3) (Case (i)) and (3, 8) must now be included in the Hamiltonian cycle by condition 2. In Case ii, the choice of (2, 7) and (5, 4) will allow edges (2, 3), (4, 9), and (5, 10) to be deleted by condition 4. The edges (3, 4) and (3, 8) must now be included in the Hamiltonian cycle by condition 2. Finally, in Case iii, the choice of edges (2, 3) and (5, 4) will allow edges (5, 10) and (2, 7) to be deleted by condition 4. The edge (3, 4) can be deleted by condition 5. The edges (4, 9) and (3, 8) must now be included in the Hamiltonian cycle by condition 2. The graphs remaining in which Hamiltonian cycles might be completed are shown in Figure 6.13.

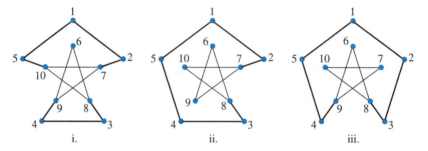

Figure 6.13   Graphs for finishing a Hamiltonian cycle. a. Case i. b. Case ii. c. Case iii.

In each of these graphs, a choice of a second edge incident to vertex 8 will lead to a violation of condition 5.

Case ii: This proof is left as an exercise for the reader.

(b) By symmetry in the graph, we need only show the result for the subgraphs formed by deleting one of the outer vertices and one of the inner vertices. Figure 6.14 shows Hamiltonian cycles in the resulting graphs.

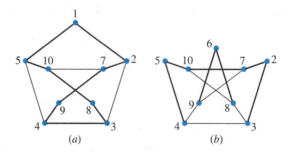

Figure 6.14   Hamiltonian cycles in vertex-deleted subgraphs. a. Inner vertex deleted. b. Outer vertex deleted.

6.4 Graph Isomorphism

For two graphs given abstractly as vertex and edge sets, we need to be able to determine whether they are the same or different. Even when representations of two graphs are drawn, it is not always clear whether the graphs are the same. The next definition details precisely what we mean when we say that two graphs are the same.

Definition 7. Two graphs $G_1 = (V_1, E_1)$ and $G_2 = (V_2, E_2)$ are the **same,** or **isomorphic,** if there is a bijection F from V_1 to V_2 such that $(u, v) \in E_1$ if and only if $(F(u), F(v)) \in E_2$. F is referred to as an **isomorphism.**

Figure 6.15 shows two graphs and displays an isomorphism between them.

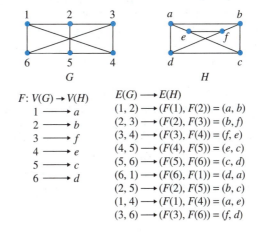

$$F: V(G) \to V(H)$$

1	$\to$ a
2	$\to$ b
3	$\to$ f
4	$\to$ e
5	$\to$ c
6	$\to$ d

$$E(G) \to E(H)$$
$$(1, 2) \to (F(1), F(2)) = (a, b)$$
$$(2, 3) \to (F(2), F(3)) = (b, f)$$
$$(3, 4) \to (F(3), F(4)) = (f, e)$$
$$(4, 5) \to (F(4), F(5)) = (e, c)$$
$$(5, 6) \to (F(5), F(6)) = (c, d)$$
$$(6, 1) \to (F(6), F(1)) = (d, a)$$
$$(2, 5) \to (F(2), F(5)) = (b, c)$$
$$(1, 4) \to (F(1), F(4)) = (a, e)$$
$$(3, 6) \to (F(3), F(6)) = (f, d)$$

Figure 6.15   Graph isomorphism.

An interesting class of graphs are those that are isomorphic to their complement. A graph that is isomorphic to its complement is called a **self-complementary graph.** Examples of self-complementary graphs on four and five vertices are shown in Figure 6.16. Exercises 38 and 39 in Section 6.6 explore this class of graphs.

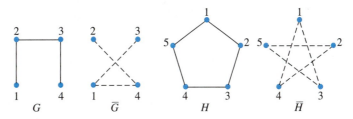

Figure 6.16   Two self-complementary graphs.

As important as it may be to show that two graphs are the same, it is often just as important to show that two graphs are not. Many algorithms for graphs need to distinguish among graphs to complete the processing both correctly and efficiently. The two graphs shown in Figure 6.17 are not isomorphic; can you prove it?

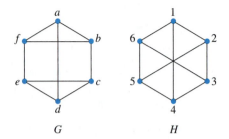

Figure 6.17   Two nonisomorphic graphs.

One way to prove that these two graphs are not isomorphic is to show that they do not share a property, called an **invariant,** that is preserved by an isomorphism. For example, two isomorphic graphs will have the same number of even vertices. A triangle in a graph G must be mapped to a triangle in any graph to which G is isomorphic. Since the graph G in Figure 6.17 contains a triangle but the graph H in Figure 6.17 has no triangle, the graphs are not isomorphic.

The exercises in Section 6.6 include problems about determining properties that are preserved by an isomorphism. These results will be useful when you face the problem of deciding whether two graphs are the same. Chemists use graph isomorphism to determine if two graphical representations of a chemical are, in fact, the same chemical. Graph algorithms, especially those dealing with coloring problems, often need to distinguish among a set of graphs to determine appropriate processing steps. An algorithm must be able to determine exactly which graph, up to isomorphism, is to be processed at a particular step of the algorithm.

6.5   Representation of Graphs

Two methods commonly are used for representing graphs (other than drawing a picture or just listing the vertices and edges). Both of these methods lead to widely used computer representations of a graph. In some applications, the order of the complexity of the algorithm will even depend on which representation is used.

6.5.1   Adjacency Matrix

For positive integers n and m, a matrix A of size $n \times m$ is a rectangular array of values organized as n rows, each with m entries called columns. The value at a location row i and column j is denoted as $A(i, j)$. Let $G = (V, E)$ be a graph with n vertices named $1, 2, \ldots, n$. An $n \times n$ matrix A is an **adjacency matrix** for G if and only if for $1 \le i, j \le n$,

$$A(i, j) = \begin{cases} 1 & \text{for } (i, j) \in E \\ 0 & \text{for } (i, j) \notin E \end{cases}$$

Since $A(i, j) = A(j, i)$ for $1 \le i, j \le n$, an adjacency matrix is symmetric. The diagonal entries of an adjacency matrix are zero, since a graph has no edge with both ends the same. Figure 6.18 shows a graph and its associated adjacency matrix.

	1	2	3	4	5	6	7	8	9
1	0	1	0	0	0	1	0	0	0
2	1	0	1	0	0	1	0	0	0
3	0	1	0	1	0	0	0	0	0
4	0	0	1	0	1	0	0	0	0
5	0	0	0	1	0	1	0	0	0
6	1	1	0	0	1	0	0	0	0
7	0	0	0	0	0	0	0	1	0
8	0	0	0	0	0	0	1	0	1
9	0	0	0	0	0	0	0	1	0

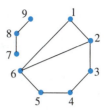

Figure 6.18 A graph and its adjacency matrix.

Other adjacency matrices for this graph can be formed by labeling the rows and columns of the matrix differently. The number of paths joining any pair of vertices in a graph can be found using an adjacency matrix. Exercise 43 in Section 6.6 will explore some properties of adjacency matrices.

6.5.2 Adjacency Lists

An **adjacency list** representation of a graph $G = (V, E)$ with n vertices only stores information about the edges that are in a graph. An adjacency list representation for a graph consists of a list of its vertices $v_1, v_2, \ldots, v_n$ together with a separate list for each vertex that contains all the vertices adjacent to v_i for $1 \le i \le n$. Figure 6.19 shows the graph of Figure 6.18 with an adjacency list representation.

Vertices	List of Adjacencies		
1	2	6	
2	1	6	3
3	2	4	
4	3	5	
5	4	6	
6	1	5	2
7	8		
8	7	9	
9	8		

Figure 6.19 A graph and its adjacency list.

Other adjacency list representations can be constructed both by listing the vertices in a different order and by listing the adjacencies of a vertex in a different order. An algorithm that examines all the edges of a graph as a computation step can have different measures of complexity if the graph is represented by an adjacency matrix rather than by an adjacency list. Neither representation introduced, however, is always better than the other.

6.6   Exercises

1. Find a graph with 12 edges having six vertices of degree three and the remaining vertices of degree less than three.

2. Give an example of a graph with at least four vertices, or prove that none exists, such that:
   (a) There are no vertices of odd degree.
   (b) There are no vertices of even degree.
   (c) There is exactly one vertex of odd degree.
   (d) There is exactly one vertex of even degree.
   (e) There are exactly two vertices of odd degree.

3. (a) Construct a graph with six vertices and degree sequence 1, 1, 2, 2, 3, 3.
   (b) Construct a graph with six vertices and degree sequence 1, 1, 3, 3, 3, 3.
   (c) Can you find at least two graphs with each of these degree sequences?

4. Construct all degree sequences for graphs with four vertices and no isolated vertex.

5. Determine all possible degree sequences for graphs with five vertices containing no isolated vertex and six edges.

6. Determine all possible degree sequences for graphs with five vertices containing no isolated vertex and eight edges.

7. Let $d_1, d_2, \ldots, d_n$ be a nondecreasing sequence of non-negative integers representing the degrees of the vertices of some graph. Prove that $\sum_{i=1}^{n} d_i$ is even. Is the converse to this result true?

8. Show that the sequence 2, 2, 2, 3, 3, 4, 5, 5, 6 is graphical. Build an answer from a graph with degree sequence 1, 1, 1, 1, 1, 1.

9. For $n = 2, 3, 4, 5$, determine a relationship between the number of edges and the number of vertices in an n-regular graph with p vertices where $p = 1, 2, 3, \ldots$. Construct all 3-regular graphs on four and six vertices.

10. Let G be a graph. Prove that G is bipartite if and only if G contains no odd cycle.

11. Draw a graph with 16 vertices labeled with the elements of $\{0, 1\} \times \{0, 1\} \times \{0, 1\} \times \{0, 1\}$ and edges that correspond to the edges in Q_4.

12. Prove that Q_n where n is some integral power of 2 has 2^n vertices and $n \cdot 2^{n-1}$ edges.

13. Prove that Q_n is bipartite for $n = 2, 3, 4, \ldots$.

14. Prove that for any graph G on six vertices, either G or $\bar{G}$ contains a triangle.

15. Construct $\overline{C_5}$, $\overline{K_{3,3}}$, and $\overline{K_{2,4}}$.

16. For the graphs

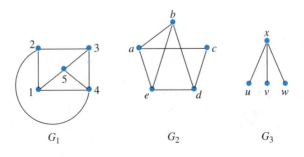

identify:

(a) the degree of each vertex
(b) a path of length greater than four, if any exists
(c) a cycle of size greater than four, if any exists
(d) a trail of length six, if one exists
(e) one circuit with more edges than the number of vertices in the graph, if one exists

17. For the graph

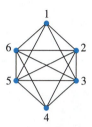

do the following:

(a) Find subgraphs without isolated vertices that are n−regular for $n = 2, 3, 4, 5$.
(b) Find one path for each of the lengths three, four, five, and six.
(c) Find one cycle of each of the sizes three, four, five, and six.
(d) Find one trail for each of the lengths 6, 8, 10, and 12.
(e) Find the induced subgraphs determined by the sets of vertices $\{1, 3, 4\}$, $\{2, 3, 5, 6\}$, $\{2, 4, 6\}$, and $\{1, 3, 4, 5, 6\}$.

The solutions for parts (a), (b), (c), and (e) should be drawings of graphs with the vertices labeled as in the figure. For part (d), list the vertices as they occur in the trail.

18. Prove that if a graph G has an n-circuit with n odd and $n > 3$, then G has an odd cycle.

19. Show that 1, 2, 2, 3, 4 is graphical but that 1, 3, 3, 3 is not. Prove the theorem of Havel-Hakimi that for $n \geq 1$, the sequence $d_1, d_2, \ldots, d_n$ is graphical if and only if $d_1, d_2, \ldots, d_{n-d_n}, d_{n-d_n+1} - 1, \ldots, d_{n-1} - 1$ is graphical.

20. Let $d_1, d_2, \ldots, d_n$ be a sequence of distinct integers where $n \geq 1$ and $0 \leq d_i \leq n - 1$ for $1 \leq i \leq n$. Prove that such a sequence is not graphical.

21. Let $G = (V, E)$ be a bipartite graph with vertex partition $V = A \cup B$. Prove that if $|A| \neq |B|$, then G is not Hamiltonian.

22. Find a Hamiltonian cycle in Q_3.

23. Find a Hamiltonian cycle in the following graphs:

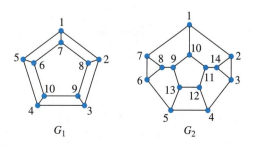

G_1 G_2

24. Find a Hamiltonian cycle in the following graph:

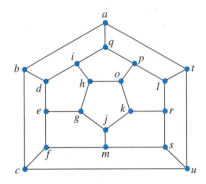

25. Complete the proof of Example 5 in Section 6.3.1.
26. Complete the proof of Case ii of Example 6 in Section 6.3.1.
27. Show that the function $F(a) = 3$, $F(b) = 1$, $F(c) = 4$, and $F(d) = 2$ is an isomorphism between the graphs G and H as shown:

28. Prove that for two graphs G and H that G is isomorphic to H if and only if $\overline{G}$ is isomorphic to $\overline{H}$.
29. Let $G = (V, E)$ and $H = (V_1, E_1)$ be isomorphic graphs. Prove the degrees of the vertices of G are exactly the degrees of the vertices of H. Show that $|V| = |V_1|$ and $|E| = |E_1|$ alone do not imply that G and H are isomorphic.
30. Prove that G and H as shown are isomorphic:

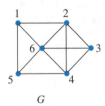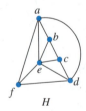

31. Prove that G and H as shown are not isomorphic:

32. Prove that G and H as shown are isomorphic:

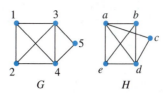

33. Prove that no pair of G_1, G_2, and G_3 as shown are isomorphic:

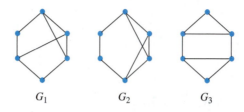

34. Prove that G and H as shown are isomorphic:

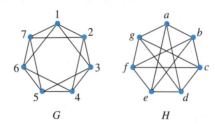

35. For each graph shown in Exercise 16, construct both an adjacency matrix and an adjacency list representation.

36. Construct two different adjacency matrices and two different adjacency lists for C_4.

37. Prove that if $G = (V, E)$ is isomorphic to $\overline{G}$, then $|V| \equiv 0, 1 \ (mod \ 4)$.

38. Given a self-complementary graph with $4m$ vertices for some $m > 0$, construct a self-complementary graph on $4m + 1$ vertices that contains the self-complementary graph on $4m$ vertices as an induced subgraph.

39. Construct all self-complementary graphs on four and five vertices.

40. Let $G = (V, E)$ be a graph with $|V| \geq 3$. Prove that if the degree of each vertex in G is at least $|V|/2$, then G is Hamiltonian.

41. Let $G = (V, E)$ be a graph. Prove that if for any two nonadjacent vertices v and w of G we have $deg(v) + deg(w) \geq |V|$, then G is Hamiltonian.

42. Let $\mathcal{G}$ be a set of graphs. For all $G, H \in \mathcal{G}$, define the relation R as $G \ R \ H$ if and only if G and H are isomorphic. Prove that R is an equivalence relation.

43. Let $A^n = [a_{ij}^{(n)}]$ be the nth power of the adjacency matrix of G. Prove that:

    (a) $a_{ij}^{(2)}$, $i \neq j$, is the number of $i - j$ paths of length 2 in G.

    (b) $a_{ii}^{(2)} = deg(i)$.

    (c) $(1/6)\sum a_{ii}^{(3)}$ is the number of 3-cycles in G.

    (d) For each graph given below by its adjacency matrix representation, verify parts (a) through (c). For each pair of vertices, determine how many paths of length 3 are joining them. List all the 3-cycles in each of the graphs (3-cycle a–b–c is counted as being different from the 3-cycle a–c–b.)

Adjacency Matrix			
	1	2	3	4
1	0	1	0	1
2	1	0	1	1
3	0	1	0	0
4	1	1	0	0

Adjacency Matrix					
	1	2	3	4	5	6
1	0	0	0	1	1	1
2	0	0	0	1	1	1
3	0	0	0	1	1	1
4	1	1	1	0	0	0
5	1	1	1	0	0	0
6	1	1	1	0	0	0

6.7   Connected Graphs

How do we characterize the obvious difference between graphs G and H shown in Figure 6.20?

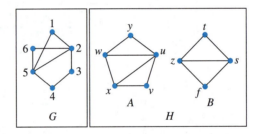

Figure 6.20   Two graphs.

This is an important question, because many algorithms that use graphs as models first find the pieces of a graph, like A and B in H, and then solve the problem for each piece separately. The notion we will use to resolve this question is based on the simple idea of asking whether a pair of vertices can be the ends of a trail. Extensions of this simple idea even play a role in efficient algorithms for numerical analysis involving large systems of equations.

Section 6.8 also deals with the first problem of graph theory. The problem simply asks whether you can stroll across the bridges over a river and return to your starting point after having crossed each bridge exactly once. The result is interesting, because there is a complete characterization of those graphs that model such a stroll.

6.7.1   The Relation *CONN*

The important building block for this section is the relation based on the idea of whether two vertices can be the ends of a trail. This relation is an equivalence relation that allows us to find the pieces of a graph, such as shown for H in Figure 6.20, in a unique way.

Definition 1.   A graph for which each pair of (not necessarily distinct) vertices is joined by a trail is **connected.** A graph that is not connected is **disconnected.** A **connected component** of a graph is a subgraph that is connected but not contained in any connected subgraph with more vertices or edges.

For the graphs drawn in Figure 6.20, it is easy to identify G as a connected graph and the subgraphs A and B as the connected components of the disconnected graph H.

The key notion for designing an algorithm to decompose a graph into connected components is the following relation defined on the vertices of a graph.

Definition 2.   Let $G = (V, E)$ be a graph. For all $v, w \in V$, define v *CONN* w if and only if there is a trail in G from v to w.

Example 1.   Find the connected components of the graphs G and H:

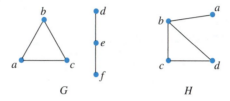

$$G \qquad\qquad\qquad H$$

Solution.   There is no trail in G from a to e. Therefore, G is disconnected. The two components of G are

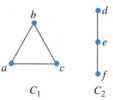

$$C_1 \qquad\qquad C_2$$

The graph H is connected, as can be seen by considering the trail $a - b - c - d$. H has one connected component, itself!   ■

To begin the process of decomposing a graph into its connected components, we will show that the relation *CONN* is an equivalence relation on the vertices of the graph. Once we know that the relation is an equivalence relation, we can use the fact that the distinct equivalence classes form a partition. It will follow that the elements of the partition are the vertex sets of the distinct connected components of the graph. The actual components of the graphs are the subgraphs induced by the elements of this partition.

Theorem 1.   *CONN* is an equivalence relation on the vertices of a graph.

Proof.   The proof consists of showing that the relation *CONN* is reflexive, symmetric, and transitive. Let $G = (V, E)$ be a graph. For each $v \in V$, the vertex itself is a trail from v to v. Thus, v *CONN* v holds for each $v \in V$, and *CONN* is reflexive. For vertices $v, w \in V$ such that v *CONN* w, there is a trail in G from v to w. Such a trail can be reversed to form a trail from w to v, showing that w *CONN* v. Therefore, *CONN* is symmetric. Now, to show that *CONN* is transitive let v, w and x be vertices of G such that v *CONN* w and w *CONN* x. Since v *CONN* w and w *CONN* x, there are trails in G from v to w and from w to x. By following the trail from v to w and then following the trail from w to x, a trail from v to x is formed in G provided no edge occurs in both trails. If an edge occurs in both trails, it is easy to modify this sequence of vertices and edges to make it into a trail from v to x. This shows that v *CONN* x. Therefore, the relation *CONN* is transitive. Since *CONN* has been shown to be reflexive, symmetric, and transitive, *CONN* is an equivalence relation. ∎

Since *CONN* is an equivalence relation, the vertex set of any graph is uniquely partitioned into equivalence classes by means of the relation (see Theorem 2 in Section 3.6.1). For a graph $G = (V, E)$, let $C \subseteq V$ such that C is an equivalence class of V determined by *CONN*. The induced subgraph $<C>$ contains all the edges of G that have both their ends in C. The subgraph $<C>$ is a connected component of G. Similarly, the subgraphs induced by the other distinct equivalence classes of G relative to *CONN* determine all the other connected components of G. Figure 6.21 shows an example of how *CONN* is used to find the connected components of a graph.

$$G = (\{1, 2, 3, 4, 5, 6, 7, 8, 9, 10\}, \{(1, 2), (2, 4), (2, 6), (1, 4),$$
$$(5, 6), (3, 5), (3, 4), (7, 8), (8, 9)\,(8, 10), (10, 7)\})$$

Equivalence Classes of CONN

$\{1, 2, 3, 4, 5, 6\}$ $\{7, 8, 9, 10\}$

Components

$<\{1, 2, 3, 4, 5, 6\}>$ $<\{7, 8, 9, 10\}>$

Figure 6.21   Connected components.

If a graph is connected, it contains a single equivalence class with respect to the relation *CONN*. In any case, each vertex of a graph is contained in exactly one connected component of a graph.

In solving problems such as finding the connected components of a graph, we need an efficient method for examining all the vertices and edges of a graph. Two methods that are very efficient are depth first search and breadth first search.

6.7.2   Depth First Search

A **depth first search** of a connected graph $G = (V, E)$ is a recursive algorithm to examine the vertices and the edges of a graph. To begin a depth first search, choose an adjacency list representation for G. Different adjacency list structures will give rise to examining the vertices and edges in different orders, but in any representation, each edge and vertex will be visited.

First, mark all the vertices of G as "unvisited." For the next step, choose any vertex v, and mark it as "visited." Suppose the first vertex on the adjacency list for v is w. If w is marked "visited," do nothing with the edge that this entry represents, and proceed to examine the next vertex in the list of vertices adjacent to v. If w is marked "unvisited," use w as a new starting point for this process. When all the edges incident to w are examined, return to v, and examine the next unexamined vertex on v's list of adjacent vertices. The process terminates when the procedure returns to the starting vertex and all the edges incident to this vertex have been examined.

An example of a depth first search is shown in Figure 6.22. A depth first search encounters two kinds of edges. Some edges, called *tree edges,* join the vertex being examined to a vertex that has not already been visited. All other edges incident to the current vertex join the current vertex to an already-visited vertex. In a depth first search, the tree edges are shown as solid lines, and all other edges are shown as dashed lines.

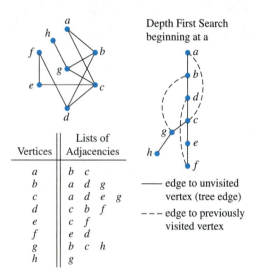

Figure 6.22 Depth first search starting at *a*.

In Figure 6.22, the search has examined the edges in the order in which they are listed in the adjacency list.

The edges in a graph can be partitioned into two sets by the search procedure. One set of edges consists of all the edges (v, w) for which both ends of the edge have $Visited[\, *\,] =$ "*visited*" when the edge is examined for the first time. The other set of edges consists of all the edges (v, w) with one end—say, v—having $Visited[\, v\,] =$ "*visited*" and the other end w having $Visited[w] =$ "*unvisited*" when the edge is examined for the first time. All the edges in this second set form a special subgraph called a **depth first search tree.** Trees are formally defined and discussed in more detail later in this chapter, but for the search just shown in Figure 6.22, the depth first search tree is shown in Figure 6.23.

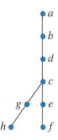

Figure 6.23 Depth first search tree.

If a graph is not connected, then the process is repeated for each connected component. The algorithm given here is intended to deal with connected graphs.

Algorithm: Depth First Search (Dfs)

INPUT: Connected graph $G = (V, E)$ and start vertex v_0 and vertices numbered $1, \ldots, |V|$

RESULT: Each vertex and each edge is examined

Create an array *Visited* with $|V|$ positions initialized to *"unvisited"*
$Dfs(G, v_0, Visited)$ /* The initial call uses start vertex v_0 */

    $Dfs(G, v, Visited)$/* The recursive procedure */
        $Visited[v] =$ *"visited"*
        for each vertex w adjacent to v do
            if ($Visited[w] =$ *"unvisited"*) then
                $Dfs(G, w, Visited)$

To prove *Dfs* is correct, we must prove that every vertex and every edge of a connected graph G is examined when *Dfs* is called for any vertex of the graph. Since every adjacency of a vertex v is examined when $Dfs(G, v, Visited)$ is executed, we can prove that every vertex and every edge is examined if we can prove that *Dfs* is executed for each vertex of G.

Suppose for some vertex $w \in V$ that $Dfs(G, w, Visited)$ is not executed and $Visited[w] =$ *"unvisited"* after execution of *Dfs*. Without loss of generality, we can assume w is adjacent to a vertex v that had $Dfs(G, v, Visited)$ executed, because the graph is connected and $Dfs(G, v, Visited)$ is executed. When the adjacencies of v were examined, however, $Visited[w]$ would have been *"unvisited."* Therefore, $Dfs(G, w, Visited)$ would have been executed, and $Visited[w]$ would have been asssigned *"visited."* This contradicts the assumption that a vertex such as w exists. Therefore, the algorithm is executed for each vertex of G, and the algorithm is correct.

6.7.3   Complexity of *Dfs*

The analysis of the complexity of this search method is restricted to the case of connected graphs. The generalization to disconnected graphs is left for the reader. Also, assume that a graph has an adjacency list representation. The analysis for the case using an adjacency matrix representation for the graph is also left to the reader.

In proving the correctness of *Dfs*, we showed that for a graph $G = (V, E)$, the procedure *Dfs* is called $|V|$ times, once for each vertex of G. We can calculate the complexity of *Dfs* by adding up the time taken by each of these $|V|$ calls to *Dfs*. There are two parts to each call, the marking process and the for loop. The marking process takes a constant amount of time—say, c_1. Each time through the loop, the for loop takes a constant amount of time to test the condition—say, c_2—and a constant amount of time—say, c_3—as an upper bound for the time to execute the body of the loop (that is, to test the condition and make the recursive call to *Dfs* when the condition is *TRUE*). For each vertex, the loop will be executed once for each entry on the vertex's adjacency list. The total time to execute the loop for a single vertex v is bounded by $c_1 + (c_2 + c_3)deg(v)$. Summing the time taken by each of the calls to *Dfs* gives an upper bound on the complexity of *Dfs*. Since $\sum_{v \in V} deg(v) = 2 \cdot |E|$ by Theorem 2 in Section 6.2, we have

$$\text{Running time of } Dfs \text{ on } G \leq \sum_{v \in V}(c_1 + (c_2 + c_3)deg(v))$$

$$\leq c_1 \cdot |V| + (c_2 + c_3)\sum_{v \in V} deg(v)$$

$$\leq c_1 \cdot |V| + (c_2 + c_3) \cdot 2 \cdot |E|$$

$$= \mathbf{O}(|V| + |E|)$$

Observe that this argument does not just count operations. It uses results about graphs to provide a more careful analysis.

6.7.4   Breadth First Search

A **breadth first search** is one of the simplest algorithms for searching a graph. The range of effective applications of this search method is quite different from the range of applications typically used with a depth first search.

A typical problem using a breadth first search involves finding an optimal path between two vertices. When the edges have weights, you might be asking how much information can flow from one computer to another if the information passes through a number of computers with different capacities. If the vertices represent plants or warehouses or retail stores, you might be asking if you can supply each store with all the product they can sell. When researching a family genealogy, this search could be used to print out the members of the family tree, one generation at a time. For example, the family tree shown in Figure 3.2 (see Section 3.1) would have its members printed in the order Mary, John, Peter, Elaine,

Maude, Harold, George, and Elizabeth if a breadth first search was used. Even the Job Assignment Problem introduced in Section 6.1 can be effectively solved using this search procedure as part of the algorithm.

When a depth first search starts at a vertex v and finds a vertex w with $Visited[w] = FALSE$, no other adjacency of v will be examined until all the adjacencies of w have been examined. In contrast, a breadth first search examines all the vertices adjacent to a vertex before any vertex not adjacent to that vertex is examined. A temporary storage area is used to store vertices for later processing. This storage area is managed as a list, with entries being added at the end and removed from the front (this is formally known as a **queue,** or a **first-in-first-out** list, which we only need to deal with intuitively here).

To implement a breadth first search of a graph $G = (V, E)$, examine the vertices and edges of G as follows. For getting started, first mark all the vertices as *"unvisited."* Choose any vertex—say, v. Mark v as *"visited,"* and put it on the end of the list. Now, for each vertex on the list, examine in some order all the vertices adjacent to v. When all the *"un-visited"* vertices adjacent to v have been marked as *"visited"* and put on the end of the list, remove the vertex from the front of the list and repeat this procedure for any of its adjacencies that are marked as *"unvisited."* Continue this procedure until the list is empty.

A breadth first search will be performed on the graph used in the depth first search example so that it will be easier to see the differences between the search procedures. As with a depth first search, a **breadth first search tree** can be defined. One way to see the differences between these two search procedures is to notice the difference between the structure of the depth first search tree shown in Figure 6.23 and the corresponding breadth first search tree shown in Figure 6.24. The same convention for tree edges and all other edges explained for a depth first search are used for a breadth first search.

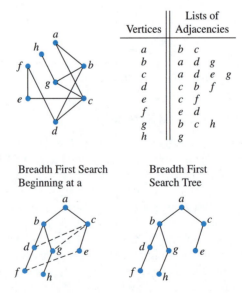

Figure 6.24   Breadth first search and breadth first search tree.

Algorithm: Breadth First Search (*Bfs*)

INPUT: Connected graph $G = (V, E)$, and start vertex v_0
RESULT: Each vertex and each edge is examined

Create an array *Visited* with $|V|$ positions initialized to *FALSE*
Create a queue Q that is initially empty
          /* Q holds the discovered vertices waiting to be processed */
Bfs(G, v_0, *Visited*, Q)   /* Start the search at v_0 */

Bfs(G, v, *Visited*, Q)   /* The recursive procedure */
          Visited[v] = *TRUE*
          Put v at the end of Q
          while $Q \neq \emptyset$
                    Remove the vertex w at the front of Q
                    for each vertex u adjacent to w do
                              if (*Visited*[u] = *FALSE*) then
                                        Visited[u] = *TRUE*
                                        Put u at the end of Q

By very similar arguments to the ones used for the depth first search, the reader can prove that this algorithm is correct and that the complexity of *Bfs* is also $\mathbf{O}(|V| + |E|)$.

6.7.5   Finding Connected Components

One use of connectedness in many graph algorithms is to find the components of a graph and then solve a problem for each of the components separately. Since the components can be found in $\mathbf{O}(|E| + |V|)$ time, this operation does not affect the overall complexity of an algorithm that must typically examine each vertex and each edge in a graph to determine an answer.

The algorithm presented for finding connected components uses a depth first search of the graph to find the components one at a time. The adjacency list representation of a graph makes the depth first search particularly easy to implement. The procedure to find the Connected Components that calls *DfsComp* uses a counter (*CompNumber*) to keep track of how many connected components are found in the graph being processed. If all the vertices are visited as a result of the first call to *DfsComp*, then this counter will have a value of 1 after the process is finished. In this case, the graph is connected.

Algorithm: Connected Components

INPUT: Graph $G = (V, E)$
OUTPUT: For each vertex, a number labeling its connected component

Create an array *Visited*[1..|V|] with |V| positions each initialized
      to *FALSE*
Create an array *Comp* with |V| positions each initialized to 0
      /* *Comp* [v] holds the integer label of v's component */
CompNumber = 0   /* No components discovered yet */
for v = 1 to |V| do
      /* Make sure each vertex gets visited */
      if (*Visited* [v] = *FALSE*) then   /* New component! */
          Add 1 to *CompNumber*
              /* *CompNumber* is the number of the new component */
          DfsComp(G, v, *Visited*, *Comp*, *CompNumber*)
              /* Explore the new component starting at v */

DfsComp(G, v, *Visited*, *Comp*, *CompNumber*)
          /* The recursive procedure */
      Comp [v] = *CompNumber*
          /* Record the number of v's component */
      Visited [v] = *TRUE*
      for each vertex w adjacent to v do
          if (*Visited* [w]) = *FALSE* then
              DfsComp(G, w, *Visited*, *Comp*, *CompNumber*)

The procedure *DfsComp* is a slight modification of the earlier procedure called *Dfs*. The difference is that this algorithm ensures every vertex in the graph is eventually examined, even if the graph is not connected.

A better understanding of the roles of the arrays and variables in *DfsComp* will help to show how the procedure operates. The variables *Visited*, *Comp*, and *CompNumber* implement one way to keep track of what is learned by *DfsComp* at a vertex. At the end of the algorithm's execution, all vertices with the same value in *Comp* have been identified as belonging to the same connected component. The array *Visited* keeps track, as the algorithm is executed, of whether a vertex has been examined. If *Visited* has a value *TRUE* for a vertex, then it follows that *Comp* has been assigned a value for that vertex. Because *Visited* only has its value changed once for each vertex, *Comp* is assigned exactly one value for each vertex. The variable *CompNumber* is used to make sure that the same number is assigned to each vertex in a particular component and that vertices in different components are identified by different numbers.

A breadth first search could also be used to find the connected components of a graph. The complexity of such an algorithm would be the same as the algorithm that uses a depth first search. The design of an algorithm for finding connected components using a breadth first search is left to the reader.

6.8   The Königsberg Bridge Problem

In the eighteenth century, the city of Königsberg, Prussia, had seven bridges that crossed the Pregel River. The bridges connected the two islands in the river with each other and with the opposite banks. Figure 6.25 shows the configuration of the bridges.

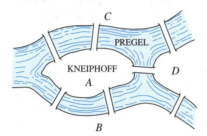

Figure 6.25   Königsberg bridges on the Pregel River.

The town's folk had wondered if the following problem, called the **Königsberg Bridge Problem,** had a solution: Is there a continuous walk starting at C that crosses each of the seven bridges exactly once and returns the walker to C? The problem was initially solved by the famous mathematician Leonhard Euler (1707–1783, b. Switzerland), who modeled the problem with a graph and then found necessary and sufficient conditions for the graph constructed to represent a walk of the required kind.

To define a graph that represents the problem, let each piece of land be a vertex. Each bridge defines an edge joining the vertices that represent the pieces of land at the ends of the bridge. For the Königsberg Bridge Problem, the resulting graph is shown in Figure 6.26.

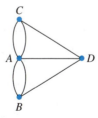

Figure 6.26   Königsberg bridge graph.

Properly speaking, the model in Figure 6.26 is not a graph, because there are pairs of vertices with more than a single edge joining them. This extension of the notion of a graph is called a **multigraph.** The extension of Euler's theorem to multigraphs would not be difficult, but it would require that the terminology of multigraphs be explained. The notions are obvious extensions of the ideas used with graphs and can be formulated by the reader. The main point is to treat the multiple edges as distinct entities in the set of edges rather than as multiple listings of the same element in a set. Euler's theorem will be proved for graphs as defined in Section 6.1.1.

By trial and error, it will become clear that no circuit in the graph shown in Figure 6.26 starts and ends at the same vertex and contains each edge once. Hence, no walk of the required kind exists in Königsberg. When there is a circuit (trail) in a graph that includes each edge exactly once, the circuit (trail) is called an **Eulerian circuit (trail).** Euler's theorem gives necessary and sufficient conditions for the existence of an Eulerian circuit in a graph.

Theorem 1.   (Königsberg Bridge Theorem) (Euler, 1736)   Let G be a connected graph. G has an Eulerian circuit if and only if each vertex is even.

Proof. $(\Rightarrow)$ Let G be a graph that has an Eulerian circuit C, and show that every vertex of G is even.

Let C be $v_1, v_2, \ldots, v_n$ where a vertex v_i for $1 \le i \le n$ may occur more than once in the list. We want to show that each vertex v_i has even degree.

Before continuing the proof, we consider an example that motivates how this part of the proof will go.

Motivation for the proof: In Figure 6.27, the circuit

$$C = 1, 2, 3, 1, 4, 5, 1, 6, 7, 8, 6, 2, 9, 1$$

with the edges

$(1, 2), (2, 3), (3, 1), (1, 4), (4, 5), (5, 1), (1, 6), (6, 7), (7, 8), (8, 6), (6, 2), (2, 9), (9, 1)$

is an Eulerian circuit. In Figure 6.27, you can follow the Euler circuit by traversing the edges by going from vertex to vertex in the following order: 1-2-3-1-4-5-1-6-7-8-6-2-9-1.

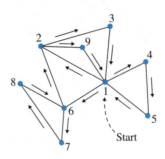

Figure 6.27   Graph with an Eulerian circuit.

For each of the vertices $2, 3, \ldots, 9$, we can see that each time the vertex occurs as the end of an edge, it is the beginning of the next edge. Thus, each time one of these vertices occurs in the circuit, we can add 2 to the running total of the degree of that vertex. For the vertex 1, the same computation works for all its occurrences except for the first and the last in the circuit. Both the first and the last occurrence of vertex 1 contribute one to the degree of vertex 1 so there is an addition of two to the degree of vertex 1 for these occurrences. A summary for these calculations is given in Table 6.2.

Table 6.2 Computing Degrees in Traversing an Euler Circuit

Traversal of an Euler Circuit Add to Current Total—1 When Entering a Vertex and 1 When Leaving a Vertex— Table shows running totals																
Euler Circuit		(1,2)(2,3)(3,1)(1,4)(4,5)(5,1)(1,6)(6,7)(7,8)(8,6)(6,2)(2,9)(9,1)														
Vertex	Start	1-	2-	3-	1-	4-	5-	1-	6-	7-	8-	6-	2-	9-	1	Degree
1	0	1			3			5							6	6
2	0		2										4			4
3	0			2												2
4	0					2										2
5	0						2									2
6	0								2			4				4
7	0									2						2
8	0										2					2
9	0													2		2

Continuation of the proof: First, consider the case where $v \neq x_1$; then, also, $v \neq x_n$. Vertex v_i occurs in C some number t of times. In each case, there are edges in G from v_i to the preceding and succeeding vertices of G—two edges per occurrence of v_i. Moreover, no edge is allowed to occur twice in the circuit, so in all, there are $2t$ edges incident to v_i.

If $v = x_1$, then the argument is analogous, except that one pair of edges is (x_1, x_2) and (x_{n-1}, x_n).

($\Leftarrow$) Conversely, suppose that every vertex of a connected graph G has even degree, and prove that G contains an Eulerian circuit. Select a vertex v in G, and begin a trail T at v. Continue this trail as long as possible until a vertex w is encountered that has all the edges incident to it in T. To prove that a circuit has been formed, show that $v = w$. Suppose $v \neq w$. Each time w is encountered on T, except for the last time, one edge is used to enter w and another to leave w. The last time w is encountered in T, only one edge is used—the one to enter w. Therefore, an odd number of edges incident to w occur in T. Since the degree of w is even, there must be at least one edge incident with w that does not belong to T. Therefore, T can be continued, which is a contradiction, and so $w = v$. If T contains all the edges of G, then T is an Eulerian circuit of G.

Before continuing the proof, we consider an example that motivates how the rest of this proof will go.

Motivation for the proof: Suppose the circuit T does not contain all the edges of G. An example of such an occurrence is shown in Figure 6.28. (This graph is known as Mohammed's Scimitar.)

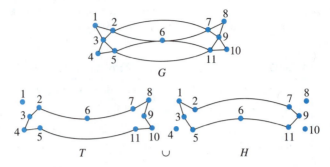

Figure 6.28 Circuit that is not Eulerian.

Continuation of the proof: Define H to be the spanning subgraph of G with edge set $E(G) - E(T)$. Since every vertex in T has even degree, every vertex in H has as degree a value that is the difference of two even numbers and, hence, is an even number. Choose an edge (u, v) that is in H and that has at least one endpoint, v, in T. Such a choice is possible because G is connected. Let H_1 be the connected component of H that contains the vertex v.

Begin a trail U at v in H_1, and continue it as long as possible. As before, U will end at its starting vertex v and be a circuit. Now, form a circuit C in G that has more edges than T by inserting U at v. The larger circuit is formed when v is reached while traveling around T, by taking a side trip around U (splicing U to T) before completing the rest of T.

Motivation for the proof: An example of this operation is shown in Figure 6.29.

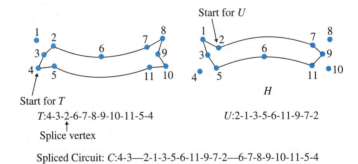

Figure 6.29 Splicing circuits.

Continuation of the proof: If T now contains all the edges of G, then T is an Eulerian circuit. If T still does not contain all the edges of G, then form the spanning subgraph with edge set $E(G) - E(T)$, and repeat the process outlined for U to find a circuit in this graph that can be spliced onto T. Repeat this process until an Eulerian circuit for G is formed. ∎

No wonder the people of Königsberg could not find a good walk: Every vertex of the graph in Figure 6.26 is odd!

6.8.1   Graph Tracing

In directing a plotting device, commands are given to indicate where a pen should be positioned to begin to draw a portion of the diagram and how the point of the pen should move on the paper. When the pen is being repositioned to draw a different part of the diagram, the write head is in the up position, or off the paper. If the diagram is composed of many different parts, then a considerable amount of plotting time may be taken up with repositioning the pen. It would be helpful to know ahead of time how many times the pen must be repositioned and to schedule the plotting to use only as many pen motions in the nondrawing position as are absolutely necessary.

An application of Euler's theorem is to determine whether all the edges of a graph can be traced by a plotter exactly once, starting and ending at a given vertex. The next result will answer the question about whether a graph could be drawn (passing over each edge exactly once) without lifting the plotting pen. Figure 6.30 shows a graph that cannot be drawn without lifting the pen.

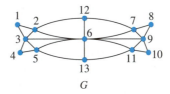

Figure 6.30   A graph to plot.

Clearly, the pen could be lifted and repositioned for each edge in Figure 6.30. The real question is this: What is the fewest number of times that a pen must be lifted to complete the tracing of a graph? In particular, how many times must the pen be lifted to trace the graph shown in Figure 6.30? (Assume that at the end of the plot, the pen is lifted so the paper can be removed from the plotter but that this lifting is not counted when answering the question.) Theorem 3 gives an answer to this question.

Theorem 2.   Let $G = (V, E)$ be a connected graph with $2 \cdot k > 0$ odd vertices for some integer k. Then, there are k edge-disjoint trails

$$Tr_1, Tr_2, \ldots, Tr_k$$

in G such that

$$E(G) = E(Tr_1) \cup E(Tr_2) \cup \cdots \cup E(Tr_k)$$

Proof.   Let $v_1, v_2, \ldots, v_{2 \cdot k-1}, v_{2 \cdot k}$ be the odd vertices of G where $k > 0$. Let H be the graph formed from G by adding the k edges

$$(v_1, v_2), (v_3, v_4), \cdots, (v_{2 \cdot k-1}, v_{2 \cdot k})$$

(You may be joining pairs of vertices that are already adjacent and need Euler's Theorem in the context of multigraphs. As long as you allow the edge "set" of the graph to contain and treat as different multiple edges joining the same pair of vertices, this proof is valid.) Every vertex in H has even degree, so H has an Eulerian circuit. List the edges of an Eulerian circuit of H beginning with the newly added edge (v_1, v_2). Deleting the edge (v_1, v_2) from the list results in a trail from v_2 back to v_1. Now, delete the added edge (v_3, v_4) from this trail. The result will be two trails of the form v_2 to v_3 and v_4 to v_1 or v_2 to v_4 and v_3 to v_1.

After deleting the remaining edges that were added to G to form H, k edge-disjoint trails containing all the edges of G are formed. ∎

Corollary 1:   Let G be connected and have two vertices of odd degree. There is a Euleria trail in G that begins at one of the odd vertices and ends at the other.

For the proof of Corollary 1, it is possible that the pair of odd vertices to which you add an extra edge are already connected. Keeping track of such an edge as separate from the one originally in the graph is all that is needed to make the proof valid.

For any decomposition of a graph into pairwise edge disjoint trails, each vertex of odd degree must be the end of at least one trail. Therefore, since each trail has two ends, a graph with $2 \cdot k$ odd vertices with $k > 0$ cannot be decomposed into fewer than k trails. In terms of the plotting problem, interpret this result to say that the pen will have to be lifted and moved at least $k - 1$ times if the plot graph contains $2 \cdot k$ vertices of odd degree where $k > 0$. For the graph shown in Figure 6.30, it would be necessary to lift the pen once, because there are four vertices of odd degree. The two part tracing of the graph shown in Figure 6.31 accomplishes the tracing while lifting the pen a minimum number of times.

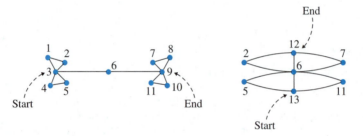

Figure 6.31   Minimum tracing of a graph.

Figure 6.32 shows other possible trail decompositions of the graph shown in Figure 6.30. Theorem 3 mentions nothing about the uniqueness of the decomposition. Each different way of adding edges to pairs of odd vertices may give a different set of trails.

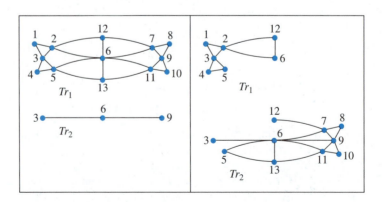

Figure 6.32   Other trail decompositions.

6.9   **Exercises**

1. Construct connected graphs of the following sorts:
   (a) All graphs of five vertices with at least seven edges
   (b) All cubic graphs with at most eight vertices
   (c) One 4-regular graphs of six vertices
   (d) Three 5-regular graphs of eight vertices

2. In a graph, the edges are pairs of vertices, with no ordering—unlike our formalization with relations as sets of ordered pairs. So, here treat the edges as ordered pairs as follows: For any graph $G = (V, E)$, let $\hat{E}$ be the set of all *ordered pairs* (v_1, v_2) and (v_2, v_1) where $(v_1, v_2) \in E$. Show that *CONN* is the reflexive and transitive closure of $\hat{E}$.

3. Find a graph G such that $\overline{G}$ is not connected.

4. Let G be a graph. Prove that if G is disconnected, then $\overline{G}$ is connected.

5. Let $G = (V, E)$ be a connected graph with at least two vertices. Prove that if $|V| > |E|$, then G has a vertex of degree one.

6. Suppose that $2n$ computers are networked so that each computer is connected to n other computers. Prove that it is always possible for any pair of computers to pass messages to one another. (*Hint:* See Exercise 40 in Section 6.6.)

7. Let G be a connected graph with average degree greater than two. Prove that G contains at least two cycles.

For problems involving depth first and breadth first searches, assume the adjacency lists are formed by listing the adjacencies of each vertex in increasing order.

8. Carry out a depth first search on the graph in Figure 6.22 starting at the vertex f. Display the result of the search process as was done in Figure 6.22.

9. Carry out a depth first search on the graph in Figure 6.22 starting at the vertex e. Display the result of the search process as was done in Figure 6.22.

10. Carry out a breadth first search on the graph in Figure 6.24 starting at the vertex f. Display the result of the search process as was done in Figure 6.24.

11. Carry out a breadth first search on the graph in Figure 6.24 starting at the vertex g. Display the result of the search process as was done in Figure 6.24.

12. Let the graph G be given by the following adjacency list:

Vertices	List of Adjacencies			
1	2	6	7	8
2	1	3	4	
3	2	4	5	6
4	2	3		
5	3			
6	1	3		
7	1	8	9	10
8	1	7	9	10
9	7	8		
10	7	8		

Draw the depth first search tree or the breadth first search tree that results from the following:

(a) A depth first search on G beginning at vertex 1
(b) A depth first search on G beginning at vertex 6
(c) A breadth first search on G beginning at vertex 1
(d) A breadth first search on G beginning at vertex 6

13. For each of the following graphs:

    (a) Carry out a depth first search starting at vertex b.
    (b) Carry out a breadth first search starting at vertex b.
    (c) Repeat parts (a) and (b) using vertex g as the starting vertex.

 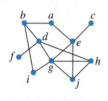

In each case display the result of the search process as was done in Figures 6.22 through 6.24.

14. Perform a depth first search and a breadth first search on the graph given by the following adjacency list:

Vertices	List of Adjacencies		
1	9	2	4
2	1	3	5
3	2	4	
4	3	1	
5	6	2	
6	7	9	5
7	8	6	
8	7		
9	6	1	

Start each search at vertex 3. Repeat the problem starting with vertex 5. Draw the depth first search tree and the breadth first search tree for each case.

15. Write an algorithm for finding a shortest path between any two vertices in a connected graph.

16. Let $G = (V, E)$ be a graph. Prove that if the minimum degree of G is greater than $(|V| - 1)/2$, then G is connected.

17. Let $G = (V, E)$ be a graph. Prove that if $|E| < |V| - 1$, then G is disconnected.

18. Find connected graphs with at least eight vertices such that:

    (a) G is Eulerian and not Hamiltonian.
    (b) G is not Eulerian but is Hamiltonian.
    (c) G is not Eulerian and not Hamiltonian.
    (d) G is Eulerian and Hamiltonian.

19. The Hall of Horrors at an amusement park challenges you to enter at the Start door and find your way to the Escape door. After passing through a door, the door closes and locks. How many doors can you leave closed as you find your way from the Start door to the Escape door? The layout of the Hall of Horrors is shown below.

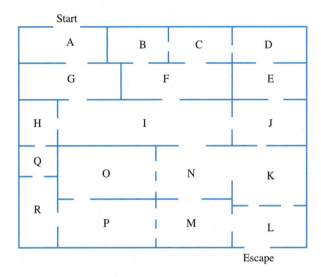

20. A real estate agent wants to show the three houses shown. Lay out a tour that goes through each door.

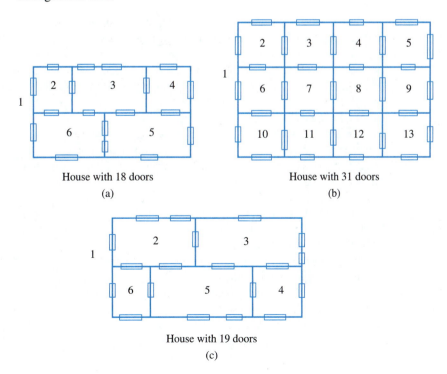

House with 18 doors
(a)

House with 31 doors
(b)

House with 19 doors
(c)

21. Find a tracing that minimizes the number of pen lifts for the graph shown.

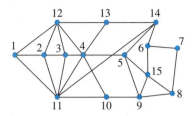

22. Prove that any Eulerian graph can be decomposed into a set of pairwise edge-disjoint cycles.

6.10 Trees

Historically, the study of trees was independently initiated by Arthur Cayley (1821–1895 b. England) and Gustav Kirchoff (1824–1887, b. Prussia). Cayley's work in the 1850s used these graphs as a tool for counting the number of saturated hydrocarbons $C_n H_{2n+2}$ containing n carbon atoms and $2n + 2$ hydrogen atoms. Kirchoff's work in 1847 used this same family of graphs to solve a system of simultaneous linear equations that give the current in each branch and around each circuit of an electrical network. Figure 6.33, on page 371, shows examples of typical graphs representing the problems Cayley and Kirchoff dealt with.

In Cayley's work, trees are used to represent the structure of hydrocarbons. Hydrocarbons contain atoms of hydrogen, oxygen, carbon, and nitrogen. The letters in Figure 6.33(a) represent the atoms, and a line between a pair of atoms represents a chemical bond. Methane, for example, is composed of four hydrogen atoms that are bonded to one carbon atom. Atoms of hydrogen always correspond to vertices of degree 1; atoms of carbon always correspond to vertices of degree #1.

In Figure 6.33(b), the top figure represents an electrical network with resistances, condensers, and inductances. This information was replaced with the graph G. Kirchoff showed that it was not necessary to consider every cycle in the associated graph of an electrical network separately to solve the system of equations in which he was interested. He showed that all the needed information was contained in any tree, such as T.

In more recent applications, trees are used to solve problems regarding the most efficient way to link a number of locations, whether by roads or by computer networks. In computer science, trees are used to represent hierarchical structures, such as file systems, as well as arithmetic expressions.

This section provides several different characterizations of trees. In Section 6.11, attention will focus on the problem of finding a "smallest" connected spanning subgraph in a graph. After solving this problem, we will focus in Section 6.12 on rooted trees and the application of these trees to sorting and searching. We will show how trees can provide a model for determining a lower bound on the complexity of any sorting algorithm that is based on the comparison of two elements at a time.

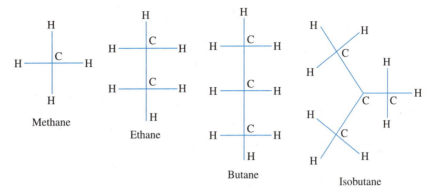

(a) Cayley's graphs for hydrocarbon compounds

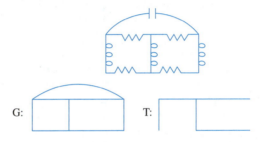

(b) Kirchoff's graphs for electrical network

Figure 6.33   Cayley's and Kirchoff's graphs. a. Cayley's graphs for hydrocarbon compounds. b. Kirchoff's graphs for an electrical network.

6.10.1   Definition of Trees

A good way to deal with a new mathematical structure is to construct examples that satisfy the definition. You should try this approach with the next definition, in which we will make use of the term *acyclic* (see Definition 6 in Section 6.3).

Definition 1.   A **tree** is a connected acyclic graph. A graph in which each connected component is a tree is called a **forest.** A vertex of degree one in a tree is called a **leaf,** or a **terminal vertex.** A vertex of degree greater than one in a tree is called an **interior vertex.**

For the usual definition of trees, the vertex set may not be empty. Later, a special application will be facilitated by relaxing this restriction on the definition of a tree.

In Figure 6.34, we see all the trees with fewer than six vertices. Can you deduce a relationship between the number of vertices and the number of edges in a tree from these examples?

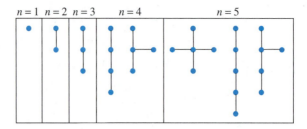

Figure 6.34   Trees with five or fewer vertices.

6.10.2   Characterization of Trees

Because trees have been "discovered" in so many different contexts, such as mathematics, chemistry, and electrical engineering, several seemingly different definitions of trees have been given. The next theorem shows that a number of these definitions are equivalent.

To prove that each of the statements in this theorem describes the same objects, we must prove that each pair of properties is equivalent. For two properties—say, A and B—we would do this by proving that property A implies property B and then that property B implies property A. In the next theorem, we want to prove that four different properties are equivalent. To prove each pair to be equivalent as above would require 12 proofs. We can, however, reduce the number of proofs required by using the fact that if property A implies property B and property B implies property C, then property A implies property C. This makes a direct proof that property A implies property C unnecessary. To show that n properties are equivalent, we form a circle of implications—that is, property 1 implies property 2, property 2 implies property 3, ..., property n implies property 1. Now, for example, to complete a proof that property i implies property j for any i and j, we simply start at the implication that property i implies property $i + 1$ and follow the circle of implications until the conclusion is property j. Such a sequence of implications and the transitive property of implication lead to the desired conclusion. More formally, what we observe is that the formula

$$(p_1 \leftrightarrow p_2) \wedge (p_1 \leftrightarrow p_3) \wedge (p_1 \leftrightarrow p_4) \wedge (p_2 \leftrightarrow p_3) \wedge (p_2 \leftrightarrow p_4) \wedge (p_3 \leftrightarrow p_4)$$

is logically equivalent to the formula

$$(p_1 \rightarrow p_2) \wedge (p_2 \rightarrow p_3) \wedge (p_3 \rightarrow p_4) \wedge (p_4 \rightarrow p_1).$$

The proof of Theorem 1 uses a circle of implications. In Theorem 1, the number of proofs is reduced to four from 12 by using this proof technique.

Theorem 1.   (Tree Characterization)   Let T be a graph with n vertices and m edges. The following statements are equivalent:

1. T is a tree.
2. T is connected and $m = n - 1$.
3. T is acyclic and $m = n - 1$.

4. T is acyclic and the addition of any edge joining two nonadjacent vertices creates a cycle.

Proof.

(1 ⇒ 2)   The proof is an induction on n, the number of vertices in T. Let $n_0 = 1$ and $\mathcal{T} = \{n \in \mathbb{N} : \text{all trees on } n \text{ vertices are connected and } m = n - 1\}$.

(Base step)   If $n = 1$, then the only tree to consider has one vertex and zero edges, so the result is true for $n = 1$.

(Inductive step)   Now, let $n \geq n_0$. Show that if $n \in \mathcal{T}$, then $n + 1 \in \mathcal{T}$. Let T be a connected and acyclic graph with $n + 1$ vertices. Since T is connected, we only have to prove that $m = n - 1$.

Choose a path W in T of maximum length, and let v be an endpoint of W. Vertex v has degree 1 in T, because W is a maximum length path and T is acyclic. Form the subgraph T_1 of T consisting of the vertices of T, except for v, and the edges of T, except for the edge incident with v. T_1 is connected and acyclic and contains n vertices. By induction, T_1 contains $n - 1$ edges. Therefore, T contains n edges.

By the Principle of Mathematical Induction, $\mathcal{T} = \{n \in \mathbb{N} : n \geq 1\}$.

(2 ⇒ 3)   Suppose T is connected and $m = n - 1$. We must prove that T is acyclic and $m = n - 1$. Since it is given that $m = n - 1$, it only remains to prove that T is acyclic.

Assume T has a cycle of length c, where $c \leq n$. There are c vertices and c edges in the cycle. If $c = n$, then $m \geq n$, which is a contradiction. For $c < n$, each of the $n - c$ vertices not on the cycle is incident to an edge on a shortest path from that vertex to a vertex of the cycle. Each of these edges is different. Therefore, T contains

$$m \geq c + n - c = n$$

edges, which is a contradiction.

(3 ⇒ 4)   Assume T is acyclic and $m = n - 1$, and prove that the addition of an edge joining two nonadjacent vertices creates a cycle.

Since T is acyclic, every component of T must be a tree. Suppose T had $k \geq 1$ components. Each component has one more vertex than edge, so $n = m + k$. However, $n - m = 1$, so $k = 1$ and T is connected.

Let u and v be any two nonadjacent vertices in T. Since T is connected, there is a path from u to v. When the edge (u, v) is added to T, a cycle is formed since there are two distinct paths from u to v in T.

(4 ⇒ 1)   Suppose T is acyclic and the addition of any edge joining two nonadjacent vertices creates a cycle. The proof will be completed by showing that T is connected, since it is given that T is acyclic.

Suppose T is not connected. T must have at least two connected components. Let u be a vertex in one component and v be a vertex in a different component. Adding the edge (u, v) to T does not form a cycle, which is a contradiction. Therefore, T is connected. ■

When we study a special class of trees used to represent family trees and various sets with order relations, we will use the following characterization of trees. In this theorem, we use only the original definition of a tree in the proof.

Theorem 2.   Let T be a graph. T is a tree if and only if any two vertices of T are joined by a unique path.

Proof.

($\Rightarrow$)   Suppose T is a graph such that every pair of vertices is joined by a unique path. T is connected, since each vertex pair has a path joining them. Furthermore, T is acyclic. For example, suppose T were not acyclic. Let A and B be vertices on an existing cycle. A and B have different paths between them, contradicting the assumption, so T must therefore be acyclic.

($\Leftarrow$)   Let T be a tree that satisfies the hypotheses. Clearly, T is connected, since there is a path between any pair of vertices. If T is not acyclic, there are at least two vertices A and B for which there are at least two paths from A to B, which is a contradiction.   ∎

Additional characterizations of trees will be given in the exercises (see Exercise 8 in Section 6.13). Having so many different ways to think about trees can make some proofs a lot easier: You can pick a characterization of trees that is convenient for the situation at hand rather than having to use the original definition of a tree.

6.11   Spanning Trees

Recall that for a graph $G = (V, E)$, a **spanning subgraph** is a subgraph $H = (V_1, E_1)$ of G with $V_1 = V$. Obviously, every graph has a spanning subgraph—namely, the graph itself. The subgraph of a graph consisting of just the vertex set and no edges is also a spanning subgraph. Obviously, no "smaller" subgraph, in terms of the number of edges in the subgraph, can be a spanning subgraph than this subgraph with zero edges. The problem of more interest is to find the smallest, in terms of the number of edges, connected spanning subgraph of a connected graph. Clearly, such a graph must be a tree. This smallest connected spanning subgraph, a **spanning tree,** can be found using the algorithm of Kruskal.

6.11.1   Kruskal's Algorithm

Kruskal's algorithm proceeds by examining the edges of the graph one at a time, in any particular order. As an edge is examined, the algorithm determines whether a spanning tree exists that contains the edge and all the previously chosen edges. It turns out that the only condition the edge must satisfy to be added to the edges already chosen is that it does not form a cycle with any subset of the edges already chosen. If an edge does not form a cycle with any subset of the previously chosen edges, then it is included in the chosen edge set. If the edge does form a cycle with some of the previously chosen edges, then it is not chosen, and the next edge is examined. When all the edges of a graph have been examined, the edges in the subgraph consisting of the chosen edges form a spanning tree.

Kruskal's algorithm is an example of the family of algorithms that try at each stage to advance toward the goal as far as possible in a single step. The algorithms of this family are known as **greedy algorithms,** and they sometimes—but not always—find the best possible solution. As we have just seen, in the case of Kruskal's algorithm, the best possible outcome, a spanning tree, is always found.

Algorithm: Spanning Tree—Kruskal

INPUT: Connected graph $G = (V, E)$
OUTPUT: A spanning tree T of G

$V(T) = V(G)$ /* Initialize */
$E(T) = \emptyset$
for each $e \in E(G)$ do
    if $(\{e\} \cup E(T)$ is acyclic) then
        $E(T) = E(T) \cup \{e\}$

6.11.2   Correctness of Kruskal's Algorithm

To show that Kruskal's algorithm is correct, it must be shown that the subgraph T formed by the algorithm is a spanning subgraph for G and that T is both connected and acyclic. T is clearly a spanning subgraph, because $V(T)$ is defined to be $V(G)$. Since no edge is ever added to T that forms a cycle, T is acyclic, but it remains to be shown that T is connected. Suppose T is not connected. Let u and v be vertices that lie in different components of T. Since G is connected, there must be a path P in G from u to v. Traveling along P from u toward v, let e be the first edge that leaves the component of T containing u and enters some other component of T. Since e bridges two components of T, e does not belong to T. Since the endpoints of e are not joined by any path in T, $E(T) \cup e$ is acyclic. Thus, when e was considered in the execution of the algorithm, $E(T) \cup e$ was acyclic; therefore, e was added to $E(T)$. This contradicts $e \notin E(T)$. Since assuming T is not connected led to a contradiction, T must be connected.

The algorithm will be clarified by applying the proof to the example shown in Figure 6.35. The edges listed in each part of the diagram indicate the tree edges (solid lines) and the edges that form cycles (dashed line) and, thus, are not included in T.

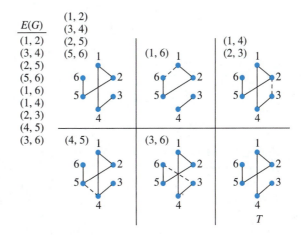

Figure 6.35   Finding a spanning tree.

6.11.3   Kruskal's Algorithm for Weighted Graphs

Figure 6.36 shows a graph that models a communication network where the length of the communication link is a measure of the cost associated with joining the two vertices. The Communication Network Problem is to find a set of edges from this graph such that all the cities are connected and the sum of the lengths of the edges is a minimum. To make this problem clear, we introduce the notion of a value or a weight for a graph.

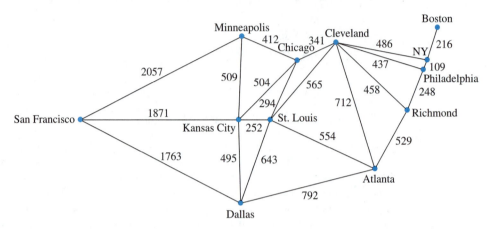

Figure 6.36   Communication network.

Definition 2.   Let $G = (V, E)$ be a graph. A **weighted graph** is a graph G together with a function $F : E \to \mathbb{R}$. The **weight**, or **cost**, or **value**, of an edge e is just $F(e)$. The weight of a subgraph H of G is $W(H) = \sum_{e \in E(H)} F(e)$.

Example 1.   Let the values shown on the edges of G be their weights.

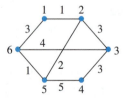

Determine the weight of the 4-cycle with vertices $1, 2, 5,$ and 6. Calculate the weight of G.

Solution. $W(1 - 2 - 5 - 6) = 1 + 2 + 1 + 3 = 7$, whereas $W(G) = 1 + 3 + 3 + 5 + 1 + 3 + 2 + 4 = 22$. ■

Since the cost of constructing a communication link is often proportional to the distance between the sites being connected, a weighted graph, with weights equal to the distance between the locations represented by the ends of an edge, is a good model for a network when deciding which links to construct. The problem is to find a good spanning tree. To measure how good a spanning tree is, one sums the weights of all the edges included in the spanning tree. By the previous definition, this is the weight of the tree. When the edge weights represent costs, the smaller the weight (cost), the better the tree. Thus, the problem is to find a spanning tree of minimum weight for the network. A spanning tree of minimum weight is called a **minimum cost spanning tree (MCST).**

Kruskal's algorithm to find a spanning tree can be viewed as the special case of finding an MCST in a graph with each edge having weight one. The modifications needed to find an MCST where the edges have arbitrary weights are included in the next algorithm.

Algorithm: Minimum Cost Spanning Tree—Kruskal

INPUT: Connected graph $G = (V, E)$ with weighted edges
OUTPUT: A minimum cost spanning tree T of G

$V(T) = V(G)$ /* Initialize */
$E(T) = \emptyset$
Sort $E(G)$ in order of increasing weight
for each $e \in E(G)$ in increasing order of edge weights do
    if $(e \cup E(T)$ is acyclic) then
        $E(T) = E(T) \cup \{e\}$

Figure 6.37 shows MCST in the graph from Figure 6.36.

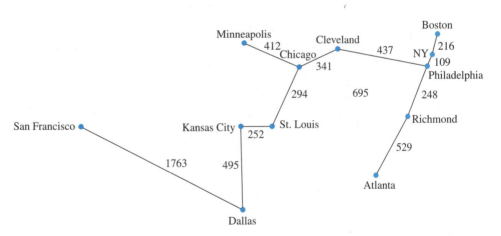

Figure 6.37 Minimum cost spanning tree.

Several approaches are used to determine which edge to examine next. One method is to sort the edges of the graph in increasing order by weight and then examine the edges in that order. A second approach involves repeating a process whereby only the next edge to be examined is identified rather than preprocessing all the edges by sorting them. An effective data structure for this second possibility is explored in Exercise 26 of Section 6.13. No matter what procedure is used to determine which edge to examine next, however, the procedure can be stopped as soon as number of edges chosen is one less than the number of vertices in the graph.

6.11.4   Correctness of Kruskal's Weighted Graph Algorithm

Since the order in which the edges are considered is the only difference between this version of Kruskal's algorithm and the earlier version for unweighted graphs, the algorithm correctly finds a spanning tree.

It remains to be proven that this algorithm generates an MCST. Let $G = (V, E)$ be a weighted, connected graph with weight function c. Let T be a weighted spanning tree generated by Kruskal's algorithm. Let T_1 be an MCST for G. Let $E(T)$ be the edges of T and $E(T_1)$ be the edges of T_1. If $E(T) = E(T_1)$, then T is also an MCST. If $E(T) \neq E(T_1)$, let e be a minimum cost edge such that $e \in E(T)$ and $e \notin E(T_1)$. Observe that every edge of T with a weight strictly less than the weight of e is included in $E(T_1)$ because of the way that e is chosen.

The graph $E(T_1) \cup \{e\}$ contains a unique cycle (see Exercise 6 in Section 6.13)—say,

$$e_1, e_2, e_3, \ldots, e_k, e$$

Let e_j be an edge on this cycle such that $e_j \notin E(T)$. The edge e_j exists or else the cycle

$$e_1, e_2, \ldots, e_k, e$$

would also be contained in T, which is a contradiction. If $c(e_j) < c(e)$, then e_j would have been included in T by the way that e was chosen. Therefore, $c(e_j) \geq c(e)$.

Now, consider the graph T_2 with edge set $E(T_1) \cup \{e\}$. Removal of any edge of the cycle

$$e_1, e_2, \ldots, e_k, e$$

will result in a spanning tree. In particular, removing e_j results in a spanning tree T_3 with a cost no more than the cost of T_1 since $c(e_j) \geq c(e)$. Since T_1 is an MCST, T_3 must also be an MCST.

The argument that T_3 is an MCST is worth a more careful look, because it is useful in other contexts. We started with T_1 being an MCST with weight $wt1$. We then showed that the weight of T_3—say, $wt2$—has the property that $wt2 \leq wt1$. Now, if $wt2 < wt1$ we would be contradicting the fact that T_1 is an MCST. Therefore, $wt1 = wt2$, and T_3 has the same weight as T_1, which makes T_3 an MCST. Notice that we did not claim T_1 was the only MCST, just one of the possible choices.

After at most $|V(T_1)| - 1$ transformations of the sort described, all the edges of T_1 not contained in T will be replaced by edges of T. Hence, T is an MCST.

6.12   Rooted Trees

In a family tree, such as that shown in Figure 3.2 (see Section 3.1), there is a clear interpretation of the edges and vertices as being on paths from the root. The relation being represented precludes thinking of edges as "going both ways." The special tree used to represent such relations is called a **rooted tree.** A rooted tree consists of a tree together with one vertex distinguished as a **root.** For any vertex in a rooted tree, other than the root, there is a unique path from the root to the vertex. We view the edges as directed along the path from the root to a vertex (see Theorem 2 in Section 6.10.2). With this interpretation of direction, a drawing of a rooted tree normally has the root vertex at the top and the remainder of the tree displayed below the root. Depth first search trees and breadth first search trees are both important examples of rooted trees. In those cases, the root of the tree is the vertex at which the search begins. As another example, Figure 6.38 shows how rooted trees can be used to represent a hierarchical structure, such as a file system for a computer support group.

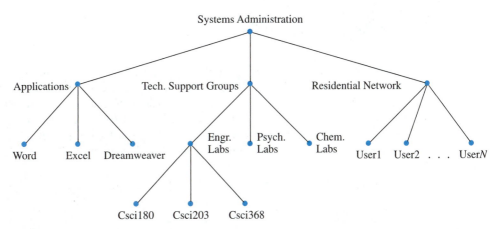

Figure 6.38   Computer center file structure.

As with ordinary trees, vertices of degree one, except the root, in a rooted tree are called leaves, or terminal vertices, and the other vertices in a rooted tree are called internal or interior vertices. The maximum length of any path from a vertex to a leaf beneath that vertex is the **height** of the vertex. The height of the tree is the height of the root. The length of a path from the root to a vertex is the **level** of the vertex. These notions are illustrated in Figure 6.39.

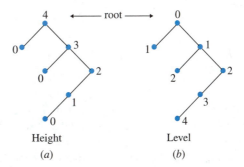

Figure 6.39   Height and level in a rooted tree. a. Height. b. Level.

Another widely used terminology for rooted trees draws on the analogy with genealogy. For any directed edge (v, w), the vertex v is the **parent** of w, and w, is the **child** of v. If v and w are any two vertices in a rooted tree for which there is a directed path from v to w, then the vertex v is an **ancestor** of w, and w is a **descendant** of v. Vertices with the same parent are **siblings.** Figure 6.40 shows examples of these notions.

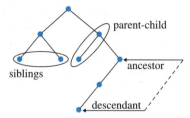

Figure 6.40   Relationships between vertices.

In discussing graphs, the notion of a subgraph clarified what is meant by a substructure. In the case of a rooted tree, a **rooted subtree** is the analogue. To define a rooted subtree, let v be a vertex of the rooted tree T. The rooted subtree of T rooted at v is the induced subgraph of T defined by the vertex set consisting of v together with all its descendants in T. Figure 6.41 shows two subtrees of the rooted tree shown in Figure 6.38. (A similar notion was introduced with expression trees in Section 2.1.2.)

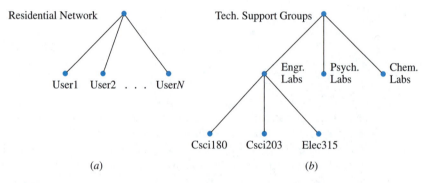

Figure 6.41   Support area for the computer center. a. Residental network. b. Technical support groups.

6.12.1   Binary Trees

A vertex in a rooted tree may have any number of children. The children of a vertex in a rooted tree are not arranged in any order with respect to their parent or with respect to each other. The two rooted trees in Figure 6.42 are isomorphic rooted trees. In addition to the usual properties of an isomorphism between two graphs, an isomorphism between rooted trees must map the root of one tree to the root of the other.

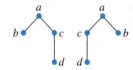

Figure 6.42   Two isomorphic rooted trees.

Of special interest in computer science is the family of rooted trees in which each vertex has at most two children and the children are ordered by being designated as either the left child or the right child of the vertex.

Definition 3. A **binary tree** is either a tree with no vertices or a rooted tree for which each vertex has at most two children. Each child of a vertex in a binary tree is designated as the **left child** of the vertex or the **right child** of the vertex (but not both).

Normally, a tree may not have an empty vertex set, but in the case of binary trees, the notion of an empty tree is useful. The leaves of a binary tree are considered to have empty subtrees for both the left and the right child. This convention gives a convenient way to determine when a leaf in a binary tree has been reached. The definition of isomorphism for binary rooted trees includes more than the conditions for isomorphism between two graphs. For two binary rooted trees T_1 and T_2 to be isomorphic, we require, as with any rooted tree, that the root of T_1 be mapped to the root of T_2. We also require one additional property of the isomorphism of binary rooted trees: If $F : V(T_1) \rightarrow V(T_2)$ is an isomorphism for two rooted binary trees T_1 and T_2, and if w is a left (right) child of v in T_1, then $F(w)$ must be a left (right) child of $F(v)$ in T_2. The fact that each child of a vertex is designated as either the left or the right child of the vertex means the two binary trees in Figure 6.43(a) and 6.41(b) are different.

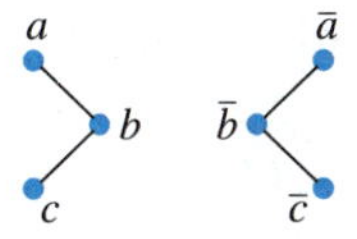

Figure 6.43 Two distinct binary trees.

All binary trees with fewer than five vertices are pictured in Figure 6.44. That is, every binary tree with fewer than five vertices is isomorphic to one of these binary trees.

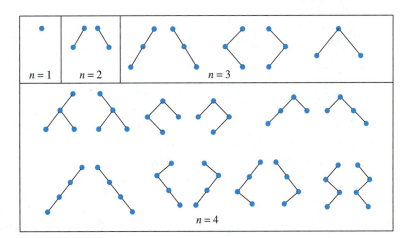

Figure 6.44 Binary trees with fewer than five vertices.

6.12.2   Binary Search Trees

In addition to directing the edges away from the root in a binary tree, we can assign values to the vertices in a special way. This new kind of tree is called a binary search tree. A typical application of a binary search tree includes searching an ordered set for an element.

Definition 4.   Let X be a linearly ordered set. A binary tree $T = (V, E)$ with vertices labeled by elements of X is a **binary search tree** if and only if:

(a) The value of the label of each vertex in the left subtree of a vertex v and not equal to v is less than the value of the label at v.
(b) The value of the label of each vertex in the right subtree of v and not equal to v is greater than the value of the label at v.
(c) no element of X occurs as a label of more than one vertex of T.

We have several choices for managing data when it must be stored in alphabetical order for ease of access in applications. Certainly, the data can be kept in order in a list. The data can also be stored as labels of the vertices of a binary search tree. If the binary search tree can be constructed with a relatively small height, then the search algorithm will be very efficient.

When the data stored in a binary search tree are a collection of words or names, the ordering relation that is used to determine the location of a label is simply the familiar dictionary ordering of words (see Example 9 in Section 3.8.2). The tree pictured in Figure 6.45 is a binary search tree with respect to the usual dictionary ordering of words.

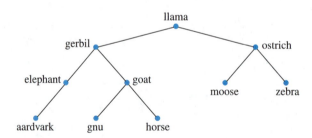

Figure 6.45   Binary search tree.

The search for a word in a set of words labeling a binary search tree can be implemented with the algorithm on the facing page.

Since the labels for the vertices of the binary search tree come from a linearly ordered set and no two vertices have the same label, each comparison of the elements sought with the label of a vertex results in one of two outcomes: You find the element, or you restrict the continuation of the search to a smaller subtree. Since each leaf has an empty subtree for both its left and right subtrees, the search will terminate either by finding the element as the label of a vertex or by exhausting the possibility of finding the element by coming to a leaf of the tree.

Algorithm: Search Using a Binary Search Tree

INPUT: Binary search tree T with root R and an element *item* of the type stored at the vertices of T
OUTPUT: *TRUE* if *item* is stored in T and *FALSE* if not

BinSrchTree$(R, item)$ /* Begin the search at root R of T */

BinSrchTree$(v, item)$ /* The recursive procedure */
    if $(v = \emptyset)$ then
        return *FALSE*
    else
        if (*item* equals the element stored at v) then
            return *TRUE*
        else
            if *item* is less than the element at v then
                BinSrchTree(*item*, left child of v)
            else
                if *item* is greater than the element at v then
                    BinSrchTree(*item*, right child of v)

As an example of the use of this algorithm, we will show how *zebra* is found on the binary search tree shown in Figure 6.46.

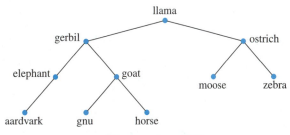

Compare zebra with llama
Continue search in light subtree with root ostrich

Compare zebra with ostrich
Continue search in light subtree with root zebra

zebra

Compare zebra with zebra

Figure 6.46   Search using a binary search tree.   SUCCESS!

The efficiency of searching a binary search tree depends on the form of the tree. In Figure 6.47, we see extreme cases for a binary search tree with five vertices.

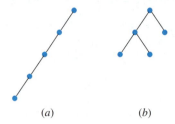

(a)   (b)

Figure 6.47   Two binary search trees with five vertices.

The difficulty with the binary search tree shown in Figure 6.47(a) is that its height is four. On the other hand, the height of the binary search tree in Figure 6.47(b) is only two. These two facts say that a search using the tree shown in Figure 6.47(a) can require five comparisons while a search using the tree shown in Figure 6.47(b) will require at most three comparisons in any search. To be efficient, algorithms using a binary search tree must deal with the problem of keeping the height of the tree as small as possible. This is called "balancing" the tree, and it is often a major implementation problem. One technique for dynamic balancing of a rooted tree involves operations called **rotations.** The family of **A-V-L trees** use rotations to keep the tree "balanced." An A-V-L tree rebalances a vertex when its left subtree has height two greater than the height of its right subtree. An example of this procedure is shown in Figure 6.48.

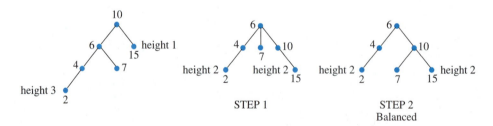

Figure 6.48   Rebalancing an A-V-L tree.

When building a binary search tree one element at a time, the detection of a vertex having a left subtree with height two greater than the height of the other subtree causes a balancing action to take place. In this example, vertex 10 has a left subtree of height three and a right subtree of height one. Step 1 in Figure 6.48 is to move the vertex that is unbalanced (10) down one level in the tree and its left child (6) up one level to become its parent. The problem now is that vertex 6 has three children! We observe, however, that when vertex 10 was moved down one level, it lost its left child! Therefore, we can now make vertex 7 the left child of vertex 10 (see Step 2 in Figure 6.48), and the tree becomes a binary search tree with no vertex having a left subtree with height more than one different from the height of its right subtree.

In general, if we suppose a binary search tree is built from a list of n randomly ordered elements. The first element on the list creates the root; the second element creates a left or right child of the root, depending on whether it is greater than or less than the first element; and so on. It is possible to prove that the height of such a tree, on average, is $O(\log_2(n))$. This bound is independent of any balancing operation.

6.12.3   Tree Traversals

A binary search tree can be built for a sequence of elements by adding the elements to the tree one by one. For example, consider the sequence D, G, B, F, H, A, E, C. The elements have a natural linear ordering, the alphabetical ordering, but they do not appear in alphabetical order in the sequence. The first element, D, becomes the root of the tree. The next element, G, is greater than D in alphabetical order, so G becomes the right child of D. The third element, B, is less than D, so it becomes the left child of D. The fourth element, F, is greater than D, so it belongs to the subtree rooted at the right child, G, of D. Since $F \neq G$ and $F < G$, F becomes the left child of G. Continuing in this way creates the binary search tree shown in Figure 6.49.

A binary search tree can be traversed in a way that makes it possible to print out its labels in their proper order. The directed trail in Figure 6.49 shows how this is done. Traveling along the trail, which starts at the root, output the label of an internal vertex that has a left child when the vertex is encountered for the *second* time. Output the labels of leaves and internal vertices with no left child the *first* time they are encountered. (Recall that the root is an internal vertex.) For the tree in Figure 6.49, the labels are encountered along the trail in the order $D, B, A, B, C, B, D, G, F, E, F, G, H, G, D$. Following the procedure will list the letters A, B, C, D, E, F, G, H in alphabetical order.

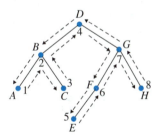

Figure 6.49   Listing the labels of the vertices. The numbers indicate the order in which labels are ordered.

There are actually three ways to list the labels at the vertices of a binary tree using the traversal shown. The listings that result are determined by choosing when to print out the label at a vertex: before the labels in its subtree, after the labels in its subtree, or between the labels in its left subtree and the labels in its right subtree. For a nonempty tree T with root r, left subtree T_1, and right subtree T_2, the formal description of these procedures follows.

VERTEX LABEL LISTINGS

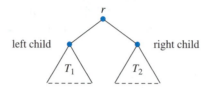

PREORDER

1. List the label at the root vertex r of T.
2. If T_1 is nonempty, list the labels of the vertices of T_1 in preorder.
3. If T_2 is nonempty, list the labels of the vertices of T_2 in preorder.

INORDER

1. If T_1 is nonempty, list the labels of the vertices of T_1 in inorder.
2. List the label of the root vertex r of T.
3. If T_2 is nonempty, list the labels of the vertices of T_2 in inorder.

POSTORDER

1. If T_1 is nonempty, list the labels of the vertices of T_1 in postorder.
2. If T_2 is nonempty, list the labels of the vertices of T_2 in postorder.
3. List the label of the root vertex r of T.

Figures 6.49 and 6.50 illustrate **inorder** listings.

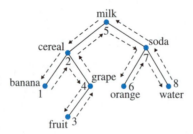

Figure 6.50   Inorder traversal.

In Figure 6.50, the tree shown is a binary search tree. The word *fruit* is listed by the inorder traversal before *grape,* because the inorder process lists the label of the left subtree before **listing** the label at the vertex itself and, finally, lists the labels in the right subtree of that vertex. In this example, the vertex labeled *grape* has no right subtree, so this finishes the listing of the labels in the right subtree labeled *cereal* (already listed). This in turn completes the listing of the labels in the left subtree of the vertex labeled *milk*. So, the next label listed is *milk*.

To carry out a preorder or a postorder listing, follow the same trail alongside the tree. For a preorder listing, however, record the label on the vertex the first time the vertex is encountered. For a postorder listing, record the label on the vertex the last time it is passed, as the traversal moves up to the vertex's parent. Examples of these listings of the labels are shown in Figure 6.51. Notice that for **preorder** and **postorder** listing, the labels at the vertices are not listed in alphabetical order.

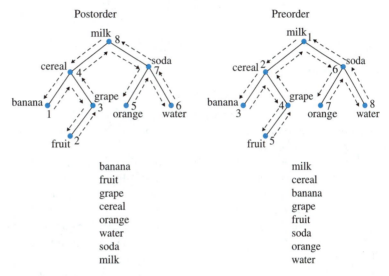

Figure 6.51   Tree traversals.

6.12.4   Application: Decision Trees

To find a lower bound for the complexity of a class of algorithms, it is necessary to find a model for all possible algorithms of the class. The lower bound for sorting algorithms based on using a comparison of two elements as the fundamental operation can be found by using a decision tree model. A **decision tree** is a rooted binary tree with **labeled vertices** and edges.

Each step of a comparison-based sorting algorithm involves the comparison of two distinct elements (if two elements were the same, we would be reduced to a smaller case) using the linear order on the set of elements being sorted. These comparisons are used to label the vertices of the tree. The label $a : b$ at a vertex indicates that a and b are to be compared. The left edge from a vertex represents what is known about the order of the elements when $a < b$ is true. The right edge represents what is known about the order of the elements when $b < a$ is true. Each leaf represents the fact that the questions used to get to that leaf have completely identified the order among the elements of the input. The discovered order is then presented in a box. Figure 6.52 shows a tree that represents one way to sort three elements.

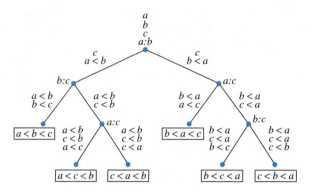

Figure 6.52   Decision tree for sorting three elements.

The number of vertices with boxes must be at least $n!$, since there are that many possible orderings of n elements. Obviously, a sorting algorithm must be able to identify each of these orderings.

The level of any vertex turns out to be the number of comparisons that must be made to get from the starting question to the vertex—that is, the number of edges traversed in going from the root of the tree to the vertex. Since each comparison results in one of two possible answers, after k comparisons at most 2^k different vertices can be labeled with orderings.

Let $S(n)$ be the minimum number of comparisons used by any sorting algorithm that correctly sorts any input of n elements. The decision tree representing any sorting-by-comparisons algorithm must have at least $S(n)$ levels.

Thus, the decision tree representing any sorting-by-comparisons algorithm must have at least $S(n)$ levels. Since at most $2^{S(n)}$ orderings can be represented by vertices in a tree with $S(n)$ levels, we must have

$$2^{S(n)} \geq n!$$
$$S(n) \geq \log_2(n!)$$

The problem is to find a good approximation for $\log_2(n!)$.

Example 2. $\mathbf{O}(\log_2(n!)) = \mathbf{O}(n \log(n))$.

Solution.   We must prove that (a) $\log(n!) \in \mathbf{O}(n \log(n))$ and (b) $n \log(n) \in \mathbf{O}(\log(n!))$.

(a)

$$\begin{aligned}
\log_2(n!) &= \log_2(1) + \log_2(2) + \log_2(3) + \cdots + \log_2(n-1) + \log_2(n) \\
&\leq \log_2(n) + \log_2(n) + \log_2(n) + \cdots + \log_2(n) + \log_2(n) \\
&= n \log_2(n)
\end{aligned}$$

Now, let $c = 1$ and $N_0 = 1$. Since $n \log_2(n) \leq c \log_2(n!)$ for all $n \geq N_0$, we conclude that $\log_2(n!) \in \mathbf{O}(n \log_2(n))$.

(b)

$$\log_2(n!) = \log_2(n) + \log_2(n-1) + \cdots + \log_2(n/2) + \log_2(n/2 - 1) + \cdots$$
$$+ \log_2(4) + \log_2(3) + \log_2(2) + \log_2(1)$$
$$> (n/2)\log_2(n/2) + (n/2)\log_2(2)$$
$$= (n/2)\log_2(n) - (n/2)\log_2(2) + (n/2)\log_2(2)$$
$$= (n/2)\log_2(n)$$
$$= (1/2)(n\log_2(n))$$
$$2\log_2(n!) > n\log_2(n)$$

Now, let $c = 2$ and $N_0 = 1$. Since $n\log_2(n) \leq c\log_2(n!)$ for all $n \geq N_0$, we conclude $n\log_2(n) \in \mathbf{O}(\log_2(n!))$.   ∎

From Example 1, it follows that any sorting algorithm based on the comparison of a pair of elements at each step will have the complexity given by a function at least as large as $c(n\log_2(n))$ where n is the size of the input set and $c \in (0, \infty)$.

6.13 Exercises

1. Construct all trees on six vertices. Find an algorithm for constructing all possible trees on six vertices if you know all possible trees on five vertices.
2. If the sequence $d_1, d_2, \ldots, d_n$ of nonnegative integers represents the degrees of the vertices of a tree with n vertices, then $\sum_{i=1}^{n} d_i = 2(n-1)$. Show that the converse is false.
3. Let G be a tree with all vertices having odd degree. Prove that G contains an odd number of edges. Show that this is not true if G is not a tree.
4. Prove that a tree is a bipartite graph.
5. Use a graph to represent the possible paths through the maze shown. Use the graph to find a path from A to N.

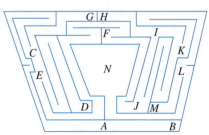

6. Let T be a tree. Prove that if an edge e joins two nonadjacent vertices in T, then $T \cup \{e\}$ contains a cycle.
7. Prove that any tree with two or more vertices has at least two vertices of degree one.
8. Prove that a graph T is a tree if and only if T is connected but the deletion of any edge disconnects T.

9. Prove that a cycle and the complement of any spanning tree must have at least one edge in common.

10. Let T_1 and T_2 be two spanning trees of a graph G. Prove that if a is any edge in T_1, then there exists an edge b in T_2 such that the graph obtained from T_1 by replacing a with b is a spanning tree of G.

11. Prove that a graph G is connected if and only if G contains a spanning tree.

12. Find a MCST in the graph shown:

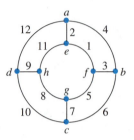

13. For the following graphs:

   (a) Find a MCST in each of the graphs G_1, G_2, and G_3.
   (b) Repeat part (a) for graph G_1 if it must include $(3, 5)$ and $(1, 7)$ in the MCST.
   (c) Repeat part (a) for graph G_2 if it must include $(3, 10)$, $(6, 11)$, and $(8, 9)$ in the MCST.
   (d) Repeat part (a) for graph G_3 if it must include $(2, 9)$, $(6, 8)$, and $(4, 5)$ in the MCST.

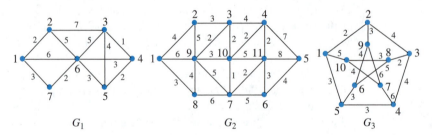

14. Find a MCST in each of the graphs G and H:

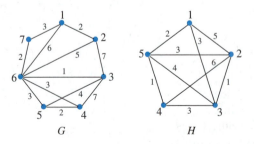

   (a) Is the MCST for graph G unique? If not, find all examples. Find an MCST in the graph if $(6, 3)$ may not be included. Is such an MCST unique? If not, find all examples.

(b) Is the MCST in graph H unique? If not, find all examples. Find an MCST in the graph if (3, 4) and (2, 5) must be included. Is such an MCST unique? If not, find all examples.

15. A university has eight buildings that need to be connected so that each building's computer network is accessible to the networks in the other buildings. The distance between buildings is given in units of 1000 yards. These distances between buildings are given in the table that follows. The distance from building i to building j is the same as the distance from building j to building i.

	1	2	3	4	5	6	7	8
1	—	1.6	1.4	0.5	1.2	1.5	1.8	2.3
2		—	0.9	1.8	1.2	2.6	2.3	1.1
3			—	2.6	1.7	2.5	1.9	1.0
4				—	0.7	1.6	1.5	0.9
5					—	0.9	1.1	0.8
6						—	0.6	1.0
7							—	0.5
8								—

Which pairs of buildings should be directly connected to connect all the buildings with a minimum total network length? What is the length of a minimum network? What are the different possible minimum networks?

16. Assume that all the weights on the edges of a graph are positive. Prove that if no two edges in a graph have the same weight, then the MCST is unique. (*Hint:* Assume G has more than one minimal spanning tree. Order the edges, and consider the smallest subscript k of an edge that belongs to some, but not all, minimal spanning trees.)

17. Find an algorithm for determining an MCST that is based on removing the maximum cost edge from a cycle at each step.

18. Let $F : T \rightarrow S$ be an isomorphism between binary trees T and S. Let $v \in T$ be a vertex at level k for $k \geq 0$. Prove that $F(v)$ is at level k in S.

19. For each of the following sets of words, find the binary search tree that is formed when the underlined word is the label of the root and the remaining words are added to the tree in left-to-right order as they occur in the sentence.

   (a) The time has <u>come</u> for all good.
   (b) The time <u>has</u> come for all good.
   (c) The time has come for <u>all</u> good.
   (d) Since all the words in this sentence are <u>different</u>, it is possible to represent them with a binary search tree.

20. Write a recursive algorithm to traverse a binary search tree and list its labels in increasing order. Write the procedure so that it can be extended easily to a program in some programming language.

21. Write an algorithm for the postorder traversal of any rooted tree.

22. Design a minimum depth decision tree to solve each of the following problems:

    (a) Take the water in an eight-gallon jug, and divide it into two four-gallon portions using a three-gallon jug and a five-gallon jug that initially are empty.
    (b) Take the water in a six-gallon jug, and end with two gallons in a four-gallon jug using a four-gallon jug and a three-gallon jug that initially are empty.

23. Draw a minimal depth decision tree that represents sorting four elements from a linearly ordered set. Only five levels are required.

24. Let $G = (V, E)$ be a connected graph. The distance between two vertices $v, w \in V$, denoted $D(v, w)$, is the length of a shortest path joining v to w. A vertex u of G is in the center of G if $\max_{v \in V(G)} d(u, v)$ is as small as possible. Prove that a tree has either one or two vertices in its center.

25. A binary tree is said to be complete if all its leaves occur at the lowest level. For a complete binary tree of height h, determine the number of vertices at level i where $1 \le i \le h$. Prove that the number of leaves is one more than the number of internal vertices.

Definition 1.   A **heap** of height h is an arrangement of numbers at the nodes of a binary tree such that the following hold:

(a) All the leaves are either at level h or at level $h - 1$.
(b) There are 2^i vertices at level h for $h = 0, 1, \ldots, h - 1$.
(c) All the leaves at level h are at the leftmost possible positions.
(d) The number at any internal node is larger than the number at either of its two children.

26. Given a heap, what can you say about the number at its root? By removing the number at the root of a heap and replacing it with the number in the rightmost leaf, the property of being a heap is destroyed. Devise an algorithm to restore the property of being a heap. Use this algorithm to devise a sorting algorithm, and then determine the complexity of this sorting algorithm.

6.14   Directed Graphs

The detection of a bottleneck in allocating resources to computer programs, the scheduling of jobs that depend on the completion of other jobs, and the design of one-way street grids are all problems that are modeled by graphs. An important facet in these problems, however, is not reflected in the structure of a graph. Often, a natural notion of direction is associated with the connection between vertices. For instance, one can associate the vertices of a graph with the intersections of streets and the edges with pairs of vertices that indicate the way that traffic is allowed to flow on the streets in a section of a town with only one-way streets. We first introduce directed graphs, which deal with graphs that have a direction associated with each edge. We then use this terminology to model deadlock in an operating system. An analogue of a depth first search in an undirected graph will be used with directed graphs to solve scheduling problems for sets of prioritized events. A distinction between the notion of connectedness for directed graphs and for undirected graphs will be explored. Finally, an analogue of Euler's theorem will be presented for directed graphs.

6.14.1   Basic Definitions

To understand the differences between directed and undirected graphs, we must supply definitions of directed graphs and of their subgraphs and induced subgraphs.

Definition 1.   A **directed graph,** or **digraph,** $D = (V, E)$ consists of a finite, nonempty set V, the elements of V are the vertices of D, and a finite set E of ordered pairs of distinct elements of V, the elements of E are the **directed edges** of D. For an edge $(a, b) \in E$, the vertex a is the **tail** of the edge, and b is its **head.** Both a and b are **incident** to the directed edge (a, b). The vertex a is **adjacent** to b, but b is not adjacent to a unless $(b, a) \in E$. Two directed edges, (a, b) and (c, d), are adjacent if $b = c$ or $d = a$.

The direction of an edge in a drawing of the graph is normally indicated by attaching an arrowhead to the head of the edge. For a vertex a in a directed graph, the **indegree** of a, denoted $indeg(a)$, is the number of edges with a as the head, and the **outdegree** of a, denoted $outdeg(a)$, is the number of edges with a as the tail. When we think of two objects as being adjacent—say, a is adjacent to b—we feel that we can just as well say that b is adjacent to a. For undirected graphs, that is indeed true. For directed graphs, however, the word has a more restricted meaning.

Subgraphs and induced subgraphs of a directed graph are defined very similarly to the way they were defined for undirected graphs.

Definition 2.   Let $D = (V, E)$ be a directed graph. A directed graph $D_1 = (V_1, E_1)$ is a **directed subgraph** of D provided that $V_1 \subseteq V$ and $E_1 \subseteq E$ and, for each $e \in E_1$ that both ends of e are in V_1. An edge in E_1 will have the same head and tail as that edge when considered as an edge of D. D_1 is an **induced subgraph** of D if D_1 is a subgraph of D such that E_1 consists of all the edges of D with both head and tail in V_1.

Figure 6.53 illustrates these notions in a directed graph.

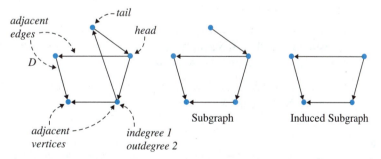

Figure 6.53   Directed subgraph and directed induced subgraph.

With each directed graph is an associated undirected graph, called the **underlying graph,** that is formed from the same vertices and edges but considers all the edges to be undirected. The underlying graph contains just one edge joining a to b when the directed graph contains both the edge (a, b) and the edge (b, a) for some pair of vertices a and b. A directed graph is called **bipartite** if the underlying graph is bipartite.

6.14.2   Directed Trails, Paths, Circuits, and Cycles

The definitions for a directed trail, a directed path, and a directed cycle in a directed graph are straightforward generalizations of the corresponding notions for undirected graphs.

Definition 3.   Let $D = (V, E)$ be a directed graph. A **directed trail** in D is a sequence of not necessarily distinct vertices $v_1, v_1, v_2, \ldots, v_k$ such that $(v_i, v_{i+1}) \in E$ for $1 \leq i \leq k - 1$ and the edges are distinct. If all the vertices in a directed trail are distinct, then the directed trail is a **directed path.** A directed trail for which $v_1 = v_k$ is a **directed circuit.** A directed trail for which all the vertices $v_1, v_2, \ldots, v_{k-1}$ are distinct and $v_1 = v_k$ is a **directed cycle.** The length of a directed trail is the number of edges it contains.

Figure 6.54 gives examples of these notions.

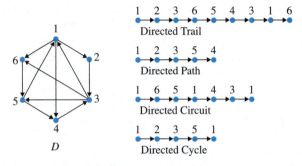

Figure 6.54   Directed trail, path, circuit, and cycle in D.

6.14.3   Directed Graph Isomorphism

The generalization of the notion of an isomorphism to the case of directed graphs gives a formal definition of what it means to say that two directed graphs are isomorphic regardless of how the vertices and edges are labeled.

Definition 4.   Let $D_1 = (V_1, E_1)$ and $D_2 = (V_2, E_2)$ be directed graphs. D_1 and D_2 are the **isomorphic directed graphs** if and only if there is a bijection $F : V_1 \rightarrow V_2$ such that $(a, b) \in E_1$ if and only if $(F(a), F(b)) \in E_2$.

Remember that in a directed graph, the edge directed from a to b and denoted (a, b) is not the same as the edge denoted (b, a).

6.15   Application: Scheduling a Meeting Facility

A small meeting facility consists of three scheduleable resources: a meeting room for large groups, a lobby area for small-group meetings, and a slide projector for use in either room. Two organizations are planning to use the facility on the same afternoon, so a schedule for the resources is needed. The first organization would like to start in the lobby for a slide presentation for its program committee, then move into the large meeting room for a slide presentation to its full membership. The second organization is planning a slide presentation to its full membership in the large meeting room. The second organization would like to begin its meeting at the same time the first organization is completing its program committee meeting.

At any point in time, we can build a model of the resource requests and allocation for this meeting facility using a directed graph. An edge (a, b) where

$$a \in \{Organization\ One,\ Organization\ Two\}$$

and

$$b \in \{Meeting\ Room,\ Slide\ Projector,\ Lobby\}$$

indicates that group a has requested resource b, whereas (b, a) indicates that group a has been allocated resource b. Assuming that Organization Two has scheduled the large meeting room and Organization One is completing its use of the slide projector in the lobby, a model at the time the meeting of the second organization is about to begin is shown in Figure 6.55.

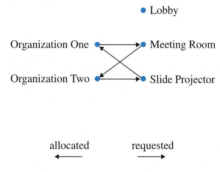

Figure 6.55 Model of meeting facility.

From examining the model, it is clear that neither organization can continue its meeting, because a resource needed by each organization is assigned to the other organization, which has no plan to release it.

The graph of the meeting facility's resource allocation clearly is dependent on the particular time the model represents. For example, the meeting facility would be represented by the directed graph shown in Figure 6.56 for the case that Organization One is completing its meeting just as Organization Two is scheduled to begin its meeting.

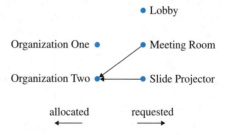

Figure 6.56 A second model of the meeting facility.

Although this model is given in terms of organizations and meeting facilities, the same problem arises in an operating system when two programs request the same printer, tape drive, or disk drive at the same time.

6.15.1 WAITFOR Graphs

To model the general situation of handling the request for and the allocation of resources involving many users and many resources, we formally define a special type of directed graph.

Definition 1. A **WAITFOR graph** for the allocation status of a set of resources is a directed bipartite graph $D = (V, E)$ where V_1 and V_2 are a bipartition of V. The elements of $V_1 = \{P_1, P_2, \ldots, P_n\}$ represent users of the resources, and the elements of $V_2 = \{R_1, R_2, \ldots, R_m\}$ represent resources. For $P \in V_1$ and $R \in V_2$, there is an edge $(P, R) \in E$ if user P requests resource R and has not been granted resource R and an edge $(R, P) \in E$ if user P has been allocated the resource R where $P \in V_1$ and $R \in V_2$.

A WAITFOR graph can be thought of as taking a snapshot of the status of requests for and allocations of resources at a fixed point in time. Different points in time may generate different WAITFOR graphs for the same users and the same resources.

In the operation of a computer, it is important to know at each point in time if any of the processes being executed by an operating system cannot progress toward completion. Two or more processes waiting indefinitely for an event that can only be completed by one of the waiting processes are said to be in a **deadlock state,** or **deadlock.** A deadlock state will depend on the scheme that is used by the operating system to allocate resources such as input or output devices or files. Regardless of the allocation scheme, an operating system must deal with the problems of detecting a deadlock state and deciding what to do when a deadlock state is detected. One allocation scheme for an operating system is to allocate resources so that any resource is assigned completely to one process at a time but can be reassigned when one process has completed its use. Typically, printers, tape drives, and files are allocated in this way. Such an allocation scheme is called a **single-unit resource scheme.**

Figure 6.57 shows a WAITFOR graph that represents a deadlock state for an operating system using the single-unit resource allocation scheme. User 1 needs both a tape drive and a printer to continue. The tape drive is allocated to User 1, but the printer is allocated to User 3. At the same time, User 3 needs both a tape drive and a printer to continue. User 3 has been allocated the printer but must wait for the tape drive. The problem is that neither User 1 nor User 3 can proceed until one of the devices allocated to the other is freed. Since neither can proceed until all their requested resources are allocated, the processing comes to a deadlock state. The operating system could break the deadlock by taking the tape drive away from User 1 or by taking the printer away from User 3. The reader should draw the WAITFOR graphs that result from these drastic actions to see whether they are deadlock free. Figure 6.57 also shows a second problem that an operating system must solve. In the case, if User 1 finishes with the tape drive or the operating system takes it away from User 1, should User 2 and User 3 be given access to that resource next? The WAITFOR graph gives no help with this question. A key property about WAITFOR graphs is summarized in the next theorem.

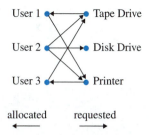

Figure 6.57   Deadlock.

Theorem 1.   A WAITFOR graph represents a deadlock state for a single unit resource scheme if and only if the digraph contains a directed cycle.

6.16   Finding a Cycle in a Directed Graph

The theorem that characterizes a WAITFOR graph that represents deadlock states is useful, because there is an algorithm for detecting a directed cycle in a directed graph. The algorithm proceeds by first recognizing a vertex v with outdegree zero as a vertex that cannot be contained in any directed cycle. Then, all the edges with v as head can be identified as not being in a directed cycle through v and eliminated from further consideration. This follows because each of these edges leads to v, from which no continuation is possible. It then follows that the process can be repeated for the subgraph formed by deleting all the edges with v as head and putting v in the set of vertices that cannot be in a directed cycle. This process can be repeated until no vertex remains or every vertex remaining is contained in a directed cycle. An example of using this procedure is shown in Figure 6.58.

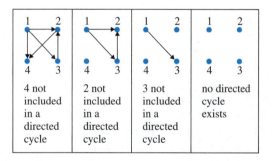

Figure 6.58   Cycle-free directed graph.

If all the vertices of the directed graph can be identified as not being in a directed cycle, then no directed cycle was present in the original directed graph.

6.16.1   Directed Cycle Detection Algorithm

A number of data structures are assumed to be available to the algorithm. The algorithm uses *NotInCycle* as the initial set of vertices known to be unable to lie in any directed cycle—that is, those vertices of outdegree zero. Q is the queue of vertices in *NotInCycle* waiting to be processed. Array entry *NumExits*[v] is initialized as the current maximum for the number of ways a directed cycle can exit vertex v. Initially, this is just the outdegree of a vertex.

Algorithm: Directed Cycle Detection

INPUT: Directed graph $D = (V, E)$
OUTPUT: *TRUE* if D contains a directed cycle and
FALSE if it does not

Initialize *NotInCycle* and Q to contain the vertices
    with outdegree 0 and *NumExits*[$1 .. |V|$] to the outdegrees of the vertices of D
while $Q \neq \emptyset$
    Remove the vertex v at the head of Q
    for each $w \in V$ such that $(w, v) \in E$ do
        Remove (w, v) from E
        NumExits[w] = *NumExits*[w] $- 1$
        if (*NumExits*[w] $= 0$) then
            NotInCycle = *NotInCycle* $\cup \{w\}$
            Put w at the end of Q
if (*NotInCycle* $= V$) return *FALSE*
    else return *TRUE*

Initially, there is no way to know which vertices, other than vertices of outdegree zero, cannot be in a directed cycle. The vertices with outdegree zero give the algorithm a place to start. As the algorithm proceeds, any vertex v that is identified as not being in a directed cycle has the edges with it as the head identified as edges not belonging to a directed cycle. For all vertices w that are the tail of an edge not in a directed cycle, *NumExits*[w] is decremented by one. If *NumExits*[w] becomes zero for some vertex w through this process, then the vertex must have all the edges with it as the head the recognized as not being in a directed cycle. The set *NotInCycle* merely collects the vertices that are identified as not belonging to any directed cycle. If at the end of the procedure every vertex in the directed graph is in *NotInCycle*, then the digraph does not contain a directed cycle. If not every vertex is an element of *NotInCycle* when the procedure terminates, there is a directed cycle in D contained in the induced subgraph $< V - NotInCycle >$.

6.16.2   Correctness of Directed Cycle Detection

Let us argue that the algorithm terminates and produces the correct answer. If no vertex has outdegree zero, then the directed graph has a directed cycle by Exercise 2 in Section

6.20, and the correct result is returned. This follows because $Q = \emptyset$ and the outer loop is never entered. At the end of the procedure, $NotInCycle = \emptyset \neq V$.

If some vertex has outdegree zero, then at least one vertex is put in both *NotInCycle* and Q. The outer loop is executed, since $Q \neq \emptyset$ is true. The inner loop may identify additional vertices that cannot be in a directed cycle. All such additional vertices, as they are recognized, are included in the set *NotInCycle* and put on Q. Each instance of the inner loop will terminate, since each vertex is adjacent only to finitely many other vertices. The outer loop executes at most once for each vertex in V. This is because the first time a vertex is put on Q, the vertex has outdegree 0 in the current graph. Hence, it can never serve as the tail of an edge in the remaining graph and, so, cannot be put on Q a second time.

Therefore, the outer loop terminates after at most $|V|$ iterations. Finally, the procedure determines whether there is a directed cycle in the directed graph by asking whether all the vertices have been put in *NotInCycle*. If $V = NotInCycle$, then the directed graph does not contain a directed cycle. If $V \neq NotInCycle$, then it can be shown that the subgraph induced by the vertices in $V - NotInCycle$ contains a directed cycle by using Exercise 3 in Section 6.20.

6.17 Priority in Scheduling

In addition to being able to determine whether a set of processes is in deadlock, an operating system must be able to schedule all the processes in a system so that each process has its required input available before it is executed. When a complicated expression is evaluated, each operator must have access to the values of its operands at the time it is carried out, so the operators in the expression must be "scheduled."

One additional example involves a manufacturing assembly process. In manufacturing a complex product, assembly is decomposed into a sequence of subassemblies that must be scheduled so that required subassemblies are completed before their output is needed by another subassembly. Figure 6.59 shows a directed graph that models the relationships among a set of subassemblies in a manufacturing process. The directed graph has a directed edge from subassembly A to subassembly B if subassembly A must be completed before subassembly B can begin. A practical problem to solve is how to schedule all these processes sequentially so that whenever a subassembly begins, all the subassemblies that it requires have been completed.

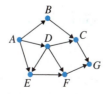

Figure 6.59   Related subassemblies.

Before examining the general problem, consider the smaller scheduling problem represented by the directed graph shown in Figure 6.60.

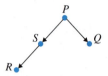

Figure 6.60   A small scheduling problem.

Clearly, P must be scheduled first. For the remaining processes Q, R, and S, any order in which S precedes R is acceptable. The possible schedules are

$$P \rightarrow Q \rightarrow S \rightarrow R \qquad P \rightarrow S \rightarrow Q \rightarrow R \qquad P \rightarrow S \rightarrow R \rightarrow Q$$

What we have accomplished is to take the partial ordering relation represented by a directed graph in Figure 6.60 and find a linear order of the vertices. The linear order has the property that for each edge, we have the head of the edge in the directed graph occurring after the tail of that edge in the final linear order. Check that each of the edges (P, Q), (P, S), and (S, R) satisfy the property.

More formally, this process is known as **embedding a partial order** in a linear order (see Section 3.8.6). The first problem is to know when a directed graph representing a set of constrained events can be scheduled sequentially so that all the constraints are satisfied. A second problem is to find a schedule if one is possible.

If the directed graph contains a directed cycle, then clearly, no schedule can satisfy all the constraints. On the other hand, if the digraph has no directed cycle, then it is always possible to schedule the events by a process called a **topological sort.**

Whether a digraph contains a directed cycle can be determined using the algorithm presented in Section 6.16.1 (Directed Cycle Detection). Directed acyclic graphs are often called **dags.**

One algorithm for finding a topological sort of a dag uses a depth first search on a directed graph. The only difference between a depth first search for a directed graph and for an undirected graph is that in the case of a directed graph, the edge (i, j) has j occurring on the adjacency list for i but does not have i occurring on the adjacency list for j unless the edge (j, i) is also included in the directed graph.

6.17.1   Algorithm for Topological Sort

An algorithm for topological sorting will be given followed by an argument showing its correctness. Finally, an example of using the algorithm will be presented.

As an intuitive aid to understanding the algorithm, think of the vertices as representing jobs and the edges as representing precedence constraints between jobs. At each step, the list *Stack* correctly orders the vertices (jobs) considered so far.

Algorithm: Topological Sort

INPUT: A directed acyclic graph $D(V, E)$
OUTPUT: An ordering of the vertices so that if D contains a
directed path from v to w, then $v < w$

Initialize the $|V|$ positions *Visited* $[1..|V|]$ to *FALSE*
Stack $= \emptyset$
for each $v \in V$ do
 if (*Visited* $[v]$ = *FALSE*) then
  TopSort(v)
print the elements in *Stack* from top to bottom

TopSort(v)   /* The recursive procedure */
 Visited $[v]$ = *TRUE*
 for each vertex w at the head of an edge (v, w) leaving v do
  if (*Visited* $[w]$ = *FALSE*) then
   TopSort(w)
 Add v to the top of *Stack*

6.17.2   Correctness of Topological Sort Algorithm

We will use induction on the number of vertices that are topologically sorted to prove that this algorithm is correct.

Theorem 2.   The output of the Topological Sort algorithm is correct.

Proof.   Let $n_0 = 1$ and $\mathcal{T} = \{n \in \mathbb{N}$: the output of Topological Sort algorithm is correct for every dag on n vertices$\}$.

(Base step)   The base case is $n_0 = 1$. For a dag with one vertex, the algorithm is correct, since the single vertex is printed.

(Inductive step)   Choose n such that $n > n_0$, and suppose that for all dags with k vertices where $1 \le k < n$ that $k \in \mathcal{T}$. Now, let D be a dag with n vertices. Since the vertices are initially marked *FALSE*, call *TopSort*(1) to start at vertex 1. The procedure is just a depth first search with the added feature that the vertex passed to *TopSort* is put on a first-in-last-out list, called a **stack,** before *TopSort* is exited. Any vertex that can be reached from vertex 1 by a directed path will be visited before *TopSort*(1) is completed and, hence, will be put on the list before 1 is put on the stack. Thus, 1 will be closer to the top of the stack than vertices reachable from 1 by directed paths. These vertices reachable from 1 in D and appearing below 1 on the stack are just the vertices that represent processes that must be executed after 1. Therefore, when *TopSort*(1) is completed, all the vertices that can be reached from 1 will be on the list below 1. After completing the call to *TopSort*(1), there will be at most $n - 1$ vertices marked *FALSE*.

The induced subgraph D_1 formed from the set of vertices still marked *FALSE* that remain will be a dag with fewer than n vertices. Therefore, by the induction hypothesis, this dag will be topologically sorted correctly. It remains to observe that no vertex in D_1—say, v—can be the head of an edge—say, (w, v)—such that w is already on the list. This is true because if w occurred on the list, then v would have been visited by the depth first search when w was encountered and, consequently, v would have been put on the stack before w. This means that each vertex in D_1 can be placed on the stack after all the vertices already on the stack without violating any of the constraints represented by D. Therefore, $n \in \mathcal{T}$.

By the Strong Form of Mathematical Induction, $\mathcal{T} = \{n \in \mathbb{N} : n \geq 1\}$. ■

Figure 6.61 shows the resulting order for the dag shown where the depth first search begins at A and the vertices adjacent to A are put on its adjacency list in the order B, E, D.

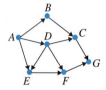

Figure 6.61   *TopSort* on a dag. Topological sort order: *A, D, E, F, B, C, G*.

6.18   Connectivity in Directed Graphs

The notion of connectedness for a directed graph is not as simple as the notion of connectedness for undirected graphs. The reason is that directed edges represent an adjacency between the tail and the head of the edge, but not one between the head and the tail of the edge. The simplest notion of connectedness for directed graphs, called **weakly connected,** holds whenever the underlying graph (that is, the undirected graph formed by omitting the directions on the edges of the directed graph and eliminating multiple edges if both (a, b) and (b, a) are in the directed graph for some pair of vertices a and b) is connected. A directed graph that is not weakly connected is called **disconnected.**

6.18.1   Strongly Connected Directed Graphs

A second notion of connectedness for directed graphs depends on the existence of directed paths from v to w and from w to v for each pair of vertices v and w. A directed graph with such paths for each pair of vertices is called **strongly connected. A strongly connected component** of a digraph is a directed subgraph that is strongly connected and is not contained in any larger strongly connected subgraph. The graph shown in Figure 6.62(a) is strongly connected, and the graph shown in Figure 6.62(b) is not. The graph shown in Figure 6.62(b) actually has three strongly connected components.

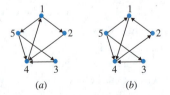

(a)   *(b)*

Figure 6.62   Digraphs with and without the path property.

The property of having two directed paths joining a pair of vertices, one path in each direction, defines a relation on the vertices of a directed graph. This relation helps us to understand strong connectivity in directed graphs.

Definition 1.   Let $D = (V, E)$ be a directed graph. For all $v, w \in V$ define v *STCONN* w if and only if there is a directed path in D from v to w and a directed path in D from w to v.

Although *STCONN* may seem to be an abstract way to approach the notion of connectedness for directed graphs, use of this relation makes it easier to find the strongly connected components of a directed graph—that is, the largest strongly connected subgraphs.

Theorem 1.   *STCONN* is an equivalence relation.

Proof.   Show that the relation *STCONN* is reflexive, symmetric, and transitive. Let $D = (V, E)$ be a directed graph. For each $v \in V$, the vertex itself is a directed path from v to v. Thus, v *STCONN* v for each $v \in V$, and the relation is reflexive. For vertices $v, w \in V$ such that v *STCONN* w, there is a directed path in D from v to w and a directed path from w to v. Since there is a directed path from w to v and a directed path from v to w, it follows that w *STCONN* v. This proves that *STCONN* is symmetric.

Finally, for vertices $u, v, w \in V$ such that u *STCONN* v and v *STCONN* w, it will follow that u *STCONN* w. Since u *STCONN* v, there are directed paths in D from u to v and from v to u. Since v *STCONN* w, there are directed paths in D from v to w and from w to v. Using the directed paths from u to v and v to w, form a directed trail that starts at u and ends at w. If a vertex occurs twice in this path, it is easily eliminated so that a path results. Similarly, form a directed trail from w to u. Delete any directed cycles from these newly formed directed trails to form directed paths. Therefore, u *STCONN* w, and the relation *STCONN* is transitive.

Since *STCONN* is reflexive, symmetric, and transitive, *STCONN* is an equivalence relation.   ■

The induced subgraphs determined by the equivalence classes of D relative to *STCONN* are the strongly connected components of D. The example in Figure 6.63 shows that the strongly connected components may not include all the edges of the directed graph. In particular, (h, g) and (h, i) are such edges. A directed graph that has exactly one strongly connected component is called strongly connected.

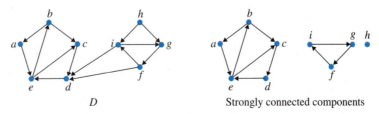

D   Strongly connected components

Figure 6.63   Strongly connected components.

6.18.2   Application: Designing One-Way Street Grids

To alleviate traffic problems on narrow, crowded streets, one strategy is to allow traffic to flow in only one direction. One obvious condition for a system of one-way streets to satisfy is that traffic can somehow get from any intersection to any other intersection. The first step in designing a system of one-way streets with this property is to describe a suitable model. For this purpose, define a graph with vertices representing street intersections and an edge joining a pair of vertices if the intersections they represent are joined directly. This gives a good model for the two-way street traffic pattern. To make this into a one-way street pattern, place directions on the edges to represent the direction traffic may flow on that one-way street.

Definition 2.   Let G be an undirected graph. An **orientation** O for G is an assignment of directions to the edges of G.

An orientation is simply a way to make an undirected graph into a directed graph. Obviously, there are many ways to orient the same graph. Three different ways to orient the same graph are shown in Figure 6.64.

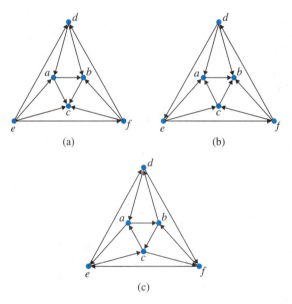

Figure 6.64   Three orientations of a graph.

In Figure 6.64(a), the orientation does not permit any directed path starting at one of a, b, or c to reach any of d, e, or f. In addition, no directed path from d can reach either e or f. In Figure 6.64(b), directed paths beginning at d can reach e and f as well as a, b, and c. The graph in Figure 6.64(c) has the property that for any two vertices v and w, there is a directed path from v to w and a directed path from w to v where $v, w \in \{a, b, c, d, e, f\}$. The property of having directed paths joining each pair of vertices is just the property that the directed graph is strongly connected. This property, as formally defined in Definition 3, is just the property needed in any grid of one way streets.

Definition 3. Let G be an undirected graph and O an orientation for the edges of G. G is **orientable** if G with orientation O is strongly connected.

In Exercise 9 of Section 6.20, the reader is asked to prove that a directed graph is orientable if and only if each undirected edge is contained in a cycle. This theorem tells when a one-way street pattern with each intersection accessible from all other intersections can be established. The theorem does not tell how good the traffic flow will be. Figure 6.65 shows two examples of orientations that make the graph orientable. Notice, however, that the directed path from v to w in G is much shorter than the directed path from v to w in H.

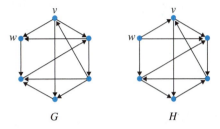

$$G \qquad\qquad\qquad H$$

Figure 6.65   Two orientations that make a graph orientable.

6.19   Eulerian Circuits in Directed Graphs

For a directed graph, we define a **directed Eulerian circuit** to be a directed circuit that contains each edge in the graph exactly once. The necessary and sufficient conditions for a directed graph to have a directed Eulerian circuit can be stated in terms of the indegree and outdegree of each vertex in the graph

Theorem 1. Let $D = (V, E)$ be a connected digraph. D contains a directed Eulerian circuit if and only if $indeg(v) = outdeg(v)$ for each $v \in V$.

The proof of this theorem can be constructed by a straightforward modification of the proof of the necessary and sufficient condition for the existence of an Eulerian circuit in an undirected graph. The necessary and sufficient condition for the existence of an Eulerian circuit in the undirected case is that every vertex has even degree. In the Euler Theorem for directed graphs, this hypothesis translates into the condition that there are as many directed edges with a vertex as head as there are directed edges with the same vertex as tail. The adaptation of the proof of the Euler Theorem in the undirected case to the directed case involves observing that each time the directed Eulerian circuit passes through a vertex, it accounts for one of the edges contributing to the indegree of the vertex and for one of the edges contributing to the outdegree of the vertex. Thus, each time an edge is directed into a vertex in the directed Eulerian circuit, an edge is also directed out of the vertex and can be used to continue the directed Eulerian circuit—unless this vertex happens to be the start of the circuit being constructed. In the case of the vertex at the start of the directed Eulerian circuit, the outdegree contribution from the start vertex is matched with the indegree contribution of the vertex at the end.

Information Links on the Web

The pages accessible from a Web browser can be represented by a graph in which the vertices are the pages and there are directed edges between pages that are linked. A query will determine the importance of a page by incorporating information about the importance or relevance of the information in pages pointed to and from the page being examined. When you make a Web query, the pages associated with a given page are as important as the page itself. The final listing of pages of interest for a query are determined by searching the graph that represents the linkages of a page. Additional mathematical techniques are used to discern the final selection of pages to present to the user. Optimization techniques involving linear algebra and probability are used to rank pages that are found by the search procedure.

6.20   Exercises

1. In each of the following graphs:

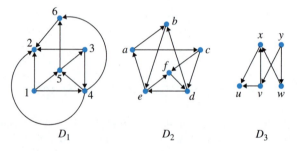

$$D_1 \qquad\qquad D_2 \qquad\qquad D_3$$

   (a) Determine the indegree of each vertex.
   (b) Determine the outdegree of each vertex.
   (c) Find a directed cycle of length four or greater, if any exists.
   (d) Find a directed path of length four or greater, if any exists.
   (e) Find a directed circuit of length at least six.

2. Let D be a directed graph with an outdegree of each vertex of at least one. Prove that D contains a directed cycle. Show that the same result holds if the hypothesis is that each node has an indegree of at least one.

3. Prove that any directed acyclic graph contains at least one vertex with an indegree of zero. Use this result to devise a different algorithm to do a topological sort.

4. Verify that the following graph D is a dag. Perform a topological sort on the vertices of the graph.

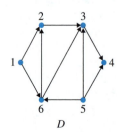

$$D$$

5. Verify that the following graph D is a dag. Perform a topological sort on the vertices of the graph.

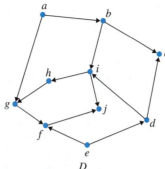

D

6. Prove that if $C_1, C_2, \ldots, C_k$ is a sequence of cycles in a directed graph D such that every two consecutive cycles have at least one common vertex, then the subgraph determined by the union of these cycles is strongly connected.
7. Prove that if a graph G contains an Eulerian circuit, then G is orientable.
8. Prove that each of the following graphs is orientable:

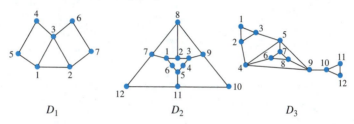

D_1 D_2 D_3

9. Prove that a connected undirected graph is orientable if and only if each edge is contained in a cycle.
10. Show that any graph with a Hamiltonian cycle is orientable.
11. Prove that if a directed graph is Eulerian, then it is strongly connected.
12. Prove that a directed graph $D = (V, E)$ is strongly connected if and only if for any partition V_1, V_2 of V there is an edge (x, y) with $x \in V_1$ and $y \in V_2$.
13. Find a directed graph that is not Eulerian but for which the underlying graph is Eulerian.
14. Devise an algorithm to find an Eulerian circuit in a directed graph, if one exists. Modify the algorithm to find all Eulerian circuits in a graph.
15. *Challenge:* A complete directed graph is a directed graph whose underlying graph is a complete graph. Show that the sum of the squares of the indegrees over all vertices is equal to the sum of the squares of the outdegrees over all vertices in any directed complete graph.

Definition 1. A **tournament** T_n is a directed graph on n vertices such that each pair of vertices v, w is joined by one and only one of the directed edges (v, w) or (w, v). The score of a vertex of a tournament is its outdegree. The **score sequence,** or **ranking,** of a tournament is the sequence formed by arranging the scores of its vertices in nondecreasing order. A tournament is **transitive** if the existence of edges (a, b) and (b, c) imply the existence of the edge (a, c). The **complement** of a tournament is formed by reversing all the directions on the edges of a tournament.

16. For the tournament shown, find a ranking of the players:

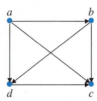

17. Let $D = (V, E)$ be a tournament. Prove that if a has maximum score, then for any vertex y, either there is an edge (a, y) or a vertex w and edges (a, w) and (w, y).

18. Prove that a tournament with no cycles is transitive.

19. Prove that if D is a tournament, then it has a directed Hamiltonian path.

20. Show that a transitive tournament can have its vertices ordered so that if a precedes b, then the score of a is greater than or equal to the score of b.

21. Show that the figure

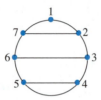

    can be used to schedule a "tournament" for seven players. The tournament has a directed edge (i, j) if and only if i beats j. (*Hint:* Rotate the edges but not the vertices.)

Representing a Relation.   A digraph can be used to represent a relation. For a relation R on a set A, define the digraph $D = (V, E)$ as follows: Let $V = A$. There is an edge $(i, j) \in E$ if and only if $(i, j) \in R$. The definition of a digraph needs to be extended slightly to allow representation of elements of the form $(x, x) \in R$. A loop in a digraph is an edge with both ends the same. We think of a loop as starting and ending at the same vertex. For example, we can represent the relation R on $\{1, 2, 3, 4, 5\}$ with elements $\{(1, 1), (2, 3), (3, 4), (4, 1), (2, 5)\}$ as

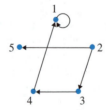

22. Represent by a digraph the partial order defined on $\mathcal{P}(\{1, 2, 3, 4\})$ where the relation is set inclusion.

23. Represent by a digraph the partial order *divides* defined on the integers 0 through 11.

24. Let R be the relation on $\{1, 2, 3, 4, 5\}$ with elements $\{(1, 1), (2, 1), (3, 2), (2, 3), (1, 4), (3, 5), (5, 2)\}$. Represent R as a digraph.

25. Let R be the relation on $\{1, 2, 3, 4, 5\}$ with elements $\{(1, 1), (2, 1), (3, 2), (2, 3), (1, 4), (3, 5), (5, 2)\}$. Represent the reflexive closure of R as a digraph.

26. Let R be the relation on $\{1, 2, 3, 4, 5\}$ with elements $\{(1, 1), (2, 1), (3, 2), (2, 3), (1, 4),$ $(3, 5), (5, 2)\}$. Represent the symmetric closure of R as a digraph.
27. Let R be the relation on $\{1, 2, 3, 4\}$ with elements $\{(1, 1), (2, 1), (3, 2), (2, 3), (1, 4)\}$. Represent the transitive closure R^* of R as a digraph.

6.21 Chapter Review

Definitions and notation are the first hurdles in studying graph theory. Vertices, edges, incidence, adjacency, and degrees are the core terms. Using these terms, one learns to identify parts of a graph, such as a subgraph, a spanning subgraph, an induced subgraph, a trail, a path, a circuit, and a cycle. The idea of a graph isomorphism is needed to distinguish between graphs that are truly different and graphs that are simply labeled differently. The standard representations of graphs by adjacency matrices and adjacency lists finish the introduction to the basic vocabulary. The next major topic involves determining whether a graph is connected and, if not, what its connected components are. Two fundamental search procedures, depth first search and breadth first search, are explained. The complexity of the algorithms for carrying out these search procedures as well as the determination that the algorithms are correct concludes this discussion. The search procedures are also used in an algorithm to identify the connected components of a graph. These algorithms are proven to be correct, and their complexity is determined. The chapter then examines two special kinds of graphs: trees, and directed graphs. Various definitions of trees are proven to be equivalent. Minimal connected spanning subgraphs are introduced. Kruskal's algorithm for finding an MCST is presented, with correctness and complexity issues being explained. A second class of trees, called rooted trees, is then introduced and used in searching procedures. Finally, with directed graphs, the vocabulary is introduced, as are two further algorithms. The first detects a directed cycle, and the second carries out a topological sort. The directed form of Euler's Theorem is presented, as is the notion of a strongly connected directed graph.

Applications include developing a process to determine if a graph has a Hamiltonian cycle. Euler's Theorem is used to determine an effective way to draw a graph. Decision trees are used to find a lower bound for the complexity of any sorting algorithm based on comparing pairs of elements. Directed graphs are used to characterize deadlock in an operating system that uses a single resource allocation scheme. The notion of being strongly connected is used to describe effective grids of one-way street patterns.

6.21.1 Terms, Theorems, and Algorithms

6.1–6.5 Summary

TERMS

acyclic	bipartite graph	complete bipartite graph
adjacency list	bipartition	complete graph
adjacency matrix	circuit	cubic graph
adjacent	clique	cycle
binary representation	complement	degree

degree sequence	intersection	regular graph
distance	invariant	same
edges	isolated vertex	self-complementary
even (odd)	isomorphic	graph
graph	isomorphism	spanning subgraph
graphical	k-cycle	subgraph
Hamiltonian cycle	length	trail
Hamiltonian graph	n-cube (Q_n)	triangle
hypercube	neighbros	union
incident	n-regular	vertex
induced subgraph	path	vertices

THEOREMS

Handshaking theorem   In any graph the number of odd vertices is even.

6.7–6.8   Summary

TERMS

breadth first search	depth first search tree	Königsberg Bridge
breadth first search tree	disconnected	Problem
connected	Eulerian circuit	multigraph
connected component	Eulerian trail	queue
depth first search	first-in-first-out list	

THEOREMS

Conn is an equivalence relation.
Königsberg Bridge Problem

ALGORITHMS

Breadth First Search (*Bfs*)   Connected Components   Depth First Search (*Dfs*)

6.10–6.13   Summary

TERMS

ancestor	height	right child
A-V-L tree	inorder	root
binary search tree	interior vertex	rooted subtree
binary tree	leaf	rooted tree
child	left child	rotation
cost	level	sibling
decision tree	Minimum Cost Spanning	spanning subgraph
descendant	tree (MCST)	spanning tree
forest	parent	terminal vertex
greedy algorithm	postorder	tree
heap	preorder	value

vertex label   weight
vertex label listing   weighted graph

THEOREM

Tree Characterization

ALGORITHMS

Minimum Cost Spanning Tree—Kruskal   Spanning Tree—Kruskal
Search Using a Binary Search Tree

6.14–6.19   Summary

TERMS

adjacent edges	embedding a partial order	score sequence
bipartite	head	single-unit resource
complement	incident	scheme stack
dag	indegree	STCONN
deadlock	induced subgraph	strongly connected
digraph	isomorphic directed	strongly connected com-
directed circuit	graphs	ponent
directed cycle	linear order	tail
directed edge	orientable	topological sort
directed Eulerian circuit	orientation	tournament
directed graph	outdegree	transition
directed path	partial order	underlying graph
directed subgraph	ranking	WAITFOR graph
directed trail	schedule	weakly connected

THEOREM

STCONN is an equivalence relation

ALGORITHMS

Directed Cycle Detection   Topological Sort

6.21.2   Starting to Review

1. A trail may contain

   (a) Repeated occurrences of vertices and edges
   (b) Repeated occurrences of vertices but no repeated occurrences of edges
   (c) No repeated occurrences of vertices, and no repeated occurrences of edges
   (d) None of the above

2. Let $G = (V(G), E(G))$ and $H = (V(H), E(H))$ be graphs. If G is isomorphic to H, what may we correctly conclude?

   (a) $|V(G)| = |V(H)|$ and $|E(G)| = |E(H)|$
   (b) $|V(G)| = |V(H)|$ only.
   (c) $|E(G)| = |E(H)|$ only.
   (d) None of the above

3. Let $D = (V, E)$ be a directed graph. An edge $(a, b) \in E$:
    (a) Contributes one to the indegree of a
    (b) Contributes one to the outdegree of a
    (c) Contributes one to the indegree of a and one to the outdegree of b
    (d) All of the above

4. The longest directed trail in the graph shown contains how many edges?

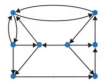

    (a) 8
    (b) 7
    (c) 6
    (d) 5

5. Construct a graph with degree sequence 1, 2, 2, 3, 4, 5, or prove that none exists.

6. Answer the following questions for the following tree

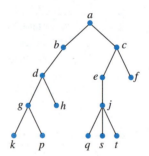

    (a) What vertices are leaves?
    (b) What vertex is the root?
    (c) What vertex is the parent of g?
    (d) What vertices are the descendants of c?
    (e) What vertices are the siblings of s?
    (f) What is the level number of the vertex f?
    (g) What vertices are at level four?
    (h) What is the height of the tree?

7. Find an MCST in G. Show all the steps of the algorithm you use.

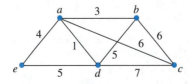

8. List the vertices in the tree shown when they are visited in a preorder traversal and in a postorder traversal.

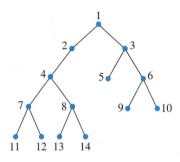

9. Find an Eulerian circuit in G. Show all steps of the process.

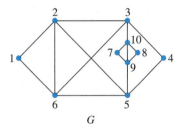

G

10. Topologically sort D.

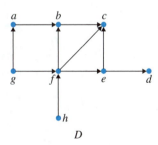

D

6.21.3   Review Questions

1. Let $G = (V, E)$ be a graph for which $deg(v) \geq r$ with $r \geq 2$ for every $v \in V$. Prove that G contains a cycle of length at least $r + 1$. Use this result to show that if everyone at a party knows at least n others, then it is possible to seat $n + 1$ of the guests at a circular table so that everyone know the persons seated on their left and their right.

2. Prove that G has no Hamiltonian cycle that contains the edge (e, j).

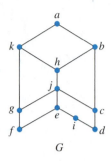

G

3. Prove that the G and H are isomorphic.

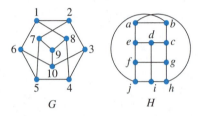

G H

4. Let G and H be graphs, and let $v \in V(G)$. Let $F : G \rightarrow H$ be an isomorphism between G and H. Let edges (v, w), (w, y), (y, x), and (x, v) form a 4-cycle in G. Prove that the images of these edges under F form a 4-cycle in H.

5. Construct the *Dfs* and *Bfs* trees for the following graph starting at vertex 1: Assume the adjacency lists are formed by listing the adjacencies of each vertex in increasing order.

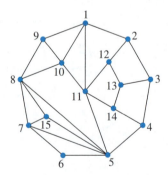

6. Let $G = (V, E)$ be a graph. Prove that if $\delta(G)$ and $\Delta(G)$ represent the minimum and the maximum degrees of all the vertices of G, respectively, then

$$\delta(G) \leq 2 \cdot |E|/|V| \leq \Delta(G)$$

7. Prove by induction that the sequence $1, 1, 2, 2, 3, 3, \ldots, n-1, n-1, n, n$ is graphical for all n such that $n \geq 1$.

8. Let $G = (V, E)$ with $|V| \geq 2$ be a graph with all vertex degrees occurring exactly once except for a single degree that occurs twice. Find the value(s) for the repeated

degree value. (*Hint:* If $|V|$ is even, then the repeated degree can have two possible values. If $|V|$ is odd, then the repeated value is unique.)

9. Prove that a graph is connected if and only if every two-part partition of the vertices of the graph has an edge with ends in both parts of the partition.

10. Using the picture of the bridges connecting the two islands in the middle of the Seine in Paris, describe a trail that takes you over each bridge exactly once. See the comment about *multigraphs* in Section 6.8.

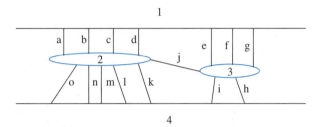

Paris bridges joining the islands in the Seine.

11. Prove that if a tree has a vertex of degree p, then it has at least p vertices of degree one.

12. Agents A, B, C, D, E, F, G, and H are embroiled in a conspiracy called the Bucknell-gate Affair. To coordinate their cover-up efforts, each conspirator must be able to communicate with each other either directly or indirectly (through another conspirator). A table of risk factors is given for each possible direct communication link. What is the least total possible risk while still meeting the requirements for communication?

Agent Pairs	AB	AC	AE	AF	AG	BC	BF	CD	CF	CG	CH	DE	DH	EH
Risk Factor	6	3	7	4	13	6	5	4	6	7	9	2	8	5

13. Prove that D_1 and D_2 are not isomorphic.

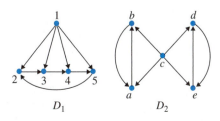

14. Topologically sort D.

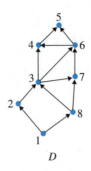

D

15. Prove that G contains no Hamiltonian cycle.

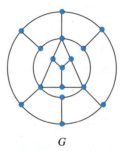

G

16. Let A be the adjacency matrix of a graph $G = (V, E)$ with $V = \{1, 2, \ldots, n\}$. Prove that the sum of the entries in any column is the degree of the corresponding vertex. Prove that the (i, j) entry of A^n, $n \geq 1$, is the number of different trails between i and j of length n in G. A^n denotes matrix multiplication of A with itself n times. For A, a 3×3 matrix, define $A \cdot A$ to have $(i, j)^{th}$ entry equal to $\sum_{k=1}^{3} a_{ik} \cdot a_{kj}$.

6.21.4 Using Discrete Mathematics in Computer Science

1. A vertex in a directed graph that has an indegree of zero is called a *transmitter.* Identify the transmitters in D. Devise a strategy for spreading a rumor so that all the nodes of the graph, which can be thought of as people on the ends of telephone lines, know the rumor.

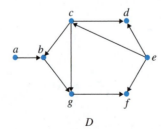

D

2. *Prim's Algorithm for MCST:* Carry out the following procedure for the graph shown in Figure 6.34 in Section 6.11.3. Let $G = (V, E)$ be a graph. Maintain two sets: E_1 and V_1 where E_1 is initially empty and V_1 is a subset of V initially containing any vertex of the graph. Repeat $|V| - 1$ times: Find an edge $e \in E$ of least cost that joins a vertex $v \in V_1$ to a vertex $w \notin V_1$. Let $E_1 = E_1 \cup \{e\}$ and $V_1 = V_1 \cup \{w\}$.

3. A printer has one printing machine and one binding machine. Let p_i and b_i denote the printing time and binding time for book i, respectively. For any two books i and j, either $b_i \geq p_j$ or $b_j \geq p_i$. Show that it is possible to specify an order in which the books are printed (and then bound) so that once the first book is printed, the binding machine will be kept busy until all the books are bound. As an example, find a schedule for four books with printing time and binding time given as an ordered pair. (The first coordinate is the printing time, and the second coordinate is the binding time.) $D = \{(2, 3), (3, 5), (4, 1), (3, 3)\}$.

4. Eight computers must be networked together. Each computer must be accessible from every other computer, but the connection does not have to be direct in all cases. The computers $(A-H)$ are represented by the vertices of the following graph, and the label on an edge is an estimate of the cost of running the network cable. Find the set of cables connecting all the computers with the minimum cost.

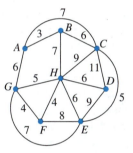

5. A bank plans to use special data lines to connect each of its branch offices with the main office. The data line from a branch office need not be connected directly to the main office. A branch office may be connected indirectly to the main office by connecting to another branch office that is connected (directly or indirectly) to the main office. Every branch must be connected to the main office by some route. The distances between pairs of offices are as follows:

	1	2	3	4	5	6
1	X	160	270	75	70	190
2	160	X	310	80	210	50
3	270	310	X	175	120	215
4	75	80	175	X	150	240
5	70	210	120	150	X	100
6	190	50	215	240	100	X

Determine which pairs of offices should be connected by data lines so as to connect every branch office (directly or indirectly) to the main office with a minimum total cost. Repeat the procedure if there must be a data link between branches 2 and 5. Repeat the procedure if branches 1 and 4 cannot be connected by a data line.

6. Suppose that a group of computers were linked together in a simple ring as shown:

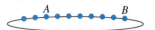

In this arrangement, each computer can communicate directly with the computers on either side of it. Since the communication lines are all separate, each computer can pass information to the computer to the right of it *all at the same time.* To pass a message from computer A to computer B, just pass it along the line from computer to computer until it reaches its destination. The communication lines are supposed to be bidirectional, so the message can be passed in seven steps if it is passed to the right and in three steps if passed to the left.

   Some obvious graph-theoretic data are important to how quickly the computers can share information and to how much it costs to connect the computers together. For example, consider:

   (a) *The Degree of the Graph:* The maximum degree of any vertex. For example, in the graph shown, each computer has a degree of two, so the degree of the entire graph is two. Here, the communication lines are fairly cheap to build. If the degrees of each vertex were large, or if there were many interconnections then the network of computers would be much more expensive to build.

   (b) *The Diameter of the Graph:* The maximum distance between any two vertices. For example, in a path with n vertices, the diameter is n. For such a graph, it can take a relatively long time to send information from one computer to another. This also means that a programmer, trying to get maximum programming speed out of such a computer configuration, will have to program very cleverly to avoid communication taking too long.

   For each of the following graphs, compute the degree and diameter of the graph:

   (a) A complete graph with n vertices.

   (b) A hypercube with n vertices. (Of course, here, n has to be a power of 2.)

   (c) A complete binary tree (see Exercise 25 in Section 6.13) with n vertices where $n = 2^m - 1$ for some non-negative integer m:

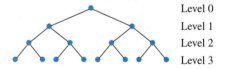

Level 0
Level 1
Level 2
Level 3

   (d) A square grid with n vertices (pictured with 16 vertices):

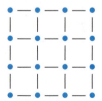

(Another very important feature is how many "bottlenecks" there are in a graph. For example, in the tree, all information from the left side to the right side must flow through the root, making the root a bottleneck in the communication.)

7. Use Fleury's algorithm to find an Eulerian circuit in each of the following graphs given by the following adjacency list representation:

Vertices	Lists of Adjacencies					
1	2	3	4	5	6	7
2	3	4	5	6	7	1
3	4	5	6	7	1	2
4	5	6	7	1	2	3
5	6	7	1	2	3	4
6	7	1	2	3	4	5
7	1	2	3	4	5	6

Vertices	Lists of Adjacencies					
1	2	3	4	6		
2	1	10	3	5		
3	1	2	4	5		
4	1	3	6	7	5	8
5	2	3	4	9	10	8
6	1	4	7	11		
7	4	6	8	11		
8	4	5	7	9		
9	5	8	10	11		
10	2	5	9	11		
11	6	7	9	10		

Fleury's algorithm is as follows:

(a) Choose an arbitrary vertex v and set $W = v$.
(b) Suppose that the trail $W_i = v, v_1 \ldots, v_i$ has been chosen. Then, choose edge e_{i+1} from $E - \{e_1, e_2, \ldots, e_i\}$ in such a way that
i. e_{i+1} is incident to v_i.
ii. Unless there is no alternative, the removal of the edge e_{i+1} does not disconnect the graph remaining after isolated vertices are removed.
(c) Stop when part (b) can no longer be implemented.

8. A Huffman code is a technique for compressing messages based on the frequency of the characters that occur in the message. The procedure for creating a Huffman tree starts by choosing the two characters with the lowest frequencies and creating a subtree with them as leaves. A character not in the message is used to name the tree with the two chosen characters from the message as leaves. The special character is given a frequency that is the sum of the frequencies of its leaves. The process is repeated with the remaining characters augmented by this new character. Draw the Huffman tree associated with the character with frequencies given in the table:

α 0.45
β 0.07
γ 0.12
δ 0.08
ϵ 0.15
ζ 0.13

9. Recall the discussion of satisfiability in propositional logic from Chapter 2 in Section 2.3.1. One of the cases in which satisfiability is easy to check is the case of 2-CNF formulas—conjunctions such as

$$\phi = (\neg p \vee \neg q) \wedge (\neg p \vee q) \wedge (p \vee \neg q) \wedge (p \vee q)$$
$$\chi = (\neg p \vee q) \wedge (p \vee \neg q) \wedge (p \vee q)$$

where each clause has just two literals. That is, each clause is of the form $\lambda_1 \vee \lambda_2$, where each λ_i is either a proposition letter x_i or its negation $\neg x_i$.

In this problem, the negation of a literal λ_i is the negation of λ_i, denoted $\neg\lambda_i$, and the *negation* of $\neg\lambda_i$ is λ_i. That is, when negating $\neg\lambda_i$, first form $\neg\neg\lambda_i$, and then drop the double negation sign.

Given a 2-CNF formula ϕ_0, construct a digraph as follows: The vertices are all literals λ_i and $\neg\lambda_i$ where λ_i is a propositional variable in the program. For each clause $\lambda_i \vee \lambda_j$, add directed edges from vertex $\neg\lambda_i$ to vertex λ_j and from vertex $\neg\lambda_j$ to λ_i.

(a) Draw the digraph constructed for the sample formulas ϕ and χ given previously.
(b) Show that for literals λ_1 and λ_2, if there is a directed trail from vertex λ_1 to vertex λ_2, then $\phi \models \lambda_1 \rightarrow \lambda_2$.
(c) Call a strongly connected component of the graph *self-contradictory* if it includes both λ_i and $\neg\lambda_i$, for some literal λ_i. Show that if the graph has a self-contradictory, strongly connected component, then ϕ is unsatisfiable.
(d) Use part c to show that formula ϕ_0 is unsatisfiable.
(e) For a directed graph G, define a binary relation $\preceq$ on the strongly connected components of the graph as follows: $C_1 \preceq C_2$ if there are vertices $a_1 \in C_1$ and $a_2 \in C_2$ where there is a directed trail in G from a_1 to a_2. Show that if $C_1 \prec C_2$, then for *every* vertex $a_1 \in C_1$ and every vertex $a_2 \in C_2$, there is a directed trail in G from a_1 to a_2. (Remember that there is always a directed trail, of length zero, from any vertex to itself.)
(f) Continue part e: Show that $\preceq$ is a partial ordering of the strongly connected components of G.
(g) Return to the graph defined from the 2-CNF formula χ. Since the formula has only finitely many proposition letters, it has only finitely many literals, and the graph has only finitely many strongly connected components. For the partial ordering $\preceq$ defined in part e, pick a strongly connected component C_0 that is minimal in the $\preceq$ ordering.

Assume that C_0 is not self-contradictory. Start constructing a truth assignment for ϕ by setting each literal in C_0 to *FALSE*—that is, if $x_i \in C_0$, set x_0 to *FALSE*, and if $\neg x_i \in C_0$, set x_i to *TRUE*. Show that if c is any clause of ϕ, either this assignment sets one of the literals in c to *TRUE* or this assignment does not assign truth values to any literal in the clause. (Such a partial truth assignment is called an *autark* assignment.)
(h) *Challenge:* Prove the converse of part c. (*Hint:* Use induction.)
(i) *Challenge:* Earlier, we presented an algorithm to check whether a 2-CNF formula is unsatisfiable: Form the graph, find the strongly connected components, and check whether any strongly connected component is self-contradictory. Analyze the complexity of this algorithm.

CHAPTER 7

Counting and Combinatorics

How many passwords can be formed using five letters and two special characters? How long would it take a computer to generate all such passwords as a way of compromising computer security? How many ways can 25 people be arranged for a group photo? How many ways can the manager of a tour of European capitals order the cities to be visited? How many ways can $100 be divided among four students? How many lottery tickets are needed to cover all the possibilities? Counting techniques introduced in this chapter provide tools for answering questions like these.

Counting techniques are also used in the study of probability to determine the number of events in a sample space and the number of successful outcomes for an experiment. Probability theory commonly uses these counts to assess the likelihood of a particular event. In computer science, elementary counting methods provide useful tools and techniques for dealing with problems such as the enumeration of all possible states that must be considered in proving that a program is correct. More generally, counting methods are used at several stages in the development of a correct and efficient program. For example, a first decision about which algorithm to use in a particular application is often based on knowing how the various alternative algorithms compare with respect to running time or storage use. A step in the process of determining the running time or the storage use of a program often includes one or more counting arguments.

In this chapter, we discuss four major topics. First, we introduce the two fundamental counting principles: the Multiplication Principle and the Addition Principle. Second, we introduce permutations and combinations, which deal with the ways of selecting subsets of a set in which the order of selection of the elements may—or may not—be important. With these preliminaries, we introduce the final two major topics: The first involves counting the number of selections possible from collections with repeated elements; the second involves combinatorial identities, such as the binomial theorem. In addition, Pascal's triangle is presented. The first example of this chapter involves determining the complexity of an algorithm to solve a well-known problem in computer science.

7.1 Traveling Salesperson's Problem

The Traveling Salesperson's Problem (TSP) is deceptively easy to state but deceptively difficult to solve. A salesperson's territory includes *n* cities that must be visited on a regular basis. Between each pair of cities, air service is available. The problem is to schedule a sequence of flights that visits each city exactly once before returning to the starting point so that the total time spent flying is minimized.

A naive algorithm for the solution of the problem uses three steps.

Algorithm: Traveling Salesperson's Problem (*TSP*)

STEP 1: Find all possible routes.
STEP 2: Find the travel time for each route found in Step 1.
STEP 3: Choose a route with travel time equal to the minimum
of the travel times calculated in Step 2.

Example 1 carries out the steps of this algorithm for a set of four cities.

Example 1.   Find a best route for visiting the four cities with the travel times given. The entry in row I and column J is the time to travel either from city I to city J or from city J to city I. Edges indicate direct air service between cities. The number on an edge gives the distance for a flight between the two cities that are the ends of the edge.

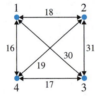

$I \backslash J$	1	2	3	4
1	—	18	30	16
2		—	31	19
3			—	17
4				—

Solution.

Step 1: Find all possible routes.

$$1\text{-}2\text{-}3\text{-}4\text{-}1$$
$$1\text{-}2\text{-}4\text{-}3\text{-}1$$
$$1\text{-}3\text{-}2\text{-}4\text{-}1$$
$$1\text{-}4\text{-}3\text{-}2\text{-}1$$
$$1\text{-}3\text{-}4\text{-}2\text{-}1$$
$$1\text{-}4\text{-}2\text{-}3\text{-}1$$

Step 2: Calculate the travel time for each route.

$$1\text{-}2\text{-}3\text{-}4\text{-}1 : 18 + 31 + 17 + 16 = 82$$
$$1\text{-}2\text{-}4\text{-}3\text{-}1 : 18 + 19 + 17 + 30 = 84$$
$$1\text{-}3\text{-}2\text{-}4\text{-}1 : 30 + 31 + 19 + 16 = 96$$
$$1\text{-}4\text{-}3\text{-}2\text{-}1 : 16 + 17 + 31 + 18 = 82$$
$$1\text{-}3\text{-}4\text{-}2\text{-}1 : 30 + 17 + 19 + 18 = 84$$
$$1\text{-}4\text{-}2\text{-}3\text{-}1 : 16 + 19 + 31 + 30 = 96$$

Step 3: Choose a minimum-distance route.

$$1\text{-}2\text{-}3\text{-}4\text{-}1 : 18 + 31 + 17 + 16 = 82$$ ■

A more general description of a solution for n cities gives insight regarding the number of operations required to find a solution. Assume that n cities need to be visited and that one city is designated as the starting point for all routes. There are $n - 1$ cities that remain to be visited immediately after the starting city. Thus, there are $n - 1$ ways to choose the second city to visit. For each of these $n - 1$ partial routes consisting of city 1 and the choice of a second city, there are $n - 2$ remaining unvisited cities. In all, we have $(n - 1) \cdots (n - i)$ partial routes starting at city 1 and containing $i + 1$ cities. The total number of routes starting at one city and not returning to that city until all the other cities have been visited is $(n - 1) \cdots (n - (n - 1)) = (n - 1)!$. Therefore, $(n - 1)!$ routes exist for visiting each of the n cities exactly once provided that one city is designated as the starting and ending city.

As the second step in this algorithm, the travel time for each of the $(n - 1)!$ possible routes must be calculated. Calculation of the traveling time for each possible route requires adding the n numbers that represent the travel times of all the flights of a route. Each addition will require at most some constant number k of such elementary operations. This means that when n additions are performed, they are carried out by at most kn elementary operations for some positive natural number k. Therefore, the number of elementary operations in calculating the length of all the routes is at most $(n - 1)!\, kn$ (or $\mathbf{O}(n!)$) operations.

For the third step, finding the minimum for a set of $(n - 1)!$ numbers requires $(n - 1)! - 1$ comparisons. Suppose the comparison of two numbers consists of some finite number l of elementary operations. This means that when $(n - 1)! - 1$ comparisons are performed, there are at most $l\,((n - 1)! - 1)$ elementary operations for some fixed constant l. Putting the analysis of these three steps together gives the complexity of this solution for the TSP. This solution will require at most $kn! + l \cdot ((n - 1)! - 1)$ operations. The order of the algorithm is $\mathbf{O}(n!)$.

The upper bound on the running time of our naive algorithm for the TSP looks very bad, but in fact, it is not known whether the TSP can be solved by an algorithm with the complexity of a polynomial in n. The problem has many practical applications, and at times, some algorithm to solve an instance of this problem must be used, however much time it will take.

This TSP example has provided a first illustration of the major counting principles that we now want to introduce more formally.

7.2 Counting Principles

Two principles of counting form a foundation for most counting techniques. The first principle, called the Multiplication Principle, is used for counting the number of elements arising from several choices that are made independently. This situation occurs, for example, in counting the number of ways to order a meal consisting of an appetizer, a main course, and a desert. The second principle, called the **Addition Principle,** is used to count the number of elements in a set that can be partitioned into disjoint subsets. This situation occurs, for example, in taking a census by adding the contributions from different regions. We discuss the Multiplication Principle first and then the Addition Principle.

Suppose that Fast Lease, Inc., has three brands of microcomputers available for lease. Each microcomputer is leased together with one of four different software packages. To have leases available for customers to sign regardless of the options they choose, how many different leases should be drawn up and available at any time?

The solution of this problem uses the following kind of analysis: First, a customer must choose a microcomputer from among the three possibilities. Second, after choosing a microcomputer, the customer must choose an appropriate software package, which can be done in four ways no matter which microcomputer was chosen. Consequently, the total number of choices is

$$(\# \text{ Choices for microcomputer}) \cdot (\# \text{ Choices for software package}) = 3 \cdot 4$$
$$= 12$$

Fast Lease should therefore have 12 different leases available at all times.

The choices can be explicitly displayed using the tree shown in Figure 7.1. Each interior vertex of the tree—that is, each vertex with an edge to its right in the figure—represents a point at which a choice can be made in a number of ways. Since a customer has three choices for choosing a microcomputer, three branches of the tree emanate from the starting point that is represented by the leftmost vertex. After choosing one of the three options for a microcomputer, a customer must then choose one of the four software packages, giving the 12 choices enumerated by the vertices at the right edge of the tree.

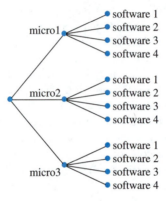

Figure 7.1   Choices for micros and software packages.

7.2.1   The Multiplication Principle

The problems of counting the number of possible tours for the TSP and of counting the number of leases are among the counting problems solved using the Multiplication Principle.

The Multiplication Principle

Let $m \in \mathbb{N}$. For a procedure of m successive distinct and independent steps with n_1 outcomes possible for the first step, n_2 outcomes possible for the second step, . . . , and n_m outcomes possible for the mth step, the total number of possible outcomes is

$$n_1 \cdot n_2 \cdots n_m$$

Example 2.   A university offers six sections of a physics course. Each student registered for the physics course also registers for 1 of 11 lab sections and 1 of 12 problem session sections. Any combination of lecture, lab, and problem session is possible. How many different ways can a single student be registered for the physics course?

Solution. To use the Multiplication Principle, we need to identify the steps of the procedure and the number of outcomes possible for each step. We also must convince ourselves that the process for each step does not depend on the process at any other step. This independence will be obvious. The steps proceed as follows: Choose a section, choose a lab, and choose a problem session. There are 6 possible outcomes for the first step, 11 for the second, and 12 for the third. Therefore, the Multiplication Principle gives

$$(\# \text{ Ways to register for physics course}) = (\# \text{ Choices for section}) \cdot (\# \text{ Choices for lab})$$
$$\cdot (\# \text{ Choices for problem session})$$
$$= 6 \cdot 11 \cdot 12$$
$$= 792$$

Example 3. Single characters displayed on a digital watch's display are formed by turning on some of the areas in a rectangular grid. Figure 7.2 shows the seven lines that a typical digital watch uses to form characters. How many different characters can be formed?

Figure 7.2 Figures on a digital watch's display.

Solution. The digits can be represented as shown in Figure 7.3.

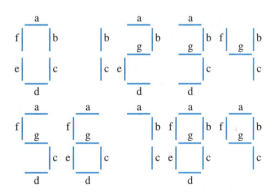

Figure 7.3 Representation of the digits.

Notice that a total of seven different line segments are used in forming the different numerals. You can form a pattern of lines by indicating for each of the seven line segments whether it is "on" or "off." Since there are seven line segments and each can be on or off independently of whether any other line segment is on or off, the Multiplication Principle says that $2^7 = 128$ characters can be represented by this grid.   ■

It is easy enough to represent the digits using seven segments, but the reader should think about how many segments, and what pattern of segments, could be used to represent the letters of the alphabet as well as the other characters on a keyboard.

Example 4.   Show that the number of subsets of a finite set with n elements is 2^n.

Solution.   Let the elements of an n-element set be $a_1, a_2, \ldots, a_n$. Each subset of these elements can be uniquely associated with an n-tuple of 0's and 1's, with 1 in the ith position indicating that the element a_i is in the subset and 0 in the ith position indicating the element a_i is not in the subset. In this procedure, for $1 \leq i \leq n$, the ith step has one of two possible outcomes: Either the ith element of the set is an element of the subset and the entry is 1, or the ith element of the set is not an element of the subset and the entry is 0. Therefore, by the Multiplication Principle, there are 2^n such n-tuples and, consequently, 2^n possible subsets of an n-element set.   ■

The reader should compare this proof with the one in Theorem 2 in Section 1.7.4 That proof is far more complicated, because it does not use the Multiplication Principle. (It also makes the induction very explicit. Indeed, if one wanted to prove the Multiplication Principle from "first principles," then the proof might look rather like the one in Section 1.7.4

7.2.2   Addition Principle

Another counting principle arises when choices are made from sets of mutually exclusive options. For example, suppose that Fast Machine Repair, Inc., offers two categories of service contracts. The first is a full-service contract that provides parts and services as needed for a fixed annual fee. The second is a service-on-demand contract that charges a small annual fee and then charges separately for services by the hour and for parts as they are needed. To satisfy different kinds of customers, Fast Machine Repair offers three levels of full-service contracts and two levels for service-on-demand service. These two sets of choices are disjoint. How many different contracts should be available so that no matter what option the customer may choose, the appropriate contract is ready for signing? Since there are three choices for the full-service option and two-choices for the hourly charge option, and because these two sets of choices are disjoint, the total number of choices is

$$(\# \text{ Service options}) = (\# \text{ Full-service options}) + (\# \text{ Hourly charge options})$$
$$= 3 + 2$$
$$= 5$$

Rather than representing these choices by a tree as was done with the Multiplication Principle, the possible choices here are represented by a collection of disjoint sets. Each set represents a collection of options in the same category. The options available in this problem are represented by the two disjoint sets pictured in Figure 7.4.

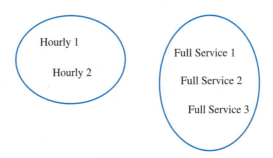

Figure 7.4 Set representation of choices.

The total number of possible choices is just the sum of the number of elements in each of the disjoint sets. This counting problem is one in which the Addition Principle is used to find a solution.

The Addition Principle

For a collection of m disjoint sets with n_1 elements in the first, n_2 elements in the second, ..., and n_m elements in the mth, the number of ways to choose one element from the collection is

$$n_1 + n_2 + \cdots + n_m$$

As an example, suppose we would like to know the number of elements between 1 and 100 that are divisible by 11 or 13. $Set11 = \{11, 22, 33, 44, 55, 66, 77, 88, 99\}$, and $Set13 = \{13, 26, 39, 52, 65, 78, 91\}$. Since $Set11 \cap Set13 = \emptyset$, the answer is found by simply adding together $|Set11|$ and $|Set13|$ to get 16 as the answer.

Two steps need to be completed before the Addition Principle can be applied. The first is to verify that the sets being considered are disjoint. This often proves to be more difficult than you might expect. The second is to count the number of elements in each of the sets as defined in the first step.

Example 5. How many ways can you choose a 9, a red card with value greater than 9, or a black card with a value less than 6 from a standard deck of cards? (A standard deck of cards is defined in Section 3.1. In this example, an Ace is a high card.)

Solution. The choices can be put into three disjoint sets as shown in Figure 7.5 on page 428. The Addition Principle says that there are

$$4 + 10 + 8 = 22$$

possible choices.

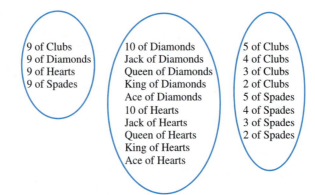

Figure 7.5 Card choices.

7.3 ## Set Decomposition Principle

Counting is not done using just the Multiplication Principle or the Addition Principle. In fact, many problems are solved by breaking the problem into **subproblems,** each of which can be solved without reference to any of the other subproblems. After solving each of the subproblems, you combine these results to solve the original problem. The solution of each subproblem may use the Addition Principle, the Multiplication Principle, or some combination of both. Since the problem is broken down into subproblems that are solved independently, the last step involves using the Addition Principle to add the number of solutions of the subproblems to solve the original problem.

Example 6. Student registration for computer science courses in the spring semester involves choosing two courses from Comp240, Comp225, and Comp326. If six sections of Comp240, seven sections of Comp225, and five sections of Comp326 are offered, then how many different registrations are possible? Assume that course conflicts are not an issue.

Solution. Break the problem into three different subproblems that represent the three pairs of ways that courses can be chosen. Then, count the number of solutions for each of these three subproblems separately using the Multiplication Principle. The final answer uses the Addition Principle and sums the number of ways that each of the three subproblems can be solved.

Subproblem 1: Choose Comp240 and Comp225:

(# Ways to choose Comp240) · (# Ways to choose Comp225) = 6 · 7

Subproblem 2: Choose Comp240 and Comp326:

(# Ways to choose Comp240) · (# Ways to choose Comp326) = 6 · 5

Subproblem 3: Choose Comp225 and Comp326:

(# Ways to choose Comp225) · (# Ways to choose Comp326) = 7 · 5

Final Result:

$$(\# \text{ Choices}) = (\# \text{ Choices for subproblem 1}) + (\# \text{ Choices subproblem 2})$$
$$+ (\# \text{ Choices subproblem 3})$$
$$= 6 \cdot 7 + 6 \cdot 5 + 7 \cdot 5$$
$$= 107$$

Example 6 introduces a very useful strategy for counting problems. The strategy involves breaking a problem into a number of subcases, each of which can be solved independently. The insights that are needed to see how to break a problem into subproblems can be developed by examining the solutions of similar problems and then attempting to solve other similar problems.

7.3.1   Counting the Complement

To count the number of objects in a subset of a class of objects that have a common property, you can begin by counting all the objects in the class. Then, you can count the number of objects in the class that do not have the required property. The difference between these two counts is the number of objects in the subset that have the property in question.

Example 7. How many 3-strings or words with three letters have a letter repeated if the words are formed from {a, b, c, d, e}?

Solution. It is easier to determine this count indirectly. We will count the total number of 3-strings possible and the total number of 3-strings with no repeated letters. The answer will be the difference between these two counts.

There are five possibilities for each of the letter positions in forming an arbitrary word. By the Multiplication Principle, there are 5^3 words in all. For the words with distinct letters, the first position can be filled with any of the five letters, the second position with any of the four remaining letters after the first letter is chosen, and the third position with any of the three remaining letters after the first two choices are made. By the Multiplication Principle, this gives $5 \cdot 4 \cdot 3$ possible words with all letters distinct:

$$(\# \text{ Words with repeated letters }) = (\# \text{ Words}) - (\# \text{ Words with distinct letters})$$
$$= 5^3 - 5 \cdot 4 \cdot 3$$
$$= 65$$

The same idea of counting the **complement** will tell us how many functions are not *1–1*. It would not be as easy to do this count directly as it would have been with the previous example.

Example 8. Let $S = \{a, b, c, d, e\}$ and $T = \{1, 2, 3, 4, 5, 6\}$. How many functions from S to T are not *1–1?*

Solution. The general class of objects to count is the number of functions from S to T. The second count that is needed determines how many of these functions are *1–1*. The answer is the difference between these two counts:

$$(\# \text{ Functions}) = (\# \text{ Possible images of } a) \cdot (\# \text{ Possible images of } b)$$
$$\cdots (\# \text{ Possible images of } e)$$

$$= 6 \cdot 6 \cdot 6 \cdot 6 \cdot 6$$
$$= 7776$$

(# *1–1* functions) = (# Possible images of a) $\cdot$ (# Possible images of b)

(excluding the image of a) $\cdots$ (# Possible images of e)

(excluding the images of a, b, c, d)

$$= 6 \cdot 5 \cdot 4 \cdot 3 \cdot 2$$
$$= 720$$

Therefore, the total number of functions that are not *1–1* is

$$7776 - 720 = 7056$$ ■

7.3.2   Using the Pigeon-Hole Principle

The **Pigeon-Hole Principle** (see Section 4.6) states that if m objects are to be put in n locations, where $m > n > 0$, then at least one location must receive at least two objects. Thus, to prove that a set of objects has at least two elements with the same property, first count the number of distinct properties of objects in the set, and then count the number of distinct elements. If the total number of elements is larger than the number of distinct properties of objects, then the Pigeon-Hole Principle implies that at least two of the elements have the same property. The next example is an illustration of this type of argument.

Example 9.   A local bank requires customers to choose a four-digit code to use with an ATM card. The code must consist of two letters in the first two positions and two digits in the other two positions. The bank has 75,000 customers. Show that at least two customers choose the same four-digit code.

Solution.   First, use the Multiplication Principle to calculate the number of distinct codes possible:

(# Four-symbol codes) = (# Choices of letter 1) $\cdot$ (# Choices for letter 2)

$\cdot$ (# Choices for digit 1) $\cdot$ (# Choices for digit 2)

$$= 26 \cdot 26 \cdot 10 \cdot 10$$
$$= 67,600$$

Now, apply the Pigeon-Hole Principle. Since there are 75,000 customers and only 67,600 codes, the Pigeon-Hole Principle implies that at least two of the customers choose the same code.   ■

The Pigeon-Hole Principle gives very sketchy information about what could actually happen when the 75,000 customers choose four-digit codes. One possibility is that the same code is chosen by all customers. Another possibility is that all codes are chosen at least once and at most twice. In Example 4, the Pigeon-Hole Principle only says that at least two customers choose the same code.

The Generalized Pigeon-Hole Principle can give us answers that will tell us more than simply that a single value is repeated at least once.

Example 10.   Suppose a group of vacationers is split into 159 teams. How many leagues must be formed if a league should contain at most 8 teams? 10 teams? 12 teams?

Solution.   The Generalized Pigeon-Hole Principle tells us that the answers are

$$\left\lceil \frac{159}{8} \right\rceil = 20 \quad \left\lceil \frac{159}{10} \right\rceil = 16 \quad \left\lceil \frac{159}{12} \right\rceil = 14$$

It remains for the organizers to determine which size of a league is most manageable.   ■

In many applications, the Pigeon-Hole Principle is used to prove that a problem cannot be solved. A typical example involves deciding what letter and digit patterns to use on a state's license plates. Problems that must be considered when designing license plates include how long the plates will be used and what scheme of letters and digits should appear on a plate so that every car that is registered during the period can have a distinct letter-digit pattern. The counting problems for license plates are good, simple examples of using the Multiplication Principle, because each position must contain a fixed kind of symbol. (Later, we will consider problems for which we do not know explicitly that a letter or a digit will occur in a particular place in a pattern, just that letters and digits will occur.)

Example 11.   Determine the number of different license plates that are possible using each of the following schemes, where D represents one of the digits $0, 1, 2, \ldots, 9$ and L represents one of the letters A, B, C, ..., Z. The leading digit must not be zero (the digit in the leftmost position).

(a) $DDDDDD$
(b) $DDDDDL$
(c) $DDDLDDD$
(d) $DDDLLL$

Solution.

(a) There are nine choices for the first digit, since it cannot be zero. There are 10 choices for each of the other five digits. By the Multiplication Principle, the number of possibilities is

$$9 \cdot 10^5 = 900{,}000$$

(b) There are nine choices for the first digit, since it cannot be zero. There are 10 choices for each of the other four digits. The letter can be chosen in 26 ways. By the Multiplication Principle, the number of possibilities is

$$9 \cdot 10^4 \cdot 26 = 2{,}340{,}000$$

(c) There are nine choices for the first digit, since it cannot be zero. There are 10 choices for each of the other five digits. There are 26 choices for the letter. By the Multiplication Principle, the number of possibilities is

$$9 \cdot 10^2 \cdot 26 \cdot 10^3 = 23{,}400{,}000$$

(d) There are nine choices for the first digit, since it cannot be zero. There are 10 choices for each of the other two digits. There are 26 choices for each of the letter positions. By the Multiplication Principle, the number of possibilities is

$$9 \cdot 10^2 \cdot 26^3 = 15{,}818{,}400$$ ■

Once the number of possibilities is known for a particular license plate scheme, it can be decided whether enough different license plates will be available for all the cars to be licensed. The Pigeon-Hole Principle will tell which schemes cannot be used. For example, if there are three million cars to license before the scheme is changed or new plates are issued, only alternatives (c) and (d) in Example 6 may be used.

7.3.3   Application: UNIX Logon Passwords

The UNIX operating system requires that users prove their identity before accessing system resources. A user typically begins a session by providing a username and a secret **password** to a login program. This program then verifies the password using a systemwide password file. A program called *crypt* takes a user's password and turns it into an entry in the system's password file. To crack a system by guessing a password, you must figure out what sequence of characters gives rise to the image of the *crypt* operation. A fast workstation with specially designed software can try more than 200,000 words a second to see if any trial word gives rise to the output of *crypt*. The question is whether it is reasonable to try all possible combinations of symbols that can be valid passwords to discover a legal password.

The UNIX operating system has the following requirements for the user forming a password:

- Each password must have at least six characters. Only the first eight characters are significant.
- Each password must contain at least two alphabetical characters and at least one numeric or special character. In this case, *alphabetical* refers to all uppercase or lowercase letters.
- Each password must differ from both the user's login name and any reverse or circular shift of that login name. In the application of this rule, an uppercase letter and the corresponding lowercase letter are considered to be the same; for example, user johnson could not have password hnsOnjO.
- A new password must differ from the old one by at least three characters. In the application of this rule, an uppercase letter and its corresponding lowercase letter are considered to be the same.

Example 12.   How many passwords can be formed in the UNIX system? How long would it take a user to try all possible combinations with the *crypt* program to discover a valid password?

Solution.   We are going to simplify the problem a bit: First, assume that all passwords are six characters long. (We shall remove this restriction in an exercise at the end of the chapter.) Second, temporarily assume that the required letters in the password occur as the first two symbols of the password as we solve the problem. Also, temporarily assume that the required special character occurs as the last symbol. (In Section 7.5, we will see how to remove these two temporary restrictions.) We will find that these two restrictions add a factor of 60 to the count. We will just add that factor without explanation at this time. We will also assume that all 32 special characters on a standard keyboard count as possible special characters.

$$(\# \textit{Passwords}) = 60 \cdot (\# \text{ Possible at position 1}) \cdots (\# \text{ Possible at position 6})$$
$$= 60 \cdot 52 \cdot 52 \cdot 84 \cdot 84 \cdot 84 \cdot 32$$
$$= 3{,}077{,}129{,}502{,}720$$

If we use the rate estimate for *crypt* operations of 200,000 per second, then it would take about 15,385,647 seconds—about 4273 hours, which is almost half a year—to try all possible passwords. ■

7.4   Exercises

1. How many license plates can be made using two uppercase letters followed by a 3-digit number?

2. How many ways can you draw a club or a heart from an ordinary deck of cards? A spade or an ace? An ace or a jack? A card numbered 3 through 9? A numbered card (Aces are not numbered cards) or a king?

3. How many ways can one choose one right glove and one left glove from six pairs of different gloves without obtaining a pair?

4. In planning a round trip from Cleveland to Dover by way of New York, a traveler decides to do the Cleveland–New York segments by air and the two New York–Dover segments by steamship. If six airlines operate flights between Cleveland and New York and four steamship lines operate between New York and Dover, in how many ways can the traveler make the round trip without using the same company twice?

5. Given the digits 1, 2, 3, 4, and 5, find how many 4-digit numbers can be formed from them:

   (a) If no digit may be repeated
   (b) If repetitions of a digit are allowed
   (c) If the number must be even, without any repeated digit
   (d) If the number must be even

6. How many odd numbers between 1000 and 10,000 have no digits repeated?

7. How many natural numbers greater than or equal to 1000 and less than 5400 have the properties:

   (a) No digit is repeated.
   (b) The digits 2 and 7 do not occur.

8. How many 7-digit numbers are there such that the digits are distinct integers taken from $\{1, 2, \ldots, 9\}$ and the integers 5 and 6 do not appear together in either order?

9. How many 6-digit numbers can be formed using $\{1, 2, 3, \ldots, 9\}$ with no repetitions such that 1 and 2 do not occur in consecutive positions?

10. Answer the following questions about 9-digit natural numbers with no repeated digits (leading zeros are not permitted):

    (a) How many such 9-digit numbers exist?
    (b) How many are divisible by 2?
    (c) How many are divisible by 5?
    (d) How many are greater than 500,000,000?

11. How many positive integers less than 1,000,000 can be written using only the digits 7, 8, and 9? How many using only the digits 0, 8, and 9?

12. Find the number of ways that flipping five coins can give at least three heads. Use the Multiplication Principle and symmetry to conclude that the answer is 2^4.

13. A palindrome is a string that reads the same forward and as it reads backward. An example (if blanks and punctuation are ignored) is: *A man, a plan, a canal, Panama.* How many n-letter palindromes can be formed using the alphabet $\{0, 1\}$?

14. In the United States and Canada, a telephone number is a 10-digit number of the form *NXX–NXX–XXXX* where $N \in \{2, 3, \ldots, 9\}$ and $X \in \{0, 1, 2, \ldots, 9\}$. How many telephone numbers are possible? The first three digits of a telephone number are called an area code. How many different area codes must a city with 23,000,000 phones have? A previous scheme for forming telephone numbers required a format of *NYX-NXX-XXXX* where N and X are defined as above and Y is either a 0 or a 1. How many more phone numbers are possible under the new format than under the old format?

15. How many ways can a computer system be configured if there are k input devices, m processors, and n output devices. A configuration consists of an input device, a processor, and an output device connected for use together. If $k = 3$, $m = 6$, and $n = 4$, draw three possible system configurations if every processor must be connected to at least one input device and two output devices.

16. The Omnibus Society has four officers: chair, secretary, treasurer, and editor. The by-laws for holding office state that one person shall be eligible to hold two, but not more than two, offices concurrently. If the Society has 1000 members, in how many ways can the officers be selected?

17. A flag is to consist of six vertical stripes in yellow, green, blue, orange, brown, and red. It is not necessary to use all the colors. The same color may be used more than once. How many possible flags are there with no two adjacent stripes the same color?

18. A convex polygon is a polygon such that any line segment joining two points inside the polygon lies entirely inside the polygon. If no 3 of the 15 diagonals of a convex, six-sided polygon intersect at a point common to all three, into how many line segments are the diagonals divided by their intersection points? Can you conjecture and prove a general result for an n-sided convex polygon?

19. How many sequences of length n can be formed using the alphabet $\{0, 1\}$? Using the alphabet $\{0, 1, 2\}$? Using the alphabet $\{1, 2, \ldots, k\}$ for $k \in \mathbb{N}$? How many possible words are there in the English language of length 13 at most? If a dictionary contains 500,000 words of length less than or equal to 13, what percentage of all words of length less than or equal to 13 does it contain?

20. How many ways can three integers be selected from $3n$ consecutive integers so that the sum is a multiple of 3? Here, n is a positive integer. What if the three chosen integers must be distinct?

21. A three out of five series is a competition between two teams consisting of at most five games and ending as soon as one of the two competing teams wins three games. How many different three out of five series are possible? Two series are "different" if the sequence of winners and losers in one series is not the same as in the other series. Draw a tree to represent the possibilities.

22. A four out of seven series is a competition between two teams consisting of at most seven games and ending as soon as one of the two competing teams wins four games. How many different four out of seven series are possible? Two series are "different" if the sequence of winners and losers in one series is not the same as in the other series. Draw a tree to represent the possibilities.

23. How many injective functions are there from S to T if $|S| = n$ and $|T| = m$ where $n \leq m$?

24. A survey asks the respondent to order by importance 10 properties of a car. How many orderings are possible? How many orderings are there if the first and last property are given?

25. Find the number of paths from A to F in the following diagram with six letters. A path can only go through letters that are consecutive, either horizontally or vertically, and it goes only to the right or up at each step.

```
F
E  F
D  E  F
C  D  E  F
B  C  D  E  F
A  B  C  D  E  F
```

Prove that a similar path with n letters has 2^{n-1} paths from the lower left corner to any letter in the rightmost position in a row.

Internet Addresses: IPv4 and IPv6. The Internet requires an address for each machine that is connected to it. The address space of the addressing architecture of Internet Protocol version 4 (IPv4) consists of a 32-bit field. Since not every combination of bits can be used as an address, plans are underway to change the address space to a 128-bit field in IPv6. The 32-bit IPv4 addresses are usually written in a form called *dotted decimal*. The 32-bit address is broken up into four 8-bit bytes, and these bytes are then converted to their equivalent decimal form and separated by dots. For example,

$$10000000\ 00000011\ 00000010\ 00000011$$

is written as 128.3.2.3, which is obviously more readable. The 128-bit IPv6 addresses are divided into eight 16-bit pieces. Each 16-bit piece is converted to its equivalent hexadecimal value (each sequence of 4 bits is converted to one hexadecimal digit). The eight four-character hexadecimal strings are separated by colons. It is not practical to list 128 bits and show the conversion to the final IPv6 address form. As an example of what you might end up with, however, we show one IPv6 address: FEDC:BA98:7654:3210:FEDC:BA98:7654:3210.

26. How many IPv4 addresses are possible?
27. Write the following IPv4 address in dotted decimal format:

$$01001010\ 11001010\ 10001000\ 11011101$$

where $x_1x_2x_3x_4x_5x_6x_7x_8$ as a binary number is the decimal number

$$x_1 \cdot 2^7 + x^2 \cdot 2^6 + x_3 \cdot 2^5 + x_4 \cdot 2^4 + x_5 \cdot 2^3 + x_6 \cdot 2^2 + x_7 \cdot 2^1 + x_8 \cdot 2^0$$

28. How many IPv6 addresses are possible?
29. What would the string

$$01001000\ 11101010\ 01101001\ 01110111$$

look like as the first part of an IPv6 address?

Permutations and Combinations

In scientific experiments used to determine how well several nutrients interact to stimulate plant growth, a sequence of different treatments may be proposed. The order in which the treatments are performed may make a difference. To show whether the order of application for the nutrients or the nutrients themselves is the important factor, all possible orders of application must be known and tried. On the other hand, when a court system selects a number of citizens to be in a jury pool, it is not important in what order citizens were chosen, only that they were chosen. There are many fewer ways to select a jury pool of a given size than there are ways to schedule the sequential application of the same number of treatments in a scientific experiment. In this section, the Multiplication Principle is used to count the number of possibilities for both kinds of problems.

7.5.1 Permutations

The first problem to consider here is how to count the number of possibilities when the order of choice is important. We also consider that each of the elements being chosen is distinguishable from all the other elements. Recognizing whether the elements that are being arranged or counted or chosen are distinguished from one another is a key step in deciding what formula is applicable.

Definition 1.   Let $n, r \in \mathbb{N}$. A **permutation** of an n-element set is a linear ordering of the n elements of the set. For $n \geq r \geq 0$ an **r-permutation** of an n-element set is a linear ordering of r elements of the set.

Example 1.   List all permutation of the elements a, b, and c.

Solution.   The permutations are abc, acb, bac, bca, cab, and cba.   ■

Let $P(n, r)$ denote the number of r-permutations of an n-element set. We define $P(n, 0) = 1$ for all $n \in \mathbb{N}$.

Theorem 1.   Let $n \geq r > 0$. Then, $P(n, r) = n \cdot (n - 1) \cdot (n - 2) \cdots (n - r + 1)$.

Proof.   The proof is by induction on n. Let $n_0 = 1$. Let $\mathcal{T} = \{n \in \mathbb{N} : P(n, r) = n \cdot (n - 1) \cdots (n - r + 1)$ for $n \geq r > 0\}$.
(**Base step**) There is only one way to arrange the one element of a one element set. Thus, $P(1, 1) = 1$, and $1 \in \mathcal{T}$.
(**Inductive step**) Let $n \geq n_0$, and assume $n \in \mathcal{T}$. Now, prove $n + 1 \in \mathcal{T}$. That is, $P(n + 1, r) = (n + 1) n \cdots ((n + 1) - (r + 1))$ for all r such that $n \geq r > 0$.
Let X be any set of $n + 1$ elements for which we must form an ordering of r of the elements. Choose one element from X—say, x—to be the first element of the ordering. This can be done in $n + 1$ ways. The problem remaining is to form an ordered arrangement of $r - 1$ of the remaining n elements. By the inductive hypothesis, this can be done in $P(n, r - 1)$ ways. The total number of ways to form an ordered r-arrangement of a set with $n + 1$ elements is $(n + 1) \cdot P(n, r - 1) = P(n + 1, r)$. Therefore, $n + 1 \in \mathcal{T}$.
By the Principle of Mathematical Induction, $\mathcal{T} = \mathbb{N} - \{0\}$.   ■

Using the factorial function, we can express $P(n, r)$ as

$$P(n, r) = n \cdot (n - 1) \cdots (n - r + 1)$$
$$= \frac{(n \cdot (n - 1) \cdots (n - r + 1))((n - r)(n - r - 1) \cdots 2 \cdot 1)}{((n - r)(n - r - 1) \cdots 2 \cdot 1)}$$
$$= \frac{n!}{(n - r)!}$$

This formula includes the case $r = 0$. If $n = r$, then $P(n, n) = n!$ (*Remember:* $0! = 1$.)

7.5.2 Linear Arrangements

An arrangement of n books on a shelf can be associated with a permutation of the integers $1, 2, 3, \ldots n$, where each number represents one of the books. The permutation indicates the order in which the books are stored on the shelf from left to right. Two arrangements of three books are shown in Figure 7.6.

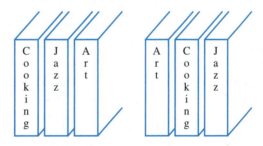

Figure 7.6   Arrangements of three books on a shelf.

Example 2.

(a) How many ways can eight different books be arranged on a shelf?
(b) How many ways can four of eight different books be arranged on a shelf?
(c) How many ways can eight different books be arranged on two shelves so that each shelf contains four books?

Solution.

(a) The answer is the number of ordered ways of arranging the books on the shelf. That is,

$$P(8, 8) = 8! = 40{,}320$$

(b) The number of ways to arrange four of the eight books is

$$P(8, 4) = 1680$$

(c) The answer is the product of the number of ways to put four books on one shelf and the number of ways to put the remaining books on the second shelf. The number of ways to arrange four books on the first shelf is $P(8, 4)$, and the four remaining books can be arranged in $P(4, 4)$ ways on the second shelf. Therefore, the total number of arrangements will be

$$(\text{\# Arrangements of books on two shelves}) = (\text{\# Arrangements on first shelf})$$
$$\cdot (\text{\# Arrangements on second shelf})$$
$$= P(8, 4) \cdot P(4, 4)$$
$$= (8!/4!) \cdot (4!/0!)$$
$$= 8!$$
$$= 40{,}320$$

The solution to part (c) is the same as the one found for part (a). To see that the answer for part (c) is the same as the answer for part (a) just put the last four books in any arrangement for (a) on the second shelf. Similarly, for any arrangement for part (c), move the books on the second shelf up to the first shelf, placing them after the books that are already there. ■

In some instances, certain objects must be placed in specified positions. The next example shows how such a counting problem can be solved.

Example 3.   A collection of eight books consists of two books on artificial intelligence, three books on operating systems, and three books on data structures.

(a) How many ways can the books be arranged on a shelf so that all books on a single subject are together?
(b) How many ways can the books be arranged on a shelf so that the three books on operating systems are together?
(c) How many ways can the books be arranged on a shelf so that the two books on artificial intelligence occur at the right end of the arrangement?

Solution.

(a) Identify the books on each subject with a single "book". Let BK1 represent the two books on artificial intelligence, BK2 represent the three books on operating systems, and BK3 represent the three books on data structures. The problem becomes one of arranging three "books" called BK1, BK2, and BK3. The first step is to determine the number of ways these three "books" can be arranged, these books can be arranged 3! ways. The second, third, and fourth steps involve finding the number of ways to arrange the books represented by BK1, BK2, and BK3, respectively. Wherever the two books on artificial intelligence occur, they can be arranged in 2! ways. The three books on operating systems and the three books on data structures can each be arranged in 3! ways. Therefore, by the Multiplication Principle, the total number of arrangements is

$$(\text{\# Ways to arrange the three categories of books})$$
$$\cdot (\text{\# Ways to arrange the artificial intelligence books (BK1)})$$
$$\cdot (\text{\# Ways to arrange the operating systems books (BK2)})$$
$$\cdot (\text{\# Ways to arrange the data structures books (BK3)})$$
$$= 3! \cdot 2! \cdot 3! \cdot 3!$$
$$= 432$$

(b) This problem can be considered the same as the problem of arranging six "books" on a shelf where one "book" represents the three books on operating systems that must occur together. After the six books are arranged, the location for the three books on operating systems that are considered as a single "book" to this point of the arrangement process can be arranged in 3! ways. The number of arrangements is

$$\text{(\# Ways to arrange six books)} \cdot \text{(\# Ways to arrange the operating systems books)}$$

$$= 6! \cdot 3!$$
$$= 4320$$

(c) In this case, there are six books to arrange, and for each of these arrangements, the two artificial intelligence books must follow in some order. The number of such arrangements is

$$\text{(\# Ways to arrange six books)} \cdot \text{(\# Ways to arrange two artificial intelligence books)}$$

$$= 6! \cdot 2!$$
$$= 1440$$

■

7.5.3 Circular Permutations

The permutations considered to this point assume that the ordering is a linear ordering. A slight variation of this occurs when one tries to count the number of ways to seat n people at a round table. Three possible arrangements of five people seated at a round table are shown in Figure 7.7.

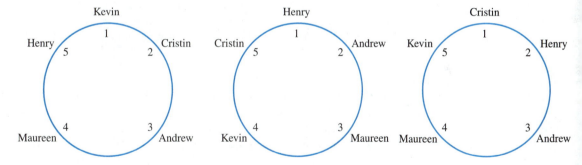

Figure 7.7 Different seating arrangements.

For the seating arrangements shown in Figure 7.7(a) and 7.7(b), each person does not have the same pair of neighbors. By convention, two seating arrangements are considered to be equivalent if one arrangement can be transformed into the other by a clockwise rotation of all the people by the same number of seats around the table. The seating arrangements shown in Figure 7.7(b) and 7.7(c) are considered to be the same seating arrangement, since everyone in the seating arrangement shown in Figure 7.7(b) can be shifted one place around the table in a clockwise direction to form the seating arrangement shown in Figure 7.7(c). Thus, for any fixed seating arrangement, n seating arrangements are considered to be the same; that is, we move all the people i seats in a clockwise direction for any $i \in \{1, 2, \ldots, n\}$.

To count the number of possible seating arrangements at a round table, consider the case of 12 people. Suppose that one of the people is named Henry and that Henry's seat is chair number 1. We number the remaining chairs clockwise around the table from Henry's chair. There are $P(11, 11) = 11!$ ways to seat the remaining 11 people in chairs 2 through 12. Counting the ways of seating people relative to Henry avoids counting arrangements that are considered to be the same. By a similar argument, it can be shown that the number of **circular permutations** of n people is just $P(n - 1, n - 1) = (n - 1)!$.

7.5.4   Combinations

A card game uses only Aces, Kings, and Queens from a deck, with four of each. Each card of a given value is from one of four suits, called Clubs, Diamonds, Hearts, and Spades. At the beginning of the game, each player is dealt three cards. How many different three-card combinations can be dealt?

We solve this problem by answering two questions and then combining the answers to those questions to find the final answer. The first question is how many ordered arrangements are comprised of 3 of 12 cards? The answer to this question is $P(12, 3)$. The second question is how many ordered arrangements are generated by a single set of three cards? For example, the two sets of choices of cards shown in Table 7.1 cannot be distinguished from one another.

Table 7.1   Different Orders for Choosing Cards

	Choice 1	Choice 2
Card 1	Ace of Spades	King of Hearts
Card 2	Queen of Diamonds	Ace of Spades
Card 3	King of Hearts	Queen of Diamonds

The answer to this question is $P(3, 3)$. Since the answer to the second question tells how many times each set of three cards occurs as a different ordered arrangement, the answer to the original question is $P(12, 3)/P(3, 3)$.

We now define more formally the general notion illustrated in this example for counting elements that are indistinguishable from one another.

Definition 2.   Let $n, r \in \mathbb{N}$ such that $n \geq r \geq 0$. An unordered selection of r elements from an n element set is called a **combination.**

Example 4.   List all the combinations of the set $\{a, b, c\}$.

Solution.   The combinations will be of sizes 0, 1, 2, and 3. All combinations are $\emptyset$, $\{a\}, \{b\}, \{c\}, \{a, b\}, \{a, c\}, \{b, c\}$, and $\{a, b, c\}$. ∎

Let $C(n, r)$ denote the number of combinations of r elements selected from a set of n elements. Another common notation for $C(n, r)$ is $\binom{n}{r}$. Note that $C(n, 0)$ is defined to be 1 for all $n \in \mathbb{N}$. $C(n, r)$ is also called a **binomial coefficient,** because these numbers occur as coefficients in the expansion of powers of binomials, such as $(x + y)^n$. We will prove this in Theorem 5 in Section 7.9.1.

Theorem 2. For $n \geq r \geq 0$,

$$C(n, r) = \frac{P(n, r)}{P(r, r)} = \frac{n!}{(n - r)! r!}$$

Proof. The number of r-permutations of an n-element set is $P(n, r)$. An r-permutation can be formed by choosing an r-element set without regard to order and then arranging these elements in $P(r, r)$ different ways. Therefore,

$$P(n, r) = C(n, r) \cdot P(r, r)$$
$$C(n, r) = \frac{P(n, r)}{P(r, r)}$$
$$= \frac{n!}{(n - r)! r!}$$

∎

Corollary 1: $C(n, r) = C(n, n - r)$.

Proof. $C(n, r)$ is the number of ways of choosing r elements from an n-element set. Each choice of r-elements determines a unique subset of $n - r$ elements—namely, the complement of the set of elements chosen. Therefore, the function $F(A) = \overline{A}$ with A ranging over the r-element subsets of an n-element set is well defined. It is easy to show this function is *1–1* and onto using the fact that $\overline{\overline{A}} = A$ (see Theorem 7 in Section 1.3.2). Therefore, $C(n, r)$ and $C(n, n - r)$ count the same number of elements. ∎

You can also formulate a direct argument to prove Corollary 1 by using the identity

$$\frac{n!}{(n - r)! r!} = \frac{n!}{(n - (n - r))!(n - r)!}$$

7.5.5 Poker Hands

A class of interesting examples that uses counting techniques involves poker hands. A poker hand is a set of five cards dealt from a deck of 52 cards, divided into four suits (ordered as Clubs, Diamonds, Hearts, and Spades in increasing order), with each suit containing 13 cards with the values 2, 3, . . . , 10, Jack, Queen, King, and Ace (listed in increasing order for purposes of poker). The Ace can be either the highest or the lowest card in a hand. The order in which the cards are dealt does not make a difference. For example, the two hands shown in Table 7.2 are the same poker hand.

Table 7.2 Poker Hands

Order Dealt	Name of Card	Name of Card
1	Ace of Hearts	Jack of Hearts
2	King of Spades	3 of Diamonds
3	4 of Clubs	King of Spades
4	Jack of Hearts	Ace of Hearts
5	3 of Diamonds	4 of Clubs

Basic questions about hands for card games include how many hands of a certain kind exist and what percentage of the total number of hands are of a certain kind.

Example 5.   How many different poker hands are there?

Solution.   This answer is just the number of ways of choosing five cards from the 52-card deck:

$$C(52, 5) = \frac{52!}{47!\,5!} = 2{,}598{,}960$$ ■

Example 6.   What percentage of poker hands contain four of a kind? A hand that contains four of a kind has four cards with the same value. For example, a hand with four Queens and a fifth card having any other value is of this type.

Solution.   The total number of poker hands was determined in Example 5. It remains to be determined how many different poker hands contain four of a kind.

Since there are 13 different card values in a deck and four of a kind consists of four cards of the same value, there are 13 possibilities for these four cards. It remains to be determined in how many ways the hand can be completed. Since four of a kind uses four of the 52 cards in the deck, there are 48 cards remaining from which the last card in the hand can be chosen. Use these two observations and the Multiplication Principle to find the total number of such hands:

$$(\text{\# Hands with four of a kind}) = (\text{\# Possibilities for four of a kind})$$
$$\cdot\ (\text{\# Ways to complete the hand})$$
$$= 13 \cdot C(48, 1)$$
$$= 624$$

The percentage of such hands is $624/2{,}598{,}960 \approx 0.02\%$.   ■

Example 7.   How many poker hands contain exactly one pair and three other cards, no two of which have the same value?

Solution.   As an example, there are four Jacks. Any two of these can be chosen for the pair, so there are six pairs of Jacks. There are $C(4, 2) = 6$ pairs for any card value. Since there are 13 values for cards and we want to choose one of these values to represent two of the cards in the poker hand, there are $C(13, 1) \cdot C(4, 2)$ ways to choose two cards with the same value. The remaining three cards have values different from the value of the pair, and at the same time, no two of the three have the same value. Since there are 12 values for cards not in the pair, there are $C(12, 3)$ ways to choose the values for the three other cards. After choosing the three values for the other cards, the suit of each of these cards can be chosen in $C(4, 1)$ ways, since there are four cards from different suits in the deck with each value:

$$(\text{\# Hands with 1 pair and 3 unmatched cards}) = (\text{\# Pairs}) \cdot (\text{\# Ways to complete the hand})$$
$$= C(13, 1) \cdot C(4, 2) \cdot C(12, 3) \cdot C(4, 1)^3$$
$$= 1{,}098{,}240$$

The percentage of such hands out of the total number of hands possible is $1{,}098{,}240/2{,}598{,}960 \approx 42\%$.   ■

We now look at a noncard problem that uses the same technique for counting.

Example 8. An examination consists of 20 questions, of which the student must answer any 12.

(a) How many different ways can a student choose questions to answer?
(b) The 20-question exam is split into three parts. There are 6 questions in the first part, 10 in the second part, and 4 in the third part. A student must choose three from the first part, eight from the second part, and one from the third part. How many ways can a student choose questions to answer?

Solution.

(a) The answer is just the number of different 12-element subsets of a 20-element set, or $C(20, 12) = 125{,}970$.
(b) By the Multiplication Principle, the answer will be the product of the number of ways to make choices in each category:

$$(\# \text{ Possible choices}) = (\# \text{ Choices for part 1}) \cdot (\# \text{ Choices for part 2})$$
$$\cdot (\# \text{ Choices for part 3})$$
$$= C(6, 3) \cdot C(10, 8) \cdot C(4, 1)$$
$$= 3600$$

∎

7.5.6   Counting the Complement

We return to the idea of counting the complement introduced earlier. This time, it is a useful technique, because direct counts often can involve detailed analysis to determine all the cases. An illustration of this is given in Example 9.

Example 9. How many poker hands consist of five cards of the same suit, but not in sequence? Such a poker hand is called a *flush.*

Solution. First, count the number of hands with all five cards from one suit. This count is just the number of ways to pick five of the 13 possible values, and there are $C(13, 5)$ such hands. A poker hand with all five values in sequence is called a *straight.* A straight with all the cards from the same suit is called a *straight flush.* We want to subtract the number of straight flushes in this suit from $C(13, 5)$ to get the number of flushes for the suit.
   To find the number of straight flushes in a suit, note that all five cards in a sequence are determined by the lowest card in the sequence. In the case of a straight, the rule is that the Ace can be used as either the highest or the lowest card in the sequence. The card values are arranged as

$$\text{Ace } \underline{2}\ \underline{3}\ \underline{4}\ \underline{5}\ \underline{6}\ \underline{7}\ \underline{8}\ \underline{9}\ \underline{10}\ \text{Jack Queen King Ace}$$

The smallest card in a straight can be any one of the card values underlined. Therefore, there are $C(10, 1)$ straight flushes for a suit, so there are $C(13, 5) - C(10, 1)$ flushes for a suit.

The answer to the original question is then found by multiplying this answer by the number of ways to choose the suit of the *flush*.

(# Hands with all cards from a given suit − # Straight flushes in that suit)

· (# Possible suits)

$$= (C(13, 5) - C(10, 1)) \cdot C(4, 1)$$
$$= 5108$$

The percentage of such hands is $5108/2,598,960 \approx 0.2\%$. ∎

7.5.7 Decomposition into Subproblems

Many problems can be solved by decomposing the original problem into a set of subproblems. Two examples of this technique include finding how many different pizzas with a certain number of toppings can be ordered in five years and finding how many poker hands contain more Aces than Kings.

Example 10. Pat's Pizza claims that it stocks enough topping ingredients so that you can order a pizza with a different combination of up to five ingredients every night for five years. What is the least number n of topping ingredients that must be available to make this claim true?

Solution. Split the problem into six cases, depending on how many ingredients are chosen. The possibilities are to choose from zero to five ingredients for a pizza. The number of different pizzas is

(# Pizzas with 0 ingredients) + (# Pizzas with 1 ingredient)
+ (# Pizzas with 2 ingredients) + (# Pizzas with 3 ingredients) + (# Pizzas with
4 ingredients) + (# Pizzas with 5 ingredients)
$$= C(n, 0) + C(n, 1) + C(n, 2) + C(n, 3) + C(n, 4) + C(n, 5)$$

This sum must be greater than $5(365) + 2$ (the maximum number of days in five consecutive years, allowing for at most two leap years). The answer to the original question involves finding the smallest n that makes the following inequality true:

$$C(n, 0) + C(n, 1) + C(n, 2) + C(n, 3) + C(n, 4) + C(n, 5) \geq 1827$$

which is the same as

$$1 + n + \frac{n(n - 1)}{2!} + \frac{n(n - 1)(n - 2)}{3!} + \frac{n(n - 1)(n - 2)(n - 3)}{4!}$$
$$+ \frac{n(n - 1)(n - 2)(n - 3)(n - 4)}{5!} \geq 1827$$

The answer is $n = 13$, since the expression has the value 2380 for 13 and only 1586 for $n = 12$. ∎

In Example 10, it was not difficult to see how to break up the elements to count into six cases. In the next example, however, the count for each of the cases will be more complex.

Example 11. How many poker hands have more Aces than Kings?

Solution. First determine all the different cases that are possible, and then count the number of hands for each separate case. Finally, add up the counts for each case to get the total count required. The different cases are listed in Table 7.3.

Table 7.3   Distribution of Aces and Kings

# of Aces	Possible # of Kings
0	No hand possible
1	0
2	0 or 1
3	0 or 1 or 2
4	0 or 1, since the hand has only five cards

There are eight subcases for which the number of hands must be counted. The count for each of these cases is given in Table 7.4.

Table 7.4   Number of Hands with Given Aces and Kings

# of Aces	# of Kings	Aces	Kings	Other Cards	# of Hands
1	0	$C(4, 1) \cdot$		$C(44, 4)$	$= 543{,}004$
2	0	$C(4, 2) \cdot$		$C(44, 3)$	$= 79{,}464$
2	1	$C(4, 2) \cdot$	$C(4, 1) \cdot$	$C(44, 2)$	$= 22{,}704$
3	0	$C(4, 3) \cdot$		$C(44, 2)$	$= 3784$
3	1	$C(4, 3) \cdot$	$C(4, 1) \cdot$	$C(44, 1)$	$= 704$
3	2	$C(4, 3) \cdot$	$C(4, 2)$		$= 24$
4	0	$C(4, 4) \cdot$		$C(44, 1)$	$= 44$
4	1	$C(4, 4) \cdot$	$C(4, 1)$		$= 4$

The answer is the sum of the number of hands in each of the eight subcases, which is 649,732.   ■

In Example 11, we saw how the Multiplication Principle was first used in each case and then how the Addition Principle was used to find the total number for all cases. This analysis proceeded fairly directly, because we first split all the cases into disjoint subsets so that each of the possible hands being counted arose in one and only one of the subcases.

Using the Principle of Inclusion-Exclusion

In many problems, it is not particularly easy to verify that a set of subsets are pairwise disjoint. If you want to count the number of elements in the union of such subsets, the Principle of Inclusion-Exclusion is often useful as a proof technique.

Let $X = \{x_1, x_2, \ldots, x_n\}$ be a set. A permutation $y_1\, y_2, \ldots, y_n$ of the elements of X has a **fixed point,** or **fixes,** x_i if $y_i = x_i$ for some i where $1 \leq i \leq n$. A **derangement** of X is a permutation of X with no fixed point.

Example 12.   Find the number of derangements of a set with three elements.

Solution.   Let $X = \{1, 2, 3\}$. Let $A_i = \{$permutations of X that fix $i\}$ where $1 \leq i \leq 3$. To compute $|A_1 \cup A_2 \cup A_3|$, we need to compute $|A_1|, |A_2|, |A_3|, |A_1 \cap A_2|,$

$|A_1 \cap A_3|$, $|A_2 \cap A_3|$, and $|A_1 \cap A_2 \cap A_3|$. $A_1 = \{1\,2\,3, 1\,3\,2\}$. $A_2 = \{1\,2\,3, 3\,2\,1\}$. $A_3 = \{1\,2\,3, 2\,1\,3\}$. $|A_1 \cap A_2| = |A_1 \cap A_3| = |A_2 \cap A_3| = |\{123\}| = 1$. $|A_1 \cap A_2 \cap A_3| = 1$. We first count the number of permutations that fix at least one element:

$$\begin{aligned}
|A_1 \cup A_2 \cup A_3| &= |A_1| + |A_2| + |A_3| - |A_1 \cap A_2| - |A_1 \cap A_3| - |A_2 \cap A_3| \\
&\quad + |A_1 \cap A_2 \cap A_3| \\
&= 2 + 2 + 2 - 1 - 1 - 1 + 1 \\
&= 4
\end{aligned}$$

The number of derangements is just the difference between the number of permutations and the number of permutations that fix at least one element: $6 - 4 = 2$. ■

The reader should compare this solution to the Hat Check Problem in Section 1.5.3.

7.6   Constructing the *k*th Permutation

Computer simulations often use randomly generated permutations of sets of n elements as test data. If all the permutations on n elements are listed in some order, then by generating a random number k such that $1 \leq k \leq n!$, it is possible to choose the kth permutation of n elements at random.

An alternative to storing all the $n!$ permutations of an n-element set is to find a method for constructing the **kth of $n!$ permutations** on n elements one element at a time without constructing any of the other $n! - 1$ permutations.

The kth of $n!$ permutations of n elements must be found relative to some **ordering of the permutations.** For the method shown here, assume that the permutations are in **lexicographical** or **dictionary order** (see Example 9 in Section 3.8.2). This ordering for the permutations on three and four elements is shown in Figure 7.8. Observe that the permutations are numbered beginning with 0, not with 1.

3 Elements				4 Elements											
#	Permutation			#	Permutation				#	Permutation				
0.	1	2	3	0.	1	2	3	4	12.	3	1	2	4
1.	1	3	2	1.	1	2	4	3	13.	3	1	4	2
2.	2	1	3	2.	1	3	2	4	14.	3	2	1	4
3.	2	3	1	3.	1	3	4	2	15.	3	2	4	1
4.	3	1	2	4.	1	4	2	3	16.	3	4	1	2
5.	3	2	1	5.	1	4	3	2	17.	3	4	2	1
				6.	2	1	3	4	18.	4	1	2	3
				7.	2	1	4	3	19.	4	1	3	2
				8.	2	3	1	4	20.	4	2	1	3
				9.	2	3	4	1	21.	4	2	3	1
				10.	2	4	1	3	22.	4	3	1	2
				11.	2	4	3	1	23.	4	3	2	1

Figure 7.8   Lexicographical Ordering of Permutations.

As a result of examining the dictionary ordering of the permutations of three and four elements, a predictable pattern is found for determining which element occurs at each position in the permutation. For the case of four elements, observe that 1 is the first element of the first 3! = 6 permutations, 2 is the first element of the second 3! = 6 permutations, 3 is the first element of the next 3! = 6 permutations, and finally, 4 is the first element of the last 3! = 6 permutations. Now, observe the pattern of occurrence of the elements in the second position. For example, the pattern in the second elements following the six occurrences of 2 in the first position has each of the remaining elements repeated 2! times. The element 1 occurs as the second element in the first 2! = 2 of these permutations, 3 occurs as the second element in the next 2! = 2 of these permutations, and finally, 4 occurs as the second element in the last 2! = 2 of these permutations. The elements occurring in the second position occur in increasing order. The next observation is that the two elements that follow 2-3 in permutations 8 and 9, for example, are the elements 1 occurring 1! times and then 4 occurring 1! times. Again, the elements that did not occur in earlier positions occur in increasing order with frequency 1!.

In any sequence of $(n - 1)!$ permutations with the same element in all the first positions, the second positions will have $(n - 2)!$ occurrences of the smallest element not in position 1, followed by $(n - 2)!$ occurrences of the next smallest element not in position 1, and so on, for each of the $n - 1$ elements not in position 1. Similarly, in any sequence of $(n - 2)!$ permutations in which the elements in positions 1 and 2 do not change, the third position will have $(n - 3)!$ occurrences of the smallest element not in positions 1 or 2, followed by $(n - 3)!$ occurrences of the next smallest element not in positions 1 or 2, and so on. Any sequence of $(n - k)!$ permutations for which the elements in positions 1 through k do not change begins with a permutation whose $(k + 1)$-st position is occupied by the smallest element not in positions 1 through k.

A procedure for generating a particular permutation element by element is straightforward to implement using these ideas—provided that divisions by $(n - 1)!, (n - 2)!, \ldots, 2!$ can be done. The next example illustrates the process.

Example 1.   Find the 79th permutation of the elements 1, 2, 3, 4, and 5 relative to the lexicographical order.

Solution.   There are 120 permutations of five elements. Imagine ordering them in lexicographical order and then numbering them from 0 to 119. Permutations 0 to 23 begin with 1, permutations 24 to 47 with 2, permutations 48 to 71 with 3, and permutations 72 to 95 with 4. Consequently, the first element of the 79th permutation is 4. To find the second element of the permutation, imagine decomposing permutations 72 to 95 into sets of six permutations. Permutations 72 to 77 have 1 as a second element, and permutations 78 to 83 have 2 as a second element. Therefore, the second element of the 79th permutation is 2.

To find the third element of the permutation, imagine decomposing permutations 78 to 83 into three sets of two permutations. Permutations 78 and 79 have 1 as the third element. Thus, the third element of the 79th permutation is 1. The two elements 3 and 5 are in the fourth position in the 78th and 79th permutation following the 1 in the third position of these permutations. Since we are constructing permutation 79, the next element is 5. Finally, the only element not yet found is 3, and that occurs in position 5 of the 79th permutation.

Therefore, the 79th permutation of 1, 2, 3, 4, and 5 relative to the lexicographical order is 4-2-1-5-3.   ■

Exercises

1. A "word" is a string of one or more lowercase letters. How many words can be formed using all the letters of the word *hyperbola*? In how many words will *h* and *y* occur together? In how many will *h* and *y* not occur together?

2. A bookshelf contains three novels, six books of poetry, and four reference books. In how many ways can these books be arranged so that the books of each type are together?

3. A shelf has room for 10 books.

    (a) Given an inventory of 25 books, how many years will it take to display all combinations of 10 books if the display is changed once a week?

    (b) How many years will it take if the display is changed five times a week?

4. A passenger train consists of two baggage cars, four day coaches, and three parlor cars. In how many ways can the train be made up if the two baggage cars must be in the front and the three parlor cars must be in the rear? Assume that the baggage cars can be told apart, that the day coaches can be told apart, and that the parlor cars can be told apart.

5. The English alphabet contains 26 letters, including five vowels. In each case determine how many words of length five are possible provided that:

    (a) Words contain at most two distinct vowels

    (b) Words contain at most one letter that is a vowel

    (c) Words contain at least four distinct vowels

6. How many five-letter words formed with *a, b,* and *c* have at least one letter missing?

7. A student has four examinations to write, and there are 10 examinations periods available. How many ways are there to schedule the examinations?

8. A student must complete the following sequence of courses: Two of four lab science courses, one of two literature courses, two of three mathematics courses, and one of seven physical education courses. Assume that none of these courses is a prerequisite for any other.

    (a) How many ways can courses be chosen if the possibility of time conflicts is disregarded?

    (b) How many ways can courses be chosen if two different lab courses are scheduled at the same time as one of the literature courses?

    (c) How many ways can courses be chosen if all the physical education courses are offered at the same time as one of the literature courses?

9. A student must answer 8 out of 10 questions on an exam.

    (a) How many choices does the student have?

    (b) How many choices does a student have if the first three questions must be answered?

    (c) How many choices does a student have if exactly four out of the first five questions must be answered?

10. How many ways can you list the 12 months of the year so that May and June are not adjacent?

11. How many permutations are there for the 26 letters of the alphabet if the five vowels occur together?

12. How many n-bit binary numbers have exactly r 1's assuming $n \geq r$? Prove by induction that for $n \geq 1$, exactly half the n-bit binary strings have an even number of 1's.

13. Twelve-tone music requires that the 12 notes of the chromatic scale be played before any tone is repeated. How many different ways can the 12 tones be played? How long will it take to play all possible sequences of 12 tones if one sequence can be played in four seconds?

14. Find the sum of all four-digit numbers that can be obtained by using the digits 1, 2, 3, 4, and 5. Repeats are not allowed. Explain your reasoning.

15. A raffle has three prizes to award to 10,000 ticket holders. How many different ways can the prizes be distributed if no one can win more than one prize? If one person can win more than one prize?

16. Thirty contestants, including the local champion, enter a competition. When the first six places are announced:

    (a) How many different announcements are possible?
    (b) How many different announcements are possible if the local champion is assured of a place in the first six?

17. How many ways are there to seat eight people at a round table? How many ways if Smith and Jones cannot be seated next to each other?

18. How many ways can 12 people be seated at a round table if a certain pair of individuals refuse to sit next to one another?

19. How many ways can n men and n women be seated at a round table if no two women are seated next to each other?

20. A party has n guests. Two of the guests do not get along well with each other. In how many ways can the guests be seated in a row so that these two persons do not sit next to each other?

21. Determine the number of five-card poker hands with the following patterns:

    (a) Four deuces (2's) and one other card
    (b) Four of a kind and one other card
    (c) Two pairs (but not four of a kind) and one card with a different value
    (d) Three cards of one value and two cards of a second value (this is called a *full house*)
    (e) A *straight flush* (a straight with all the cards from one suit)
    (f) A hand with five different card values that is not a straight and is not a flush
    (g) Three of one kind and two other cards with different values

22. A bridge hand consists of 13 of the 52 cards from a standard deck of cards. How many bridge hands contain no cards in one or more suits?

23. There are six points in a plane, no three of which are collinear. In how many ways can you draw a pair of triangles with the six points as vertices.

24. For an even integer n, prove that $C(n, n/2)$ does not have polynomial order. Interpret this in terms of implementing an algorithm that examines this number of subsets of a set.

25. Winning a state lottery is based on trying to guess which six randomly picked numbers in the set $\{1, 2, \ldots, 30\}$ will be chosen. No repeats are allowed. Winning a second state

lottery is based on trying to guess which six randomly picked numbers from the set $\{1, 2, \ldots, 38\}$ will be chosen. Winning a third state lottery is based on trying to guess 7 of 11 randomly picked numbers from the set of $\{1, 2, \ldots, 80\}$. How many possible winning combinations are there for each of these lotteries?

26. Given $1, 2, \ldots, 11$, select a subset of five elements from this set and a second subset with two of these elements. In how many ways can these groups be formed if:

   (a) There are no restrictions.
   (b) Each group contains all even or all odd integers.
   (c) No repetitions are allowed, and the smallest member of the second group is larger than the largest member of the first group. Show that it does not matter whether the two-element set or the five-element set is chosen first.

27. How many ways can a committee of three men and two women be chosen from six men and four women? What if Adam Smith and Abigail Smith will not serve on the same committee?

28. How many ways can a committee of three be chosen from four teams of two with each team consisting of a man and a woman if:

   (a) All are equally eligible.
   (b) The committee must consist of two women and one man.
   (c) A man and a woman from the same team cannot serve on the committee.

29. Two committees of five persons each must be chosen from a group of 375 people. If the committees must be disjoint, in how many ways can the committees be chosen? If the committees need not be disjoint, in how many ways can this be done?

30. A series of articles about 26 athletic teams will appear over 26 consecutive weeks. How many ways can the articles be ordered? What if the first week's article must be about the current champion? What if team A and team B must be featured in consecutive weeks?

31. Solve the problem in Example 10 if the claim is that a different pizza can be ordered every day for three years.

32. Magic Ice sells ice cream sundaes. With m flavors of ice cream and k possible toppings, how large should m and k be so that a customer can have a different sundae every night for five years? A sundae has just one kind of ice cream and at most three types of toppings.

33. How many ways are there for a person to travel from the southwest corner to the northeast corner of an $m \times n$ grid? Enumerate all the ways possible if the grid is 5×3. How many ways are there if the grid is 10×10 and no move may take the person below the main diagonal (those positions that are k steps over and k steps up from the starting point where $1 \le k \le 10$).

34. Five students (A, B, C, D, and E) are scheduled to present papers in class.

   (a) How many ways can this be arranged?
   (b) How many ways can this be arranged without B speaking before A?
   (c) How many ways can this be arranged if A speaks immediately before B?

35. How many ways can 12 black pawns be placed on the black squares of an 8×8 chess board? How many ways can 12 black pawns and 12 white pawns be placed on the black squares of an 8×8 chess board? Half the 64 squares are black and half are red. No black (red) square shares an edge with a black (red) square.

36. The Old Town Softball League has 16 teams arranged in four groups of four teams each. How many different ways can these groups be made up?

37. How many ways can a committee be selected consisting of two Independents, two Republicans, and two Democrats if the choices are made from seven Independents, nine Republicans, and eight Democrats?

38. A team of 11 players is to be chosen from a group of 15 candidates.

    (a) How many different teams can be chosen?
    (b) How many teams can be chosen if one player is designated captain and must play on the team?

39. How many seven-digit sequences can be formed using the symbols $\{0, 1, 2, 3\}$?

40. How many four-person teams can be formed from three men and five women if at least one man and at least one woman are on each team?

41. (a) Which permutation of $\{1, 2, 3, 4, 5\}$ follows 3-1-5-2-4 in the lexicographical ordering of the permutations of five elements?
    (b) Repeat the question for 4-6-1-3-7-5-2 as a permutation of $\{1, 2, \ldots, 7\}$.

42. (a) Construct the permutation numbered 39 in the dictionary ordering of the permutations of the letters $\{1, 2, 3, 4, 5\}$. Remember this is actually the 40th permutation, since the numbering of permutations starts with 0.
    (b) Construct the permutation numbered 387 in the permutations of $\{1, 2, 3, 4, 5, 6\}$.
    (c) Construct the permutation numbered 3764 in the permutations of $\{1, 2, 3, 4, 5, 6, 7\}$.
    (d) Construct the permutation numbered 27,459 in the permutations of $\{1, 2, 3, 4, 5, 6, 7, 8\}$.

43. What is the 311th permutation of $\{1, 2, 3, 4, 5, 6, 7\}$ relative to the lexicographical ordering? (Remember that the 311th is numbered 310, since the first element is numbered 0). What is the 2374th?

44. A football team of 11 players is to be selected from a set of 15 players. Five players in this set can only play in the backfield, eight can only play on the line, and two can play either in the backfield or on the line. A team has seven players on the line and four players in the backfield. How many different teams can be selected?

45. A classroom has two rows of eight seats. There are 14 students in the class. Five students always sit in the front row, and four always sit in the back row. In how many ways can the students be seated?

46. How many ways are there to roll 10 dice so that all six different faces show?

7.8 Counting with Repeated Objects

Although the word *abracadabra* has 11 letters, there are not 11! permutations that result in different words. For example, the first two occurrences of *a* can be interchanged, and no different word would be apparent.

In the word *abracadabra* the letters *a, b,* and *r* occur more than once. If we permute the letters of this word, then it is impossible, for example, to recognize two anagrams as being different when they simply interchange different pairs of occurrences of a repeated letter. A generalization of results about permutations and combinations is needed when some objects are identical and not distinguishable, as in this case. Sections 7.8–7.10 explain methods for handling such situations as well as introducing combinatorial identities. In Section 7.10, a representation of the binomial coefficients known as Pascal's triangle is also introduced and used to prove several useful combinatorial identities.

7.8.1   Permutations with Repetitions

A typical problem that motivates how to count permutations of objects not necessarily all distinct—or permutations with repetitions—is the problem of counting the number of permutations of the letters in a word with repeated letters. As an example, let us determine the number of distinguishable permutations of the letters in the word *abracadabra*. The letter *a* is repeated five times, the letter *b* two times, the letter *r* two times, and the letters *c* and *d* just once each. A permutation of these 11 letters can be constructed as follows: Pick 5 of the 11 positions for the letters of this word, and assign them the letter *a*. This can be done in $C(11, 5)$ different ways. Of the remaining six letter positions, choose two of these positions for the occurrences of the letter *b*. This can be done in $C(6, 2)$ ways. Now, choose two of the remaining letter positions for the occurrences of the letter *r*. This can be done in $C(4, 2)$ ways. Choose one of the two remaining letter positions for the occurrence of the letter *c*. This can be done in $C(2, 1)$ ways. Put *d* in the letter position remaining. By the Multiplication Principle, the total number of arrangements will be the product of the number of ways of assigning each of the different letters.

$$
\begin{aligned}
&(\# \text{ Permutations of the letters of } abracadabra) \\
&= C(11, 5) \cdot C(6, 2) \cdot C(4, 2) \cdot C(2, 1) \cdot C(1, 1) \\
&= \frac{11!}{6!\,5!} \cdot \frac{6!}{4!\,2!} \cdot \frac{4!}{2!\,2!} \cdot \frac{2!}{1!\,1!} \cdot \frac{1!}{1!\,0!} \\
&= \frac{11!}{5!\,2!\,2!\,1!\,1!} \\
&= \frac{11!}{5!\,2!\,2!} \\
&= 83{,}160
\end{aligned}
$$

When the binomial coefficients are replaced with their factorial equivalents, many of the factors cancel. The terms remaining in the denominator represent the number of times that each letter is repeated in the permutation.

A question that should arise here is whether it makes any difference in what order letters are assigned places. For example, suppose the *abracadabra* problem is solved by placing the occurrences of *r* first, *c* second, *a* third, *b* fourth, and *d* fifth. The answer would be

$$
\begin{aligned}
&(\# \text{ Permutations of the letters of } abracadabra) \\
&= C(11, 2) \cdot C(9, 1) \cdot C(8, 5) \cdot C(3, 2) \cdot C(1, 1) \\
&= \frac{11!}{9!\,2!} \cdot \frac{9!}{8!\,1!} \cdot \frac{8!}{3!\,5!} \cdot \frac{3!}{2!\,1!} \cdot \frac{1!}{1!\,0!} \\
&= \frac{11!}{5!\,2!\,2!\,1!\,1!} \\
&= \frac{11!}{5!\,2!\,2!} \\
&= 83{,}160
\end{aligned}
$$

We see that the answer is the same in both cases. In general, the order in which letters are assigned locations will not affect the answer, since we only use the number of such cases.

This example is a special case of the theorem stated below that gives the number of permutations of n letters with repetitions allowed. In this case, we have sets of distinguishable objects, but the objects within each set are indistinguishable from one another.

Definition 1. Let $m_1, m_2, \ldots, m_k$ be distinct symbols. Let a set of n symbols consist of r_i copies of m_i for $1 \le i \le k$ such that $\sum_{i=1}^{k} r_i = n$. The number of **permutations** with repetitions using these n symbols is denoted as $P(n; r_1, r_2, \ldots, r_m)$.

In the notation given by Definition 1, the terms r_i for $1 \le i \le k$ are not assumed to be in any particular order. Observe that $P(n, r)$ and $P(n; r)$ represent very different ideas.

Theorem 1. Let $n = r_1 + r_2 + \cdots + r_k$ be any sum of positive integers. The number of ways to arrange r_1 objects of type 1, r_2 objects of type 2, $\ldots$, and r_k objects of type k is given by

$$P(n; r_1, r_2, \ldots, r_k) = C(n, r_1) \cdot C(n - r_1, r_2) \cdots C(n - r_1 - \cdots - r_{k-1}, r_k)$$

$$= \frac{n!}{r_1! r_2! \cdots r_k!}$$

Proof. The proof is by induction on n. Let $n_0 = 1$ and $r_1 + r_2 + \cdots + r_k = n$. Define

$$\mathcal{T} = \{n : P(n; r_1, r_2, \ldots, r_k) = \frac{n!}{r_1! r_2! \cdots r_k!}\}$$

(Base step) The base case is 1. We leave that case to the reader.

(Inductive step) Let $n \ge n_0$. Assume that for all m where $n_0 \le m < n$, $m \in \mathcal{T}$. Now, prove that $n \in \mathcal{T}$.

The number $P(n; r_1, r_2, \ldots, r_k)$ is, by definition, the number of ways that n letters can be arranged if r_i of the letters are the same for $1 \le i \le k$.

Choose r_1 of these locations for occurrences of the first letter. This can be done in

$$C(n, r_1) = \frac{n!}{r_1! (n - r_1)!}$$

ways. After any choice of r_1 locations, there are $n - r_1$ locations remaining to be filled by the $k - 1$ other letters. There are $P(n - r_1; r_2, r_3, \ldots, r_k)$ ways to arrange the remaining letters, and, since $n_0 \le n - r_1 < n$, by the inductive hypothesis

$$P(n - r_1; r_2, r_3, \ldots, r_k) = \frac{(n - r_1)!}{r_2! \cdots r_k!}$$

It follows from the Multiplication Principle that

$$P(n; r_1, r_2, \ldots, r_k) = C(n, r_1) \cdot P(n - r_1; r_2, r_3, \ldots, r_k)$$

$$= \frac{n!}{r_1! (n - r_1)!} \cdot \frac{(n - r_1)!}{r_2! \cdots r_k!}$$

$$= \frac{n!}{r_1!, r_2! \cdots r_k!}$$

Thus, $n \in \mathcal{T}$. Therefore, by the Strong Form of Mathematical Induction, $\mathcal{T} = \{n \in \mathbb{N} : n \ge 1\}$. ∎

The result just given is a generalization of the case in which all n letters of the permutation are distinguishable, since in that case, the denominator consists of 1! occurring n times.

The following examples will show how to count the number of permutations when some elements are not distinguishable among themselves, like repeated letters in a word.

Example 1.   How many permutations are there of the letters in the word *excellent?*

Solution.   The letter *e* occurs three times, and the letter *l* occurs twice. Each of the remaining letters x, c, n, and t occurs once. By Theorem 1, the number of such permutations is

$$P(9; 3, 2, 1, 1, 1, 1) = \frac{9!}{3!\, 2!\, 1!\, 1!\, 1!\, 1!}$$
$$= 30{,}240 \qquad \blacksquare$$

Example 2.   How many four-letter words can be formed using the letters a, a, a, b, b, c, c, c, c, d, d?

Solution.   The first step in finding the required count is to decompose the problem into a number of disjoint subcases. Theorem 1 can be used on each subcase, and the final answer is found by using the Addition Principle with the answers for the subcases. A description of the subcases and the count for each subcase are given in Table 7.5.

Table 7.5   Subcases for Example 2

	Kinds of Letter	No. of Choices for Letters	No. of Arrangements for Letters
	Subcases		
(a)	4—same	$C(1, 1)$	$P(4; 4)$
(b)	3—same	$C(2, 1)$	$P(4; 3, 1)$
	1—different	$C(3, 1)$	
(c)	2—pairs	$C(4, 2\ 2)$	$P(4; 2, 2)$
(d)	2—same	$C(4, 1)$	$P(4; 2, 1, 1)$
	2—different	$C(3, 2)$	
(e)	4—different	$C(4, 4)$	$P(4; 1, 1, 1, 1)$

\# Using the Addition Principle, the final answer is

$$\begin{aligned}
\# \text{ Words} &= C(1, 1) \cdot P(4; 4) + C(2, 1) \cdot C(3, 1) \cdot P(4; 3, 1) + C(4, 2) \cdot P(4; 2, 2) \\
&\quad + C(4, 1) \cdot C(3, 2) \cdot P(4; 2, 1, 1) + C(4, 4) \cdot P(4; 1, 1, 1, 1) \\
&= 1 + 2 \cdot 3 \cdot 4 + 6 \cdot 6 + 12 + 1 \cdot 2 \cdot 3 \cdot 4 \\
&= 229 \qquad \blacksquare
\end{aligned}$$

When the partitions consist of sets of fixed size and the sizes are not all distinct, we must take into account that the elements of the partition can be permuted among themselves without changing the count that is needed.

Example 3.   In how many ways can 12 examinations be split into three sets of four examinations each?

Solution.   By Theorem 1, we get

$$\frac{12!}{4! \cdot 4! \cdot 4!}$$

This count does not take into account, however, the fact that the three sets can be permuted among themselves, so the final answer is

$$\frac{12!}{4! \cdot 4! \cdot 4! \cdot 3!}$$ ■

This last example shows how to count partitions of indistinguishable elements in sets that may themselves be indistinguishable from one another.

7.8.2   Combinations with Repetitions

The methods developed so far have involved counting the objects directly. Another technique used for counting **combinations with repetitions** is to find a bijection from the instances of a given problem to a subset of the instances of a standard problem. The number of objects in the subset of the standard problem is then just the number of instances in the original problem.

As an example of this technique, suppose that 10 identical marbles are to be distributed among three youngsters. The question is how many ways this can be done. Begin by representing each marble by an X as shown:

$$XXXXXXXXXX$$

Suppose the first and the second youngsters will receive three of the marbles while the third will receive four. To denote this, put a | following the X that represents the third marble. Now, put a second | following the sixth X to represent the fact that the second youngster also received three marbles. The remaining X's in locations 7 through 10 represent the four marbles received by the third youngster. Thus, one partition of the 10 marbles consisting of two sets of size three and one set of size four can be represented as

$$XXX \mid XXX \mid XXXX$$

If one youngster is to receive no marbles, then two of the dividing marks would be placed next to each other. Thus, the technique handles both the case in which each set of the partition is nonempty as well as the case in which some of the sets in the partition may be empty.

We can generalize the process for this problem of marbles and youngsters to solve other problems as well. We proceed as follows. For an arbitrary partition of n identical elements into k subsets, first display $n + k - 1$ positions. Choose $k - 1$ of these locations for the occurrence of the symbol |. This choice can be made in $C(n + k - 1, k - 1)$ ways. The $k - 1$ locations chosen will completely define k sets as follows: (1) the elements before the first mark, (2) the elements between the first and the second mark, ..., and (k) the elements following the $(k - 1)$-st mark. Each choice for locating the $k - 1$ marks will determine a different partition of the n elements into k subsets. This example is a special case of the next theorem.

Theorem 2.   The number of partitions of n identical objects into k sets is $C(n + k - 1, k - 1)$, where $n \geq k > 0$.

Proof.   Associate with the problem a set of $n + k - 1$ marks arranged as

$$- - - \quad \cdots \quad - - -$$
$$\underbrace{- - - - - - - n + k - 1 - - - - -}$$

Choose $k - 1$ of these locations to designate the boundaries between the k sets. The objects are placed in the positions that are not designated as boundary positions.   ∎

Example 4.   Four members of a soccer team are working together to sell 100 raffle tickets. In how many different ways can members contribute to this effort?

Solution.   The number of ways to split 100 into four subsets will represent the number of ways the players could have sold the tickets. To solve the problem, use Theorem 2, with $n = 100$ and $r = 4$. The answer is

$$C(103, 3) = 176{,}851$$ ∎

One more variation of the problem of distributing identical objects into disjoint sets is to count the number of ways to distribute n identical balls into k distinct urns when $n \geq k$. Although this problem is posed in terms of balls and urns, this is simply the problem of distributing n identical objects into k distinct sets. The answer to this question is $C(n + k - 1, k - 1)$. As a related question, in how many ways can k urns be filled with n balls provided that each urn contains at least one ball? This number will give the number of partitions of n elements into k nonempty sets. When a certain number of the balls must be in particular urns, just put this number of balls where required and count the number of ways to distribute the remaining balls into the original number of urns. In the case that each urn is required to be nonempty, put one ball in each urn, and then distribute the remaining $n - k$ balls in $C((n - k) + k - 1, k - 1) = C(n - 1, k - 1)$ ways. Put the balls occurring before the first mark into the first urn, the balls between the first and second mark in the second urn, and so on, until the $n - k$ balls are put into the k urns that already contain one ball each.

Example 5.   A data set contains 500 observations. Analysis of the data is carried out by three programs that together process the 500 observations such that each program processes at least 100 observations. If the partition of the 500 observations for use by the three programs is done by arbitrarily choosing the observations for each program, in how many ways can the data be processed?

Solution.   Think of the programs as urns and the observations as balls. The problem asks in how many ways 500 balls can be put into three urns, with each urn containing at least 100 balls. The answer is

$$C(500 - 300 + 3 - 1, 3 - 1) = C(202, 2) = 20{,}301$$ ∎

Example 6.   How many ways can the equation

$$k_1 + k_2 + \cdots + k_r = n$$

for $r \leq n$ be solved with integers $k_i \geq 0$ for $r \geq i \geq 1$?

Solution.   First, restate the problem in terms of urns and balls. How many ways can n balls be put into r urns? This number is just $C(n + r - 1, r - 1)$. The number of balls in urn i is just k_i for $i = 1, 2, \ldots, r$. ∎

Example 7.   How many ways can you solve

$$k_1 + k_2 + k_3 + k_4 = 18$$

provided that k_1, k_2, k_3, and k_4 are integers and $k_1, \ k_2 \geq 0, \ k_3 \geq 3, \ k_4 \geq 2$?

Solution.   First, put the five required values in k_3 and k_4. Then, ask how many ways there are to solve

$$k_1 + k_2 + k_3 + k_4 = 13$$

with $k_1, \ k_2, \ k_3, \ k_4 \geq 0$. This number is just

$$C(13 + 4 - 1, 4 - 1) = C(16, 3)$$
∎

7.9   Combinatorial Identities

In this section, a representation of the binomial coefficients known as Pascal's triangle is introduced and used to prove several useful combinatorial identities. Typical arguments for solving counting problems can be algebraic or combinatorial in nature. We will show examples of how to prove combinatorial identities using both types of arguments. **Combinatorial arguments** for proving **combinatorial identities** usually involve counting the same objects in two different ways. Since the same objects are being counted, the two expressions for the count must be equal. Often, a combinatorial argument restates the problem in a context with an obvious interpretation for the two sides of the identity. Theorems 3, 4, and 5 give examples of this kind of a proof. An **algebraic argument** normally involves straightforward algebraic manipulations to turn the expression on one side of the identity into the expression on the other side. Pascal's Triangle is also used to prove a number of combinatorial identites.

The next two theorems have two proofs each, one an algebraic proof and the other a combinatorial proof.

Theorem 3.   (Newton's Identity)   Let $n \geq k \geq m \geq 1$. Then,

$$C(n, k) \cdot C(k, m) = C(n, m) \cdot C(n - m, k - m)$$

Proof.   **(Combinatorial)**   The left-hand side first counts the number of k-element subsets of an n-element set. The left-hand side then counts how many m-element subsets are contained in an arbitrary k-element subset. The right-hand side first counts the number of m-element subsets of an n-element set. The right-hand side then determines how many ways an m-element subset could have elements added to form a k-element subset of the original set. Since the first step chooses m elements, the augmentation for that m-element set is done by choosing from the $(n - m)$ elements of the original set that were not chosen. In both cases, the result is the number of k-element subsets of an n-element set with m of the k elements distinguished. Therefore, the result follows.

(Algebraic)

$$C(n, k) \cdot C(k, m) = \frac{n!}{k! \, (n-k)!} \cdot \frac{k!}{m! \, (k-m)!}$$

$$= \frac{n!}{k!(n-k)!} \cdot \frac{k!(n-m)!}{m!(k-m)!(n-m)!}$$

$$= \frac{n!}{m! \, (n-m)!} \cdot \frac{(n-m)!}{(k-m)! \, (n-m-(k-m))!}$$

$$= C(n, m) \cdot C(n-m, k-m) \qquad \blacksquare$$

Corollary 1:

$$C(n, k) = \frac{n}{k} \cdot C(n-1, k-1)$$

Proof. By Theorem 3, we have

$$C(n, k) \cdot C(k, m) = C(n, m) \cdot C(n-m, k-m)$$

Substitute $m = 1$ in this identity to get

$$C(n, k) \cdot C(k, 1) = C(n, 1) \cdot C(n-1, k-1)$$

$$C(n, k) \cdot k = n \cdot C(n-1, k-1)$$

$$C(n, k) = \frac{n}{k} \cdot C(n-1, k-1) \qquad \blacksquare$$

To calculate a particular binomial coefficient, Corollary 1 leads to a good method. When Pascal's Triangle is introduced, you will see a way to calculate all the binomial coefficients for $1, 2, \ldots, n$ for any $n \in \mathbb{N}$.

The next identity is credited to Pascal (1623–1662, b. France). This result is used to compute the entries in Pascal's triangle, which is a way of displaying binomial coefficients.

Theorem 4.   (Pascal's Identity)   Let $n \geq k \geq 1$. Then,

$$C(n, k) = C(n-1, k) + C(n-1, k-1)$$

Proof.   (Combinatorial)   The left-hand side counts the number of k-element subsets of an n-element set. The right-hand side can be interpreted as follows. Let A be an n-element set, and pick any $x \in A$. Then, $C(n-1, k)$ k-element subsets of A do not contain x; the k elements must be chosen from the remaining $n-1$ elements of A. On the other hand, the number of k-element subsets of A that do contain x is $C(n-1, k-1)$, since $k-1$ elements must be added to x to get such a k-element subset. Since these two collections of k-element subsets of the original set are disjoint, the Addition Principle gives $C(n, k) = C(n-1, k) + C(n-1, k-1)$.

(Algebraic)

$$C(n-1, k) + C(n-1, k-1) = \frac{(n-1)!}{k! \, (n-1-k)!} + \frac{(n-1)!}{(k-1)! \cdot (n-k)!}$$

$$= \frac{(n-1)! \, (n-k)}{k! \, (n-k)!} + \frac{(n-1)! \, k}{k! \, (n-k)!}$$

$$= \frac{(n-1)! \, (n-k+k)}{k! \, (n-k)!}$$

$$= \frac{n!}{k!\,(n-k)!}$$
$$= C(n, k) \qquad \blacksquare$$

In the combinatorial proof that the two sets of subsets are disjoint, we are claiming that no set in one collection of k sets belongs to the other collection of k sets. This is not the same as claiming that each k set in one collection is disjoint from each k set in the other collection, which is clearly not the case.

7.9.1 Binomial Coefficients

The **binomial coefficients** derive their name from the role they play as coefficients in the expansion of a binomial, such as $(x + y)^7$. For small exponent n—say, $n = 1, 2, 3$—the coefficients of $(x + y)^n$ can be remembered, because they are often used. For larger n, the problem of expanding $(x + y)^n$ is not as simple. However, by giving these numbers a combinatorial interpretation, it is quite easy to write down the coefficients for each term in the expansion of $(x + y)^n$ for any n.

Example 8. Expand $(x + y)^3$.

Solution. The product is found by choosing, in all possible ways, one term from each of the three factors and then multiplying those choices together. To see these choices, we have numbered the elements in each term in the first two steps of the computation.

$$
\begin{aligned}
(x + y)^3 &= (x_1 + y_1)(x_2 + y_2)(x_3 + y_3) \\
&= x_1 x_2 x_3 + x_1 x_2 y_3 + x_1 y_2 x_3 + x_1 y_2 y_3 + y_1 x_2 x_3 + y_1 x_2 y_3 \\
&\quad + y_1 y_2 x_3 + y_1 y_2 y_3 \\
&= xxx + 3xxy + 3xyy + yyy \\
&= C(3, 0)x^3 + C(3, 1)x^2 y + C(3, 2)xy^2 + C(3, 3)y^3. \qquad \blacksquare
\end{aligned}
$$

We now prove a theorem that shows how to find all coefficients for the expansion of a binomial.

Theorem 5. (Binomial Theorem) Let $n \in \mathbb{N}$. For all x and y,

$$
\begin{aligned}
(x + y)^n &= C(n, 0)x^n + C(n, 1)x^{n-1}y + C(n, 2)x^{n-2}y^2 + \cdots + C(n, n)y^n \\
&= \sum_{i=0}^{n} C(n, i)x^{n-i}y^i
\end{aligned}
$$

Proof. The proof is by induction on n. Let $n_0 = 0$ and $\mathcal{T} = \{n \in \mathbb{N} : \text{the identity is true}\}$.

(Base step) For $n = 0$, the formula is

$$
\begin{aligned}
(x + y)^0 &= \sum_{i=0}^{0} C(0, i)\, x^{0-i}\, y^i \\
&= C(0, 0)\, x^0\, y^0 \\
&= 1
\end{aligned}
$$

which is true.

(Inductive step)   Choose $n \geq n_0$, and assume $n \in T$. Now, prove that $n + 1 \in T$. Begin by setting

$$(x + y)^{n+1} = (x + y)(x + y)^n.$$

By the inductive assumption, this is

$$(x + y)^{n+1} = (x + y) \cdot \sum_{i=0}^{n} C(n, i)\, x^{n-i}\, y^i$$

$$= x\left(\sum_{i=0}^{n} C(n, i)\, x^{n-i}\, y^i \right) + y\left(\sum_{i=0}^{n} C(n, i)\, x^{n-i}\, y^i \right)$$

$$= C(n, 0)\, x^{n+1} + \sum_{i=1}^{n} C(n, i)\, x^{n-i+1}\, y^i$$

$$+ \sum_{i=0}^{n-1} C(n, i)\, x^{n-i}\, y^{i+1} + C(n, n) y^{n+1}$$

Next, replace i by $i - 1$ in the third term, giving

$$\sum_{i=0}^{n-1} C(n, i)\, x^{n-i}\, y^{i+1} = \sum_{i=1}^{n} C(n, i - 1)\, x^{n-i+1}\, y^i$$

This makes the two sums start at the same value so that they can be added term by term to get

$$\sum_{i=1}^{n} (C(n, i) + C(n, i - 1))\, x^{n-i+1} y^i$$

By Pascal's Identity, the sum becomes

$$\sum_{i=1}^{n} C(n + 1, i)\, x^{n-i+1}\, y^i$$

Combining these results gives

$$(x + y)^{n+1} = y^{n+1} + \sum_{i=1}^{n} C(n + 1, i)\, x^{n-i+1}\, y^i + x^{n+1}$$

$$= \sum_{i=0}^{n+1} C(n + 1, i)\, x^{n+1-i}\, y^i$$

Therefore, $n + 1 \in T$.

By the Principle of Mathematical Induction, $T = \mathbb{N}$.   ∎

Knowing the form of the coefficients of $(x + y)^n$ allows some interesting results to be proven.

Theorem 6.   For $n \in \mathbb{N}$,

$$2^n = C(n, 0) + C(n, 1) + \cdots + C(n, n)$$

Proof. Expand $(x + y)^n$ with $x = y = 1$. ■

Corollary 2: The number of subsets of an n-element set is 2^n.

For different arguments to prove Corollary 2, see Theorem 2 in Section 1.7.4 and Example 4 in Section 7.2.1.

Theorem 7. For $n \geq 1$,

$$\sum_{i=0}^{n} C(n, i)(-1)^i = 0$$

Proof. The idea of the proof is to use the Binomial Theorem and expand $(x + y)^n$ with $x = 1$ and $y = -1$. By the Binomial Theorem with $n \geq 1$, we have $(x + y)^{n+1} = \sum_{i=0}^{n+1} C(n, i)x^{n-i}y^i$. Let $x = 1$ and $y = -1$. The identity then becomes

$$(1 + (-1))^{n+1} = \sum_{i=0}^{n+1} C(n, i)1^{n-i}(-1)^i$$

$$0 = \sum_{i=0}^{n+1} C(n, i)(-1)^i$$

■

Corollary 3: For n even,

$$\sum_{i=0}^{n/2} C(n, 2i) = \sum_{i=0}^{n/2-1} C(n, 2i + 1)$$

Corollary 4: For n odd,

$$\sum_{i=0}^{\lfloor n/2 \rfloor} C(n, 2i) = \sum_{i=0}^{\lfloor n/2 \rfloor + 1} C(n, 2i + 1)$$

The corollaries to Theorem 7 can be interpreted as saying that for any n-element set, there are as many subsets with an even number of elements as there are subsets with an odd number of elements. An obvious result of the corollaries is that

$$\sum_{i=0}^{n/2} C(n, 2i) = \sum_{i=0}^{\lfloor n/2 \rfloor} C(n, 2i)$$

Since it is also true that

$$\sum_{i=0}^{n/2} C(n, 2i) + \sum_{i=0}^{\lfloor n/2 \rfloor} C(n, 2i) = 2^n$$

it follows that

$$\sum_{i=0}^{n/2} C(n, 2i) = \sum_{i=0}^{\lfloor n/2 \rfloor} C(n, 2i) = 2^{n-1}$$

A second technique for proving identities using binomial coefficients involves recognizing that the expression

$$(1+x)^n = \sum_{i=0}^{n} C(n, i)x^i$$

can be viewed as a polynomial identity. Since this expression is an identity, its derivative will also be an identity, so

$$n(x+1)^{n-1} = \sum_{i=1}^{n} i C(n, i) x^{i-1}$$

This fact is used in evaluating the next sum.

Example 9.   Show that $\sum_{i=1}^{n} i C(n, i) = n \, 2^{n-1}$.

Solution.   Compute the derivative of

$$(x+1)^n = \sum_{i=0}^{n} C(n, i)x^{n-i}$$

getting

$$n(x+1)^{n-1} = \sum_{i=1}^{n} i C(n, i) x^{i-1}.$$

Now, substituting $x = 1$ gives

$$n2^{n-1} = \sum_{i=1}^{n} i C(n, i)$$

∎

In addition to differentiating an identity to produce another identity, it is also possible to multiply an identity by a power of the variable to produce another identity. Such techniques will be explored in the exercises (see Section 7.11).

7.9.2   Multinomials

The computation of the coefficients of terms in expansions of a **multinomial** such as $(a+b+c)^6$ is an obvious generalization of the problem of finding the coefficients for the expansion of powers of a binomial. These multinomial coefficients can be computed using the result about counting permutations with repeated letters. For example, the coefficient of a^2bc^3 in

$$(a+b+c)^6 = (a+b+c)(a+b+c)(a+b+c)(a+b+c)(a+b+c)(a+b+c)$$

is determined by choosing, in all possible ways, a from two factors of this product, b from a different factor, and c from the remaining three factors. The number of ways to do this is just

$$\frac{6!}{2!1!3!}$$

as was proved in Theorem 1 of Section 7.8.1. These coefficients are called **multinomial coefficients**.

7.10   Pascal's Triangle

Pascal's Identity (see Theorem 4 in Section 7.9) can be used to calculate the binomial coefficients for any n and m such that $0 \leq m \leq n$ using the boundary conditions and the binomial coefficients for $n - 1$. The terms pointed to by a pair of arrows in Figure 7.9 are found by adding the two values at the head of the arrows.

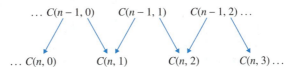

$$\ldots C(n-1, 0) \qquad C(n-1, 1) \qquad C(n-1, 2) \ldots$$

$$\ldots C(n, 0) \qquad C(n, 1) \qquad C(n, 2) \qquad C(n, 3) \ldots$$

Figure 7.9   Finding terms in Pascal's Triangle.

There is a long history for the binomial coefficients and an extensive list of identities they satisfy. The display of these numbers in triangular form in Figure 7.10 is called **Pascal's triangle** even though these numbers were displayed in that fashion centuries before Pascal's time.

$$1$$
$$C(0, 0)$$

$$1 \qquad 1$$
$$C(1, 0) \quad C(1, 1)$$

$$1 \qquad 2 \qquad 1$$
$$C(2, 0) \quad C(2, 1) \quad C(2, 2)$$

$$1 \qquad 3 \qquad 3 \qquad 1$$
$$C(3, 0) \quad C(3, 1) \quad C(3, 2) \quad C(3, 3)$$

$$1 \qquad 4 \qquad 6 \qquad 4 \qquad 1$$
$$C(4, 0) \quad C(4, 1) \quad C(4, 2) \quad C(4, 3) \quad C(4, 4)$$

Figure 7.10   Five rows of Pascal's triangle.

The rows of Pascal's triangle represent the coefficients of the terms in $(x + y)^n$. By representing the numbers in Pascal's triangle in a slightly different form, as seen in Figure 7.11 on page 464, it is natural to define **rows, columns,** and **diagonals** for Pascal's triangle.

The next theorem gives a formula for the sum of elements in a row, consecutive elements from the beginning of a column, or elements on a diagonal of Pascal's triangle.

Theorem 8.   Let $n \geq r \geq 0$. Then,

$$\text{Row Sum} \quad : C(n, 0) + C(n, 1) + \cdots + C(n, n) = 2^n.$$
$$\text{Column Sum} \quad : C(r, r) + C(r + 1, r) + \cdots + C(n, r) = C(n + 1, r + 1).$$
$$\text{Diagonal Sum} \quad : C(n, 0) + C(n + 1, 1) + \cdots + C(n + r, r) = C(n + r + 1, r)$$

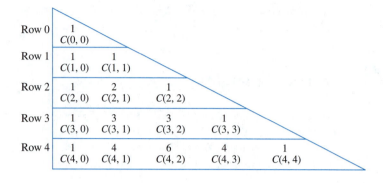

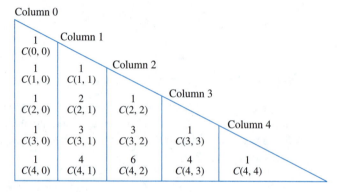

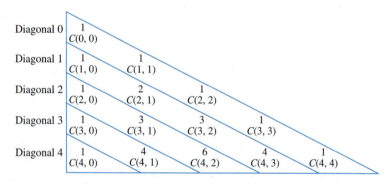

Figure 7.11 Rows, columns, and diagonals in Pascal's triangle.

Proof. The Row Sum is the content of Theorem 6 in Section 7.9.1. The proofs of the other identities are left as exercises for the reader. ∎

By recognizing how to represent powers of integer variables as linear combinations of binomial coefficients, the row, column, and diagonal sum identities can be used to evaluate finite sums of such powers.

Theorem 9. (**Sums of Powers**) Let $n \geq 1$. Then,

(a) $1 + 2 + 3 + \cdots + n = n(n+1)/2$

(b) $1^2 + 2^2 + 3^2 + \cdots + n^2 = \frac{(2n+1)\,n\,(n+1)}{6}$

Proof.

(a) Since $C(k, 1) = k$, the sum is found using the column sum formula.

$$\sum_{k=1}^{n} k = \sum_{k=1}^{n} C(k, 1) = C(n+1, 2) = \frac{n\,(n+1)}{2}$$

(b) The formula for the column sum is used twice to evaluate the sum of the first n squares. First, write $k^2 = k\,(k-1) + k$. Since $k\,(k-1)$ can be written as $2\,C(k, 2)$ and k as $C(k, 1)$, the sum now becomes

$$\sum_{k=1}^{n} k^2 = \sum_{k=2}^{n} 2\,C(k, 2) + \sum_{k=1}^{n} C(k, 1)$$
$$= 2C(n+1, 3) + C(n+1, 2)$$
$$= \frac{(2n+1)\,n\,(n+1)}{6}$$ ∎

7.11   **Exercises**

1. How many permutations are there for the letters of the name *Bathsheba? Solomon? Ahab?* your own name?

2. How many arrangements are possible for the letters of the following words:

   (a) *Tennessee*
   (b) *Mississippi*
   (c) *Kansas*
   (d) *Oregon*
   (e) *Manitoba*
   (f) *Visiting*

3. How many words or strings of 12 letters can be formed from the symbols

   $$a, a, a, a, b, b, b, b, b, b, b, b$$

   provided that no two a's can occur together?

4. Find the number of arrangements of the word *engineering*.

   (a) In how many of these are the three e's together?
   (b) In how many of these are exactly two e's together?

5. How many ways can five identical advertisements be placed in three mailboxes if each mailbox receives at least one advertisement? How many ways if a mailbox may receive none? (The order in which a messenger delivers a message is immaterial).

6. A word consisting of five letters is just a string of five letters with no meaning required. For example, *xqzrp* is a five-letter word. How many five-letter words or strings can be formed using an alphabet consisting of 35 letters if no repetition of letters is allowed? How many repetitions of a letter are allowed?

7. How many ways can you choose eight letters from

aaaaa   bbbbbb   cccccccc

with at least one a, one b, and two c's?

8. How many integer solutions are there for the following equations?

(a) $x + y + z = 8$ where $x \geq 0$, $y \geq 0$, and $z \geq 0$
(b) $x + y + z + t = 18$ where x, y, z, and t are each greater than zero
(c) $x + y + z + t = 12$ where $x \geq 1$, $y \geq 2$, and $t \geq 1$

9. How many ways can two booksellers divide between themselves 300 copies of one book, 200 copies of another, and 100 copies of a third if neither bookseller is to get all the copies of any one of the books?

10. A bookbinder is to bind 10 different books using red, green, or blue cloth for each book. How many ways can this be done if each color is to be used for:

(a) At least one book
(b) At least two books
(c) With no restriction on the number of times a color may be used

11. How many ways can six candy bars be distributed among three children if every child is to receive at least one candy bar?

12. How many ways can a class of 25 students be assigned to three different lab sections if each lab section has at least 5 students?

13. A king is placed on the bottom left-hand square of an 8×8 chess board and is to move to the top right-hand corner square. If the piece can move only up or to the right, how many possible paths does it have?

14. An XYZ-3000 is a front-end processor to five mainframe computers at RST U. There are 64 incoming phone lines to the XYZ-3000. In how many ways can the front-end processor assign lines to computers so that 8 are directed to C_1, 14 to C_2, 17 to C_3, 16 to C_4, and the remaining to C_5? A program called TUNE monitors the performance of a computer system. Suppose each user is assigned to one of the 64 memory areas when first logged onto the system. TUNE samples memory areas or partitions when a user is first logged on to the system to decide how to assign the new user memory. How many can this be done if TUNE samples 17 of 64 system partitions? How many if the one fixed partition Z is always excluded from the sample? How many if two fixed partitions are always chosen?

15. Three first-year, three second-year, and three third-year students are to be seated in a row. The students in each class are indistinguishable. How many ways can they be seated so that no three students of the same class sit together?

16. How many ways can an examiner assign 90 points to 12 questions with each question getting at least four points?

17. A shop sells six flavors of ice cream. Each ice cream cone holds one, two, or three scoops of ice cream. How many ways can four ice cream cones be made such that:

(a) All the cones have a different flavor, and each cone has a single flavor for each of its scoops.
(b) Not necessarily all the cones have a different flavor.
(c) The cones contain only two or three flavors of ice cream.
(d) The cones contain three different flavors.

18. A six-person committee is to be chosen from 16 university students, (4 from each class—first, second, third, and fourth years). Determine how many committees are possible if:

    (a) Each class is represented.
    (b) No class has more than two representatives, and each class has at least one representative.

19. How many ways can an eight-person committee be chosen from a group of 10 new members and 15 old members if the committee is composed of:

    (a) Four members from each group
    (b) More new members than old members
    (c) At least two new members

20. A string consisting of 0's and 1's has even parity if 1 occurs an even number of times; otherwise, the string has odd parity. How many strings of length n have even parity? How many strings of length n have odd parity?

21. A domino is made of two squares, each of which is marked with one, two, three, four, five, or six spots or is left blank. A set of dominoes consists of dominoes with all possible pairs showing in the two squares. How many different dominoes are there in a set?

22. A bridge hand consists of 13 cards dealt from the 52-card deck. Bridge involves four players named North, East, South, and West. How many ways can the cards be dealt so that the game can be played?

23. Prove $\sum_{k=0}^{n} C(n,k)^2 = C(2n, n)$.

    (a) Prove the identity using the fact that $(1+x)^{2n} = (1+x)^n (1+x)^n$.
    (b) Give a combinatorial proof of the identity in part a.
    (c) Find the number of 14-digit binary sequences for which the number of 1's in the first seven digits is the same as the number of 0's in the last seven digits of the sequence. Enumerate all such sequences of length six.

24. Prove that $n(n+1)2^{n-2} = \sum_{k=1}^{n} k^2\, C(n,k)$.

25. Sum $1^3 + 2^3 + \cdots + m^3$ using the fact that m^3 can be represented as

    $$m^3 = aC(m, 3) + bC(m, 2) + cC(m, 1)$$

    where a, b, and c are rational numbers. This problem should not be solved using a proof by induction.

26. Prove that $1 \cdot 2 + 2 \cdot 3 + \cdots + n \cdot (n+1) = n(n+1)(n+2)/3$. This problem should not be solved using a proof by induction.

27. Prove that
    $$1 \cdot 2 \cdot 3 + 2 \cdot 3 \cdot 4 + \cdots + n \cdot (n+1) \cdot (n+2) = n\,(n+1)\,(n+2)(n+3)/4.$$
    This problem should not be solved using a proof by induction.

28. Given that $(2n)!/n! = 2(1 \cdot 3 \cdot 5 \cdot \ldots \cdot (2n-1))$, conclude that $2 \cdot 6 \cdot 10 \cdot \ldots \cdot (4n - 6) \cdot (4n-2) = (n+1)(n+2)(n+3)(n+4)\cdots(2n-1)2n$.

29. Construct the first 10 rows of Pascal's triangle.

30. Using the Binomial Theorem, expand:

    (a) $(x+y)^5$
    (b) $(x+y)^6$

31. Expand $(1+x)^8$ using the Binomial Theorem.

32. Expand $(2x - y)^7$ using the Binomial Theorem.
33. In the expansion of $(3x - 2y)^{18}$, what are the coefficients of:

    (a) $x^5 \cdot y^{13}$
    (b) $x^3 \cdot y^{15}$

34. Expand $(a + b + c)^2$.
35. Use the Binomial Theorem to prove that

$$3^n = \sum_{k=0}^{n} C(n, k) 2^{n-k}.$$

Write out what the identity says for $n = 4$.
36. Use the Binomial Theorem to prove that

$$2^n = \sum_{k=0}^{n} (-1)^k C(n, k) 3^{n-k}.$$

Write out what the identity says for $n = 4$.
37. Theorem 4 in Section 7.9 proves that $C(n, m) C(m, k) = C(n, k) C(n - k, m - k)$. Use this result to prove that $C(n, k) C(n - k, m) = C(n, m) C(n - m, k)$.
38. Prove the Diagonal Sum formula of Theorem 8 in Section 7.10.
39. Prove that $C(n, r) > C(n, r - 1)$ if $r < n/2$ for each row of Pascal's triangle.
40. Prove that $C(2n, n) + C(2n, n - 1) = (1/2) C(2n + 2, n + 1)$.
41. Find a closed form for $\sum_{k=1}^{n} k C(n, k)$.
42. Count the number of triples (x, y, z) where $z > \max\{x, y\}$ and $1 \le x, y, z \le n + 1$. Deduce that $1^2 + 2^2 + \ldots + n^2 = C(n + 1, 2) + 2C(n + 1, 3)$.
43. Evaluate $(1/4) \sum_{k=1}^{5} k (6 - k)$ and deduce the number of points of intersection for the diagonals of an octagon if no three diagonals meet at a point.
44. Find the following multinomials:

    (a) The coefficient of $x^2 y^3 z^5$ in the expansion of $(x + y + z)^{10}$
    (b) The coefficents of $x^3 y^4 z^2$ and $x^3 y^3 z^3$ in the expansion of $(3x + 2y - z)^9$

45. Show that there are $(3^n + 1)/2$ strings of length n consisting of the letters a, b, and x in which a occurs an even number of times.
46. For $n = 1, 2, 3, \ldots$, write

$$[x]_t = x(x - 1)(x - 2) \cdots (x - t + 1)$$

for $0 \le t \le n$. We can represent $[x]_t$ as a linear combination of powers of x. The coefficients for this expansion are denoted as $s(n, t)$ and are known as the **Stirling numbers of the first kind.** Thus, for any n, we can write

$$[x]_t = \sum_{t=0}^{n} s(n, t) x^t$$

The numbers $s(n, t)$ can be defined as $s(n, 0) = 0$ for $n = 1, 2, 3, \ldots$; $s(n, n) = 1$ for $n = 0, 1, 2, \ldots$; and

$$s(n, t) = s(n - 1, t - 1) - (n - 1)s(n - 1, t)$$

for $t = 1, 2, \ldots, n - 1$. Make a table of the Stirling numbers of the first kind for $n = 1, 2, 3, 4, 5, 6$.

47. For a positive integer t, define $[x]_t = x(x - 1) \cdots (x - t + 1)$. We can represent x^n as a linear combination of $[x]_t$, where $n = 1, 2, 3, \ldots$, and $t = 0, 1, 2, \ldots, n$. The coefficients for this expansion are denoted as $S(n, t)$ and are known as the **Stirling numbers of the second kind.** Thus, for any n, we can write

$$x^n = \sum_{t=0}^{n} S(n, t)[x]_t$$

The numbers $S(n, t)$ can be defined for $n = 1, 2, 3, \ldots$ as $S(n, 0) = 0$; $S(n, n) = 1$; and

$$S(n, t) = t S(n - 1, t) + S(n - 1, t - 1)$$

for $1 \leq t \leq n - 1$. Make a table of the Stirling numbers of the second kind for $n = 1, 2, 3, 4, 5, 6$.

48. Still another application of the Principle of Inclusion-Exclusion: Prove that

$$\sum_{r=0}^{m} (-1)^r C(m, r) (m - r)^n$$

counts the number of surjections from a set of size n to a set of size m. Use this result to determine the number of ways to discharge eight people who get on an elevator at the ground floor of a four-story building. The elevator discharges its last passenger on the fourth floor.

49. Still another application of the *Principle of Inclusion-Exclusion:* Let

$$x_1 \, x_2 \, \cdots \, x_{n-1} \, x_n$$

be a permutation of $\{1, 2, 3, \ldots, n - 1, n\}$ such that for $1 \leq i \leq n$, there is no $x_i = i$. A permutation with this property is called a **derangement.** We denote the set of **derangements** on n elements as D_n. Find a formula for the number of **derangements** on $1, 2, \ldots, n - 1, n$ for any $n \in \mathbb{N}$. Evaluate this formula for $n = 2, 3, 4, 5, 6$. Construct all **derangements** for $n = 2, 3$. (*Hint:* see example 12 in Section 7.5.7.)

7.12 Chapter Review

Most counting results are based on two fundamental principles, the Multiplication Principle and the Addition Principle. Permutations (orderings of elements) and combinations (sets of elements) provide a foundation for analyzing problems. Solution techniques for counting problems include counting the complement, decomposing a problem into disjoint subproblems, and using the Pigeon-Hole Principle. Counting problems are quite different when the objects are not distinct. Results about counting permutations and combinations involving repeated elements are presented. A more practical problem for experimentation is to choose a random permutation of n elements from the $n!$ possible permutations. A technique for constructing a random permutation is given. In quite a different context, it is necessary to expand both powers of binomials and powers of multinomials. Both of these problems are solved using the Binomial Theorem and techniques for counting permutations with repeated objects. The Binomial Theorem leads to a number of combinatorial identities that are useful in a variety of counting problems. Pascal's triangle is introduced and used to prove a number of important combinatorial identities.

Besides solving problems that count the number of ways to arrange objects and design letter patterns, counting is used as a step in determining the complexity of an algorithm that solves the TSP. Counting the number of ways to distribute identical objects among other objects is often seen in probability theory. Even counting the number of integer solutions for a linear equation with several variables is an application of counting combinations with repeated objects. The chapter focuses on presenting the foundations for approaching counting problems rather than focusing on particular problems.

7.12.1   Terms, Theorems, and Algorithms

7.1–7.3   Summary

TERMS

Addition Principle
complement
crypt
Multiplication Principle

password
Pigeon-Hole Principle
subproblems

ALGORITHMS

Traveling Salesperson's Problem (TSP)

7.5–7.6   Summary

TERMS

binomial coefficient
$C(n, r)$
circular permutation
combination
derangement
dictionary ordering
fixed points

fixes
k-th of $n!$ permutations
ordering of permutations
$P(n, r)$
permutation
r-permutation

7.8–7.11   Summary

TERMS

algebraic argument
columns
combinations with repeated elements
combinations with repetitions
combinatorial arguments
combinatorial identities
diagonal
multinomial

multinomial coefficients
permutations with repetitions
$P(n; r_1, r_2, \ldots, r_m)$
Pascal's triangle
permutations with repeated letters
rows
Stirling numbers of the first kind
Stirling numbers of the second kind

THEOREMS

Binomial Theorem   Pascal's Identity
Newton's Identity   Sums of Powers

7.12.2 Starting to Review

1. What are the values of $C(5, 3)$ and $P(5, 3)$?
   (a) 60, 10
   (b) 60, 20
   (c) 60
   (d) None of the above
2. There are 12 roads to Merced, 8 roads from Merced to Planter, and 13 roads from Planter to San Francisco. How many possible ways are there to get to San Francisco from Merced?
3. How many ways can you roll two dice and get a total of six appearing on the top faces?
4. Names on Ork are formed according to the rules that a name is six characters long and has two vowels. The vowels may not be in either the first or the last positions, and the two vowels may not occur in adjacent locations. How many people can live on Ork with each person having a different name?
5. To graduate, Sally needs two courses to complete the general education requirement. Courses in anthropology and economics will satisfy the requirement. If Sally has satisfied the prerequisites for 18 anthropology and 21 economics courses, how many schedules are possible?
6. A bakery sells six different kinds of pastries. How many different dozens of pastry can you buy? What if you buy at least one of each kind?
7. There are 10 geography books, 12 chemistry books, and 18 detective novels. How many ways can you pick two books from each of two different groups of books?
8. Using the Binomial Theorem, compute 11^4. Show all work.
9. Find n such that $C(n, 0) + C(n, 1) + \cdots + C(n, n) = 128$.
10. Compute the coefficient of $x^3 y^2 z w^2$ in the expansion of $(x - y + 2z - 2w)^8$.

7.12.3 Review Questions

1. Find the number of ways in which nine 3's and six 5's can be placed in a row so that no two 5's are together.
2. Determine the number of sequences of length r for any $r \in \mathbb{N}$ if the first portion is comprised of the letters $a, b, c,$ and d with the remainder being comprised of the Greek letters $\{\alpha, \beta, \gamma, \delta\}$.
3. In the interest of efficiency, spelling rules for words have been revised. The word *relief* can be spelled in the ways described by the following rules:
   - The number of letters must not exceed six.
   - The word must contain at least one *l*.
   - The word must begin with an *r* and end with an *f*.
   - There is just one *r* and one *f*, and only the letters *e, i,* and *l* may occur in the middle positions.

   How many ways can *relief* be spelled?

4. How many $r - digit$ sequences composed of 0's, 1's, 2's, and 3's are there in which each of 1, 2, and 3 appears at least once?

5. How many solutions are there for $x_1 + x_2 + x_3 + x_4 = 30$ with $-10 \le x_i \le 20$ for $1 \le i \le 4$.

6. Find the number of solutions for $x_1 + x_2 + x_3 + x_4 = 18$ with $x_i \le 7$ for $1 \le i \le 4$.

7. How many solutions in integers are there for

$$x_1 + x_2 + x_3 = 14$$

if $-2 \le x_1 \le 4$; $x_2 \le 5$; and $2 \le x_3 \le 6$?
(*Hint:* Let U be the set of solutions satisfying the conditions $x_1 \ge -2$; $x_2 \ge 0$; $x_3 \ge 2$. Treat the -2 as if you could put -2 elements in box 1, and then solve the equation for 16. The upper bounds are handled by the Principle of Inclusion-Exclusion using the sets $A_1 = \{(x_1, x_2, x_3) \in U : x_1 > 5\}$, $A_2 = \{(x_1, x_2, x_3) \in U : x_2 > 6\}$, and $A_3 = \{(x_1, x_2, x_3) \in U : x_3 > 7\}$.

8. Give a combinatorial proof of $P(n, r) = rP(n - 1, r - 1) + P(n - 1, r)$.

9. Prove that $\sum_{k=0}^{r} C(m, k) \cdot C(n, r - k) = C(m + n, r)$. This is called Van Dermonde's Identity.

10. Prove that $C(n, m) \cdot C(m, k) = C(n, k) \cdot C(n - k, m - k)$. This is called Newton's Identity. Prove that $C(n, k) \cdot C(n - k, m) = C(n, m) \cdot C(n - m, k)$.

11. Count the number of ways to choose 10 elements from among three pears, four apples, and five bananas. (*Hint:* A_i = the set of 10-combinations with more than three pears. A_2 = the set of 10-combinations with more than four apples. A_3 = the set of 10-combinations with more than five apples.)

12. How many ways can the 26 letters of the alphabet be permuted so that none of the patterns *car, dog, pun,* or *byte* occurs?

13. Prove that $\binom{n}{r-1} + \binom{n}{r} = \binom{n+1}{r}$ both algebraically and combinatorially.

14. How many derangements are there of the set $\{1, 2, 3, 4\}$? List all these permutations.

15. A "word" is a string of one or more lowercase letters. How many words can be formed from the letters of the word ... In how many words will p and i occur together? In how many will h and y not occur together? How many words can be formed from the letters in the word *zigzag* where no two consecutive letters are the same?

7.12.4   Using Discrete Mathematics in Computer Science

1. In Section 7.3.3, we counted the number of UNIX passwords of length six, except that we left a factor of 60 underived.

   (a) Derive the factor of 60.

   (b) Count how many UNIX passwords there are of length six, seven, or eight.

   (c) At a speed of 200,000 words a second, how long would it take to try all the passwords you counted in part (b)?

2. Assume that a computer stores each file at *consecutive* locations on a hard disk, with files written on both sides of the disk. When a large file is erased and a smaller file is written in its place, only some of the space is used. If the original disk had files written on both sides, a small block of space is left unused. After many changes to the disk— that is, after a sequence of additions and deletions mixed together—there may be many

available storage locations, but these may be scattered across the disk in pieces too small to use. This sort of problem is called *fragmentation*.

   Suppose a hard disk can store a total of 2^{30} bytes of data and you need to store a file that is 2^{12} bytes long. Suppose each file currently on the disk is 2^{10} bytes long. What is the minimum number of files that could be on the disk so that there is no room to add the new file (in 2^{12} consecutive bytes).

3. Suppose that in a computer program we want an easy way to list—and to access—all the edges of the complete graph K_n, with each edge listed only once and no unused space. We can use a part of the adjacency matrix:

$$
\begin{array}{c|ccccccccc}
n-1 & n & & & & & & & \\
n-2 & n-1 & n & & & & & & \\
\cdots & \cdots & & & & & & & \\
3 & 4 & 5 & 6 & 7 & \cdots & n & & \\
2 & 3 & 4 & 5 & 6 & 7 & \cdots & n & \\
1 & 2 & 3 & 4 & 5 & 6 & 7 & \cdots & n
\end{array}
$$

where after vertex i, we list only the edges going to higher-numbered vertices. Now, we could list all the edges in a single array:

$\{n-1, n\}$	$\{n-2, n-1\}$	$\{n-2, n\}$	$\{n-3, n-2\}$	$\cdots$	$\{1, n-1\}$	$\{1, n\}$

Use combinations to express the position of edge (i, j) in the array above.

4. One method of checking for errors in data is to add "parity bits." If a number n has the form $x_1 x_2 \ldots x_n$ where x_i is either 0 or 1 is to be transmitted, an additional bit x_{n+1} is added at the end of the original n bits. The added bit is 1 if n has an odd number of 1 bits and 0 if n has an even number of 1 bits. When the number is received, the receiving computer can check the number and see whether the parity bit is correct. This way, the receiving computer can recognize any 1-bit error in transmitting the number.

   (a) A string of 0's and 1's is to be processed, from left to right, and converted to an even-parity string by adding a parity bit to the end of the string. The parity bit is initially 0. When a 0 character is processed, the parity bit remains unchanged. When a 1 character is processed, the parity bit is switched from 0 to 1 or from 1 to 0. Prove that the number of 1's in the final string—that is, including the parity bit—is always even.) (*Hint:* Consider cases).

   (b) A string consisting of 0's and 1's has even parity if 1 occurs an even number of times; otherwise, the string has odd parity. How many strings of length n have even parity? How many strings of length n have odd parity?

5. Six distinct symbols are transmitted through a communication channel. A total of 12 blanks are to be inserted between the symbols, with at least two blanks between every pair of symbols. How many ways can the symbols and blanks be arranged? How many ways can the symbols and blanks be arranged if there are 25 blanks?

6. A boolean function of n boolean variables is defined by the assignment of a value of either 0 or 1 to each of the 2^n ordered n-tuples of $0-1$ values for the variables. Recall the usual convention that 0 represents *FALSE* and 1 represents *TRUE*.

   (a) How many different boolean functions of n variables are there?

   (b) Approximate, in ordinary decimal notation, the number of boolean variables in eight variables: How many decimal digits in length is that number?

(c) A self-dual, two-valued boolean function defined on an n-bit binary number whose bits are $a_1, a_2, \ldots, a_n$ is a function F where

$$F(a_1, a_2, \ldots, a_n) = F(1 - a_1, 1 - a_2, \ldots, 1 - a_n)$$

How many self-dual boolean functions of n variables are there?

7. In Section 2.5.3, we discussed formulas in CNF. A CNF formula, such as

$$(x_{13} \vee x_{17} \vee \neg x_{23}) \wedge (\neg x_2 \vee x_5) \wedge (x_2 \vee \neg x_{13} \vee \neg x_{17})$$

is a formula that is composed of a conjunction ($\wedge$) of disjunctions ($\vee$) of literals (proposition letters x_i and their negations $\neg x_i$). Note that if we assume a $\vee$ applies to as few literals as possible, reversing normal operator precedence rules, we can unambiguously drop parentheses—for example,

$$x_{13} \vee x_{17} \vee \neg x_{23} \wedge \neg x_2 \vee x_5 \wedge x_2 \vee \neg x_{13} \vee \neg x_{17}$$

The CNF formulas are very convenient in computer representations of logic, but they also tend to get exponentially long. Show that for any fixed k, for large enough n, there are boolean functions of n variables that are not expressed by any CNF formula with less than or equal to n^k literals. (Actually, allowing any mixing of $\wedge$'s and $\vee$'s would still not solve the exponential length problem.) (*Hint:* Count the number of CNF formulas with less than or equal to n^k literals. In fact, this overcounts the number of n-ary boolean functions expressible by such CNF formulas, since several CNF formulas may express the same boolean function—that is, be logically equivalent).

8. A system encodes integers as strings of 0's and 1's, with no two consecutive 0's. (Two 0's are used to indicate the end of one number and the start of the next.) Approximate, as best you can, how many different integers can be represented in no more than 100 digits. Compute the maximum error of your approximation—and prove that your error assertion is correct.

CHAPTER 8

Discrete Probability

Expressions of chance and probability are common in everyday language and thinking. We speak of the chance of rain, the probability of getting heads when flipping a coin, and the chance of surviving a disease. We may wonder which of several moves in a board game is most likely to succeed or whether responding to a magazine's contest is worth the price of a stamp. This chapter makes these ideas mathematically precise so that we can reason about them and compute with them.

The chapter contains discussion of five major topics sections. The first gives the formal definition of a probability density function and shows how to use the frequency of occurrence for an outcome in defining a probability density function. The probability of unions and intersections of events are also discussed. The second introduces the idea of interpreting events in sample spaces as events in the product of these sample spaces. The important application to Bernoulli trial processes is discussed, and the relationship between the probabilities on individual sample spaces and the probability in the cross product of sample spaces is explored. The third deals with the important ideas of independent events and conditional probabilities. Independence tells when the probabilities of two events do not affect each other. Conditional probability gives insight regarding how probabilities can change if one knows that some outcome has occurred. For example, asking the probability of flipping a head using two fair coins is a question whose answer may change if new information is received (for example, the first coin landed tails). The last two topics deal with random variables, which are in fact functions defined on a sample space that is endowed with a probability density function. We see how a probability density function can be associated with a random variable, and we ask questions about what values can be expected from a large number of repetitions of an experiment and how likely it is that these values deviate very much from their average.

8.1   Ideas of Chance in Computer Science

Ideas of **chance** arise in computer science in at least two important and related ways. First, many systems (for example, the stock market, traffic flow in a network, ecological systems) behave in complex ways that have seemingly random aspects. Probability theory is used to create models for the computer simulation of such systems. Second, probability theory is used to design and to analyze the performance of algorithms that are run on computers. As just one example, suppose that a certain algorithm for alphabetizing a list of n names performs $c \cdot n^2$ comparisons if the input to the program gives the names in reverse order but only $c \cdot n$ comparisons if the input happens to give the names in order. Since the performance differs greatly in these two situations, we might ask how many comparisons are required on average, assuming that each of the $n!$ permutations of the names is equally likely to occur. If it is not easy to compute the average number of comparisons required, then we might try the algorithm on several randomly chosen permutations of n names to get an estimate of the algorithm's average performance. To do this, we would want to know how to generate permutations at random using a computer.

Many of the examples in this chapter will be about dice, cards, and coin-flipping experiments, because these provide simple and interesting ways to illustrate the basic concepts of probability. Familiarity with basic probability concepts applied to these situations will enable the reader to extend these ideas to other contexts, because they also arise in computing courses, such as Artificial Intelligence, Algorithm Design and Analysis, Software Engineering, Graphics, Networks, and Simulation.

Statements about probability and chance have many possible interpretations. For example, what do we mean when we say that we have a 50–50 chance of getting heads when we flip a coin? We may have in mind that there is no reason for the experiment to favor either outcome, so we assign equal likelihood to the two outcomes and arrive at a 50% chance for each. Perhaps we imagine that if we continue flipping the coin, we would get heads about 50% of the time. (Nevertheless, we would be surprised to get heads *exactly* 500 times after 1000 repetitions of the experiment.) This so-called frequency interpretation breaks down, however, when we consider an experiment that is not repeatable. If we say that a certain daredevil has a 50% chance of surviving a plunge over Niagara Falls in a barrel, do we mean that survival would be the outcome about half the time if the experiment could be repeated many times under exactly the same conditions? And what do we mean by "about half the time" and "exactly the same conditions"? Perhaps we mean that we would be willing to bet "even money" on the daredevil—that is, we would stand to lose or gain the same amount—depending on the outcome.

Probability theory has a philosophical and interpretive aspect that is not found in most other branches of mathematics. No one has been able to give a single, precise definition of probability that everyone agrees on, but it is possible to give mathematical rules for combining probabilities once they are assigned. Everyone can agree on these rules, and they form the basis for the mathematical subject of probability theory. One of the challenges of this subject is to be alert to the difference between the following: making mathematical deductions based on the axioms of probability, and reasoning based on common sense and everyday experience.

8.1.1   Introductory Examples

To begin the mathematical treatment of probability theory, we introduce some basic terminology informally. Let us start by considering a familiar situation. Suppose that we roll a pair of dice, A and B, and that we are interested in our chances of getting the sum of the **spots** (also called **pips**) on the top faces to be a certain one of the values 2 through 12. This is a typical instance of the following situation: There is a process (rolling the dice) with a set of possible results that we can enumerate. However, the result that will actually occur in any given execution of the process is not known in advance. A single execution of such a process is called an **experiment,** and each possible result is called an **outcome.** The set of all possible outcomes is called the **sample space.**

The symbol ω will denote an outcome, and Ω will denote a sample space. For $\omega \in \Omega$, we say that ω is an outcome in sample space Ω to emphasize the probability context. Of course, we can also say that ω is an element of the set Ω.

In the dice-rolling situation just described, each roll of the dice is an experiment. What are the outcomes, however, and what is the sample space? Here, we have a choice to make. On the one hand, since it is the sum of the spots that is of interest, each possible sum can be regarded as an outcome, giving a sample space

$$\Omega_1 = \{2, 3, 4, 5, 6, 7, 8, 9, 10, 11, 12\}$$

On the other hand, the outcome of a particular experiment (roll) can be regarded as an ordered pair (i, j) where i is the number of spots shown by die A and j is the number shown by die B. (We sometimes call the number of spots or pips on the top face of the die the **value** of the die.) From this point of view, the sample space is

$$\Omega_2 = \{(1, 1), (1, 2), \ldots, (6, 5), (6, 6)\}$$

which has 36 ordered pairs as elements or outcomes. Each of these sample spaces is legitimate, as are others. The choice of sample space is made according to what fits best with the questions we want to ask. The only requirement is the following:

• Whenever an experiment is run, there is exactly one outcome $\omega \in \Omega$ to describe what happens.

Also, it is usually pointless to include impossible situations in Ω, leading to a second requirement:

• There are no impossible outcomes in Ω.

Once we have decided on a sample space Ω, we next assign a **probability** $p(\omega)$ to each outcome $\omega \in \Omega$. This is done by choosing a value for $p(\omega)$ between 0 and 1 inclusive, subject only to the requirement that all the probabilities sum to 1:

$$\sum_{\omega \in \Omega} p(\omega) = 1$$

This requirement expresses the idea that we are 100% certain that exactly one outcome will occur when we roll the dice.

In the example, if we choose the sample space to be

$$\Omega_2 = \{(1, 1), (1, 2), \ldots, (5, 6), (6, 6)\}$$

then it is natural to choose

$$p(1, 1) = p(1, 2) = \cdots = p(5, 6) = p(6, 6) = \frac{1}{36}$$

because we see no reason to favor one outcome over another. It would not be wrong *mathematically* to choose other probabilities, but this would not agree with our experience. Since we believe that the 36 outcomes are equally likely, we choose $p(\omega)$ to be $1/36$ for each $\omega \in \Omega_2$.

The important point here is that mathematics does not tell us exactly how to assign probabilities, nor does it provide us with a precise definition of "experiment" or tell us what sample space to choose. In fact, mathematics does not even offer a definition for "probability of an outcome" in terms of our intuitive ideas about likelihood of occurrence of uncertain phenomena. It is entirely up to us to supply the meanings and interpretations, to define the sample spaces, and to make the probability assignments. This we do on the basis of what seems to be reasonable to us. Once we have done this, mathematics can help us to determine the logical consequences of our assumptions.

8.1.2   Basic Definitions

Here we give the mathematical definitions of the terms we have been using. The definitions make no reference to the notions of likelihood, randomness, and uncertainty that we all have in mind when we talk about probability.

Definition 1.   A **discrete sample space** is a nonempty set that has only a finite or countably infinite number of elements.

The word *discrete* in the definition refers to the fact that the set has only a finite or countably infinite number of elements. Although it is meaningful to consider processes having more than countably many outcomes, all the sample spaces in this chapter will be discrete. Hence, from now on, we will usually just say "sample space" instead of "discrete sample space." Definition 1 does not say that a sample space must be the set of possible outcomes of an experiment, but this is the interpretation we will have in mind when we elect to call a set a sample space.

Definition 2.   An **outcome** is an element of a sample space.

Definition 3.   An **event** is a subset of a sample space.

Sample spaces will be denoted by Ω's and outcomes by ω's. Events will generally be denoted by E's, although other capital letters will sometimes be used. According to Definition 3, both the entire sample space Ω and the empty set $\emptyset$ are events. An event E can also consist of a single outcome $\omega \in \Omega$, in which case we write $E = \{\omega\}$.

Example 1.   Suppose we take

$$\{\text{Saturday, Sunday, Monday, Wednesday, Friday}\}$$

as a sample space Ω. Then, Wednesday is an outcome, but Thursday is not. Wednesday can also be regarded as the event $E = \{\text{Wednesday}\}$. The weekend can also be regarded as an event in Ω, since

$$\{\text{Saturday, Sunday}\} \subset \Omega$$

(Watch out for the notation: {Wednesday} is a **subset** of Ω and denotes an event, whereas Wednesday is an **element** of Ω and denotes an outcome.) ■

Suppose a probability experiment with sample space Ω is executed. Then, an event $E \subseteq \Omega$ is said to **occur** if the outcome ω of the experiment belongs to E.

Example 2.   Suppose $\Omega = \{0, 1\}$. Then, 0 and 1 are the outcomes. Here, 0 might represent the outcome that a communication line is busy and 1 that the line is free. The experiment might be to check the status of the line. Four events are associated with Ω—namely, $\emptyset$, $\{0\}$, $\{1\}$, and $\{0, 1\}$. The event $E = \{0, 1\}$ always occurs, because the line is always either busy or free.   ■

Example 3.   Suppose Ω is the set of positive integers. Then, each positive integer is an outcome. This sample space might be used when the experiment is to flip a coin until it comes up heads. The outcome is the number of flips. The set

$$E = \{2n : n \text{ is a positive integer}\}$$

expresses the event of an even number of flips.   ■

Definition 4.   A **probability density** p on a discrete sample space Ω is a function with domain Ω satisfying

$$\text{For each } \omega \in \Omega, 0 \le p(\omega) \le 1, \text{ and } \sum_{\omega \in \Omega} p(\omega) = 1.$$

Any function satisfying these properties is, mathematically speaking, a legitimate **probability density function.** The value of the probability density function on an outcome is the **probability of the outcome.**

In the dice-rolling experiment, suppose that the sample space Ω is chosen to be

$$\Omega_1 = \{2, 3, 4, 5, 6, 7, 8, 9, 10, 11, 12\}$$

Then, the following function on Ω_1 is a legitimate probability density function:

$$p(2) = p(12) = \frac{1}{36}$$

$$p(3) = p(11) = \frac{2}{36}$$

$$p(4) = p(10) = \frac{3}{36}$$

$$p(5) = p(19) = \frac{4}{36}$$

$$p(6) = p(18) = \frac{5}{36}$$

$$p(7) = \frac{6}{36}$$

Clearly, $p(\omega) \ge 0$ for each outcome, and the sum of the probability density function over all the outcomes is 1. Hence, the two requirements of Definition 4 are satisfied.

Definition 5.   Let E be an event in a sample space Ω endowed with a probability density function $p(\omega)$. If $E \ne \emptyset$, the probability $P(E)$ of event E is

$$P(E) = \sum_{\omega \in E} p(\omega)$$

If $E = \emptyset$, there are no terms in the sum, and $P(\emptyset)$ is 0.

Example 4.   For the sample space $\{1, 2, 3, 4, 5, 6\}$ that represents the outcomes of record-ing the number of pips on the top face after rolling a fair die, define

$$p(1) = p(2) = p(3) = p(4) = p(5) = p(6) = \frac{1}{6}$$

Determine the probability that the top face shows an even number of pips, and determine the probability that the number of pips on the top face is greater than four.

Solution.   The event that the top face shows an even number of pips is $E_1 = \{2, 4, 6\}$, and its probability is $P(E_1) = p(2) + p(4) + p(6) = 1/2$. The event that the top face shows more than four pips is $E_2 = \{5, 6\}$, and its probability is $P(E_2) = p(5) + p(6) = 1/3$. ■

Note that for any event E, we have $0 \leq P(E) \leq 1$. Also, note that $P(\Omega) = 1$, and that for a singleton event $E = \{\omega\}$, we have $P(E) = p(\omega)$. (Why?)

8.1.3   Frequency Interpretation of Probability

In the preceding discussion, we defined the probability of an event, including that of a sin-gleton event. The probability was defined in terms of a probability density function that can be chosen arbitrarily, subject only to the two conditions in Definition 4 in Section 8.1.2. The usefulness in practice of computations based on the definitions depends on how well the chosen probability density function models the situation of interest. When we can imagine doing a probabilistic experiment over and over, we generally choose $p(\omega)$ to estimate the proportion of times that we think outcome ω will occur. When we compute the probability of an event E, we generally interpret it as an estimate of the proportion of times that the event should occur. This is called the **frequency interpretation** of probability.

The Frequency Interpretation of Probability

The **frequency interpretation** of probability is to take the quantity $P(E)$ as an es-timate for the proportion of times that event E will occur when an experiment is repeated over and over. The reasonableness of the estimate depends on how well the probability density function estimates the frequencies of the outcomes.

8.1.4   Introductory Example Reconsidered

To illustrate the difference between mathematical requirements and personal choices, let us reconsider the sample spaces Ω_1 and Ω_2 from the dice-throwing example in light of the definitions of the preceding subsection. In the 36-element sample space

$$\Omega_2 = \{(1, 1), (1, 2), \ldots, (6, 6)\}$$

obtaining a sum of 3 is described by the event $E = \{(1, 2), (2, 1)\}$, which breaks the situ-ation into its smallest subcases, showing exactly how the 3 can be obtained. However, for the sample space

$$\Omega_1 = \{2, 3, \ldots, 12\}$$

obtaining a sum of 3 is both an outcome and a one-element event $E = \{3\}$.

Suppose, now, that we want to represent the situation of obtaining at least one 1 on a die. This can be expressed as the 11-element event

$$E = \{(1, 1), (1, 2), \ldots, (1, 6), (2, 1), (3, 1), \ldots, (6, 1)\}$$

in Ω_2 and cannot be expressed at all as an event in Ω_1. Hence, choosing a very detailed sample space makes it possible to represent more situations as events.

What about the probability of obtaining a sum of 3? First, we must choose probability density functions p_1 and p_2 for Ω_1 and Ω_2, respectively. For Ω_2, we are comfortable with assigning $p_2(\omega) = 1/36$ for all outcomes ω. Definition 5 then leads us to calculate

$$P(sum = 3) = P(\{(1, 2), (2, 1)\}) = p_2(1, 2) + p_2(2, 1) = (2) \cdot \left(\frac{1}{36}\right) = \frac{1}{18}$$

which seems to agree with experience. If we choose p_1 by assigning the same probability density to each of the 11 outcomes in Ω_1, we get

$$P(sum = 3) = P(\{3\}) = \sum_{\omega \in \{3\}} p_1(\omega) = p_1(3) = \frac{1}{11}$$

This is not mathematically incorrect, just out of line with the experience of people who roll dice often. To repair things requires assigning a different probability density to Ω_1. What probability is obtained for this event if the probability density function following Definition 4 (in Section 8.1.2) is chosen for p_1?

What about the probability of having at least one of the two dice show one pip on the top face after a roll? Using the probability density p_2 defined on Ω_2, we compute the probability of the event

$$E = \{(i, j) : i = 1, \text{ and } 1 \leq j \leq 6\} \cup \{(i, j) : 1 \leq i \leq 6, \text{ and } j = 1\} \subseteq \Omega_2$$

to be

$$P(E) = \sum_{\omega \in E} p_2(\omega) = (11)\left(\frac{1}{36}\right) = \frac{11}{36}$$

With Ω_1, we are stuck. The situation cannot be described as an event in Ω_1, so we cannot compute a probability for it, no matter what we choose for p_1.

As this discussion illustrates, it is important to define a sample space with elements that are versatile building blocks for the events of interest. One way to do this is to choose outcomes that do not themselves decompose into subcases. In other words, the outcomes should be the most basic, elemental situations that can occur.

Terminology Summary

- A sample space Ω is a set.
- An outcome ω is an element of a sample space: $\omega \in \Omega$.
- An event E is a subset of a sample space: $E \subseteq \Omega$.
- The probability $P(E)$ of an event E is the sum of the values of the probability density function for the outcomes in the event.

> ### How to Calculate the Probability of Events
>
> 1. Describe in words the experiment and the event or events of interest.
> 2. Choose a sample space Ω so that the events of interest are easy to describe in Ω.
> 3. Define a probability density function p on Ω.
> 4. Formulate the events of interest as subsets of Ω, and calculate their probabilities by summing the probability density function on the outcomes of the events.

8.1.5   The Combinatorics of Uniform Probability Density

Choosing a sample space with outcomes that do not decompose into more detailed subcases makes it easy to express situations of interest as events in the sample space. Choosing such a sample space has the further advantage that it is often easy to define an appropriate probability density function on it. Whenever we have no reason to believe that one basic situation is any more or less likely to occur than any other, we set $p(\omega) = 1/|\Omega|$ for each $\omega \in \Omega$ provided that $|\Omega|$ is finite. (Recall that putting vertical lines on each side of the symbol for a finite set denotes the number of elements in that set.)

Definition 6.   A probability density function p such that $p(\omega) = 1/|\Omega|$ for all ω in a finite sample space Ω is called a **uniform probability density function.**

For the standard deck of cards, it is useful to think of the cards as being represented by the values 1, 2, 3, ..., 51, 52. Usually, we do not need to be concerned about which card is represented by which value. We know, for example, that 26 of these values represent red cards (hearts and diamonds) and that four of these values represent each card value. We also assume that the deck is fair—that is, no card is more likely than any other to be chosen in a random pick.

Example 5.   Define a uniform probability density function on the standard deck of cards. Determine the probability of drawing one of the 3's and the probability of drawing a face card (Jack, Queen, or King).

Solution.   For each card in the deck, $p(card) = 1/52$. The event E_1 (that the card is a 3) consists of four elements—3 of Clubs, 3 of Diamonds, 3 of Hearts, and 3 of Spades—so $P(E_1) = 1/13$. The event E_2 (that the card is a face card) is a set consisting of 12 cards, so $P(E_2) = 12/52 = 3/13$.   ∎

When a finite sample space Ω is assigned a uniform probability density p, the combinatorial counting methods of Chapter 7 can provide shortcuts to computing the probability of an event E. Since

$$P(E) = \sum_{\omega \in E} p(\omega) = \frac{|E|}{|\Omega|}$$

when Ω is finite and p is uniform, evaluating probabilities in this case is a matter of counting set sizes.

The following examples illustrate the use of counting techniques to evaluate probabilities. Throughout this chapter, unless stated otherwise, all coins, dice, and decks of cards are assumed to be **fair,** meaning that each card of a deck, face of a coin, or side of a die is equally likely to occur.

Notation.   Since we will be using the counting techniques of Chapter 7, now is a good time to recall that $C(m, n)$, denotes the binomial coefficient—that is, the number of ways of choosing n elements from a set of m elements. See Section 7.9.1 for more details.

Example 6.   A (fair) coin is tossed three times. What is the probability that at least two heads turn up?

Solution.   The experiment is to toss a coin three times. The event E of interest is that either two or three heads turn up. To make this event easy to describe, we choose a sample space Ω with outcomes ω that are ordered triples (H representing heads and T representing tails) describing what happens on each toss:

$$
\begin{aligned}
\omega_1 &= (H, H, H) & \omega_5 &= (H, T, T) \\
\omega_2 &= (H, H, T) & \omega_6 &= (T, H, T) \\
\omega_3 &= (H, T, H) & \omega_7 &= (T, T, H) \\
\omega_4 &= (T, H, H) & \omega_8 &= (T, T, T)
\end{aligned}
$$

Hence, $\Omega = \{\omega_1, \omega_2, \ldots, \omega_8\}$ and $E = \{\omega_1, \omega_2, \omega_3, \omega_4\}$.

Since the coin is fair, we choose a uniform probability density function for Ω. This assigns a probability of $p(\omega) = 1/|\Omega| = 1/8$ to each outcome ω_i for $1 \le i \le 8$. Hence, by Definition 5 in Section 8.1.2, $P(E) = |E|(1/8) = 4/8 = 1/2$.   ■

Example 7.   A coin is tossed 10 times. What is the probability that eight or more heads turn up?

Solution.   Choosing $\Omega = \{(f_1, f_2, \ldots, f_{10}) : f_i \in \{H, T\}\}$ gives a sample space with outcomes that can describe every possible result of flipping the coin 10 times. Since the coin is fair, the outcomes can be assumed to be equally likely. Since $|\Omega| = 2^{10}$, setting $p(\omega) = 2^{-10}$ for each $\omega \in \Omega$ defines a uniform probability density function on Ω. Rather than enumerate all the elements of Ω, we use counting techniques from combinatorics to compute the size of the set E that describes the situation "eight or more heads." This set E is the **disjoint union** of three other events—namely, getting exactly 8 heads, exactly 9 heads, and exactly 10 heads. These events are given by sets of size $C(10, 8) = 45$, $C(10, 9) = 10$, and $C(10, 10) = 1$, respectively. Consequently, $P(E) = (2^{-10})(45 + 10 + 1) = 56 \cdot 2^{-10}$.   ■

Example 8.   A die is rolled five times. What is the probability of obtaining exactly one 2?

Solution.   We choose the sample space with outcomes that are all the length of five sequences of die values. Each single roll results in one of six numbers, so there are 6^5 possible outcomes in Ω. Putting a uniform probability density function on Ω gives $p(\omega) = 6^{-5}$ for each sequence ω. We can count the number of length-five sequences with exactly one 2 using the Multiplication Principle. First, choose a location for the 2. This can be done in $C(5, 1)$ ways. For the other four locations, we can fill each in five ways using all the values that can occur on the top face after the roll of a die, except for 2. Therefore, the total number of such sequences is $C(5, 1) \cdot 5^4$. It follows that the probability of obtaining exactly one 2 is $C(5, 1)(5^4)(6^{-5})$.   ■

Example 9. Suppose the analysis of a sorting algorithm shows that the worst case (the one requiring the most comparisons of names) occurs when the input data lists the names in reverse order. What is the probability that this occurs if there are n names? Assume that all orderings of the names are equally likely.

Solution. There are $n!$ permutations of the names. These make up a sample space Ω. Assuming that each permutation is equally likely to occur implies that the probability of reverse order is just $(1/n!)$. ∎

Example 10. Suppose that incoming computer mail messages are equally likely to be addressed to user 1, 2, or 3. Three messages are received for delivery to these users. What is the probability that no two messages are addressed to the same person?

Solution. The sample space should reveal all the ways the messages can be addressed. This can be done by choosing

$$\Omega = \{(u_1, u_2, u_3) : 1 \leq u_1, u_2, u_3 \leq 3\}$$

Here, u_i is the user to whom message i is addressed. The sample space consists of 3^3 outcomes, which we assume are equally likely. The event E (that no two messages go to the same user) consists of all 3-tuples that are permutations of $(1, 2, 3)$. There are $3!$ of these, so $P(E) = (3!)(3^{-3}) = 2/9$. ∎

Example 11. Suppose in the previous example that there are three messages and five users. What is the probability that no two messages go to the same person?

Solution. Now, Ω has size 5^3. However, E no longer consists of permutations of $(1, 2, 3)$. Instead, it consists of permutations of the elements in sets of the form $\{u_i, u_j, u_k\}$ where u_i, u_j, u_k are three distinct integers chosen from $\{1, 2, 3, 4, 5\}$. There are $C(5, 3)$ choices for a set $\{u_i, u_j, u_k\}$, and there are $3!$ ways to permute the elements of a given three-element set. Therefore, $P(E) = C(5, 3)(3!)(5^{-3}) = 12/25$. ∎

Event Probabilities Under Uniform Density

To calculate the probability of an event E when the sample space Ω has been assigned a uniform probability density function:

1. Use counting techniques to determine $|E|$ and $|\Omega|$.
2. Compute $P(E) = |E|/|\Omega|$.

8.1.6   Set Theory and the Probability of Events

The sample spaces in Examples 6 through 11 were chosen to make it easy to formulate the situations of interest as events—that is, as subsets of the sample spaces. The combinatorial counting techniques of Chapter 7 were then used to determine the number of elements in those subsets. For finite sample spaces with uniform probability densities, this essentially determines the event probabilities. This section shows that the operations on sets introduced in Section 1.3 also help in computing the probability of events (with or without uniform probability densities).

As a simple example, suppose we know that an event E in a sample space Ω is the disjoint union of two other events A and B. To find the probability of the event E, we sum up the probabilities of the individual outcomes in E. This can be done by summing the probabilities of outcomes in A (which gives $P(A)$), summing the probabilities of outcomes in B (which gives $P(B)$), and then adding the two totals together (which gives $P(A) + P(B)$). No double counting occurs, because A and B are disjoint. Each outcome in E contributes its probability, since it belongs either to A or to B. In other words, if $A \cap B = \emptyset$, then

$$P(A \cup B) = \sum_{\omega \in A \cup B} p(\omega)$$

$$= \sum_{\omega \in A} p(\omega) + \sum_{\omega \in B} p(\omega)$$

$$= P(A) + P(B)$$

This observation extends to collections of more than two mutually disjoint sets and gives the Additive Principle of Disjoint Events.

Additive Principle of Disjoint Events

If $E_1, E_2, \ldots, E_n$ are pairwise disjoint subsets of a sample space Ω, then

$$P(\cup_{1 \leq i \leq n} E_i) = \sum_{1 \leq i \leq n} P(E_i)$$

In other words, the probability of a union of pairwise disjoint events is the sum of their probabilities. This statement remains valid for countably infinite collections of pairwise disjoint subsets of a countably infinite sample space.

We now use this principle to obtain several useful relationships among the probabilities of events.

Theorem 1. (Elementary Probability Facts) Let Ω be a sample space endowed with a probability density p, and let E and F be subsets of Ω. Then:

(a) $E \subseteq F$ implies $P(E) \leq P(F)$.
(b) $P(E) = P(E \cap F) + P(E \cap \overline{F})$ where $\overline{F} = \Omega - F$.
(c) $P(E) = \sum_i P(E \cap A_i)$ where $\Omega = \cup_{c=1}^{n} A_i$ for some $n \in \mathbb{N}$ and the set of A_i's form a partition of Ω.
(d) $P(E \cup F) = P(E) + P(F) - P(E \cap F)$.

Proof.

(a) Since $E \subseteq F$, set F can be written as the disjoint union of E and $(F - E)$. Hence, by the Additive Principle of Disjoint Events,

$$P(F) = P(E) + P(F - E)$$

Since probabilities are nonnegative, $P(E) \leq P(F)$.

(b) Since E is the disjoint union of $(E \cap F)$ and $(E \cap \overline{F})$, the Additive Principle implies statement (b).

(c) Since the A_i's form a partition of Ω, set E can be written as the union of the pairwise disjoint sets $(E \cap A_i)$. Hence, statement (c) follows from the Additive Principle.

(d) Expressing events E, F, and $(E \cup F)$ as unions of disjoint events and applying the Additive Principle gives

$$P(E \cup F) = P(E - F) + P(F - E) + P(E \cap F)$$
$$P(E) = P(E - F) + P(E \cap F)$$
$$P(F) = P(F - E) + P(E \cap F)$$

Adding together the expressions for $P(E)$ and $P(F)$ gives

$$P(E) + P(F) = P(E - F) + P(F - E) + 2P(E \cap F)$$

Comparison with the expression for $P(E \cup F)$ shows that

$$P(E) + P(F) = P(E \cup F) + P(E \cap F)$$

Statement (d) is just a rearrangement of this last equation. ∎

The proof of Theorem 1(d) is identical in form to the proof of the theorem found in Section 1.5.2. In particular, the idea of expressing a set as a union of other, pairwise disjoint sets is used in both places. The difference is that in Chapter 1, the numbers of elements in the sets are counted, whereas here, the probabilities of the elements are totaled. Thus, in Chapter 1, each element contributes a 1 to the total, whereas here, each element contributes its probability to the total.

The next example shows the usefulness of expressing an event as a disjoint union of other events.

Example 12.   What is the probability that a card drawn at random from a 52-card deck will be an Ace or a spade?

Solution.   We take Ω to be the 52-element set of cards and model drawing a card at random by assigning a uniform probability density. The event E of getting an Ace or a spade can be written as $E = A \cup S$ where A is the set of four Aces and S is the set of 13 spades. The intersection $A \cap S = \{$the Ace of Spades$\}$. Hence, by Theorem 1(d),

$$P(E) = P(A) + P(S) - P(A \cap S)$$
$$= \frac{4}{52} + \frac{13}{52} - \frac{1}{52}$$
$$= \frac{4}{13}$$ ∎

Sometimes it is easy to evaluate the probability that an event does not occur. This immediately gives the probability that it does occur, as the next result shows.

Theorem 2.   (Probability of the Complement)   Suppose E is an event in a sample space Ω, and let $\overline{E} = \Omega - E$. Then,

$$P(E) = 1 - P(\overline{E})$$

Proof. Since $\Omega = E \cup \overline{E}$ expresses Ω as a union of disjoint sets,

$$P(\Omega) = P(E) + P(\overline{E})$$

Since $P(\Omega) = 1$, $P(\overline{E}) = 1 - P(E)$. ∎

Example 13. Two parts are chosen at random from a bin containing 10 parts, three of which are defective. What is the probability that at least one of the parts chosen is good?

Solution. Consider the sample space Ω to be the set of all $C(10, 2)$ pairs of items, and give Ω a uniform probability density. Let E be the set of all pairs of parts in which at least one is good. Then, $\overline{E} = \Omega - E$ is the set of all pairs of parts in which both are bad. There are $C(3, 2)$ pairs in $\overline{E}$, so $P(\overline{E}) = C(3, 2)/C(10, 2)$. Hence,

$$P(E) = 1 - \left(\frac{C(3, 2)}{C(10, 2)} \right) = 1 - \left(\frac{3}{45} \right) = \frac{14}{15}$$ ∎

Of course, $P(E)$ could also be computed directly: E can be written as the disjoint union of A, the set of all pairs containing two good items, and B, the set of all pairs containing exactly one good item. Set A contains $C(7, 2)$ pairs, and set B contains $7 \cdot 3$ pairs. Hence,

$$P(E) = \frac{(C(7, 2) + 21)}{C(10, 2)} = \frac{(21 + 21)}{45} = \frac{14}{15}$$

In the preceding example, finding $1 - P(\overline{E})$ provided an alternative to computing $P(E)$ directly. In the next example (a classic), this idea is extremely useful.

Example 14. (The Birthday Problem) What is the probability that in a group of n people, at least two have the same birthday? (Leap years are ignored, and all combinations of birthdays are assumed to be equally likely.)

Solution. Represent the birthdays of the group by an n-tuple with components that are integers in the range 1 through 365. Take Ω to be the 365^n possible n-tuples, and put the uniform probability density on Ω. If E is the event that at least two people have the same birthday, then $\overline{E} = \Omega - E$ is the event that no two people have the same birthday. Hence, $\overline{E}$ consists of all possible n-tuples of distinct birthdays.

If $n > 365$, there are no such n-tuples. Then, $\overline{E} = \emptyset$, $P(\overline{E}) = 0$, and $P(E) = 1$.

If $n \leq 365$, there are $C(365, n)$ ways to choose n distinct birthdays and $n!$ ways to assign the chosen birthdays to the n people. Hence, $\overline{E}$ contains $C(365, n) \cdot (n!)$ n-tuples, and

$$P(E) = 1 - P(\overline{E})$$
$$= 1 - \frac{C(365, n) \cdot (n!)}{365^n}$$

Alternatively, the number of elements in $\overline{E}$ can be counted by imagining that the first person has any of 365 birthdays, the second has any of the 364 remaining possibilities, and so on. Hence, there are $365 \cdot 364 \cdot \cdots \cdot (365 - n + 1)$ n-tuples in $\overline{E}$. In fact, this is just the number of k-permutations of a 365-element set, as defined in Section 7.5.1, where $k = n$ in the present case. The reader can check that this is equal to $C(365, n) \cdot (n!)$. Thus, the two methods of calculating $| \overline{E} |$ give the same result.

Now that we have derived an expression for $P(\overline{E})$, how do we evaluate it? Trying to compute $n!$ or 365^n or $C(365, n)$ for other than the first few values of n involves numbers that are huge. However, this difficulty can be avoided by noting that the expression for $P(\overline{E})$ can be rewritten

$$\left(\frac{365}{365}\right)\left(\frac{364}{365}\right)\left(\frac{363}{365}\right)\cdots\left(\frac{365-n+1}{365}\right)$$

Using a calculator, it is easy to compute that for $n \geq 23$, $P(\overline{E}) < 0.5$ and $P(E) > 0.5$. In other words, if at least 23 students are in a class, then the chances are greater than 50% that some two or more of them have the same birthday. ∎

These examples have illustrated two ideas that aid in computing the probabilities of events, which we now highlight.

Aids to Computing the Probability of Events

- Set theory can be used to rewrite events in terms of other events. The Additive Principle of Disjoint Events and Theorem 1 in Section 8.1.6 apply this idea.
- The probability of the complement $\overline{E}$ of an event E can be used to determine the probability of E, since by Theorem 2 in Section 8.1.6,

$$P(E) = 1 - P(\overline{E})$$

8.2   Exercises

1. Suppose that Ω is a sample space with a probability density function p and that $E \subseteq \Omega$.
   (a) Prove that $0 \leq p(E) \leq 1$.
   (b) What is the probability of the singleton event $E = \{\omega\}$?
   (c) What is the probability of the event $E = \Omega$?

2. Suppose that sample space Ω_1 is chosen to model the experiment of rolling a pair of dice and that the probability density function p assigned to Ω_1 is $p(\omega) = 1/36$ for $\omega \in \Omega_1$. Under these assumptions, compute the probability of rolling a sum of 3. Compare your answer to the answers of $1/18$ and $1/11$ obtained in the text, and discuss.

3. Consider a sample space Ω consisting of five outcomes:

$$\{\omega_1, \omega_2, \omega_3, \omega_4, \omega_5\}$$

Which (if either) of the following functions p_1 and p_2 can be probability densities on Ω? Explain your answer.

ω	$p_1(\omega)$	$p_2(\omega)$
ω_1	$-1/2$	$1/3\sqrt{7}$
ω_2	$1/4$	$2/3\sqrt{7}$
ω_3	$1/4$	$1/3\sqrt{7}$
ω_4	$1/2$	$2/3\sqrt{7}$
ω_5	$1/2$	$(\sqrt{7}-2)/\sqrt{7}$

4. Two nickels and a dime are shaken together and thrown. All the coins are fair. We are allowed to keep the coins that turn up heads. Give two sample spaces together with probability density functions that reasonably describe this situation. Explain your answer.

5. A penny, a nickel, and a dime are shaken together and thrown. Someone proposes the sample space $\Omega = \{1, 5, 10\}$, explaining that the outcomes represent the values of the coins. Then, the person suggests the probability density function $p(1) = p(5) = p(10) = 1/2$, explaining that since the coins are fair, a uniform probability density is appropriate. Comment on several aspects of this situation. Do you agree with this model?

6. A penny, a nickel, and a dime are shaken together and thrown. Suppose that the nickel turns up heads twice as frequently as the penny and that the dime turns up heads half as frequently as the nickel. Define a sample space, and using the frequency interpretation, assign a reasonable probability density function based on the assumption that the penny is a fair coin.

7. Given the sample space

$$\Omega = \{0, 5, 10, 15, 20\}$$

which of the following events are in the sample space?

(a) $\{5, 10\}$
(b) $\{0, 5, 10, 15, 20\}$
(c) $\emptyset$
(d) 0
(e) $\{0\}$
(f) $\{5\}$

8. Two nickels and a dime are shaken together and thrown. We are allowed to keep the coins that turn up heads. We choose a sample space $\Omega = \{0, 5, 10, 15, 20\}$, the outcomes of which correspond to the amounts that we can keep. For each of the following situations, either describe the situation as an event in Ω by listing the elements in the appropriate subset of Ω or state that the situation cannot be described as an event in this particular sample space:

(a) No heads.
(b) All heads.
(c) Exactly one coin turns up heads.
(d) Exactly one of the nickels turns up heads.
(e) The dime turns up heads.

9. For the two nickels and the dime in Exercise 8, there are eight possible combinations of heads and tails: tails on all coins; heads on nickel 1, tails on nickel 2, tails on the dime; and so on.

   (a) Assuming that each of these eight combinations is equally likely, what probability density should be assigned to the sample space Ω of Exercise 8? Specify the probability density by giving its value on each $\omega \in \Omega$.
   (b) Invent a new sample space Ω such that each of the situations in Exercise 8 can be described as an event, and specify a reasonable probability density for the new sample space.
   (c) Describe each of the situations in Exercise 8 as an event in the new sample space from part (b).
   (d) Calculate the probability of each of the events in part (c) using the new probability density.

10. Consider a sample space $\Omega = \{a, e, i, o, u\}$ endowed with the following probability density: $p(a) = 0.22$, $p(e) = 0.35$, $p(i) = 0.13$, $p(o) = 0.20$, and $p(u) = 0.10$. Determine the probabilities of the following events:

    (a) $\{a, o\}$
    (b) $\emptyset$
    (c) The event E consisting of all those outcomes in Ω that come after the letter k in the alphabet

11. A small zoo records the proportions of visitors who prefer various animals as their favorites. Suppose that the elephants are preferred by 15%, the monkeys by 25%, the polar bears by 30%, the seals by 20%, and the boa constrictors by 10%. Suppose we are going to select a visitor at random and ask what animal that person prefers.

    (a) Set up a sample space Ω, and define a probability density on it using the given data.
    (b) Reformulate the descriptions of the following events as subsets of Ω:
        i.   The preferred animal has four legs
        ii. The preferred animal has legs
        iii. The preferred animal has either a trunk or flippers.
    (c) Calculate the probability of each event in part (b).

12. A fair coin is tossed five times. Determine the probability that:

    (a) It turns up tails every time.
    (b) It turns up heads at most three times.
    (c) It turns up heads twice in a row exactly one time.

13. Suppose that five names are drawn from a hat at random and listed in the order in which they are drawn. A name on the list is "out of order" if its position is not the same as its position after the list has been alphabetized. Determine the probability of each of the following:

    (a) Exactly two names are out of order.
    (b) At least two names are out of order.
    (c) Exactly one name is out of order.

14. What is the probability that a card drawn at random from a 52-card deck will be a heart or an even-numbered card?

15. Suppose A and B are events in a sample space such that $P(A) = 1/4$, $P(B) = 5/8$, and $P(A \cup B) = 3/4$. What is $P(A \cap B)$?

16. In a certain group of people, 50% are right-handed and wear glasses, 5% are left-handed and wear glasses, and 1% are ambidextrous and wear glasses. What is the probability that a person selected at random from this group wears glasses? Assume that *ambidextrous* means neither right-handed nor left-handed but, rather, some mixture of both. In particular, the ambidextrous people are not included in the set of right-handed people or in the set of left-handed people.

17. Four cards are dealt at random from a deck. What is the probability that at least one of them is an Ace? The answer may be given in terms of the combinatorial notation $C(a, b)$.

18. In a fierce battle, not less than 70% of the soldiers lost one eye, not less than 75% lost one ear, not less than 80% lost one hand, and not less than 85% lost one leg. What is the smallest percentage who could have lost simultaneously one ear, one eye, one hand, and one leg? This problem comes from *Tangled Tales* by Lewis Carroll, the author of *Alice in Wonderland*.

19. The waiting room of a dentist's office contains a stack of 10 old magazines. During the course of a morning, four patients, who are waiting during non-overlapping times, select a magazine at random to read. Calculate in two ways the probability that two or more patients select the same magazine.

20. What is the probability that in a group of 10 people, at least 2 have the same birthday? Assume that nobody was born on February 29th. Use a calculator to get a good, approximate answer.

21. Suppose $E_1, E_2, \ldots, E_n$ are events (not necessarily disjoint) in a sample space Ω endowed with a probability density p. Find an expression for $P(\cup_{1 \le i \le n} E_i)$, and prove that your expression is valid. (Hint: Make an analogy to the Principle of Inclusion-Exclusion of Section 1.5, but add up probabilities instead of elements.)

22. Recall that by definition, a discrete sample space may contain a countably infinite number of outcomes. This exercise gives an example of such a **countably infinite sample space.** Suppose we flip a fair coin until it comes up heads. Of course, there is no way to know in advance how many flips will be required. Design a sample space and a probability density to model this situation. Prove that the probability density you define is legitimate.

8.3 Cross Product Sample Spaces

Many probabilistic situations involve repeating an experiment over and over or combining the results of several unrelated experiments. Repeated coin flipping and the Birthday Problem are two examples of such situations. In both cases, it is appropriate to choose a sample space of n-tuples; the ith component, where i ranges from 1 to n, represents the outcome of the ith flip—or the birthday of the ith person. These are both examples of cross product sample spaces, the subject of this section. If the coin is fair, and if the days of the year are equally likely, then we assign a uniform probability density. In many situations, however, it does not make sense to do this (the coin might be biased, for example). This section explains how to assign reasonable probabilities in such situations. We study in detail situations like repeated coin flipping, and we explain how to view events in separate experiments as events in a single experiment.

8.3.1   A Multiplication Principle

Suppose that a printer is out of order (down) 8% of the time and a photocopier is out of order 10% of the time. Checking whether these machines are up or down can be regarded as an experiment with sample space $\Omega = \{(0, 0), (1, 0), (0, 1), (1, 1)\}$, where the first and second positions describe the state of the printer and the copier, respectively, and 0 and 1 denote down and up, respectively. Surely, $(1, 1)$ and $(0, 0)$ are not equally likely outcomes. Hence, we seek another method for assigning a probability density.

We might reason about this situation as follows: The status of the photocopier and the status of the printer have nothing to do with one another (assuming power failures or high office temperatures did not put them both out-of-order at the same time). Therefore, the photocopier could be down 10% of the time that the printer happens to be down and also 10% of the time that the printer happens to be up. Since the printer is down 8% of the time, and 10% of the 8% that the copier is also down, it seems to be intuitively clear that 0.8% of the time, both machines are down. The following example further illustrates this intuitive reasoning. Later, we will prove this method assigns numbers that satisfy the definition of a probability density function.

Example 1.   Suppose that you share a telephone with the other members of your household and that 30% of the time during the day, one of the others is using the telephone. Suppose further that you wish to reach a certain service number. You know that 40% of the time an incoming call to this service number is answered immediately, 35% of the time the call is placed on hold in a queue, and 25% of the time you must call again, because the queue is full and the number is busy. Based on intuitive reasoning, what is your estimate of the probability that you can reach the service number from your home telephone with no delay?

Solution.   Imagine two experiments X_1 and X_2. Experiment X_1 consists of checking the status of the telephone. We choose a sample space Ω_1 with two outcomes, 0 and 1 for busy and free, respectively, and we assign a probability density of $p_1(0) = 0.3$ and $p_1(1) = 0.7$. (The p is subscripted with a 1 to indicate that outcomes of Ω_1 are being considered.) Experiment X_2 consists of checking the status of the service number. Here, we choose $\Omega_2 = \{0, 1, 2\}$ where 0 and 1 denote busy and answered immediately, respectively, and 2 denotes hold. We assign $p_2(0) = 0.25$, $p_2(1) = 0.40$, and $p_2(2) = 0.35$. Define the sample space $\Omega = \Omega_1 \times \Omega_2$:

$$\Omega = \{(0,0), (0, 1), (0, 2), (1, 0), (1, 1), (1, 2)\}$$

Here, we have enumerated all possible pairs of outcomes, one from Ω_1 and one from Ω_2. The outcome from Ω_1 is written as the first element of the pair, and the outcome from Ω_2 is written as the second element. This product is just the product of sets defined in Section 1.3.4.

If the home telephone is free 70% of the time and the status of the service number is not related to the status of the home telephone, then we estimate that 40% of the time that the home phone is free, the service number is answered immediately. Hence, $p(1, 1)$

should be $(0.7)(0.4) = 0.28$. Continuing in this way, we assign to each ordered pair of outcomes the product of the corresponding probability densities. The tree diagram in Figure 8.1 enumerates the outcomes and computes their probabilities.

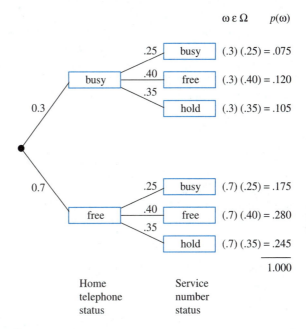

Figure 8.1   Tree diagram.

We have assigned probabilities based on an intuitive line of reasoning. Have we succeeded in defining a legitimate probability density function? The reader can check that we have. ∎

In the previous example, we multiplied the probabilities of outcomes from two unrelated experiments to obtain a probability for their combined outcome. This is an application of what we call the **Probability Multiplication Principle.**

Probability Multiplication Principle

Multiplying together the probabilities of outcomes of unrelated experiments assigns reasonable probabilities to the various combinations of outcomes.

In the Birthday Problem, the assignment of the uniform probability density function to the sample space consisting of 365^n n-tuples can be viewed as a special case of the Probability Multiplication Principle. On the one hand, we can reason as before: Each n-tuple seems to be equally likely, so each should have probability

$$1/|\Omega| = \frac{1}{365^n}$$

On the other hand, we can reason as follows: The probability that a person has a particular birthday is $1/365$, and if the birthdays of different people are unrelated, then according to the Probability Multiplication Principle, the probability of any given sequence of n birthdays is $(1/365)^n$. The two lines of reasoning assign the same probability density function.

In practice, the Probability Multiplication Principle provides good estimates for the frequencies of combined outcomes of unrelated experiments, so naturally, we want to use it to assign probability densities. Before we do, however, we must first check that the probability densities so obtained actually satisfy the two defining properties of a probability density. This check is carried out in Theorem 1.

The **Product of Sums Principle** is a simple observation about algebra that will make the proof of the theorem easier to follow. It is such a useful observation that it is worth receiving special attention.

Product of Sums Principle

The product of k sums is the same as the sum of all possible products of k terms, with each term of a product taken from a different sum.

For example, consider the following product of $k = 2$ sums:

$$(2 + 3 + 5)(4 + 9) = (2 \cdot 4 + 2 \cdot 9 + 3 \cdot 4 + 3 \cdot 9 + 5 \cdot 4 + 5 \cdot 9)$$

The right-hand side is the sum of all possible products of pairs of terms, one from $(2 + 3 + 5)$ and one from $(4 + 9)$. Similarly,

$$\left(\sum_{i=1}^{n} a_i \right) \left(\sum_{j=1}^{m} b_j \right) = \sum_{1 \le i \le n} \sum_{1 \le j \le m} a_i \cdot b_j$$

The expression on the left-hand side is written in a factored form, whereas the expression on the right-hand side is written in a "multiplied out" form. Both expressions represent the same quantity.

Theorem 1.   (Probability Density on a Cross Product Sample Space)   Let Ω_1 and Ω_2 be sample spaces with probability density functions p_1 and p_2, respectively. Let

$$\Omega = \{(\omega_1, \omega_2) : \omega_1 \in \Omega_1 \text{ and } \omega_2 \in \Omega_2\}$$

For each $\omega = (\omega_1, \omega_2) \in \Omega$, let

$$p(\omega) = p_1(\omega_1) \cdot p_2(\omega_2)$$

Then, p is a probability density function on Ω.

Proof.   Certainly, $p((\omega_1, \omega_2)) \ge 0$ for each $(\omega_1, \omega_2) \in \Omega$, because it is the product of non-negative numbers. Therefore, it remains only to verify that

$$\sum_{\omega \in \Omega} p(\omega) = 1$$

where ω ranges over the outcomes (ω_1, ω_2) in Ω. By definition of p, the left-hand side is

$$\sum_{(\omega_1,\omega_2)\in\Omega} \left(p_1(\omega_1) \cdot p_2(\omega_2) \right)$$

This expression represents the sum of all possible products $p_1(\omega_1) \cdot p_2(\omega_2)$. The product of $\sum_{\Omega_1} p_1(\omega)$ with $\sum_{\Omega_2} p_2(\omega)$, however, is the same quantity as the sum of all possible products $p_1(\omega) \cdot p_2(\omega)$ by the Product of Sums Principle. Also, p_1 and p_2 are probability densities on Ω_1 and Ω_2, respectively, so

$$\sum_{\omega_1\in\Omega_1} p_1(\omega_1) = \sum_{\omega_2\in\Omega_2} p_2(\omega_2) = 1$$

Hence,

$$\sum_{(\omega_1,\omega_2)\in\Omega} \left(p_1(\omega_1) \cdot p_2(\omega_2) \right) = \left(\sum_{\omega_1\in\Omega_1} p_1(\omega_1) \right)\left(\sum_{\omega_2\in\Omega_2} p_2(\omega_2) \right)$$
$$= 1 \cdot 1$$
$$= 1$$

This completes the proof. ∎

8.3.2 The Cross Product of Sample Spaces

The sample space Ω of Theorem 1, which is endowed with the probability density function p given by that theorem, is called a cross product sample space. This notation was introduced in Section 1.3.4, but we recall it for convenience here. Such a sample space Ω is usually denoted $\Omega_1 \times \Omega_2$ to indicate that it is based on the two sample spaces Ω_1 and Ω_2. In fact, we can form a cross product sample space from any number $n \geq 2$ of sample spaces. This construction involves taking a cross product of sets, as defined below in Definition 1, and then endowing this new set with a certain probability density, as given in Definition 2.

Definition 1. The **cross product** of n sets

$$S_1, S_2, \ldots, S_n$$

is the set of all n-tuples

$$(s_1, s_2, \ldots, s_n)$$

where $s_i \in S_i$ for $1 \leq i \leq n$.

Notation. The cross product of n sets is denoted as $S_1 \times S_2 \times \cdots \times S_n$. When the sets are all copies of the same set S, then the cross product is denoted S^n.

The cross product of a countably infinite ordered list of sets

$$S_1, S_2, \ldots$$

is the set of all sequences

$$(s_1, s_2, \ldots)$$

where $s_i \in S_i$ for $i \geq 1$. Note that the size of a cross product of finitely many finite sets is

$$|S_1 \times S_2 \times \cdots \times S_n| = |S_1| \cdot |S_2| \cdots |S_n|$$

Definition 2.   Let $\Omega_1, \Omega_2, \cdots, \Omega_n$ be sample spaces endowed with probability density functions $p_1, p_2, \ldots, p_n$, respectively. Then, the **cross product sample space** of $\Omega_1, \Omega_2, \ldots, \Omega_n$ is the cross product

$$\Omega_1 \times \Omega_2 \times \cdots \times \Omega_n = \{(\omega_1, \omega_2, \cdots, \omega_n) \ : \ \omega_i \in \Omega_i \text{ for } 1 \leq i \leq n\}$$

endowed with the probability density function p defined by

$$p(\omega_1, \omega_2, \ldots, \omega_n) = p_1(\omega_1) \cdot p_2(\omega_2) \cdots p_n(\omega_n).$$

Theorem 1 proves that p is a legitimate probability density when $n = 2$. The fact that p is a legitimate probability density function for $n > 2$ sample spaces can be proved by induction on n. All the ideas needed for the proof are contained in the proof of Theorem 1.

Example 2.   In the communication network shown in Figure 8.2, each link may be up or down. Assuming that the nodes connecting the links are always functioning and that failures of the links are not related, what is the probability that a functioning set of links is connecting node A to node C?

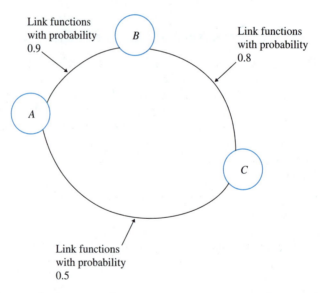

Figure 8.2   Communication network.

Solution.   This situation can be modeled by a cross product sample space. Let $\Omega_1 = \{0, 1\}$ represent the status of link AB, where 0 means down and 1 means up. Similarly, let Ω_2 and Ω_3 be sample spaces representing the status of link BC and the status of link AC, respectively. Define the probability density functions on these sample spaces as shown in Table 8.1.

The cross product sample space $\Omega_1 \times \Omega_2 \times \Omega_3$ consists of eight 3-tuples of 0's and 1's. These 3-tuples represent all possible combinations of the link conditions. For example,

Table 8.1 Link Status Sample Spaces

Sample Space	Link	Probability Density Function
Ω_1	AB	$p_1(0) = 0.1, \quad p_1(1) = 0.9$
Ω_2	BC	$p_2(0) = 0.2, \quad p_2(1) = 0.8$
Ω_3	AC	$p_3(0) = 0.5, \quad p_3(1) = 0.5$

the outcome $(0, 1, 1)$ means that AB is down, BC is up, and AC is up. This outcome is assigned the probability

$$
\begin{aligned}
p(0, 1, 1) &= p_1(0) \cdot p_2(1) \cdot p_3(1) \\
&= (0.1)(0.8)(0.5) \\
&= 0.04
\end{aligned}
$$

Figure 8.3 shows an enumeration of the entire sample space $\Omega_1 \times \Omega_2 \times \Omega_3$ and gives the values of the cross product probability density function p.

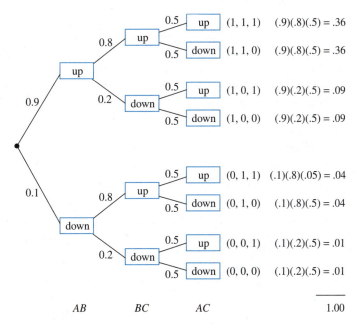

Figure 8.3 Tree diagram for the communication network.

The event "A is connected to C" occurs when AB and BC are both up; it also occurs when AC is up. The set E representing this event is

$$E = \{(1, 1, 1), (1, 1, 0), (1, 0, 1), (0, 1, 1), (0, 0, 1)\}$$

The probability $P(E)$ of E is

$$P(E) = \sum_{\omega \in E} p(\omega)$$

$$= (0.9)(0.8)(0.5) + (0.9)(0.8)(0.5) + (0.9)(0.2)(0.5) + (0.1)(0.8)(0.5)$$

$$+ (0.1)(0.2)(0.5)$$

$$= 0.36 + 0.36 + 0.09 + 0.04 + 0.01$$

$$= 0.86$$
◼

8.3.3   Bernoulli Trial Processes

An important special case of a cross product sample space occurs when an experiment with just two outcomes is repeated numerous times. Typically, the number n of repetitions can be controlled, and we want to predict what happens as a function of n.

Definition 3.   A **Bernoulli trial process** (abbreviated **Bernoulli process**) is a sequence of repetitions, called **trials,** of an experiment with a two-element sample space. It is assumed that the trials have no influence on one another. The two possible outcomes of a trial need not be equally likely.

Bernoulli processes arise in many contexts. Flipping a coin over and over can be regarded as a Bernoulli process. Another example is the sending of binary digits, or **bits,** over a communication line.

Often, the two elements of the sample space Ω of a Bernoulli trial are labeled **success** and **failure.** For example, when we flip a coin, we can label the outcome heads a success and the outcome tails a failure. Similarly, if a bit remains unchanged during transmission over a communication line, we say that a success has occurred; if the bit changes, we say that a failure has occurred.

In the context of a Bernoulli process, it is common to denote the probability of the success outcome with the letter p and the probability of the failure outcome with the letter q. Since the sample space of a Bernoulli trial has only two elements, $p = 1 - q$ where $0 \leq p, q \leq 1$.

Notation.   Since the probability density function on an arbitrary sample space is often denoted by the letter p, keep in mind that the letter p can have two usages. For example, suppose that $\Omega = \{\omega_1, \omega_2\}$ is a two-element sample space where ω_1 means success and ω_2 means failure. If we say that the probability of success is p and the probability of failure is q, then we are really defining a probability density p on Ω given by $p(\omega_1) = p$ and $p(\omega_2) = q$.

A Bernoulli process can be modeled by a cross product sample space as follows. Suppose the trials of the process have outcomes in the two-element sample space Ω. To regard a sequence of n trials as just one experiment, we form the cross product sample space consisting of n terms:

$$\Omega^n = \Omega \times \ldots \times \Omega$$

The probability density function associated with the cross product (see Definition 2) models the assumption that the trials do not influence one another.

Example 3. Consider an n-trial Bernoulli process where each trial has a probability of success p and a probability of failure $q = 1 - p$. What is the probability of obtaining exactly k successes in a run of n trials?

Solution. A sample space for the entire process can be obtained by taking the cross product of n copies of the sample space for a single trial. In other words, the sample space consists of the 2^n possible sequences of successes and failures. Each outcome in the event "exactly k successes" consists of a sequence of k successes and $n - k$ failures arranged in some order (see Figure 8.4).

(0,0,0,0,0)	(0,1,0,0,0)	(1,0,0,0,0)	(1,1,0,0,0)
(0,0,0,0,1)	(0,1,0,0,1)	(1,0,0,0,1)	*(1,1,0,0,1)
(0,0,0,1,0)	(0,1,0,1,0)	(1,0,0,1,0)	*(1,1,0,1,0)
(0,0,0,1,1)	*(0,1,0,1,1)	*(1,0,0,1,1)	(1,1,0,1,1)
(0,0,1,0,0)	(0,1,1,0,0)	(1,0,1,0,0)	*(1,1,1,0,0)
(0,0,1,0,1)	*(0,1,1,0,1)	*(1,0,1,0,1)	(1,1,1,0,1)
(0,0,1,1,0)	*(0,1,1,1,0)	*(1,0,1,1,0)	(1,1,1,1,0)
(0,0,1,1,1)	(0,1,1,1,1)	(1,0,1,1,1)	(1,1,1,1,1)

Figure 8.4 Bernoulli trial process for $n = 5$.

The outcomes in the event "exactly three successes in five trials" are indicated by an asterisk in Figure 8.4. The probability of any single such outcome, regardless of the ordering of successes and failures, is $p^k q^{n-k}$. There are $C(n, k)$ orderings, because choosing the positions for the k successes determines the positions for the failures. Hence, the probability of the event "exactly k successes" is $C(n, k) \cdot p^k q^{n-k}$. ∎

Notation. In an n-trial Bernoulli process with probability p of success on each trial, the probability of exactly k successes is denoted $b(n; k, p)$.

Probability of k Successes in a Bernoulli Process

The probability of exactly k successes in an n-trial Bernoulli process with probability p of success on each trial is

$$b(n; k, p) = C(n, k)\, p^k q^{n-k}$$

where $0 \le p \le 1$ and $q = 1 - p$.

Suppose that in an n-trial Bernoulli process we are interested exclusively in events of the form "exactly k successes" for various values of k. Then, there is an alternative to the cross product sample space—namely, a sample space with outcomes that correspond to the occurrence of exactly k successes: $\Omega = \{\omega_0, \omega_1, \cdots, \omega_n\}$ where outcome ω_k means "exactly k successes" for $0 \le k \le n$. Since the probability of the event "exactly k successes" in the cross product sample space is $b(n; k, p)$, it seems to be reasonable to define a probability density function on the alternative sample space Ω by setting $p(\omega_k) = b(n; k, p)$. The following theorem verifies that this is, indeed, a legitimate probability density function.

Theorem 2. **(Probability of k Successes in a Bernoulli Process)**   The function $p(\omega_k) = b(n; k, p)$ defines a probability density function on

$$\Omega = \{\omega_0, \omega_1, \omega_2, \ldots, \omega_n\}$$

Proof.   Clearly, $b(n; k, p) \geq 0$ for $0 \leq k \leq n$. To prove that

$$\sum_{0 \leq k \leq n} b(n; k, p) = 1$$

substitute in the values of $b(n; k, p)$:

$$\sum_{0 \leq k \leq n} b(n; k, p) = C(n, 0) \, p^0 q^n + C(n, 1) \, p^1 q^{n-1} + \cdots + C(n, n) \, p^n q^0$$

By the Binomial Theorem (see Section 7.9.1), the right side is $(p + q)^n$. Since $p + q = 1$, the result follows.   ∎

8.3.4   Events of Cross Product Form

We now leave the special case of Bernoulli processes and return to cross product sample spaces in general.

Many events in cross product sample spaces are cross products of events in the individual sample spaces. Suppose we roll a die and pick a card from a shuffled deck, associating with these experiments the sample spaces

$$\Omega_1 = \{1, 2, 3, 4, 5, 6\} \quad \text{and} \quad \Omega_2 = \{1, 2, 3, \ldots, 51, 52\}$$

with uniform probability densities. The combined experiment has an associated sample space $\Omega_1 \times \Omega_2$. Let $E_1 \subseteq \Omega_1$ be the event "an odd number shows on the die," and let $E_2 \subseteq \Omega_2$ be the event "a red card is drawn." Then, $E_1 \times E_2 \subseteq \Omega_1 \times \Omega_2$ is the event "an odd number on the die and a red card." (In the context of cross product events such as this, we will use E_i to denote an event in Ω_i.)

In the die-and-card example, the roll of the first die has no connection to the selection of the card, so it seems to be reasonable that $P(E_1 \times E_2)$ should be $P(E_1) \cdot P(E_2)$. However, we are not at liberty to assign probabilities to events. Once probabilities have been assigned to the individual outcomes in a sample space, the probability of an event, by definition, must be computed by summing the probabilities of the outcomes in the event. The next theorem shows that the probabilities of events having a cross product form do, indeed, obey a multiplication rule, just as we have speculated for this die-and-card situation.

Theorem 3. **(Probability of Events of Cross Product Form)**   In a cross product sample space $\Omega_1 \times \Omega_2 \times \cdots \times \Omega_n$, any event of the form $E_1 \times E_2 \times \cdots \times E_n$ has probability

$$P(E_1 \times E_2 \times \cdots \times E_n) = \prod P(E_i)$$

where E_i is an event in Ω_i for $1 \leq i \leq n$.

Proof.   The proof is by induction on n. Define

$$\mathcal{T} = \{n \in \mathbb{N} : P(E_1 \times E_2 \times \cdots \times E_n) = \prod P(E_i)\}$$

(Base step) The statement is trivially true for $n = 1$, so $1 \in \mathcal{T}$.

(**Inductive step**) Suppose that the statement is true for $n = k$. Now, prove that it is true for $n = k + 1$. Let p denote the usual probability density function on $\Omega_1 \times \Omega_2 \times \cdots \times \Omega_n$ (see Definition 2 in Section 8.3.2). Then,

$$P(E_1 \times E_2 \times \cdots \times E_{k+1}) = \sum p(\omega_1, \omega_2, \cdots, \omega_{k+1})$$

$$= \sum p_1(\omega_1) \cdot p_2(\omega_2) \cdots p_{k+1}(\omega_{k+1})$$

where the sums are taken over all combinations of choices of the ω_i from the E_i.

Rearranging terms so that all terms with the same choice for ω_{k+1} from E_{k+1} are grouped together yields

$$P(E_1 \times E_2 \times \cdots \times E_{k+1}) = \left(\sum_{\omega_{k+1} \in E_{k+1}} p_{k+1}(\omega_{k+1}) \right) \left(\sum p_1(\omega_1) \cdot p_2(\omega_2) \cdots p_k(\omega_k) \right)$$

where the second sum on the right-hand side is over all combinations of choices of the ω_i from the E_i for $1 \le i \le k$.

The second sum represents the probability of the event $E_1 \times E_2 \times \cdots \times E_k$ in the sample space $\Omega_1 \times \Omega_2 \times \cdots \times \Omega_k$. By the inductive hypothesis, it is equal to $P(E_1)P(E_2) \cdots P(E_k)$. Thus, summing over the outcomes ω_{k+1} in E_{k+1} gives

$$P(E_1 \times E_2 \times \cdots \times E_{k+1}) = \sum_{\omega_{k+1} \in E_{k+1}} p_{k+1}(\omega_{k+1}) \cdot P(E_1) \cdot P(E_2) \cdots P(E_k)$$

$$= P(E_1) \cdot P(E_2) \cdots P(E_k) \sum_{\omega_{k+1} \in E_{k+1}} p_{k+1}(\omega_{k+1})$$

$$= P(E_1) \cdot P(E_2) \cdots P(E_k) \cdot P(E_{k+1})$$

Therefore, $n + 1 \in \mathcal{T}$, and the result follows by the Principle of Mathematical Induction. ∎

Example 4. At the beginning of this section, we formulated the situation "an odd number on the die and a red card" as an event $E_1 \times E_2 \subseteq \Omega_1 \times \Omega_2$ where E_1 was the event "an odd number on the die" in $\Omega_1 = \{1, 2, 3, 4, 5, 6\}$ and E_2 was the event "a red card" in $\Omega_2 = \{1, 2, \ldots, 52\}$. Verify the conclusion of Theorem 3 in this case by computing $P(E_1 \times E_2)$ directly and comparing the result to $P(E_1) \cdot P(E_2)$.

Solution. Define $E_1 \times E_2$ to be the event of $\Omega_1 \times \Omega_2$ defined as

$$E_1 \times E_2 = \{(\omega_1, \omega_2) : \omega_1 \text{ is an odd number in } \Omega_1 \text{ and } \omega_2 \text{ is a red card in } \Omega_2\}$$

Each outcome (ω_1, ω_2) in $E_1 \times E_2$ has probability

$$p(\omega_1, \omega_2) = p_1(\omega_1) \cdot p_2(\omega_2) = \left(\frac{1}{6} \right) \left(\frac{1}{52} \right)$$

Hence,

$$P(E_1 \times E_2) = \left(\frac{1}{6} \right) \left(\frac{1}{52} \right) |E_1 \times E_2| = \left(\frac{1}{6} \right) \left(\frac{1}{52} \right) (3)(26) = \frac{1}{4}$$

On the other hand,

$$P(E_1) = \left(\frac{1}{6}\right) | E_1 | = \left(\frac{1}{6}\right)(3) = \frac{1}{2}$$

and

$$P(E_2) = \left(\frac{1}{52}\right) | E_2 | = \left(\frac{1}{52}\right)(26) = \frac{1}{2}$$

Hence,

$$P(E_1) \cdot P(E_2) = \frac{1}{4} = P(E_1 \times E_2)$$

∎

Example 5.   Suppose we flip a coin once and roll a die twice. Formulate the event

$$E = \text{"heads on the coin and an even number on the second roll"}$$

as a cross product of sets in a cross product sample space. Calculate the probability of the event.

Solution.   We choose sample spaces

$$\Omega_1 = \{H, T\} \quad \text{and} \quad \Omega_2 = \Omega_3 = \{1, 2, 3, 4, 5, 6\}$$

and assign each a uniform probability density. Flipping the coin and rolling the die twice is modeled by the sample space $\Omega_1 \times \Omega_2 \times \Omega_3$. Let $E_1 = \{H\}$, $E_2 = \Omega_2$, and $E_3 = \{2, 4, 6\}$. Since the outcome of the first roll is not specified, all possibilities must be taken into account. Hence,

$$E = E_1 \times \Omega_2 \times E_3$$

and

$$P(E) = P(E_1) \cdot P(\Omega_2) \cdot P(E_3) = \left(\frac{1}{2}\right)(1)\left(\frac{1}{2}\right) = \frac{1}{4}$$

∎

8.3.5   Two Ways of Viewing Events

Let

$$\Omega = \Omega_1 \times \Omega_2 \times \cdots \times \Omega_n$$

be the cross product of sample spaces $\Omega_1, \Omega_2, \ldots, \Omega_n$. This section looks at the relationship between events in Ω_i and events in Ω.

Suppose we want to specify an event in the cross product sample space Ω just by specifying what happens in one of its component sample spaces Ω_i (we do not care what happens in the other component sample spaces). We can reformulate a nonempty event E_i in Ω_i as a corresponding event E_i^* in $\Omega_1 \times \Omega_2 \times \cdots \times \Omega_n$ by setting E_i^* equal to all n-tuples in which the ith component belongs to E_i.

For example, suppose we flip a penny and roll a die. Then, $n = 2$, $\Omega_1 = \{H, T\}$, and $\Omega_2 = \{1, 2, 3, 4, 5, 6\}$. Getting an even number on the die is the event $E_2 = \{2, 4, 6\} \subseteq \Omega_2$.

However, this event E_2 is part of a larger context in which a coin is flipped. In the grand experiment $\Omega_1 \times \Omega_2$, getting an even number on the die corresponds to the event

$$\{H, T\} \times \{2, 4, 6\} = \{(H, 2), (H, 4), (H, 6), (T, 2), (T, 4), (T, 6)\}$$

because either H or T could have turned up on the coin flip. In other words, the nonempty event E_i in Ω_i corresponds to the event

$$E_i^* = \Omega_1 \times \Omega_2 \times \cdots \times \Omega_{i-1} \times E_i \times \Omega_{i+1} \times \cdots \times \Omega_n$$

in $\Omega = \Omega_1 \times \Omega_2 \times \cdots \times \Omega_n$. The following corollary of Theorem 3 shows that this reformulation to a corresponding event preserves the probability computed for the original event.

Corollary 1 to Theorem 3. $P(E_i^*) = P(E_i)$ where

$$E_i^* = \Omega_1 \times \cdots \times \Omega_{i-1} \times E_i \times \Omega_{i+1} \times \cdots \times \Omega_n$$

is an event in cross product sample space $\Omega_1 \times \Omega_2 \times \cdots \times \Omega_n$.

Proof. According to Theorem 3,

$$P(E_i^*) = P(\Omega_1) \cdot P(\Omega_2) \cdots P(\Omega_{i-1}) \cdot P(E_i) \cdot P(\Omega_{i+1}) \cdots P(\Omega_n)$$

Each Ω_j has $P(\Omega_j) = 1$, so $P(E_i^*) = P(E_i)$. ∎

Example 6. An experiment involves flipping a fair coin and rolling a fair die. Let $\Omega = \{H, T\} \times \{1, 2, 3, 4, 5, 6\}$ be a cross product sample space for the experiment. Compute the probability of rolling a 3 or a 5 as an event in Ω using Corollary 1 to Theorem 3.

Solution. Define sample spaces $\Omega_1 = \{H, T\}$ and $\Omega_2 = \{1, 2, 3, 4, 5, 6\}$ so that $\Omega = \Omega_1 \times \Omega_2$. The event of interest in Ω is $E_2^* = \{(H, 3), (T, 3), (H, 5), (T, 5)\}$. The event of interest in Ω_2 is $E_2 = \{3, 5\}$, which has probability $2/6 = 1/3$. Since $|\Omega| = 12$, $P(E_2^*) = 4/12 = 1/3$. According to Corollary 1 to Theorem 3, compute $P(E_2^*) = P(E_2)$, which is, indeed, the case in this example. ∎

We can also specify events in a cross product sample space by specifying what takes place in some collection of individual sample spaces. Suppose we single out k sample spaces $\Omega_{i_1}, \Omega_{i_2}, \ldots, \Omega_{i_k}$ from the n sample spaces $\Omega_1, \Omega_2, \ldots, \Omega_n$ where $k \leq n$, and then specify an event $E_{i_j} \subseteq \Omega_{i_j}$, in each of these k selected sample spaces. The events E_{i_j} correspond to events $E_{i_j}^*$ in the cross product $\Omega_1 \times \cdots \times \Omega_n$. Specifying that all the events $E_{i_1}, \ldots, E_{i_k}$ are to occur in the individual sample spaces is the same thing as specifying that the event $E_{i_1}^* \cap \cdots \cap E_{i_k}^*$ is to occur in the cross product sample space. The next corollary of Theorem 3 says that the probability of this event can be computed by a multiplication rule.

Corollary 2 to Theorem 3. $P(E_{i_1}^* \cap E_{i_2}^* \cap \cdots \cap E_{i_k}^*) = P(E_{i_1}) \cdot P(E_{i_2}) \cdots P(E_{i_k})$, where

$$E_{i_j}^* = \Omega_1 \times \cdots \times E_{i_j} \times \cdots \times \Omega_n$$

for $1 \leq j \leq k$ is an event in cross product sample space $\Omega_1 \times \Omega_2 \times \cdots \times \Omega_n$.

Proof.   The event $E_{i_1}^* \cap E_{i_2}^* \cap \cdots \cap E_{i_k}^*$ of the cross product sample space consists of all n-tuples in $\Omega_1 \times \cdots \times \Omega_n$ with entries in positions $i_1, i_2, \ldots, i_k$ that belong to the events $E_{i_1}, \ldots, E_{i_k}$ of $\Omega_{i_1}, \ldots, \Omega_{i_k}$, respectively.

We first define some terms to make it easier to express this set of n-tuples in the form

$$E_1 \times E_2 \times \cdots \times E_n.$$

Let $X = \{i_1, i_2, \ldots, i_k\}$ and $I_n = \{1, 2, \ldots, n\}$. Then for $1 \le i \le n$, let

$$E_i = \begin{cases} \Omega & \text{for } i \in I_n - X \\ E_{i_m} & \text{for } i = i_m \text{ for some } i_m \in X \end{cases}$$

Since $P(E_j) = 1$ whenever $E_j = \Omega_j$, the result follows from Theorem 3.   ■

We see the use of this corollary in the next example.

Example 7.   Consider rolling a die three times, and choose a sample space $\Omega = \Omega_1 \times \Omega_2 \times \Omega_3$ where $\Omega_1 = \Omega_2 = \Omega_3 = \{1, 2, 3, 4, 5, 6\}$. What is the probability of getting an odd number on the first roll and an even number on the third roll?

Solution.   Consider $E_1 = \{1, 3, 5\} \subseteq \Omega_1$ and $E_3 = \{2, 4, 6\} \subseteq \Omega_3$. In Ω, getting an odd number on the first roll is the event $E_1^* = E_1 \times \Omega_2 \times \Omega_3$; getting an even number on the third roll is the event $E_3^* = \Omega_1 \times \Omega_2 \times E_3$. Having both these events occur is the event $E_1^* \cap E_3^*$. By Corollary 2 of Theorem 3,

$$P(E_1^* \cap E_3^*) = P(E_1) \cdot P(E_3) = \left(\frac{3}{6}\right)\left(\frac{3}{6}\right) = \frac{1}{4}$$

Alternatively, observe that $E_1^* \cap E_3^*$ consists of $3 \cdot 6 \cdot 3 = 54$ ordered triples in a sample space of size 6^3 with a uniform probability density function and so has probability $54 \cdot 6^{-3} = 1/4$.   ■

We began the study of cross product sample spaces by stating the Probability Multiplication Principle, which Theorem 3 justified mathematically. Corollary 2 to Theorem 3 can be viewed as justifying the extension of the Multiplication Principle from outcomes to events. We now summarize the major results about Cross Product Events.

Probability of Cross Product Events

- The probability of an event in cross product form $E_1 \times \cdots \times E_n$ is the product of the event probabilities

$$P(E_1 \times \cdots \times E_n) = P(E_1) \cdots P(E_n)$$

- The simultaneous occurrence of k events $E_{i_1}, \ldots, E_{i_k}$ in some k of the n sample spaces composing a cross product can be regarded as a single event

$$E_{i_1}^* \cap \cdots \cap E_{i_k}^* \subseteq \Omega_1 \times \cdots \times \Omega_n$$

- The probability of this single event in $\Omega_1 \times \Omega_2 \times \cdots \times \Omega_n$ is the product of the event probabilities $P(E_{i_1}) \cdots P(E_{i_k})$.

8.4   Exercises

1. Suppose S is a set with k elements. How many elements are in S^n, the cross product $S \times S \times \cdots \times S$ of n copies of S?

2. Suppose that $\sum_{i=1}^{n} a_i = \sum_{j=1}^{m} b_j = 1$ where $0 \le a_i, b_j \le 1$. Use the Product of Sums Principle to prove that $\sum_{i=1}^{n} \sum_{j=1}^{m} a_i \cdot b_j = 1$. Does the result hold if some of the a_i and b_j can be less than zero and greater than one?

3. Suppose $\sum_{i=1}^{n} a_i = 2$, $\sum_{j=1}^{m} b_j = 3$, and $\sum_{k=1}^{l} c_k = 5$. Evaluate

$$\sum_{i=1}^{n} a_i \left(\sum_{j=1}^{m} \sum_{k=1}^{l} b_j \cdot c_k \right)$$

4. Consider the Birthday Problem, ignoring leap years. Determine the probability that two people in your class have the same birthday under each of the following circumstances:

    (a) There are 20 people in your class.
    (b) There are 30 people in your class.

    You may wish to use a calculator.

5. Suppose that people are equally likely to be born on each of the seven days of the week. In a group of n people, determine the probability that:

    (a) Two or more of them were born on Saturday.
    (b) Exactly two of them were born on Saturday.
    (c) Two or more of them were born on the same day of the week.

6. For the network described in Example 2 in Section 8.3.2, determine the probability of each of the following events:

    (a) Each node can communicate with the other two.
    (b) At least two links are down.
    (c) A is directly connected to B.

7. A coin is tossed, a die is rolled, and a card is drawn at random from a deck. Assume that the toss, roll, and draw are fair.

    (a) Describe this experiment as a cross product sample space.
    (b) With the aid of a tree diagram, define a probability density on the cross product.
    (c) Verify by direct computation that the probability density found in part (b) is legitimate.
    (d) Does it matter in what order the coin, the die, and the card are considered?

8. Two dice are rolled. One is fair, but the other is loaded: It shows the face with six spots half the time and the remaining five faces with equal frequencies.

    (a) Describe the experiment in terms of a cross product sample space.
    (b) Define a probability density on the cross product space.
    (c) Verify by direct computation that the probability density found in part (b) is legitimate.
    (d) Does it matter in what order the dice are considered? Explain your answer.

9. A coin that is twice as likely to show heads than it is tails is tossed three times.

   (a) Describe this experiment as a Bernoulli process.
   (b) Use a tree diagram to assign a probability density.

10. Suppose the coin in Exercise 9 is tossed n times where n is some arbitrary positive integer.

    (a) Sketch a tree diagram for this experiment.
    (b) Give a formula for the number of outcomes in which heads occur exactly k times.
    (c) Give a formula for the value of the probability density function on each of the outcomes described in part (b).
    (d) What would be the formula for the value of the probability density function on each of the outcomes in part (b) if the coin were fair?

11. A coin that is twice as likely to show heads than it is tails is tossed three times. Suppose we are only interested in the number of heads.

    (a) Formulate this experiment in terms of a sample space with outcomes that give the number of heads.
    (b) Assign a probability density function to this sample space.
    (c) Describe the event "at least two heads" as a set of outcomes in this sample space.
    (d) What is the probability of the event in part (c)?

12. Consider the events described in Exercise 6 regarding Example 2. Which (if any) of these events can be expressed in cross product form? Give the cross product expression if one exists.

13. For the cross product sample space found in Exercise 8(a), give the cross product formulation, and calculate the probability of each of the following events:

    (a) An even number of spots shows on each of the die.
    (b) The loaded die shows six spots.
    (c) The fair die shows either five or six spots, and the loaded die shows six spots.

14. For the coin, die, and card experiment of Exercise 7, let the coin have sample space $\Omega_1 = \{H, T\}$, the die have sample space $\Omega_2 = \{1, 2, 3, 4, 5, 6\}$, and the card deck have sample space $\Omega_3 = \{1, 2, \ldots, 51, 52\}$.

    (a) What event E_1^* in $\Omega_1 \times \Omega_2 \times \Omega_3$ corresponds to the event $E_1 = $ "heads on the coin"?
    (b) What event E_2^* in $\Omega_1 \times \Omega_2 \times \Omega_3$ corresponds to the event $E_2 = $ "4 or higher on the die?"
    (c) Formulate "getting a head on the coin and a 4 or higher on the die" as an event in $\Omega_1 \times \Omega_2 \times \Omega_3$ by indicating the outcomes in the event. Then, formulate this in terms of E_1^* and E_2^*.
    (d) Calculate the probability of the event "heads on the coin and a 4 or higher on the die" in two ways by using each of the formulations of part (c).
    (e) Formulate the event "either a head on the coin or a 4 or better on the die" in terms of E_1^* and E_2^*. What is the probability of this event?

15. Give two ways you could label the outcomes of an experiment that consists of flipping a dime three times. Determine a probability density function for each of the two ways.

16. What is the probability of flipping k heads out of m tosses ($0 < k \le m$)? What are the numerical results for five heads out of eight tosses?

17. Suppose the sex of a newborn child is viewed as an experiment with two equally likely outcomes. Assuming that each child represents an independent trial, what is the probability of a family with four children having two, three, or four girls? Suppose the probability of the child being a girl is 0.52; now what is the probability of two, three, or four boys?

18. Find the probability of getting a five exactly twice in seven rolls of a fair die.

8.5   Independent Events and Conditional Probability

Knowing that a fair coin came up heads on the first toss of a two-toss experiment does not cause us to believe that the chance of getting heads on the second toss is other than 50%. After all, the two tosses are physically unrelated. On the other hand, we now know that the chance of getting tails on both tosses is zero. Naturally, we anticipate that physically unrelated events resulting from, say, separate tosses of a fair coin do not affect one another. Sometimes, events arising from a single physical experiment also behave as though they are unrelated: Information about the occurrence of one does not shed any light on the occurrence of the others. This phenomenon is modeled by the concept of independence.

The subject of conditional probability tells how to revise probabilities in light of new information. Typical examples include how to revise the probability that one of two coins being flipped will come up heads once you know that the first coin did not. Quite a different use of conditional probability is to determine the likelihood that the result of a medical test result is a false-negative (that is, a positive) result. The result is a false negative (positive) if the true result is positive (negative) but if the test gives a negative (positive) result.

First, we give the mathematical definition of what it means for two events in the same sample space to be independent. This definition is intended to capture the notion, described above, that some two events do not seem to influence one another. Next, we propose a way to change the probability of an event A given that an event B in the same sample space has occurred. This is called the conditional probability of A given B. After defining **independence** and conditional probability, we argue that two events in the same sample space are independent precisely when no change is made to the probability of one event (according to the definition of the conditional probability) if we learn that the other event has occurred. In this way, these two mathematical definitions work together to model—and to make precise—the notion that some pairs of events do not seem to give any information about each other. Later, we will explore the properties and applications of conditional probabilities.

8.5.1   Independent Events

Now we give the mathematical definition intended to capture the notion that two events in the same sample space do not influence one another. Later, we will see why this definition does the job.

Definition 1.   A pair of events A and B belonging to the same sample space are said to be **independent** provided that $P(A \cap B) = P(A) \cdot P(B)$.

By Corollaries 1 and 2 of Theorem 3 in Section 8.3.4, pairs of events of the form E_i^* and E_j^* for $i \neq j$ in the sample space $\Omega = \Omega_1 \times \Omega_2 \times \cdots \times \Omega_n$ for some $n \in \mathbb{N}$ are examples of independent pairs of events in a cross product sample space. It is also possible for events that do not have this form to be independent.

Example 1.   Consider one roll of a fair die with $\Omega = \{1, 2, 3, 4, 5, 6\}$. Let A be the event that an even number is rolled. Let B be the event that the number rolled is at most two. Let C be the event that the number rolled is a prime. Which pairs of events are independent?

Solution. $A = \{2, 4, 6\}$, $B = \{1, 2\}$, and $C = \{2, 3, 5\}$. Events A and B are independent:

$$P(A \cap B) = P(\{2\}) = \frac{1}{6}$$

and

$$P(A) \cdot P(B) = \left(\frac{1}{2}\right)\left(\frac{1}{3}\right) = \frac{1}{6}$$

On the other hand, A and C are *not* independent:

$$P(A \cap C) = P(\{2\}) = \frac{1}{6}$$

but

$$P(A) \cdot P(C) = \left(\frac{1}{2}\right)\left(\frac{1}{2}\right) = \frac{1}{4} \qquad \blacksquare$$

It is important not to confuse the concept of independence with the concept of disjointness. In fact, **disjoint events** that have positive probabilities are never independent. As we have seen, physically unrelated experiments give rise to a cross product sample space in which nonempty events E_i on the various sample spaces Ω_i can be regarded as events E_i^* in a common sample space

$$\Omega_1 \times \Omega_2 \times \cdots \times \Omega_n$$

Furthermore, as we pointed out following Definition 1, these events are, mathematically speaking, independent. However, they are *not* disjoint:

$$E_i^* \cap E_j^* = \Omega_1 \times \cdots \times E_i \times \cdots \times E_j \times \cdots \times \Omega_n$$

The next definition extends the concept of independence from pairs to sets of more than two events belonging to the same sample space.

Definition 2.   Let $A_1, A_2, \ldots, A_k$ be subsets of the same sample space. The set of events $\{A_1, A_2, \ldots, A_k\}$ is called an **independent set of events** provided that

$$P(A_1 \cap A_2 \cap \cdots \cap A_k) = \prod P(A_i)$$

The definition will be used in Section 8.9.2.

Example 2.   A fair die is rolled one time. Let A denote the event $\{1, 2, 3\}$, B the event $\{1, 4, 5\}$, and C the event $\{1, 2, 3, 4\}$. Are A, B, and C an independent set of events? Are A and C independent? Are B and C independent?

Solution.   The sample space is $\Omega = \{1, 2, 3, 4, 5, 6\}$. $P(A) = P(B) = 1/2$, but $P(C) = 2/3$. Events A, B, and C are independent, because

$$P(A \cap B \cap C) = P(\{1\}) = \frac{1}{6}$$

and

$$P(A) \cdot P(B) \cdot P(C) = \left(\frac{1}{2}\right)\left(\frac{1}{2}\right)\left(\frac{2}{3}\right) = \frac{1}{6}$$

Events A and C are not independent, because

$$P(A \cap C) = P(\{1, 2, 3\}) = \frac{1}{2}$$

but

$$P(A) \cdot P(C) = \left(\frac{1}{2}\right)\left(\frac{2}{3}\right) = \frac{1}{3}$$

Events B and C are independent, because

$$P(B \cap C) = P(\{1, 4\}) = \frac{1}{3}$$

and

$$P(B) \cdot P(C) = \left(\frac{1}{2}\right)\left(\frac{2}{3}\right) = \frac{1}{3} \qquad \blacksquare$$

We now list several important points to remember about independence of events. (The reader is asked to prove some of these in Section 8.6.)

Independence

- Disjointness and independence are different concepts. In fact, disjoint events that have positive probabilities are never independent.
- Events in the individual sample spaces of unrelated experiments give rise to independent events in the cross product of the individual sample spaces.
- Independence is a property of a collection of two or more events, not a property of just one event.
- A collection of pairwise independent events does not always constitute an independent set of events.
- In an independent set of events, not all the pairs need be independent. (See Example 2).

8.5.2   Introduction to Conditional Probability

In Example 2 of Section 8.5.1, it was found that for the sample space $\Omega = \{1, 2, 3, 4, 5, 6\}$ the events $B = \{1, 4, 5\}$ and $C = \{1, 2, 3, 4\}$ were, mathematically speaking, independent even though they arise from the same experiment. Events B and C have nothing to do with one another in the following sense. Imagine that we can buy a chance on winning a prize if the die is rolled and C occurs—that is, if a 1, 2, 3, or 4 is rolled. We are hesitating about buying a chance when we learn that the die has just been rolled and that B occurred (a 1, 4, or 5 turned up). We are offered a last minute opportunity to buy a chance on C—that is, to bet that C also occurred when the die was rolled. Are we more tempted than before to buy a chance on C?

Here is a possible line of reasoning: We have new information, so we can revise the sample space from $\Omega = \{1, 2, 3, 4, 5, 6\}$ to $\Omega_1 = \{1, 4, 5\} = B$ to take into account the fact that B has occurred. Initially, we assumed that all outcomes in B were equally likely. Since we have no reason to believe otherwise now, we persist in this assumption. This means that our probability distribution on Ω_1 is uniform, with $p_1(1) = p_1(4) = p_1(5) = 1/3$. Now, for C to have occurred, it must be that $B \cap C = \{1, 4\}$ occurred. The revised probability is

$$P_1(B \cap C) = p_1(1) + p_1(4) = \frac{2}{3}$$

This, however, is exactly the same as our initial assessment of C (before we knew that B occurred): The original probability $P(C)$ was also 2/3. Therefore, the information that B occurred does not cause us to revise our predictions about C. In this sense, B and C are unrelated events.

On the other hand, if we had an opportunity to buy a last-minute chance on $A = \{1, 2, 3\}$, we would be *less* tempted: A occurred only if $A \cap B$ occurred. The revised probability of this is

$$P_1(A \cap B) = P_1(\{1\}) = p_1(1) = \frac{1}{3}$$

The initial probability of A was $P(A) = 1/2$.

The reasoning we just used is widely accepted and forms the basis for a standard technique to revise probabilities in light of new information. If events A and B are independent—that is, if they are events in the same sample space such that $P(A \cap B) = P(A) \cdot P(B)$—then knowledge that one of these events occurred or did not occur does not enable us to revise the probability for the occurrence of the other. To explain this, we must first say what we mean by the revised probability of an event because of the occurrence of another event. Then, we will note that the revised probability is equal to the initial probability precisely when the two events are independent. The formula we will develop for revised probability is also useful in its own right.

To devise a formula for what we mean by the probability of an event $B \subseteq \Omega$, given that an event A has occurred where $A \subseteq \Omega$ and $P(A) > 0$, we reason as follows: Since A has occurred, the sample space Ω with probability density function p can be revised to a new sample space $\Omega_1 = A$, for surely A includes all the outcomes that are possible. Next, we devise a probability density function p_1 for the outcomes in $\Omega_1 = A$. Of course, the sum of the $p_1(\omega)$'s for $\omega \in A$ must be one, and $p_1(\omega)$ must be greater than zero if p_1 is to be a density function. Also, we want the ratios of probability densities for outcomes in A to be the same for the new density as for the old. For example, if ω_i and ω_j are outcomes in A such that $p(\omega_i) = 2p(\omega_j)$, then $p_1(\omega_i)$ should be twice $p_1(\omega_j)$. We want this property, since we have no grounds for believing that the frequencies of outcomes have changed relative to one another just because A occurred. To obtain a legitimate probability density for the new sample space $\Omega_1 = A$ that preserves the original ratios of probabilities, we can simply divide the original densities for outcomes in A by $P(A)$. (Checking that this works is left as an exercise for the reader).

For an event A with $P(A) > 0$, how should we define "the probability of B given that A has occurred"? Obviously, no element of B outside A has occurred, so what we really seek is the probability of $B \cap A$ given that A has occurred. We will define this to be the sum of the revised probability density over the elements of $B \cap A$. In other words, we take $P_1(B \cap A)$ as the new, revised probability for B. The probability $P_1(B \cap A)$ can be rewritten as

$$P_1(B \cap A) = \sum_{\omega \in B \cap A} p_1(\omega)$$

$$= \sum_{\omega \in B \cap A} \frac{p(\omega)}{P(A)}$$

$$= \frac{1}{P(A)} \sum_{\omega \in B \cap A} p(\omega)$$

$$= \frac{P(B \cap A)}{P(A)}$$

This discussion motivates the following definition.

Definition 3.   The **conditional probability** $P(B \mid A)$ of B given that A has occurred is defined by

$$P(B \mid A) = \frac{P(B \cap A)}{P(A)}$$

where $P(A) > 0$.

Example 3.   Define an experiment of rolling two fair dice and recording the total number of pips on the top faces. Find the probability that the total number of pips is nine given that the first die shows five pips on its top face.

Solution. $P(\text{total } 9|\text{first die } 5) = P(\text{total } 9 \text{ and first die } 5)/P(\text{first die } 5)$. Assume the uniform probability density on the 36 pairs of possible outcomes for the two fair dice. Define $A = \{(5, 1), (5, 2), (5, 3), (5, 4), (5, 5), (5, 6)\}$ and $B = \{(3, 6), (4, 5), (5, 4), (6, 3)\}$. Then, we can write the probability as

$$P(B|A) = P(B \cap A)/P(A) = P(\{(5, 4)\})/P(A) = \left(\frac{1}{36}\right) \Big/ \left(\frac{6}{36}\right) = 1/6. \quad \blacksquare$$

To relate the notion of conditional probability with the notion of independence from Section 8.5.1, let us see what happens to the conditional probability of B given the occurrence of A if A and B are independent events:

$$P(B \mid A) = \frac{P(B \cap A)}{P(A)}$$

$$= \frac{P(A) \cdot P(B)}{P(A)}$$

$$= P(B)$$

In other words, the probability of B is not affected by the fact that A has occurred. In fact, the satisfaction of the condition $P(A \mid B) = P(A)$ can be regarded as an alternate definition for the independence of a pair of non-empty events A and B.

To further our understanding of the connections between conditional probability and the independence of events, let us revisit the notion of nonempty disjoint events. Consider two such events A and B on the same sample space Ω. We known $P(A) > 0$ and $P(B) > 0$. It follows immediately from the definition of independence that A and B are *not* independent, because

$$0 = P(A \cap B) \neq P(A) \cdot P(B) > 0$$

We can look at this a second way. Learning that event B, say, occurred tells us a lot about A—namely, that A definitely did *not* occur (the events are disjoint after all). This phenomenon can be expressed as

$$P(A \mid B) = \frac{P(A \cap B)}{P(B)} = 0$$

8.5.3 Exploring Conditional Probability

Note that the expression in Definition 3 for conditional probability can be rearranged to the form

$$P(B \cap A) = P(A) \cdot P(B \mid A)$$

In this subsection, we take advantage of this observation to obtain two further results, called Bayes' Rule and the Theorem of Total Probability. These two results, which are often used together, provide powerful tools for computing probabilities, as we shall see.

Let us begin with the Theorem of Total Probability, which is perhaps the easier result to understand.

Theorem 1.   (Total Probability)   Let sample space Ω be the disjoint union of events $E_1, \ldots, E_n$ with positive probabilities, and let $A \subseteq \Omega$. Then,

$$P(A) = \sum_{i=1}^{n} P(A \mid E_i) \cdot P(E_i)$$

Proof.   Event A can be expressed as a union of disjoint events as follows:

$$A = (A \cap E_1) \cup \ldots \cup (A \cap E_n)$$

Hence,

$$P(A) = \sum_{i=1}^{n} P(A \cap E_i)$$

From the definition of conditional probability,

$$P(A \cap E_i) = P(A \mid E_i) \cdot P(E_i)$$

Substituting the right side of this expression for $P(A \cap E_i)$ in the expression for $P(A)$ gives the result.   ∎

Example 4. There are three kinds of vending machines in canteens at the university: candy dispensers, hot-drink dispensers, and soda dispensers. Of the vending machines, 25% sell candy, 35% dispense hot drinks, and 40% dispense sodas. Suppose that servicing is needed by 1/2 the candy machines, 1/5 the hot drink machines, and 1/3 the soda machines. What is the probability that a machine chosen at random needs servicing?

Solution. Let Ω consist of all the machines, and endow Ω with a uniform probability density to model choosing a machine at random. Let E_1, E_2, and E_3 denote the candy, hot-drink, and soda machines, respectively. Then, Ω is the disjoint union of these events where $P(E_1) = 0.25$, $P(E_2) = 0.35$, and $P(E_3) = 0.40$. Let A denote the set of nonfunctioning machines. Event A is the disjoint union of the nonfunctioning candy, hot-drink, and soda machines, and

$$P(A \mid E_1) = \frac{1}{2} \quad P(A \mid E_2) = \frac{1}{5} \quad P(A \mid E_3) = \frac{1}{3}$$

By the Total Probability Theorem,

$$P(A) = \sum_{i=1}^{3} P(A \mid E_i) \cdot P(E_i)$$

$$= \left(\frac{1}{2}\right)(0.25) + \left(\frac{1}{5}\right)(0.35) + \left(\frac{1}{3}\right)(0.40)$$

$$\approx 0.328 \qquad \blacksquare$$

Now, we give the other result, Bayes' Rule, that can be obtained by playing with the expression in the definition of conditional probability. We began by noticing that

$$P(B \cap A) = P(A) \cdot P(B \mid A)$$

To warm up for Bayes' Rule, notice that interchanging the roles of A and B gives another expression for $P(B \cap A)$—namely,

$$P(A \cap B) = P(B) \cdot P(A \mid B)$$

Theorem 2. (Bayes' Rule) Let A and B be events in the same sample space. If neither $P(A)$ nor $P(B)$ is zero, then

$$P(B \mid A) = P(A \mid B) \frac{P(B)}{P(A)}$$

Proof. By the definition of conditional probability,

$$P(A \mid B) = \frac{P(A \cap B)}{P(B)} \quad \text{and} \quad P(B \mid A) = \frac{P(A \cap B)}{P(A)}$$

Consequently,

$$P(A \cap B) = P(A) \cdot P(B \mid A) = P(B) \cdot P(A \mid B)$$

so

$$P(B \mid A) = P(A \mid B) \frac{P(B)}{P(A)} \qquad \blacksquare$$

One key feature of Bayes' Rule is that it can be used in situations where the underlying probability distribution of the sample space is not known but estimates of the probabilities for some of the events are available. The next example illustrates this. The designers of computer systems that aid in medical diagnosis and other similar problems must deal with uncertain information of this kind.

Example 5. Suppose it is estimated that 10% of a population has a certain disease. Tests for this disease are being developed but are not yet perfect. In fact, an individual who has the disease may test negative. Suppose experience with a particular test shows that 5% of the results are actually false negatives—that is, the individual actually does have the disease. Also, suppose that 8% of the tests done so far have been positive. What is the probability that a sick person will receive a false-negative test result?

Solution. Let Ω consist of the population, A denote the subset of people who would test positive if they took the test, and B denote the subset of people who have the disease. Which people actually make up these subsets is not known: Not everybody may take the test, and only time will tell which people actually have the disease. Nevertheless, we can estimate that $P(A) = 0.08$, $P(B) = 0.10$, and $P(B|\overline{A}) = 0.05$. We are interested in the probability that a person who is ill tests negative. Hence, we must compute $P(\overline{A}|B)$. By Bayes' Rule,

$$P(\overline{A}|B) = P(B|\overline{A})\frac{P(\overline{A})}{P(B)} = \frac{(0.05)(0.92)}{(0.10)} = 0.46 \qquad \blacksquare$$

Because of the wide applicability of Bayes' Rule, we highlight it here.

Using Bayes' Rule

In many applications, the answer being sought can be expressed in the form of a conditional probability $P(B|A)$, and the other probabilities $P(A \mid B)$, $P(A)$, and $P(B)$ that are needed to apply Bayes' Rule can be estimated by experiment or experience:

$$P(B \mid A) = \frac{P(A \mid B) \cdot P(B)}{P(A)}$$

8.5.4   Using Bayes' Rule with the Theorem of Total Probability

We devote this subsection to examples illustrating the power of using Bayes' Rule together with the Theorem of Total Probability.

Example 6. **(Communication Channel Reliability)** Consider a noisy communication channel over which a 0 or a 1 is to be sent. Suppose that the probability the bit to be sent is a 0 is 0.4 and the probability that it is a 1 is 0.6. Also, suppose that due to noise, the probability that a 0 is changed to a 1 during transmission is 0.2 and the probability that a 1 is changed to a 0 is 0.1 (see Figure 8.5). Suppose a 1 is received. What is the probability that a 1 was sent?

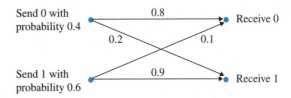

Figure 8.5 Communication channel reliability.

Solution. Let *B* denote the event that a 1 was sent, and let *A* denote the event that a 1 was received. The probability we seek is $P(B|A)$, which according to Bayes' Rule is

$$P(B\mid A) = \frac{P(A\mid B) \cdot P(B)}{P(A)} = \frac{(0.9)(0.6)}{P(A)}$$

What is $P(A)$, the probability that a 1 was received? We are not given this information directly, but we can compute it using the Theorem of Total Probability.

To apply the Theorem of Total Probability, we should check that we know what sample space Ω we are using. We can take Ω to be all 2-tuples of 0's and 1's where the first element gives the transmitted bit and the second element gives the received bit. Note that Ω is *not* a cross product sample space, because the probability of receiving a 1, say, depends on the bit that is transmitted.

Sample space Ω can be regarded as the disjoint union of events with positive probabilities as follows: Let E_1 be the event that a 1 was sent, and let E_2 be the event that a 0 was sent. Hence, $\Omega = E_1 \cup E_2$. By the Theorem of Total Probability, the probability that a 1 was received is

$$P(A) = P(A\mid E_1) \cdot P(E_1) + P(A\mid E_2) \cdot P(E_2)$$
$$= (0.9)(0.6) + (0.2)(0.4)$$
$$= 0.62$$

Now, we can complete the computation of $P(B\mid A)$, the event that a 1 was sent given that a 1 was received:

$$P(B\mid A) = \frac{(0.9)(0.6)}{P(A)} = \frac{(0.9)(0.6)}{0.62} \approx 0.87 \qquad \blacksquare$$

The communication channel reliability situation of Example 6 can be conceptualized as a two-stage process. First, a bit is selected for transmission. The bit is either transmitted correctly or incorrectly. This is not the same as flipping two coins, however, where the result of one flip has no influence on the result of the other. Here, the probability that a bit is transmitted without error depends on whether the bit is a 0 or a 1. As suggested in Example 6, we can choose $\Omega = \{(0, 0), (0, 1), (1, 0), (1, 1)\}$ where the first component of each ordered pair gives the bit transmitted and the second component gives the bit received. What probability density should we assign to this Ω? The next example shows that only one choice is consistent with the data in Figure 8.5.

Example 7. (Communication Channel Reliability Continued) Suppose that the communication channel reliability situation is modeled by the sample space Ω just suggested. Determine a probability density on Ω that is consistent with the given data.

Solution.   Each outcome $\omega = (i, j) \in \Omega$ can be regarded as an event $\{\omega\} \subseteq \Omega$, and the Theorem of Total Probability can be used. The sample space Ω is the disjoint union of two events of positive probability—namely, the event $E_1 = \{(1, 0), (1, 1)\}$ of sending a 1 and the event $E_2 = \{(0, 0), (0, 1)\}$ of sending a 0. According to the data, $P(E_1) = 0.6$, and $P(E_2) = 0.4$. Hence, the event $\{(0, 0)\}$ has probability

$$P(\{(0, 0)\}) = P(\{(0, 0)\} \mid E_1) \cdot P(E_1) + P(\{(0, 0)\} \mid E_2) \cdot P(E_2)$$
$$= (0)(0.6) + (0.8)(0.4)$$
$$= 0.32$$

Event $\{(1, 0)\}$ has probability

$$P(\{(1, 0)\}) = P(\{(1, 0)\} \mid E_1) \cdot P(E_1) + P(\{(1, 0)\} \mid E_2) \cdot P(E_2)$$
$$= (0.1)(0.6) + (0)(0.4)$$
$$= 0.06$$

Similarly,

$$P(\{(0, 1)\}) = (0)(0.6) + (0.2)(0.4) = 0.08$$

and

$$P(\{(1, 1)\}) = (0.9)(0.6) + (0)(0.4) = 0.54$$

The probabilities of these singleton events must agree with the probability density p assigned to Ω, so there is only one choice for p consistent with the data.   ■

Situations that consist of stages depending on the results of preceding stages, such as the channel reliability examples, are called **dependent trial** processes. In designing sample spaces and probability distributions, it is important to consider whether trials are independent or dependent. Sometimes, the dependency of one trial on a preceding trial is not immediately obvious (see Exercise 9 in Section 8.6).

We conclude this section with one more application of Bayes' Rule and the Theorem of Total Probability.

Example 8.   (Identifying the Source of a Bad Apple)   Suppose apples are shipped to a grocery store by three different orchards O_1, O_2, and O_3. Suppose the percentages of each shipment that are bad are 10%, 8%, and 3%, respectively. Suppose further that the percentages of the apple supply from these orchards are 20%, 30%, and 50%, respectively. Now, suppose a customer selects an apple at random and finds that it is bad. What is the probability the apple came from orchard O_2?

Solution.   Let A_i denote the event that an apple (good or bad) selected at random comes from orchard O_i for $i = 1, 2,$ and 3. Hence,

$$P(A_1) = 0.2 \quad P(A_2) = 0.3 \quad P(A_3) = 0.5$$

Let B denote the event that an apple selected at random is bad. We are to determine $P(A_2 \mid B)$. We are given the conditional probabilities

$$P(B \mid A_1) = 0.1 \quad P(B \mid A_2) = 0.08 \quad P(B \mid A_3) = 0.03$$

Hence, from Bayes' Theorem and the Theorem of Total Probability,

$$P(A_2 \mid B) = \frac{P(B \mid A_2) \cdot P(A_2)}{P(B)}$$

$$= \frac{P(B \mid A_2) \cdot P(A_2)}{P(B \mid A_1) \cdot P(A_1) + P(B \mid A_2) \cdot P(A_2) + P(B \mid A_3) \cdot P(A_3)}$$

$$= \frac{(0.08)(0.3)}{(0.1)(0.2) + (0.08)(0.3) + (0.03)(0.5)}$$

$$\approx 0.\#$$

■

8.6   Exercises

1. A sample space $\Omega = \{\omega_1, \omega_2, \ldots, \omega_6\}$ has the following probability density:

$$\begin{aligned} p(\omega_1) &= 2/5 & p(\omega_4) &= 1/10 \\ p(\omega_2) &= 1/10 & p(\omega_5) &= 1/10 \\ p(\omega_3) &= 1/5 & p(\omega_6) &= 1/10 \end{aligned}$$

Which of the following pairs A, B of events are independent? Explain your answer.

(a) $A = \{\omega_1, \omega_2\}$ and $B = \{\omega_3, \omega_5\}$
(b) $A = \{\omega_1, \omega_6\}$ and $B = \{\omega_2, \omega_6\}$
(c) $A = \emptyset$ and $B = \{\omega_3, \omega_4, \omega_5\}$
(d) $A = \{\omega_1, \omega_3\}$ and $B = \{\omega_3, \omega_4, \omega_5\}$

2. Under which of the following circumstances is the pair A, B of events in sample space Ω an independent pair? Explain your answer.

(a) A and B are disjoint, $P(A) > 0$, and $P(B) > 0$
(b) $P(A) = 0$ and $P(B) > 0$
(c) $P(A) = P(B) = 0$

3. Suppose A and B are disjoint events in a sample space Ω. Is it possible that A and B could be independent? Explain your answer.

4. A fair die is rolled, and a fair coin is tossed. The sample space is taken to be $\Omega = \Omega_1 \times \Omega_2$ where Ω_1 is the six-element sample space for the die and Ω_2 is the two-element sample space for the coin. Let $A \subseteq \Omega_1$ be the event "a 5 is rolled." Let $B \subseteq \Omega_2$ be the event "heads." Let $C \subseteq \Omega$ be the event "at most two spots on the top face of the die (with heads or tails on the coin) or at least five spots on the top face of the die together with heads on the coin." Let D be the event "at least a 5 on the die (with heads or tails on the coin)." Which of the following sets of events are independent sets? Explain your answer.

(a) $\{A, B\}$
(b) $\{A, B, C\}$
(c) $\{B, C\}$
(d) $\{B, C, D\}$

5. Suppose that $E_1, E_2, \ldots, E_k$ are events in the same sample space and that some pair E_i, E_j of these events are disjoint.

   (a) If all the events have positive probability, can the set $\{E_1, E_2, \ldots, E_k\}$ be an independent set of events? Explain your answer.

   (b) If one or more of the events has 0 probability, can the set $\{E_1, E_2, \ldots, E_k\}$ be an independent set of events?

6. A fair penny and a fair nickel are tossed. Let A be the event "heads on the penny." Let B be the event "tails on the nickel." Let C be the event "the coins land the same way."

   (a) Choose a sample space Ω, and represent A, B, and C in terms of Ω.

   (b) Which pairs of events chosen from A, B, and C are independent pairs?

   (c) Is the set of events $\{A, B, C\}$ an independent set?

7. Suppose that Ω is a sample space with a probability density function p, and suppose that $A \subseteq \Omega$. Let $P(A)$ denote the probability of A. Assume that $P(A) > 0$. Define a function p_1 on A as follows: For $\omega \in A$, $p_1(\omega) = p(\omega)/P(A)$.

   (a) Show that if $\omega_1, \omega_2 \in A$ and $p(\omega_1), p(\omega_2) \neq 0$, then

   $$\frac{p(\omega_1)}{p(\omega_2)} = \frac{p_1(\omega_1)}{p_1(\omega_2)}$$

   (b) Show that if B and C are nonempty subsets of A with elements that have positive probabilities, then

   $$\frac{P(B)}{P(C)} = \frac{P_1(B)}{P_1(C)}$$

   (c) Show that p_1 is a probability density function on $\Omega_1 = A$.

8. Suppose we have two coins. One is fair, but the other one has two heads. We choose one of them at random and flip it. It comes up heads.

   (a) What is the probability the coin is fair?

   (b) Suppose we flip the same coin a second time. What is the probability that it comes up heads?

   (c) Suppose the coin comes up heads when flipped the second time. What is the probability the coin is fair?

9. A television show features the following weekly game: A sports car is hidden behind one door, and a goat is hidden behind each of two other doors. The moderator of the show invites the contestant to pick a door at random. Then, by tradition, the moderator is obligated to open one of the two doors not chosen to reveal a goat (there are two goats, so there is always such a door to open). At this point, the contestant is given the opportunity to stand pat (do nothing) or to choose the remaining door. Suppose you are the contestant, and suppose you prefer the sports car over a goat as your prize. What do you do? (*Hint:* It may help to model this as a two-stage dependent trials process, but it may not be obvious how to do this).

(a) Suppose you decide to stand with your original choice. What are your chances of winning the car?

(b) Suppose you decide to switch to the remaining door. What are your chances of winning the car?

(c) Suppose you decide to flip a fair coin. If it comes up heads, you change your choice; otherwise, you stand pat. What are your chances of winning the car?

10. Suppose our manufacturing company purchases a certain part from three different suppliers S_1, S_2, and S_3. Supplier S_1 provides 40% of our parts, and suppliers S_2 and S_3 provide 35% and 25%, respectively. Furthermore, 20% of the parts shipped by S_1 are defective, 10% of the parts shipped by S_2 are defective, and 5% of the parts from S_3 are defective. Now, suppose an employee at our company chooses a part at random.

(a) What is the probability that the part is good?

(b) If the part is good, what is the probability that it was shipped by S_1?

(c) If the part is defective, what is the probability that it was shipped by S_1?

11. (a) Give an example that shows three pairwise independent events need not be an independent set of events.

(b) Give an example that shows three events can be independent without having the corresponding pairs of events be independent.

12. Two manufacturing companies M_1 and M_2 produce a certain unit that is used in an assembly plant. Company M_1 is larger than M_2, and it supplies the plant with twice as many units per day as M_2 does. M_1 also produces more defects than M_2. Because of past experience with these suppliers, it is felt that 10% of M_1's units have some defect, whereas only 5% of M_2's units are defective. Now, suppose that a unit is selected at random from a bin in the assembly plant.

(a) What is the probability that the unit was supplied by company M_1?

(b) What is the probability that the unit is defective?

(c) What is the probability that the unit was supplied by M_1 if the unit is defective?

13. When a roulette wheel is spun once, there are 38 possible outcomes: 18 red, 18 black, and 2 green (if the outcome is green the house wins all bets). If a wheel is spun twice, all $38 \cdot 38$ outcomes are equally likely. If you are told that in two spins at least one resulted in a green outcome, what is the probability that both outcomes were green?

14. A computer salesperson makes either one or two sales contacts each day between 1 and 2 PM. If only one contact is made, the probability is 0.2 that a sale will result and 0.8 that no sale will result. If two contacts are made, the two customers make their decisions independently of each other, each purchasing with probability 0.2 and not purchasing with probability 0.8. What is the probability that the salesperson has made two sales this hour?

15. Only 1 in 1000 adults is afflicted with a particular rare disease for which a diagnostic test has been developed. The test is such that when an individual actually has the disease, a positive result will occur 99% of the time, and an individual without the disease will show a positive test result only 2% of the time. If a randomly selected individual is tested and the result is positive, what is the probability that the individual has the disease? Draw a tree diagram for the problem.

16. Suppose three fair coins are tossed. What is the probability that precisely two coins land heads up if the first coin lands heads up and the second coin lands tails up?

Discrete Random Variables

Many probabilistic experiments have outcomes that are associated with real numbers. For example, a gambling game may involve spinning a wheel that is divided into segments that offer various payoffs or penalties (see Figure 8.6). The outcomes (wheel segments) of the experiment have real numbers (cash values) associated with them. In the case of flipping a coin n times, each outcome (n-tuple of heads and tails) can be associated with the total number of heads in that outcome. Of course, we are accustomed to using numbers, typically integers, as labels. For example, we use 1 to mean heads and 6 to mean a die that shows six spots. However, our interest in this section is to study the situation in which outcomes are associated with numerical values that are not used simply as labels for the outcome.

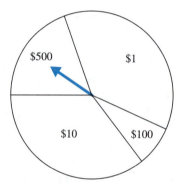

Figure 8.6   Payoffs associated with outcomes.

8.7.1   Distributions of a Random Variable

A typical question involving numerical values is how much one should be willing to pay to gamble on a game that offers various payoffs with various probabilities. The following definition explains what it means to associate real numbers with outcomes.

Definition 1.   A **discrete random variable** X is a real-valued function with a domain that is a discrete (finite or countably infinite) sample space Ω endowed with a probability density function. In other words, $X(\omega) \in \mathbb{R}$ for each $\omega \in \Omega$.

Example 1.   A game consists of rolling a fair die. You win $3 times the value on the top face after the roll if the top face is 2 or 5. The game pays $1 if the top face after the roll is 1, 3, 4, or 6. Define the random variable associated with this game.

Solution.   Let X be the random variable defined as $X(1) = 1$, $X(2) = 6$, $X(3) = 1$, $X(4) = 1$, $X(5) = 15$, and $X(6) = 1$ where $1, 2, \ldots, 6$ represent the value on the top face after rolling a fair die. The probability density for the roll of the die is the uniform probability density on $\{1, 2, 3, 4, 5, 6\}$.   ■

The term *random variable* is unfortunate, because X is not a variable in the mathematical sense and also is not random. Each $\omega \in \Omega$ is sent by X to a uniquely defined real number $X(\omega)$. In other words, X is a **function.** Nevertheless, we shall use the term *random variable,* because it is standard terminology.

Because the domain of X is a sample space endowed with a probability density function, the range of X—namely,

$$\{x : X(\omega) = x \text{ for some } \omega \in \Omega\}$$

which is a subset of $\mathbb{R}$—can easily be made into a sample space Ω_X and endowed with its own probability density. To see this, note that the preimage of any element x in the range of X is the event

$$\{\omega \in \Omega : X(\omega) = x\} \subseteq \Omega$$

Notation.   The event $\{\omega \in \Omega : X(\omega) = x\} \subseteq \Omega$ will be denoted by $(X = x)$. The probability of this event,

$$P(\{\omega \mid X(\omega) = x\})$$

will be denoted by $P(X = x)$.

The next theorem shows that the assignment of the probability $P(X = x)$ to each x defines a probability density function on the range of X.

Theorem 1.   (Probability Density Defined by a Random Variable)   Let Ω_X be the range of a random variable X defined on a sample space Ω endowed with probability density function p. Then, $p_X(x) = P(X = x)$ defines a probability density function on Ω_X.

Proof.   Since each $p_X(x)$ is the probability of an event in Ω, we have $0 \le p_X(x) \le 1$. To see that $\sum p_X(x) = 1$, we observe that the sets

$$\{\omega \in \Omega : X(\omega) = x\}$$

where x ranges over Ω_X form a partition of Ω. Hence, by the Additive Principle of Disjoint Events, the probabilities of these events sum to 1.   ∎

Definition 2.   The probability density function defined on the range of random variable X by $p_X(x) = P(X = x)$ is called the **distribution** of X.

Example 2.   Let the random variable X be defined on the value of the top face after the roll of a fair die as follows: $X(1) = 1$, $X(2) = 2$, $X(3) = 2$, $X(4) = 3$, $X(5) = 2$, and $X(6) = 3$. Determine the distribution of X.

Solution.   The range of the random variable X is $\{1, 2, 3\}$. Probability $p_X(1) = P(X = 1) = P(\{1\}) = 1/6$; $p_X(2) = P(X = 2) = P(\{2, 3, 5\}) = 1/2$; $p_X(3) = P(X = 3) = p(\{4, 6\}) = 1/3$.   ∎

8.7.2 The Binomial Distribution

A distribution is simply a density that arises in a special way—that is, from a random variable. Several distributions occur so frequently in applications that they have been given special names. Let us look at a few examples.

Suppose a coin is flipped n times where the probability of heads is p and the probability of tails is $q = 1 - p$. We consider the usual sample space Ω consisting of the 2^n possible sequences of heads and tails. Suppose a random variable X is defined on Ω by setting $X(\omega)$ equal to the number k of times heads occurs in the n-tuple ω where $1 \leq k \leq n$. Then, the range Ω_X of X is $\{0, 1, \ldots, n\}$, and the distribution p_X of X is given by

$$p_X(k) = C(n, k) \cdot p^k \cdot (1 - p)^{n-k} \text{ for } k = 0, 1, \ldots, n$$

This distribution is called the binomial distribution, and X is said to be **binomially distributed.**

Notice that the value of the binomial distribution for a particular k depends on n and p as well as on k.

Definition 3. The **binomial distribution** is the probability density function defined on $\{0, 1, \ldots, n\}$ by assigning to $k = 0, 1, \ldots, n$ the value

$$b(k; n, p) = C(n, k) \cdot p^k \cdot (1 - p)^{n-k}$$

The binomial distribution gets its name from the fact that the Binomial Theorem can be used to prove it is a valid density function. (The reader should try this.) The binomial distribution is generally thought of as arising from a random variable X that counts successes (or failures) for n independent repetitions of a two-outcome experiment. Certainly, however, nothing is wrong with assigning a density $p(k) = b(k; n, p)$ to the sample space $\{0, 1, \ldots, n\}$ without thinking of a random variable. Use of the term *distribution* instead of probability *density* simply indicates that we have a random variable in mind.

The value $b(k; n, p)$ of the binomial distribution has another interpretation: It is the probability of getting exactly k red balls if we draw a ball n times from an urn in which the proportion of red balls is p and we replace *the ball after each draw.*

8.7.3 The Hypergeometric Distribution

A second important distribution can be motivated by the process of drawing balls from an urn repeatedly without replacing a ball after it is drawn. Suppose an urn contains m balls, of which r are red and the rest are black. Thus, there are $m - r$ black balls. We will draw n balls where $n \leq m$, without replacing them. If we do not replace the ball each time, then the probability of getting a red ball on a particular draw depends on what happened on previous draws. To calculate the probability that we will draw exactly k red balls (and, hence, exactly $n - k$ black balls), imagine that we pick a handful of n balls from the urn all at once where each such handful is equally likely. We choose the sample space Ω to be the set of all possible handfuls of n balls, so each element of Ω is actually a set of n balls. We put a uniform density on Ω that consists of $C(m, n)$ outcomes. Now, let a random variable X be defined on Ω as follows:

$$X(\omega) = \text{the number of red balls in } \omega$$

Hence, X is a function from Ω to $\{0, 1, 2, \ldots, n\}$. The probability in Ω of the event $(X = k)$ is computed as follows: The number of handfuls that contain k red balls (and $n - k$ black balls) is $C(r, k) \cdot C(m - r, n - k)$, because there are $C(r, k)$ ways to choose k red balls from the total number r of red balls and $C(m - r, n - k)$ ways to choose $n - k$ black balls from the total number $m - r$ of black balls. Since Ω has size $C(m, n)$, the probability of each handful is $1/C(m, n)$. Hence, the probability $P(X = k)$ of obtaining exactly k red balls, which is the summation of $p(\omega)$ over all ω containing exactly k red balls, is

$$P(X = k) = \frac{C(r, k) \cdot C(m - r, n - k)}{C(m, n)}$$

Since $P(X = k)$ depends on n, r, and m as well as on k, we will denote $P(X = k)$ by $h(k; n, r, m)$.

Definition 4. For integer parameters k, n, r, and m satisfying $0 \le k \le n \le m$ and $k \le r$, the **hypergeometric distribution** is the probability density function defined on $\{0, 1, 2, \ldots, m\}$ by assigning to $k = 0, 1, 2, \ldots, m$ the value

$$h(k; n, r, m) = \frac{C(r, k) \cdot C(m - r, n - k)}{C(m, n)}$$

The following example will help clarify this discussion.

Example 3. Suppose that 25 of 1000 parts in a bin are defective. What is the probability that a randomly chosen set of 10 parts contains at least one defective part? (The parts are chosen without replacement.)

Solution. The probability can be obtained by subtracting from 1 the probability that none of the 10 parts is defective. This latter probability can be expressed in terms of the hypergeometric distribution. We have $m = 1000$ parts (or balls in an urn), of which $r = 25$ are defective (red). We draw a handful of $n = 10$ and ask for the probability that $k = 0$ are defective. Hence,

$$P(0 \text{ defects}) = h(0; 10, 25, 1000) = \frac{C(25, 0) \cdot C(975, 10)}{C(1000, 10)}$$

This can be rewritten as

$$1 \cdot \frac{975 \cdot 974 \cdots 966}{10!} \cdot \frac{10!}{1000 \cdot 999 \cdots 991} = \frac{975}{1000} \cdot \frac{974}{999} \cdots \frac{966}{991}$$

$$\approx 0.7754$$

The probability of at least one defective part in a handful of 10 chosen at random is therefore

$$P(\text{at least 1 defect}) = 1 - P(0 \text{ defects}) = 1 - 0.7754 = 0.2246 \qquad \blacksquare$$

When m, r, and $m - r$ are large compared to n, the binomial density $b(k; n, r/m)$ can be used to approximate $h(k; n, r, m)$. This approximation is based on the fact that the proportion of red balls changes very little as we make the first few draws. For the last example, the binomial approximation to

$$h(k; n, r, m) = h(0; 10, 25, 1000)$$

is given by

$$b(0; 10, 25/1000) = C(10, 0) \cdot (25/1000)^0 \cdot (1 - 25/1000)^{10}$$
$$= (975/1000)^{10}$$
$$\approx 0.7763$$

Being able to approximate one distribution with another can be useful, especially if the approximating distribution is easier to compute.

8.7.4   Expectation of a Random Variable

Suppose we choose a card from a shuffled deck. If we draw a spade, we must pay a $10 penalty, but if we draw a card from any other suit, we receive $5. How do we estimate our winnings in playing the game a large number n of times? If we believe that a spade will be drawn approximately $n/4$ times, then we estimate our winnings as

$$(\$5)\left(\frac{3n}{4}\right) + (-\$10)\left(\frac{n}{4}\right) = (\$1.25)n$$

and we estimate our average winning per draw to be $1.25.

In mathematical terms, the dollar amount associated with each card defines a real-valued random variable X on the sample space Ω of cards, each of which is assigned a probability of $1/52$. The range of X consists of just two values: $5, with probability $3/4$ of occurring, and $-$10, with probability $1/4$ of occurring. The sum of these payoffs, weighted by their probabilities, is

$$(\$5)\left(\frac{3}{4}\right) + (-\$10)\left(\frac{1}{4}\right) = \$1.25$$

This quantity is called the expectation of X. When we interpret probabilities as estimates of the frequency of occurrence of real-world events, then the expectation of X estimates the average value of X when the underlying experiment is repeated many times. In general, the expectation of X does not tell us what we expect to happen on any particular trial of the experiment. Only under very special circumstances, such as the ones described in the Law of Averages given at the end of Section 8.9, does the expectation tell us what to expect.

Definition 5.   The **expectation** (also called **mean** or **expected value**) $E(X)$ of a random variable X is defined by

$$E(X) = \sum x \cdot p_X(x)$$

where the sum is taken over all x in the range Ω_X of X and $p_X(x) = P(X = x)$, the probability that x occurs.

Notation.   When only one **random variable** is under discussion, the value $E(X)$ is often denoted by μ.

Example 4.   A loaded die pays $1 for an even number of spots occurring on the top face after a roll, and charges $1, $3, or $5 for showing one, three, or five spots after a roll, respectively. Suppose the probability $p(\omega)$ of getting

$$\omega = 1, 2, 3, 4, 5, 6$$

spots is, respectively,

$$\frac{1}{6}, \frac{1}{2}, \frac{1}{12}, \frac{1}{12}, \frac{1}{12}, \frac{1}{12}$$

These probabilities define a random variable X with domain

$$\Omega = \{1, 2, 3, 4, 5, 6\}$$

and range

$$\Omega_X = \{\$1, -\$1, -\$3, -\$5\}$$

Calculate the distribution $p_{\hat{X}}$ induced by X on its range $\Omega_{\hat{X}}$, and then determine $E(X)$.

Solution.   The outcomes in Ω that have a \$1 payoff are 2, 4, and 6, so

$$p_X(\$1) = p(2) + p(4) + p(6) = \frac{1}{2} + \frac{1}{12} + \frac{1}{12} = \frac{2}{3}$$

Similarly,

$$p_X(-\$1) = p(1) = \frac{1}{6}$$

and

$$p_X(-\$3) = p_X(-\$5) = p(3) = p(5) = \frac{1}{12}$$

Now that we know p_X, we can compute $E(X)$:

$$E(X) = (\$1) \cdot p_X(\$1) + (-\$1) \cdot p_X(-\$1) + (-\$3) \cdot p_X(-\$3) + (-\$5) \cdot p_X(-\$5)$$

$$= (\$1)\left(\frac{2}{3}\right) + (-\$1)\left(\frac{1}{6}\right) + (-\$3)\left(\frac{1}{12}\right) + (-\$5)\left(\frac{1}{12}\right) = -\$\frac{1}{6} \qquad \blacksquare$$

Note that the expectation of a random variable does not tell us what to "expect" the value of X to be. In the above example, we can never owe $-\$1/6$ as a result of rolling the die once. We can only get one of the four values in Ω_X. A more extreme case would be to flip a fair coin that pays \$100 for heads and charges \$100 for tails. The expected value of the payoff is zero, which is far from either of the two actual payoffs.

Since the probability $p_X(x)$ assigned to an element $x \in \Omega_X$ is the probability of the event $(X = x)$ in Ω, we can also compute the expectation of X by returning to the original sample space Ω, as the next theorem shows.

Theorem 2.   (Basic Property of Expectation)   The expectation $E(X)$ of a random variable X defined on a sample space Ω satisfies

$$E(X) = \sum_{\omega \in \Omega} X(\omega) p(\omega)$$

Proof.   By definition,

$$E(X) = \sum_{x \in \Omega_X} x \cdot p_X(x)$$

Substituting $p_X(x) = \sum p(\omega)$, where the summation is over all $\omega \in \Omega$ in the event $(X = x)$, gives

$$E(X) = \sum_{x \in \Omega_X} \left(x \cdot \sum_{\omega \in (X=x)} p(\omega) \right)$$

$$= \sum_{x \in \Omega_X} \left(\sum_{\omega \in (X=x)} x \cdot p(\omega) \right)$$

$$= \sum_{x \in \Omega_X} \sum_{\omega \in (X=x)} X(\omega) \cdot p(\omega)$$

Suppose that $\Omega_X = \{x_X, \ldots, x_n\}$. Then, the double sum above can be written out as the following sum of sums:

$$\sum_{\omega \in (X=x_1)} X(\omega) \cdot p(\omega) + \cdots + \sum_{\omega \in (X=x_n)} X(\omega) \cdot p(\omega).$$

Since the events $(X = x_1), \ldots, (X = x_n)$ partition Ω, each $\omega \in \Omega$ is contributing exactly $X(\omega) \cdot p(\omega)$ to this sum of sums. Hence,

$$E(X) = \sum_{\omega \in \Omega} X(\omega) \cdot p(\omega). \qquad \blacksquare$$

Example 5.   Compute the expectation of the random variable X in Example 4 by using Theorem 2.

Solution.   Instead of computing the probability distribution p_X induced by X on Ω_X, we work directly with the original sample space Ω.

$$E(X) = \sum_{\omega \in \Omega} p(\omega)X(\omega)$$

$$= (-\$1)p(1) + (\$1)p(2) + (-\$3)p(3) + (\$1)p(4) + (-\$5)p(5) + (\$1)p(6)$$

$$= (-\$1)\frac{1}{6} + (1)\frac{1}{2} + (-\$3)\frac{1}{12} + (\$1)\frac{1}{12} + (-\$5)\frac{1}{12} + (\$1)\frac{1}{12}$$

$$= -\$\frac{1}{6}$$

which agrees with our previous answer. $\blacksquare$

8.7.5   The Sum of Random Variables

Suppose that $X_1, \ldots, X_n$ are random variables defined on some sample space Ω. We can use these to define new random variables on Ω in various ways. For example, we can define a random variable *SUM* by setting

$$SUM(\omega) = X_1(\omega) + \cdots + X_n(\omega)$$

for each $\omega \in \Omega$.

Example 6.   Suppose we toss three fair coins. We associate with this experiment a sample space Ω of eight triples each having probability $1/8$. The first, second, and third components of each triple is 1 or 0 depending on whether the first, the second, or the third coin lands heads or tails, respectively. For a triple $\omega \in \Omega$, let $X_1(\omega)$ be 1 if the first coin lands heads up and 0 if otherwise. Define $X_2(\omega)$ and $X_3(\omega)$ in similar fashion for the second and the third coins, respectively. If the first coin lands heads and the other two land tails, what is the value of *SUM* on that outcome?

Solution.   Since $\omega = (1, 0, 0)$, we have $X_1(\omega) = 1$ and $X_2(\omega) = X_3(\omega) = 0$. Hence,

$$SUM(\omega) = 1 + 0 + 0 = 1 \qquad \blacksquare$$

Example 7.   What is the expectation of *SUM* in the preceding example?

Solution.   The range Ω_X of X is the number of heads that can show, so $\Omega_X = \{0, 1, 2, 3\}$. The probability distribution induced by X on Ω_X is

$$p_X(0) = p(\{(0, 0, 0)\}) = \frac{1}{8}$$

$$p_X(1) = p(\{(1, 0, 0), (0, 1, 0), (0, 0, 1)\}) = \frac{3}{8}$$

$$p_X(2) = p(\{(1, 1, 0), (1, 0, 1), (0, 1, 1)\}) = \frac{3}{8}$$

$$p_X(3) = p(\{(1, 1, 1)\}) = \frac{1}{8}$$

Hence,

$$E(SUM) = \sum_{x \in \Omega_X} x \cdot p_X(x)$$

$$= 0 \cdot \frac{1}{8} + 1 \cdot \frac{3}{8} + 2 \cdot \frac{3}{8} + 3 \cdot \frac{1}{8}$$

$$= \frac{3}{2} \qquad \blacksquare$$

We now establish notation for some common random variables that are defined in terms of one or more others. The symbols for the new random variables will indicate how they are computed from the old ones rather than simply being single letters like X. When we are given several random variables $X_1, \ldots, X_n$, we will assume they are all defined on the same sample space Ω.

The sum of random variables.   The random variable that sends ω to

$$X_1(\omega) + \cdots + X_n(\omega)$$

is denoted by

$$(X_1 + \cdots + X_n)$$

Its expectation is denoted by $E(X_1 + \cdots + X_n)$. Hence, the random variable *SUM* in the preceding example would be denoted $(X_1 + X_2 + X_3)$. The value of

$$(X_1 + X_2 + \cdots + X_n)$$

at a particular ω is denoted by

$$(X_1 + X_2 + \cdots + X_n)(\omega)$$

The scaling of a random variable.   Given random variable X and a real number k, the random variable that sends ω to $kX(\omega)$ is denoted by (kX), and its expectation is denoted by $E(kX)$. When $k \neq 0$, we can define a random variable (X/k) with expectation $E(X/k)$ in similar fashion.

The square of the deviation from a fixed value.   Given random variable X and a real number k, the random variable that sends ω to $(X(\omega) - k)^2$ is denoted by $(X - k)^2$. When $k = 0$, we simply write X^2.

Constant random variables.   Given a real number k, the random variable that sends ω to k is denoted by k and is called a **constant random variable.**

When it is clear which set is being summed over in a summation $\sum$, we will often omit the subscript on the summation $\sum$. Similarly, we will often omit subscripts on products $\prod$.

The next theorem relates the expectation of the sum and the scaling of random variables to the expectations of the original random variables.

Theorem 3.   (Linearity of Expectation)   Let $X_1, X_2, \ldots, X_n$, and X be random variables defined on the same sample space Ω, and let k be a real number. Then:

(a) $E(X_1 + \ldots + X_n) = \displaystyle\sum_{i=1}^{n} E(X_i)$.

(b) $E(kX) = kE(X)$, and where $k \neq 0$, $E(X/k) = \dfrac{E(X)}{k}$

(c) $E\left(\dfrac{X_1 + \ldots + X_n}{n}\right) = \dfrac{1}{n} \displaystyle\sum_{i=1}^{n} E(X_i)$.

Proof.   (a) We prove part (a) for $n = 2$ and leave the proof (by induction) for $n > 2$ as an exercise. (All unsubscripted sums $\sum$ will be taken over $\omega \in \Omega$.) From the formula for expectation given in Theorem 2, we have

$$
\begin{aligned}
E(X_1 + X_2) &= \sum ((X_1 + X_2)(\omega)) \cdot p(\omega) \\
&= \sum (X_1(\omega) + X_2(\omega)) \cdot p(\omega) \\
&= \sum (X_1(\omega) \cdot (p(\omega) + X_2(\omega)) \cdot p(\omega)) \\
&= \sum X_1(\omega) p(\omega) + \sum X_2(\omega) p(\omega) \\
&= E(X_1) + E(X_2)
\end{aligned}
$$

(b) Since division by $k \neq 0$ can be regarded as multiplication by $1/k$, the second half of part (b) follows from the first half, which can be established as follows:

$$
\begin{aligned}
E(kX) &= \sum ((kX)(\omega)) \cdot p(\omega) \\
&= \sum k \cdot X(\omega) \cdot p(\omega) \\
&= k \sum X(\omega) \cdot p(\omega) \\
&= kE(X)
\end{aligned}
$$

(c) From part (b), with

$$X = (X_1 + X_2 + \cdots + X_n)$$

and $k = n$, we have that

$$E\left(\frac{X_1 + X_2 + \cdots + X_n}{n}\right) = \frac{1}{n}E(X_1 + X_2 + \cdots + X_n)$$

so by part (a), this is

$$\left(\frac{1}{n}\right)\sum_{i=1}^{n} E(X_i)$$ ∎

8.8   Exercises

1. Consider a game based on the days of a 31-day month. A day is chosen at random—say, by spinning a spinner. The prize is a number of dollars equal to the sum of the digits in the date of the chosen day. For example, choosing the 31st of the month pays $3 + $1 = $4, as does choosing the fourth day of the month.

   (a) Set up the underlying sample space Ω and its probability density, the value of which at ω gives the reward associated with ω.
   (b) Define a random variable $X(\omega)$ on Ω with a value at ω that gives the reward associated with ω.
   (c) Set up a sample space Ω_X consisting of the elements in the range of X, and give the probability distribution p_X on Ω_X arising from X.
   (d) Determine $P(X = 6)$.
   (e) Determine $P(2 \le X \le 4) = P(\omega : 2 \le X(\omega) \le 4)$.
   (f) Determine $P(X > 10) = P(\omega : X(\omega) > 10)$.

2. Consider the following darts game: The target consists of a bull's-eye, which is a circle of radius 1, surrounded by a middle ring of outer radius 3 and inner radius 1; this region is in turn surrounded by another ring of outer radius 5 and inner radius 3. If you hit the bull's-eye, you win $10 plus the opportunity to throw again. If you hit the middle ring, you lose $2, and if you hit the outer ring, you lose $5. You must stop throwing as soon as you make a losing toss or hit three bull's-eyes. Suppose the probability that you hit a region is proportional to its area.

   (a) Set up an underlying sample space Ω and its probability density p.
   (b) Define a random variable X on Ω.
   (c) Define a sample space Ω_X and a probability distribution p_X on Ω_X.

3. Suppose we flip a fair coin four times. We are interested in counting the number of times the coin turns up heads.

   (a) Define a sample space Ω and a probability density p on Ω.
   (b) Define a random variable X on Ω to count the number of heads.
   (c) Describe the event $(X = 3)$ as a subset of Ω.
   (d) Set up a sample space Ω_X and a probability distribution p_X based on X. You may express your answer in terms of the binomial distribution.
   (e) Which number (or numbers) of heads is most likely to occur?

4. Repeat Exercise 3 for a coin that comes up heads a third of the time.
5. What is the relationship between $b(k; n, p)$ and $b(n - k; n, p)$ when $p = 1/2$? Does this relationship hold if $p \neq 1/2$?
6. Suppose we make draws from an urn containing two red balls and three black ones, replacing the chosen ball after each draw. How many draws should we make (we have to decide this number in advance) to have probability 0.5 or greater of selecting at least two red balls?
7. Suppose we draw three balls from an urn containing two red balls and three black balls. We do not replace the balls after we draw them. In terms of the hypergeometric distribution, what is the probability of getting two red balls? Compute this probability.
8. Compute the expectation $E(X)$ of the random variable X that counts the number of heads in four flips of a fair coin. (See Exercise 3.)
9. Compute the expectation $E(X)$ of the random variable X that counts the number of heads in four flips of a coin that lands heads with frequency $1/3$.
10. Suppose we make three draws from an urn containing two red balls and three black ones. Determine the expected value of the number of red balls drawn in the following situations.

   (a) The chosen ball is replaced after each draw.
   (b) The chosen ball is not replaced after each draw.

11. We have seen two ways to compute the expected value $E(X)$ of a random variable X. One way is to use the definition of expected value, summing $p_X(x)$ over the range Ω_X of X. The other way is to use Theorem 2, summing $X(\omega) \cdot p(\omega)$ over the domain Ω of X. If you have not already done so, do Exercise 10, and then compute $E(X)$ using the method you did not use the first time. In general, which method do you think will be easier to carry out, and why?
12. Game A has you roll a fair die once and receive the number of dollars that is equal to the value on the top face. Game B has you roll a fair die twice and receive the number of dollars that is the maximum of the two values that show on the top face. It costs $3 to play game A and $4 to play game B. Which game would you choose?
13. Wagga Wagga University has 15,000 students. Let X be the number of courses for which a randomly chosen student is registered. No student is registered for more than seven courses, and each student is registered for at least one course. The number of students registered for i courses where $1 \leq i \leq 7$ is 150, 450, 1950, 3750, 5850, 2550, and 300, respectively. Compute the expected value of the random variable X.

8.9   Variance, Standard Deviation, and the Law of Averages

The actual value $X(\omega)$ of a random variable X can differ drastically from its expected value $E(X)$. To measure how well the values of X tend to cluster around its expectation $E(X)$, we define in this section the notions of variance and standard deviation of a random variable.

Suppose we repeatedly run an experiment, and suppose each time we compute the average of a certain set of random variables. We do not anticipate that we will get the same number for the average each time, but we do wonder how this number will behave. Will we see wild fluctuations in the number from one run of the experiment to the next? This section studies this issue, and it concludes with a technical version of a theorem known informally as the Law of Averages. The theorem says that under the right circumstances, the values of the average of random variables computed from one run of an experiment to another will cluster around the same number; the expected value of their average. The circumstances under which the Law of Averages holds involve the notion of independent random variables, which are also introduced in this section.

8.9.1   Variance and Standard Deviation

To measure how much the actual values of a random variable cluster around its expectation, we define the variance and the standard deviation of a random variable. These measurements of clustering are very important in practice.

Definition 1.   Let X be a random variable defined on the sample space Ω. The **variance** $Var(X)$ of a random variable X is defined as

$$Var(X) = \sum_{\omega \in \Omega} \left(X(\omega) - \mu)^2 \right) \cdot p(\omega)$$

where $\mu = E(X)$.

Definition 2.   The **standard deviation** σ of a random variable X is defined as

$$\sigma = \sqrt{Var(X)}$$

Example 1.   Suppose we flip two fair coins and list a pair (i, j) for each possible outcome where $i = 1$ if the coin ends up heads and $i = 0$ if it ends up tails. Define j similarly for the second coin. The four ordered pairs determine a sample space Ω. Use the uniform probability density function on Ω. Define a random variable X on Ω that counts the number of 1's in an element of Ω. Determine the variance and the standard deviation of X.

Solution.   First, compute p_X for each value of X: $p_X(0) = 1/4$, $p_X(1) = 1/2$, and $p_X(2) = 1/4$. Therefore, $E(X) = 1$. Now,

$$V(X) = \frac{1}{4} \cdot (-1)^2 + 0^2 + 0^2 + 1^2) = \frac{1}{2}$$

with standard deviation $1/\sqrt{2}$.   ■

Theorem 1.   (Properties of Variance)   Let X be a random variable defined on sample space Ω endowed with a probability density p, and let $\mu = E(X)$. Then:

(a) $Var(X) = E((X - \mu)^2) = E(X^2) - \mu^2$.
(b) $Var(kX) = k^2 Var(X)$ for real numbers k.

Warning.   In general, $E(Y^2) \neq E^2(Y)$, and $Var(X_1 + X_2) \neq Var(X_1) + Var(X_2)$.

Proof.

(a) Computing the expectation of the random variable $(X - \mu)^2$ gives

$$E((X - \mu)^2) = \sum [(X - \mu)^2(\omega) \cdot p(\omega)] = \sum (X(\omega) - \mu)^2) \cdot p(\omega)$$

which by definition is $Var(X)$.

Regarding the random variable $(X - \mu)^2$ as the sum of the random variable X^2, the random variable $-2\mu X$, and the constant random variable μ^2 allows us to apply Theorem 3(a) **(Linearity of Expectation)** from Section 8.7.5:

$$E((X - \mu)^2) = E(X^2 - 2\mu X + \mu^2)$$
$$= E(X^2) + E(-2\mu X) + E(\mu^2)$$
$$= E(X^2) - 2\mu E(X) + \mu^2$$
$$= E(X^2) - \mu^2$$

since $E(X) = \mu$.

(b) From Theorem 3(b) in Section 8.7.5, we have $E(kX) = k\mu$. Applying part (a) of the theorem, we are now proving the random variable (kX) gives

$$Var(kX) = E((kX)^2) - (k\mu)^2 = E(k^2X^2) - k^2\mu^2$$

By Theorem 3(b) of the previous section, $E(k^2X^2) = k^2E(X^2)$, so

$$Var(kX) = k^2E(X^2) - k^2\mu^2 = k^2(E(X^2) - \mu^2) = k^2Var(X) \qquad ∎$$

Now that we have shown how expectation is calculated for various operations, we see how to use these results.

Example 2.   Suppose we toss a fair coin, associating 1 with heads and -1 with tails. Thus, we have a sample space $\Omega = \{\text{heads, tails}\}$, with $p(\text{heads}) = p(\text{tails}) = 1/2$ and a random variable X defined on Ω. The range of X is $\Omega_X = \{-1, 1\}$, with $p_X(-1) = p_X(1) = 1/2$. What is $Var(X)$?

Solution.   First, we need a value for μ, so we compute

$$E(X) = \left(\frac{1}{2}\right)(-1) + \left(\frac{1}{2}\right)(1) = 0 = \mu$$

Hence,

$$Var(X) = E((X - \mu)^2) = E(X^2)$$

However, $E(X^2)$ is not $E^2(X) = 0$. In fact,

$$E(X^2) = \sum (X(\omega))^2 \cdot p(\omega) = \left(\frac{1}{2}\right)(1)^2 + \left(\frac{1}{2}\right)(-1)^2 = 1$$

Hence, $Var(X) = 1$. $\qquad ∎$

The next theorem shows that when the variance of a random variable is small, this can be interpreted as meaning that its values tend to cluster around its expected value (mean).

Theorem 2.   (Bound on the Probability of Deviation from the Expected Value)   Let X be a random variable on sample space Ω, and let μ and σ^2 denote the mean and variance of X respectively. For $\epsilon > 0$, let $P(|X - \mu| \geq \epsilon)$ denote the probability of the event that $X(\omega)$ differs from μ by ϵ or more. Then, for all $\epsilon > 0$,

$$P(|X - \mu| \geq \epsilon) \leq \frac{\sigma^2}{\epsilon^2}$$

Proof. By definition,

$$\sigma^2 = \sum (X(\omega) - \mu) \cdot p(\omega)$$

Since the summands are all non-negative, the value of the sum certainly cannot increase if we leave out some of its terms. In particular, let

$$A = \{\omega : |X(\omega) - \mu| \geq \epsilon\}$$

and just sum over $\omega \in A$. This gives

$$\sigma^2 \geq \sum_{\omega \in A} (X(\omega) - \mu)^2 \cdot p(\omega)$$

$$\geq \sum_{\omega \in A} \epsilon^2 p(\omega)$$

$$= \epsilon^2 P(A)$$

$$= \epsilon^2 P(|X - \mu| \geq \epsilon)$$

Hence,

$$P(|X - \mu| \geq \epsilon) \leq \frac{\sigma^2}{\epsilon^2}$$ ∎

Next, we will use Theorem 2 to prove an important result—the Law of Averages.

8.9.2   Independent Random Variables

Suppose we run an experiment described by a sample space Ω, with random variables X_1 and X_2 defined on Ω. After the experiment produces some outcome $\omega \in \Omega$, someone tells us the value $x_1 = X_1(\omega)$ without telling us the outcome ω. Does this help us to guess the value $x_2 = X_2(\omega)$? The answer depends on what the random variables are. Knowing the value of one may determine the value of the other for some outcomes in Ω. On the other hand, knowing the value of one sometimes gives no information about the other. Of course, the outcomes that give rise to the values x_i for $i = 1, 2$ form events

$$\{\omega : X_i(\omega) = x_i\}$$

in Ω, so we are really asking whether those events are independent.

As usual, we denote events of the form

$$\{\omega : X_i(\omega) = x_i\}$$

by $(X_i = x_i)$ where x_i is some value in the range of X_i. We denote the event

$$\{\omega : X_1(\omega) = x_1, \ldots, X_k(\omega) = x_k\}$$

by $(X_1 = x_1, \ldots, X_k = x_k)$.

Definition 3.   Let $X_1, X_2, \ldots, X_n$ be random variables defined on a sample space Ω endowed with probability density p. Then, $X_1, X_2, \ldots, X_n$ are said to be **independent random variables** if and only if for every choice of $x_1, x_2, \ldots, x_n$ such that x_i is contained in the range of X_i for $i = i, 2, \ldots, n$, the events

$$(X_1 = x_1), (X_2 = x_2), \ldots, (X_n = x_n)$$

form an independent set. In other words (recall Definition 2 of Section 8.5.1:

$$P(X_1 = x_1, X_2 = x_2, \ldots, X_n = x_n) = \prod P(X_i = x_i)$$

Example 3. Suppose we toss a fair penny and a fair nickel. The associated sample space Ω consists of four ordered pairs, each having probability $1/4$. Let us define random variables X_1 and X_2 on Ω by

$$X_1(\omega) = \begin{cases} 1 & \text{if the coins agree} \\ -1 & \text{otherwise} \end{cases}$$

$$X_2(\omega) = \text{the number of heads in outcome } \omega$$

Are X_1 and X_2 independent?

Solution. First, we compute the probability distributions induced on $\Omega_{X_1} = \{-1, 1\}$ and $\Omega_{X_2} = \{0, 1, 2\}$:

$$p_{X_1}(1) = p_{X_1}(-1) = \frac{1}{2}$$

$$p_{X_2}(0) = \frac{1}{4}$$

$$p_{X_2}(1) = \frac{1}{2}$$

$$p_{X_2}(2) = \frac{1}{4}$$

There are $2 \cdot 3 = 6$ ways to choose $x_1 \in \Omega_{X_1}$ and $x_2 \in \Omega_{X_2}$. If X_1 and X_2 were independent, we would have to prove it by verifying that for each of the six combinations of x_1 and x_2 the events $(X_1 = x_1)$ and $(X_2 = x_2)$ are independent. However, X_1 and X_2 are not independent. To prove this requires only that we find one case for which the formula in Definition 2 of Section 8.5.1 does not hold. Consider $x_1 = 1$ and $x_2 = 2$. The event

$$(X_1 = 1, \ X_2 = 2) = \{(\text{heads, heads})\}$$

has probability $1/4$. On the other hand,

$$P(X_1 = 1) = p_{X_1}(1) = \frac{1}{2} \quad \text{and} \quad P(X_2 = 2) = p_{X_2}(2) = \frac{1}{4}$$

Hence, the random variables X_1 and X_2 are not independent:

$$P(X_1 = 1, X_2 = 2) = \frac{1}{4} \neq P(X_1 = 1) \cdot P(X_2 = 2) = \frac{1}{8} \qquad \blacksquare$$

Definition 4. Let $X_1, X_2, \ldots, X_n$ be random variables defined on the same sample space Ω. Then $X_1, X_2, \ldots, X_n$ are said to be **independent, identically distributed (i.i.d.) random variables** if and only if

(a) $X_1, X_2, \ldots, X_n$ are independent.
(b) $X_1, X_2, \ldots, X_n$ have the same range $\Omega_{X_1} = \ldots = \Omega_{X_n}$, denoted Ω_X.
(c) $X_1, X_2, \ldots, X_n$ induce on Ω_X the same distribution $p_{X_1} = \cdots = p_{X_n}$, denoted p_X.

Properties (b) and (c) imply that i.i.d. random variables share the same expectation, variance, and standard deviation. As the next example shows, Bernoulli trials give rise to i.i.d. random variables.

Example 4. Consider a Bernoulli trials experiment in which a fair coin is tossed n times. The sample space Ω consists of 2^n n-tuples ω, each having probability $(1/2)^n$. Define random variables X_i on Ω by

$$X_i(\omega) = \begin{cases} 1 & \text{if the coin lands heads on the } i\text{th toss} \\ 0 & \text{otherwise} \end{cases}$$

Each X_i has range $\{0,1\}$ and induces $p_X(0) = p_X(1) = 1/2$. Is this set of random variables independent?

Solution. Observe that the X_i's are different as functions. For example, if $n = 2$ and $\omega = $ (heads, tails), then $X_1(\omega) = 1$ whereas $X_2(\omega) = 0$. Now, choose $x_1, x_2, \ldots, x_n$ where each $x_i = 0$ or 1, and determine the probability of the event:

$$(X_1 = x_1, X_2 = x_2, \ldots, X_n = x_n)$$

Since there is exactly one ω that corresponds to any particular choice,

$$P(X_1 = x_1, X_2 = x_2, \cdots, X_n = x_n) = \frac{1}{2^n}$$

On the other hand,

$$P(X_i = x_i) = p_{X_i}(x_i) = 1/2$$

Hence,

$$P(X_1 = x_1, X_2 = x_2, \ldots, X_n = x_n) = \prod P(X_i = x_i) = \frac{1}{2^n}$$

so the X_i are independent. ∎

Sets of independent random variables have certain properties not shared by all sets of random variables. Sets of i.i.d. random variables have even more such properties. In the following discussion, do not assume that a set of random variables is independent, or i.i.d., unless this is explicitly stated. For example, an independent set of random variables is not necessarily an i.i.d. set.

Theorem 3. (Expectation of the Product of Independent Random Variables) Let X_1 and X_2 be independent random variables defined on a sample space Ω with probability density p. Then, $E(X_1 \cdot X_2) = E(X_1) \cdot E(X_2)$.

Warning. If X_1 and X_2 are not independent, this relation does not necessarily hold.

Proof. By definition,

$$E(X_1 \cdot X_2) = \sum_{x \in \Omega_{X_1 \cdot X_2}} x \cdot p_{X_1 \cdot X_2}(x)$$

where $p_{X_1 \cdot X_2}(x)$ is the probability of the event

$$(X_1 \cdot X_2 = x) = \{\omega \in \Omega | X_1(\omega) \cdot X_2(\omega) = x\} \subseteq \Omega$$

Since there may be several pairs of values $x_1 \in \Omega_{X_1}$, and $x_2 \in \Omega_{X_2}$ such that $x_1 \cdot x_2 = x$, we write this event as the disjoint union of other events in Ω. We do this as follows: For each choice of $x_1 \in \Omega_{X_1}$ and $x_2 \in \Omega_{X_2}$ such that $x_1 \cdot x_2 = x$, we put $(X_1 = x_1, X_2 = x_2) \subseteq \Omega$ into the union:

$$(X_1 \cdot X_2 = x) = \bigcup_{x_1 \cdot x_2 = x} (X_1 = x_1, X_2 = x_2)$$

where the union is taken over all pairs $x_1 \in \Omega_{X_1}$ and $x_2 \in \Omega_{X_2}$ such that $x_1 \cdot x_2 = x$. Since the events in the union are disjoint,

$$p_{X_1 \cdot X_2}(x) = \sum_{x_1 \cdot x_2 = x} P(X_1 = x_1, X_2 = x_2)$$

where the sum is taken over all pairs $x_1 \in \Omega_{X_1}$ and $x_2 \in \Omega_{X_2}$ such that $x_1 \cdot x_2 = x$. Because X_1 and X_2 are independent random variables, we can replace $P(X_1 = x_1, X_2 = x_2)$ by $p_{X_1}(x_1) \cdot p_{X_2}(x_2)$ in this equation, giving

$$p_{X_1 \cdot X_2}(x) = \sum_{x_1 \cdot x_2 = x} p_{X_1}(x_1) \cdot p_{X_2}(x_2)$$

Then, substitute for $p_{X_1 \cdot X_2}(x)$ in the expression for $E(X_1 \cdot X_2)$ to get

$$E(X_1 \cdot X_2) = \sum_{x \in \Omega_{X_1 \cdot X_2}} x \sum_{x_1 \cdot x_2 = x} p_{X_1}(x_1) \cdot p_{X_2}(x_2)$$

$$= \sum_{x \in \Omega_{X_1 \cdot X_2}} \sum_{x_1 \cdot x_2 = x} x_1 \cdot x_2 \cdot p_{X_1}(x_1) \cdot p_{X_2}(x_2)$$

This double sum consists of exactly one summand

$$x_1 \cdot x_2 \cdot p_{X_1}(x_1) \cdot p_{X_2}(x_2)$$

for each choice of a pair $x_1 \in \Omega_{X_1}$ and $x_2 \in \Omega_{X_2}$ since writing out the double sum arranges the summands so that the ones for which $x_1 \cdot x_2$ has some particular value x are grouped together. Now, consider the expression

$$\left(\sum_{x_1 \in \Omega_{X_1}} x_1 \cdot p_{X_1}(x_1) \right) \left(\sum_{x_2 \in \Omega_{X_2}} x_2 \cdot p_{X_2}(x_2) \right) = E(X_1) \cdot E(X_2)$$

When multiplied out, the expression on the left gives exactly the same set of summands. Hence,

$$E(X_1 \cdot X_2) = E(X_1) \cdot E(X_2)$$ ■

Now, we use this theorem to derive another important property of independent random variables.

Theorem 4. (Variance of the Sum of Independent Random Variables) Let $X_1, X_2, \ldots, X_n$ be independent random variables on a sample space Ω. Then,

$$Var(X_1 + X_2 + \ldots + X_n) = \sum Var(X_i)$$

Proof. We prove the statement for $n = 2$ and leave the proof by induction for $n > 2$ as an exercise. Let $\mu_1 = E(X_1)$ and $\mu_2 = E(X_2)$. By Theorem 3(a) (Linearity of Expectation) from Section 8.7.5, $E(X_1 + X_2) = \mu_1 + \mu_2$. Applying Theorem 1(a) of this section with $\mu = \mu_1 + \mu_2$ gives

$$Var(X_1 + X_2) = E((X_1 + X_2)^2) - (\mu_1 + \mu_2)^2$$

Regarding $(X_1 + X_2)^2$ as the sum of the three random variables X_1^2, $2X_1X_2$, and X_2^2 allows us to apply Theorem 3(a) from Section 8.7.5 again:

$$
\begin{aligned}
E((X_1 + X_2)^2) &= E(X_1^2 + 2X_1X_2 + X_2^2) \\
&= E(X_1^2) + 2E(X_1X_2) + E(X_2^2)
\end{aligned}
$$

Because X_1 and X_2 are independent, Theorem 3 implies that the term $2E(X_1X_2)$ can be rewritten as

$$2E(X_1X_2) = 2E(X_1) \cdot E(X_2) = 2\mu_1\mu_2$$

Replacing $E((X_1 + X_2)^2)$ by $E(X_1^2) + 2\mu_1\mu_2 + E(X_2^2)$ in the expression for $Var(X_1 + X_2)$ and simplifying gives

$$
\begin{aligned}
Var(X_1 + X_2) &= E((X_1 + X_2)^2) - (\mu_1 + \mu_2)^2 \\
&= E(X_1^2) + 2\mu_1\mu_2 + E(X_2^2) - (\mu_1 + \mu_2)^2 \\
&= E(X_1^2) + 2\mu_1\mu_2 + E(X_2^2) - (\mu_1^2 + 2\mu_1\mu_2 + \mu_2^2) \\
&= E(X_1^2) + 2\mu_1\mu_2 + E(X_2^2) - \mu_1^2 - 2\mu_1\mu_2 - \mu_2^2 \\
&= E(X_1^2) - \mu_1^2 + E(X_2^2) - \mu_2^2 \\
&= Var(X_1) + Var(X_2).
\end{aligned}
$$

∎

Suppose $X_1, X_2, \ldots, X_n$ are i.i.d. random variables with common expectation $E(X_i) = \mu$ and common variance $Var(X_i) = \sigma^2$. Consider the random variable

$$Y = \frac{(X_1 + \cdots + X_n)}{n}$$

which represents their average. By Theorem 3(c) of Section 8.7.5, we know that

$$E(Y) = \frac{E(X_1 + \cdots + X_n)}{n} = \frac{1}{n} \sum_{i=1}^{n} E(X_i)$$

Since the X_i's all have the same mean μ, the expectation of their average is

$$E(Y) = \frac{1}{n} \sum_{i=1}^{n} \mu = \mu$$

As for the variance of Y, we know from Theorem 1(b) that

$$Var(Y) = Var\left(\frac{(X_1 + X_2 + \cdots + X_n)}{n}\right) = \frac{1}{n^2} Var(X_1 + X_2 + \cdots + X_n)$$

Since the X_i's are i.i.d. random variables,

$$Var(X_1 + \cdots + X_n) = \sum Var(X_i)$$

by Theorem 1. Hence,

$$Var(Y) = \frac{1}{n^2} \sum_{i=1}^{n} Var(X_i) = \frac{\sigma^2}{n}$$

Because of their importance, we highlight these results about the mean and the variance of i.i.d. random variables.

The Mean and Variance of the Average

The average of n i.i.d. random variables with mean μ and variance σ^2 is a random variable that also has mean μ *but variance* σ^2/n.

We pointed out that the actual value $X(\omega)$ of a random variable X can differ drastically from the mean μ of X. Theorem 2 gave an upper bound for the probability that X differs from μ by ϵ or more. If we choose a small value for ϵ—say, $\epsilon \leq \sigma$—then the bound just tells us that this probability is at most some number greater than or equal to one, which we know anyway. The next theorem makes a stronger statement in the case that the random variable X happens to be the average of i.i.d. random variables $X_1, X_2, \ldots, X_n$.

Theorem 5.   **(Law of Averages, or The Weak Law of Large Numbers)**   Let $n \in \mathbb{N}$. Let $X_1, X_2, \ldots, X_n$ be i.i.d. random variables on a sample space Ω, and let μ and σ^2 denote their common mean $E(X_i)$ and variance $Var(X_i)$. Then, for $\epsilon > 0$,

$$P\left(\left| \frac{X_1 + X_2 + \cdots + X_n}{n} - \mu \right| \geq \epsilon \right) \leq \frac{\sigma^2}{n \cdot \epsilon^2}$$

Proof.   From the discussion following Theorem 4, we know that μ is also the expectation of the random variable

$$Y = \frac{(X_1 + X_2 + \cdots + X_n)}{n}$$

and that σ^2/n is the variance of Y. Applying Theorem 2 to Y to get a bound on

$$P(|Y - E(Y)| \geq \epsilon)$$

gives

$$P\left(\left| \frac{X_1 + X_2 + \cdots + X_n}{n} - \mu \right| \geq \epsilon \right) \leq \frac{(\sigma^2/n)}{\epsilon^2}$$

$$= \frac{\sigma^2}{n \cdot \epsilon^2} \qquad \blacksquare$$

The Law of Averages is particularly interesting if $X_1, X_2, \ldots, X_n$ arise in a Bernoulli trials process where each X_i is associated with the outcome of some experiment that is repeated n times. The X_i are i.i.d., there is no restriction on their common distribution, and we can choose how many times n to repeat the experiment. Furthermore, the expected value μ of the average

$$Y = \frac{X_1 + X_2 + \cdots + X_n}{n}$$

does not depend on n. Suppose we choose some small positive value for ϵ. No matter what we choose, we can make the bound

$$\frac{\sigma^2}{n \cdot \epsilon^2}$$

in Theorem 5 as small as we like by making n sufficiently large. Hence, according to the theorem, the probability is small that $Y = (X_1 + X_2 + \cdots + X_n)/n$ will differ by ϵ or more from its expected value $E(Y) = \mu$ provided that we run a large enough number n of trials. Therefore, in this special case, the expected value of the random variable Y does tell us what to expect the value of the variable to be. Repeating an experiment a large number of times increases the accuracy of estimating the average. We highlight this important interpretation below.

Interpretation of the Law of Averages

If we interpret probability as an estimate for frequency of occurrence, then the Law of Averages says that only rarely will

$$Y = \frac{X_1 + X_2 + \cdots + X_n}{n}$$

differ greatly from the expected value of Y, so the actual values of Y *do* cluster around the expected value of Y (provided that n is large).

8.10   Exercises

1. Compute the variance $Var(X)$ of the random variable X that counts the number of heads in four flips of a fair coin.

2. Compute the variance $Var(X)$ of the random variable X that counts the number of heads in four flips of a coin that lands heads with a frequency of $1/3$.

3. Define a random variable X on the sample space Ω by setting $X(\omega) = 3$ for all $\omega \in \Omega$. What is $E(X)$? $Var(X)$?

4. Suppose we flip a fair coin 100 times. Define a random variable X on the underlying sample space Ω that counts the number of heads that turn up.

   (a) What are the mean μ and the variance σ^2 of X?

   (b) Use Theorem 2 to give an upper bound for the probability that X differs from μ by 10 or more.

5. Suppose we flip a fair coin n times. Let Ω consist of n-tuples ω of H's and T's. Let $X_i(\omega) = 1$ if the ith component of ω is an H; otherwise, let $X_i(\omega) = 0$.

   (a) Do the X_i form an i.i.d. set of random variables?

   (b) Let $Y = (X_1 + \cdots + X_n)/n$. What is the mean and the variance of Y?

   (c) Suppose $n = 100$. Use Theorem 5 to give an upper bound for the probability that Y differs from its mean by 0.1 or more.

6. Continuation of Exercise 5. Record the average number of heads obtained for each run of 100 flips of a fair coin. Run the experiment many times. What proportion of these experiments produce an average number of heads that differ from the expected value for the average by 0.1 or more? Compare with the results of the Exercise 5.

7. Let a random variable X have probability density function

x	1	2	6	8
$p(X = x)$	0.4	0.1	0.3	0.2

Compute the variance and standard deviation of X with $\mu = 4$.

8. The probability density function for the random variable X defined to be the number of cars owned by a randomly selected family in Millinocket is given as

x	0	1	2	3	4
$p(X = x)$	0.08	0.15	0.45	0.27	0.05

Compute the variance and standard deviation of X.

8.11   Chapter Review

This chapter introduced the notion of a probability density function p defined on the outcomes ω of a sample space Ω. The challenge of elementary probability theory is to set up a suitable sample space and density function so that the situation of interest can be expressed in terms of events $E \subseteq \Omega$. This takes practice.

First, we showed that using set theory to express sets in terms of other sets and using counting techniques to determine the size of sets play a crucial role in calculating probabilities. As the Birthday Problem illustrated, it sometimes is extremely convenient to compute the probability of an event by determining the probability of its complement and then subtracting that from one.

The discussion of cross product sample spaces explained how to set up a sample space and compute probabilities for situations that involve repeated trials of an experiment, such as flipping a coin over and over, or combining several unrelated experiments, such as checking the status of various communication links. We proved that when k events $E_{i_1}, E_{i_2}, \ldots, E_{i_k}$ occur simultaneously in k different sample spaces $\Omega_{i_1}, \Omega_{i_2}, \ldots, \Omega_{i_k}$, then the probability of such a combined simultaneous event is the product $P(E_{i_1}) \cdot P(E_{i_2}) \cdots P(E_{i_k})$.

The material on conditional probability showed how to revise probabilities in light of new information. If events are independent, then no revision is necessary. Otherwise, Bayes' Rule, which often is used with the Theorem of Total Probability, provides a powerful tool for computing conditional probabilities.

Finally, we introduced the idea of a random variable and its expected value. Since a random variable may have a value that is very different from its expected value, we introduced the notion of variance and standard deviation of a random variable. It was shown that expectation is *linear*. For example, the expectation of a sum of random variables is just the sum of their expectations. We concluded by discussing how likely it is that the average of a set of random variables, computed for a particular trial of an experiment, differs greatly from the expected value of their average. To obtain the Law of Averages, we introduced the notion of sets of i.i.d. random variables.

8.11.1   Terms and Theorems

8.1–8.2   Summary

TERMS

chance	pips
countably infinite sample space	probability
discrete sample space	probability density
disjoint union	probability density function
element	probability of the outcome
event	sample space
experiment	spots
fair	subset
frequency interpretation	uniform probability density function
occur	value
outcome	

THEOREMS

Elementary Probability Facts
Probability of the Complement

8.3   Summary

TERMS

Bernoulli process	Probability Multiplication Principle
Bernoulli trial process	Probability of Cross Product Events
$b(n; k, p)$	Product of Sums Principle
cross product	success
cross product sample space	trials
failure	

THEOREMS

Probability Density on a Cross Product
   Sample Space
Probability of k Successes in a Bernoulli
   Process

Probability of Events of Cross Product
   Form

8.5   Summary

TERMS

conditional probability	dependent trial	independent
communication channel	disjoint events	independent set of events
reliability	independence	

THEOREMS

Bayes' Rule
Total Probability

8.7   Summary

TERMS

binomial distribution
binomially distributed
constant random variable
discrete random variable
distribution
expectation
expected value

function
hypergeometric distribution
$h(k; n, r, m)$
mean
μ
random variable

THEOREMS

Basic Property of Expectation
Linearity of Expectation

Probability Density Defined by a Random
    Variable

8.9   Summary

TERMS

independent random variables
independent, identically distributed
    (i.i.d.) random variables

σ
standard deviation
variance

THEOREMS

Bound on the Probability of Deviation
    from the Expected Value
Expectation of the Product of Independent
    Random Variables
Law of Averages, or the Weak Law of
    Large Numbers

Properties of Variance
Variance of the Sum of Independent
    Random Variables

8.11.2   Starting to Review

1. What is a sample space? An outcome? An event?
2. What conditions must a function on a sample space satisfy to be a probability density function? What is the difference between the probability of an outcome and the probability of an event?
3. What condition on a probability density function p on the sample space Ω makes p a uniform probability density function? Define a uniform probability density function on the possible results of rolling a fair die. Compute the probability that the top face after the roll of a fair die shows more than four spots, an even number of spots, and either one or five spots.
4. List the probability of each outcome in the cross product sample space formed from the sample space for flipping a fair coin and the sample space for choosing a number from the set {1, 2, 3, 4}. For each sample space, use a uniform probability density function.

5. Flip a fair coin eight times. What is the probability of getting five heads? If the coin is biased and comes up heads only 40% of the time, what is the probability of five heads out of eight flips? of six heads out of eight flips?

6. What does it mean that two events are disjoint? What does it mean that two events are independent? Are disjoint events independent? Consider two events based on rolling a fair die one time: The first event is getting an odd number of pips on the top face, and the second event is rolling either a 3 or a 6. Are these two events an independent pair of events?

7. State Bayes' Rule. Compute the probability for rolling a fair die twice and getting three pips both times, knowing that the first roll does result in three pips. What does your intuition suggest as an answer? Use Bayes' Rule to verify your intuition.

8. Define a random variable on the sample space that gives the sum of the values on the top faces after rolling two fair die. Suppose the sample space of 36 pairs has a uniform probability density defined on it. Determine $P(X = x)$ for each value x of X.

9. Define a random variable X that counts the number of tails that result from flipping a fair coin three times. Let the sample space for flipping the coin have a uniform probability density function. Compute the expectation or mean of X.

10. Choose a random number from the sample space $\{1, 2, 3, 4, 5\}$, and flip a fair coin, resulting in either heads or tails appearing on the top face. Let both sample spaces have a uniform probability density function defined on them. Let the random variable X have value twice the number drawn if heads is flipped and just the value of the number drawn if tails is flipped. The mean of the random variable is 4.5. Compute the variance and the standard deviation of X.

8.11.3   Review Questions

1. Let $\Omega = \{1, 2, 3, 4, 5, 6\}$, $A = \{1, 2, 3\}$, $B = \{3, 5, 6\}$, and $C = \{2, 4, 6\}$. Describe the following events:

   (a) At least one of the events A or B occurs.
   (b) Exactly one of the events A or B occurs.
   (c) At least one of the events A, B, or C occurs.
   (d) Exactly one of the events A, B, or C occurs.
   (e) All three of the events A, B, and C occur.
   (f) Exactly two of the events A, B, or C occur.
   (g) At least two of the events A, B, or C occur.
   (h) None of the events A, B, and C occurs.
   (i) No more than one of the events A, B, or C occurs.
   (j) No more than two of the events A, B, or C occur.
   (k) A occurs, but neither B nor C occurs.

2. An electronic system of n components is said to be a series system if failure of at least one component causes a system failure. It is called a parallel system if the system fails only when all components fail. Suppose that 15 components are connected in a parallel system and that each component has probability $1/20$ of working properly. What can be said about the probability of a system failure?

3. What is the probability that a randomly chosen integer is a member of the set of numbers divisible by 3? Not divisible by 5? Divisible by either 4 or 6?

4. Find the probability that at a deal of a hand of bridge, at least one of the four players will have 13 cards of the same suit.

5. Assign a probability density function to the possible outcomes of adding the sum of the top faces after the roll of a pair of fair dice. What is the probability that both top faces have the same value?

6. It is known that 10% of certain articles manufactured are defective. What is the probability that in a random sample of 12 such articles, at least 9 are defective?

7. A chain of home entertainment stores sells three different brands of DVD players. Fifty percent of its sales are brand 1, 30% are brand 2, and 20% are brand 3. Each manufacturer offers a one-year warranty on parts and labor. It is known that 25% of brand 1's DVD players require warranty repair work, whereas the corresponding percentages for brands 2 and 3 are 20% and 10%, respectively.

   (a) What is the probability that a randomly selected purchaser has bought a brand 1 DVD player that will need repair under warranty?

   (b) What is the probability that a randomly selected purchaser has a DVD player that will need repair while under warranty?

   (c) If a customer returns to the store with a DVD player that needs warranty repair work, what is the probability that it is a brand 1 DVD player? A brand 2 DVD player? A brand 3 DVD player?

8. Four individuals have responded to a request by a blood bank for donations. None of the four has donated before, so each person's blood type is unknown. Suppose that only type A positive is desired and that only one of the four actually has this type. If the potential donors are selected in random order for blood typing, what is the probability that at least three individuals must be typed to find a donor of type A positive?

9. Show that the three following events based on the toss of two fair coins are independent: E_1 is the event "even on the first die." E_2 is the event "even on the second die." E_3 is the event "even sum."

10. Three automatic machines produce similar automobile parts. Machine A produces 40% of the total, machine B 25%, and machine C 35%. On average, 10% of the parts turned out by machine A do not conform to specifications, and for machines B and C, the corresponding percentages are 5% and 1%, respectively. If one part is selected at random from the combined output and does not conform to the specifications, what is the probability that it was produced by machine A?

11. Let $\Omega = \{\omega_1, \omega_2, \ldots, \omega_7\}$ be a sample space that represents parcels of a large lot divided into sublots for sale. The percentage of the total area for each lot and the price for each lot is as follows:

	% Area	Cost
ω_1	5	800
ω_2	10	900
ω_3	10	1000
ω_4	10	1200
ω_5	15	800
ω_6	20	900
ω_7	30	800

Define the random variable X to be the price of the lot for $\omega_1, \omega_2, \ldots, \omega_7$. Find the expected value of X.

12. Let a random variable X have probability density function as follows:

x	3	4	5
$p(X = x)$	0.3	0.4	0.3

Compute the variance and standard deviation of X.

13. A computer store has purchased three computers of a certain type at $500 apiece. The computers then are sold for $1000 each. The manufacturer has agreed to repurchase any computers that remain unsold after one month for $200 each. Let X be the random variable that denotes the number of computers sold. Suppose the probabilities for selling i computers for $i = 0, 1, 2, 3$ are $p(0) = 0.1$; $p(1) = 0.2$; $p(2) = 0.3$; and $p(3) = 0.4$. Let $h(x)$ denote the profit from selling X units. Find the expected value of h as well as the standard deviation and the variance.

14. Student workers find that 75% of all help desk inquiries involve programs with syntax errors. Let X be the random variable that counts the number of programs with syntax errors in 10 randomly chosen consultations. Find the expected value, the variance, and the standard deviation of this random variable.

8.11.4 Using Discrete Mathematics in Computer Science

1. Consider sending a job through the series-parallel system of processors as shown:

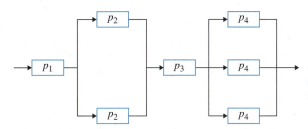

The system consists of four stages connected in sequence. Each stage i consists of identical processors connected in parallel. Each processor at stage i has probability p_i of being up. A job makes it through the system if and only if at least one processor at each stage is up. We want to determine the probability that a job makes it through the system.

(a) Model this situation with a cross product sample space $\Omega = \Omega_1 \times \Omega_2 \times \Omega_3 \times \Omega_4$. Describe in detail the component sample spaces Ω_i, their probability density functions f_i, and the cross product probability density f. (*Hint:* It may be useful to assign labels to the individual processors.)

(b) Let $A_i \subseteq \Omega_i$ be the event "at least one of the processors at stage i is up." List the elements of Ω_i belonging to A_i. What is $P(A_i)$?

(c) Let $B_i \subseteq \Omega_i$ be the event "none of the processors at stage i is up." List the elements of B_i. What is $P(B_i)$?

(d) Describe the event $A_i^* \subseteq \Omega_i$ that corresponds to $A_i \subseteq \Omega_i$. What is $P(A_i^*)$?

(e) Describe the event $B_i^* \subseteq \Omega$ that corresponds to $B_i \subseteq \Omega$. What is $P(B_i^*)$?

(f) Describe in detail the event $E \subseteq \Omega$ that at least one processor at each stage is up, and write an expression for its probability. (You may use complements of events.)

2. Suppose we double the numbers of processors at each stage of the series-parallel system in Exercise 1. Compare the probability that a job makes it through the new system with the probability that a job makes it through the old system.

3. Suppose one of the processors of stage 4 in Exercise 1 is removed and put in parallel with the processor at stage 3. Now, stage 3 has two parallel processors, one of which has probability p_3 of being up and the other of which has probability p_4 of being up, and stage 4 has two parallel processors, each with probability p_4 of being up. Answer parts (a) through (f) of Exercise 1 for this new system.

4. The probabilities of an event often can be evaluated by determining the probability density function and then summing it over the outcomes in the event. However, the theory developed in Section 8.5 often leads to simpler computations. This exercise illustrates the two approaches.

Consider the nonseries-parallel system as shown:

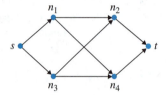

The system works correctly provided at least one of the directed paths from s to t has all its intermediate nodes working. For $i = 1, 2, 3$, and 4 let p_i denote the probability that node n_i is working, and assume that the nodes function independently of one another. Let W denote the event "the system works correctly." You will be asked to describe the situation in terms of a cross product sample space

$$\Omega = \Omega_1 \times \Omega_2 \times \Omega_3 \times \Omega_4$$

As usual, if $E_i \subseteq \Omega_i$ is an event in Ω_i, then $E_i^* \subseteq \Omega$ denotes the corresponding event in Ω. Make sure to explain your answers.

(a) Let f_i for $i = 1, 2, 3$, and 4 be the probability density function on Ω_i. Describe the situation in terms of a cross product sample space:

$$\Omega = \Omega_1 \times \Omega_2 \times \Omega_3 \times \Omega_4$$

Specify a legitimate probability density p on Ω in terms of probability densities f_i on Ω_i. (Recall that p_i is the probability that node n_i works.)

(b) Describe the event "the system works correctly" as a subset $W \subseteq \Omega$, and give an expression for $P(W)$ by summing the probability density p you defined in part a on the outcomes in W. This will be a little tedious, which is part of the point this exercise illustrates.

(c) Let $E_i^* \subseteq \Omega$ denote the event that node n_i works. How many outcomes of Ω belong to E_i^*?

(d) Compute $P(E_i^*)$ by using the theory of cross product sample spaces rather than by summing $p(\omega)$ over $\omega \in E_i^*$.

(e) Note that Ω is the disjoint union of the events E_i^* and $\overline{E_i^*}$. Use this observation and the Theorem of Total Probability to give an expression for $P(W)$.

(f) Give an expression in terms of p_2 and p_4 for the probability that nodes n_2 and n_4 both fail. Justify your answer.

(g) Note that if n_1 works, then the system works unless n_2 and n_4 both fail. Use this observation and the result of part (f) to give an expression for $P(W \mid E_1^*)$.

(h) Note that if n_1 fails, then the system can be thought of as a new, simpler system consisting of n_3 followed by the parallel pair n_2, n_4. This new system works provided n_3 and at least one of n_2, n_4 work (compare with Exercise 1). Use this observation to give an expression for $P(W \mid \overline{E_1^*})$.

(i) Refine the expression for $P(W)$ found in part (e), writing it in terms of the p_i.

5. Let p_1 denote the probability that any particular code symbol is erroneously transmitted through a communication system. Assume that on different symbols, errors occur independently of one another. Suppose also that with probability p_2, an erroneous symbol is corrected on receipt. Let X denote the number of correct symbols in a message block consisting of n symbols (count after the correction process has been carried out). What is the probability distribution of X?

6. A computer disk storage device has 10 concentric tracks (numbered $1, 2, \ldots, 10$ from outermost to innermost) and a single access arm. Let p_i be the probability that any particular request for data will take the arm to track i where $1 \leq i \leq 10$. Assume that the tracks accessed in successive seeks form an independent process. Let X be the number of tracks over which the access arm passes during two successive requests. Here, if the next track is different from the current track, X counts all the intermediate tracks plus the new track. For example, going from track 3 to track 7, the arm passes over tracks 4, 5, and 6 and then lands at track 7, so this gives $X = 4$. If two successive requests are for the same track, then $X = 0$. Hence, the possible values of X are $0, 1, \ldots, 9$. Compute the probability density function of X. (*Hint:* P(the arm is now on track i and $X = j) = P(X = j$ arm now on $i) \cdot p_i$. After writing the conditional probability in terms of $p_1, p_2, \ldots, p_{10}$ using the Law of Total Probability, the desired probability is obtained by summing over i).

7. An electronic system consisting of n components in series fails if at least one component of the system fails. An electronic system consisting of n components in parallel fails only if all n components fail. Two parallel systems of three elements each are in series as shown:

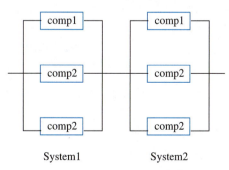

System1   System2

Let A_{ij} be the event that component i in system j fails, where $1 \leq i \leq 3$ and $1 \leq j \leq 2$. Write the event that the system fails using the terms A_{ij}.

8. A system has four components numbered 1, 2, 3, and 4 as shown:

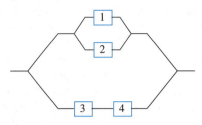

The system performs if either component 1 or component 2 performs or if both components 3 and 4 perform. Each component has probability 0.9 of performing, and all the components are independent. What is the probability the system performs?

9. Recall the 3-satisfiability problem: We are given a set of m clauses, each of which contains three boolean variables connected together by OR's; the clauses are then connected together by AND's. For example, we might want to assign truth values to the variables $X_1, X_2, \ldots, X_5$ so that the expression

$$(X_1 \ OR \ X_3 \ OR \ \neg X_5) \ AND \ (X_1 \ OR \ \neg X_2 \ OR \ X_4)$$

is true. Here, there are only $m = 2$ clauses and $n = 5$ variables. In general, for n variables (not counting their complements separately), we might have to try as many as 2^n truth assignments to the variables to determine whether the clauses can all be simultaneously satisfied. We can assume that no variable appears more than once in the same clause (whether complemented or not). If X appears with $\neg X$ in the same clause, then the clause is automatically satisfied. If X appears more than once in the same clause, that is redundant, so the clause does not truly contain three variables. (As an example of how probability theory can help to solve a problem that otherwise would seem to require examining an exponentially large number of cases, consider the following idea: For each variable, toss a fair coin. If the coin comes up heads, then set the variable to *TRUE* and its complement to *FALSE*. If it comes up tails, then make the opposite assignment). What is the expectation of the number of clauses that will be satisfied?

CHAPTER 9

Recurrence Relations

The analysis of algorithms is an area of interest in computer science, because it directly benefits the wise programmer. Program design and analysis involves making decisions about how tasks should be accomplished in a program. When a task such as sorting or searching can be accomplished in several ways, the programmer should be aware of the performance characteristics of the various alternatives. The analysis of algorithms often uses the theory of recurrence relations to make meaningful comparisons among various methods for accomplishing the same task. The chapter begins by discussing the Tower of Hanoi problem and then finding and solving a recurrence relation that describes the complexity of this problem. We then find a solution for all such recurrence relations. Next, we deal with a method for solving more general recurrence relations. Finally, we give examples and an analysis of recurrence relations that arise from divide-and-conquer algorithms.

9.1 The Tower of Hanoi Problem

The **Tower of Hanoi** problem has a long and colorful history. The problem has appeared under many names and is part of the folklore of many cultures. You start with a set of disks, each with a differents radius, stacked on a peg so that no disk is on top of another disk with a smaller radius. The problem consists in moving the stack of disks on one peg to another peg, with the use of a third peg as a temporary location so that no disk ever sits on top of a disk with a smaller radius. The difficulty arises from the requirement that at no time can a disk be placed on top of a smaller disk. It is fabled that when the problem is solved for a stack of 64 disks, the world will come to an end. A picture of an initial configuration for a stack of three disks is shown in Figure 9.1.

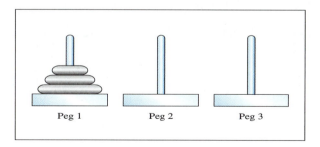

Figure 9.1 Tower of Hanoi for three disks.

In Figure 9.2, examples of legal and illegal moves are shown. The algorithm for solving the problem for n disks that is presented here involves solving a similar problem for $n - 1$ disks twice. An implementation of this algorithm involves recursive calls to the procedure.

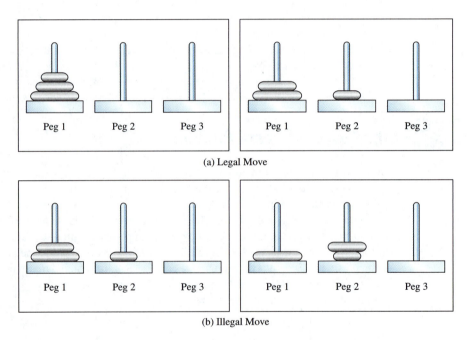

Figure 9.2 Examples of legal and illegal moves. a. Legal move. b. Illegal move.

It is instructive to follow the steps of the procedure for an initial configuration involving three disks. The steps in this recursive procedure are shown in Figure 9.3 using a tree. The actual computation order can be found by an inorder traversal of the tree (see Section 6.12.3).

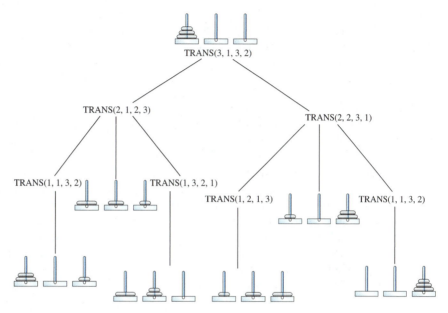

Figure 9.3 Recursion tree for Tower of Hanoi algorithm.

Algorithm: Tower of Hanoi

INPUT: Three pegs, named 1, 2, and 3, and N disks stacked by decreasing radius on peg 1, with the biggest disk at the bottom
OUTPUT: The N disks are stacked by decreasing radius on peg 3, with the biggest disk at the bottom

$TRANS(N, 1, 3, 2)$ /* Move the top N disks from peg 1 to peg 3 */

$TRANS(k, PegFrom, PegTo, PegUsing)$ /* The recursive procedure */
1. if ($k = 1$) then
          move the top disk from *PegFrom* to *PegTo*
    else
          $TRANS (k-1, PegFrom, PegUsing, PegTo)$
          move the top disk on *PegFrom* to *PegTo*
          $TRANS (k-1, PegUsing, PegTo, PegFrom)$

9.1.1   Recurrence Relation for the Tower of Hanoi Problem

The complexity of an algorithm is described by a function of n that gives the time required to execute the algorithm when n is the size of the input for the algorithm. In the case of the Tower of Hanoi problem, the amount of time that is needed to solve the problem for n disks depends directly on the number of disks that are moved. Therefore, define the complexity of the algorithm given for solving the Tower of Hanoi problem to be the function $T(n)$, the value of which is the number of disk moves that are required to solve the problem with n disks.

The algorithm consists of a single *if ... then ... else* construct. The complexity of an *if ... then ... else* construct is found by determining the complexity of the *TRUE* and the *FALSE* ranges separately and then taking the maximum of the two complexities as the complexity of the construct (see Section 5.3.2). Since it is not known if the branch that is more complex is ever actually executed, the estimate for the complexity of an *if ... then ... else* construct is usually just an upper bound for the complexity of this portion of the solution for the problem. To analyze the complexity of the *TRUE* range, observe that the *TRUE* range (line 1 of the Tower of Hanoi algorithm) consists of a single disk move. The complexity of this is 1. For the complexity of the *FALSE* range (lines 2–4), observe that line 2 and line 4 require the solution of the problem twice for a stack of $n - 1$ disks. Also, line 3 requires one additional disk move. Thus, the complexity of the *FALSE* range is $2T(n - 1) + 1$. Since the *FALSE* range of this condition is executed for $n > 1$, the function $T(n)$ is exact and not just an upper bound. The equation that gives the number of disk moves needed for the algorithm presented is

$$T(n) = \begin{cases} 2T(n-1) + 1 & \text{for } n > 1 \\ 1 & \text{for } n = 1 \end{cases}$$

The expression for $T(n)$ is called a **first-order recurrence relation,** because the value of T at any value $n > 1$ is given in terms of the value of T at a value smaller by 1—that is, by $n - 1$. The value of T at 1 is called an **initial value.** A function of n that gives the values of the recurrence is called a **solution** of the recurrence relation. In many instances, a good start for finding the solution of a first-order recurrence relation will involve a process called **back substitution,** which will be explained.

9.1.2   Solving the Tower of Hanoi Recurrence

To solve the Tower of Hanoi recurrence, begin with $T(n)$. Replace the occurrence of $T(n - 1)$ on the right-hand side with its representation in terms of $T(n - 2)$. Continue the replacement procedure for smaller and smaller values until $T(1)$ is the only value of T occurring on the right-hand side. The result of this substitution process gives $T(n)$ in terms of n and $T(1)$. Since $T(1) = 1$, this reduction process gives $T(n)$ as a function of n alone and not as an expression involving values of $T(k)$ where $k < n$. The first three steps in this process for the Tower of Hanoi recurrence relation are as follows:

$$\begin{aligned} T(n) &= 2T(n-1) + 1 \\ &= 2(2T(n-2) + 1) + 1 \quad \text{substitution for } T(n-1) \\ &= 4T(n-2) + 2 + 1 \\ &= 4(2T(n-3) + 1) + 2 + 1 \quad \text{substitution for } T(n-2) \end{aligned}$$

$$= 8T(n-3) + 4 + 2 + 1$$

$$= 2^3 T(n-3) + \sum_{i=0}^{2} 2^i$$

At this point, we want to look at the results of back substitution and see if a pattern is emerging. Seeing a pattern may suggest what will result from the remaining steps needed to reduce $T(n)$ to an expression involving n and $T(1)$.

In the case when the right-hand side contains $T(n-i)$ for $i = 1, 2,$ and 3, observe two things: the power of 2 in the expression that is the coefficient of $T(n-i)$ is 2^i, and the other constants on the right-hand side of the expression that are summed are $2^0, 2^1, \ldots, 2^{i-1}$. If this is the pattern that will continue, then it appears that after i steps, the representation for $T(n)$ would be

$$T(n) = 2^i T(n-i) + 2^{i-1} + 2^{i-2} + \cdots + 2 + 1$$

Using this pattern and continuing the back substitution for $n-2$ steps, the resulting expression is

$$T(n) = 2^{n-2} T(2) + 2^{n-3} + 2^{n-4} + \cdots + 2 + 1$$

One more back substitution step gives $T(n)$ as a function of n and $T(1)$:

$$T(n) = 2^{n-2}(2T(1) + 1) + 2^{n-3} + 2^{n-4} + \cdots + 2 + 1$$
$$= 2^{n-1} T(1) + 2^{n-2} + 2^{n-3} + \cdots + 2 + 1$$

Since the initial value for $T(1) = 1$, we replace the reference to the function T on the right-hand side and get

$$T(n) = 2^{n-1} + 2^{n-2} + \cdots + 2 + 1$$
$$= \sum_{i=0}^{n-1} 2^i$$

We compute a closed form for this sum as explained in Section 1.7.5. The computation with ratio 2 is

$$2T(n) = \qquad 2 + 4 + \cdots + 2^{n-2} + 2^{n-1} + 2^n$$
$$-T(n) = 1 + 2 + 4 + \cdots + 2^{n-2} + 2^{n-1}$$

$$T(n) = 2^n - 1$$

The back substitution has led to a function of n that we must now prove to be the solution for the Tower of Hanoi recurrence relation, because the generalization step in back substitution is not a proof. The generalization from the pattern for three terms to a pattern for all the terms was only justified by the idea that the pattern would continue as it was developing. A formal proof of a property for all values of n requires a proof by induction.

Theorem 1. For $n \geq 1$, $T(n) = 2^n - 1$ is a solution for the Tower of Hanoi recurrence relation.

Proof.   Let $n_0 = 1$. Let $\mathcal{T} = \{n \in \mathbb{N} : T(n) = 2^n - 1\}$.

(Base step)   First, show that the conclusion holds for $n = 1$. $T(1) = 1 = 2^1 - 1$ Therefore, for $n = 1$, the formula gives the correct value for $T(1)$ and $1 \in \mathcal{T}$.

(Inductive step)   Let $n \geq n_0$. Assume that $T(n) = 2^n - 1$ is a solution, and show for $n + 1$ that $T(n + 1) = 2^{n+1} - 1$ is a solution:

$$
\begin{aligned}
T(n+1) &= 2T(n) + 1 \quad &&\text{Definition of recurrence relation} \\
&= 2(2^n - 1) + 1 \quad &&\text{Induction hypothesis} \\
&= 2^{n+1} - 1
\end{aligned}
$$

Therefore, $n + 1 \in \mathcal{T}$.

By the Principle of Mathematical Induction, $T(n) = 2^n - 1$ for $n \geq 1$.   ■

After analysis of this sort, a knowledgeable comparison of this algorithm with any other algorithm that solves this problem can be made by comparing the functions that represent the complexity of each solution method.

9.2   Solving First-Order Recurrence Relations

The method of solution used for the Tower of Hanoi recurrence relation can be used to solve the general class of first-order recurrence relations. After formalizing the terms needed to describe a general first-order recurrence relation, a solution method will be given.

Definition 1.   A **first-order recurrence relation** is an expression of the form

$$
T(n) = \begin{cases} cT(n-1) + f(n) & \text{for } n > k \\ f(k) & \text{for } n = k \end{cases}
$$

defined for all natural numbers greater than or equal to some positive integer k where c is a constant and f is a function of n for $n = k, k + 1, k + 2, \ldots$.

The recurrence relation is of the **first order,** because $T(n)$ is defined as a function involving the value $T(n - 1)$ but no other values $T(l)$ where $l \leq n$. The value of T at $n = k$ is called an initial value or a **boundary value.** The initial value can be used to calculate values of $T(n)$ for increasingly larger n. For example,

$$
T(k + 1) = cT(k) + f(k + 1)
$$

will give $T(k + 1)$ in terms of the known values for $T(k)$ and $f(k + 1)$. The boundary value will often be the value $T(0)$ or $T(1)$, but it can be the value of T for any natural number. When the boundary value is not given as the value $T(0)$ but as $T(k)$ for $k > 0$, the recurrence relation may not be defined for values less than k. The function f can be any function of n, but it quite often is a constant or a polynomial function of n. The more complex the function f, the more difficult it is to put the solution of the recurrence relation in a simple form.

It is helpful to examine the Tower of Hanoi recurrence relation again and identify all its parts in terms of the new definitions. The recurrence relation is

$$
T(n) = \begin{cases} 2T(n-1) + 1 & \text{for } n \geq 2 \\ 1 & \text{for } n = 1 \end{cases}
$$

In this case, $c = 2$, $f(n) = 1$ for all $n \geq 1$, and the boundary value is given for $k = 1$. Since $T(n)$ is given as a function of $T(n-1)$, this is a first-order recurrence relation. Because the function f is not the zero function, it is called **nonhomogeneous.**

9.2.1   Solving First-Order Recurrences Using Back Substitution

After finding a solution for the general first-order recurrence relation with constant coefficient for $T(n-1)$, we will apply the result to solving some special cases.

Theorem 2.   (Solution of First-Order Recurrence Relations)   The solution of

$$T(n) = \begin{cases} cT(n-1) + f(n) & \text{for } n \geq k \\ f(k) & \text{for } n = k \end{cases}$$

where c is a constant and f is a nonzero function of n for $n \geq k$ is

$$T(n) = \sum_{l=k}^{n} c^{n-l} f(l)$$

Motivation for the Proof.   First, use back substitution to decide what the general form of the solution might be, and then prove by induction that this is the solution:

$$
\begin{aligned}
T(n) &= cT(n-1) + f(n) \\
&= c(cT(n-2) + f(n-1)) + f(n) \\
&= c^2 T(n-2) + cf(n-1) + f(n) \\
&= c^2(cT(n-3) + f(n-2)) + cf(n-1) + f(n) \\
&= c^3 T(n-3) + c^2 f(n-2) + cf(n-1) + f(n)
\end{aligned}
$$

Using back substitution one more time gives

$$T(n) = c^3 \left[cT(n-4) + f(n-3) \right] + \sum_{l=n-2}^{n} c^{n-l} f(l)$$

$$= c^4 T(n-4) + c^3 f(n-3) + \sum_{l=n-2}^{n} c^{n-l} f(l)$$

$$= c^4 T(n-4) + \sum_{l=n-3}^{n} c^{n-l} f(l)$$

If back substitution is continued until the argument of T is k—that is, for $n - k$ steps—then the expression for $T(n)$ becomes

$$T(n) = c^{n-k} T(n - (n-k)) + \sum_{l=n-k+1}^{n} c^{n-l} f(l)$$

$$= c^{n-k} T(k) + \sum_{l=n-k+1}^{n} c^{n-l} f(l)$$

Since $T(k) = f(k)$, replace the reference to T on the right-hand side of the equation, getting

$$T(n) = c^{n-k} f(k) + \sum_{l=n-k+1}^{n} c^{n-l} f(l)$$

$$= \sum_{l=n-k}^{n} c^{n-l} f(l)$$

Proof.   By induction, show that

$$T(n) = \sum_{l=k}^{n} c^{n-l} f(l)$$

Let $n_0 = k$. Let $\mathcal{T} = \{n \in \mathbb{N} : n \geq k \text{ and } T(n) \text{ is a solution}\}$.

(Base step)   First, show that

$$\sum_{l=k}^{n} c^{n-l} f(l)$$

is a solution for $n = k$ so that $k \in \mathcal{T}$.

$$\sum_{l=k}^{k} c^{k-l} f(l) = c^{k-k} f(k) = f(k) = T(k)$$

(Inductive step)   Now, assume that $T(n)$ is given by this expression for $n \geq n_0$, that is, $T(n) = \sum_{l=k}^{n} c^{n-l} f(l)$. Now prove that $T(n+1)$ is also given by this expression: In this case, prove that $T(n+1) = \sum_{l=k}^{n+1} c^{n-l} f(l)$.

$$T(n+1) = cT(n) + f(n+1) \quad \text{(Definition of recurrence relation)}$$

$$= c \sum_{l=k}^{n} c^{n-l} f(l) + f(n+1) \quad \text{(Inductive hypothesis)}$$

$$= \sum_{l=k}^{n} c^{n-l+1} f(l) + f(n+1)$$

$$= \sum_{l=k}^{n+1} c^{n+1-l} f(l)$$

This proves $n + 1 \in \mathcal{T}$.

By the Principle of Mathematical Induction, $\mathcal{T} = \{n \in \mathbb{N} : n \geq k\}$.   ■

Corollary 1.   The solution of

$$T(n) = \begin{cases} T(n-1) + f(n) & \text{for } n \geq 0 \\ f(0) & \text{for } n = 0 \end{cases}$$

is $T(n) = \sum_{k=1}^{n} f(k)$

Proof.   In the general formula, let $c = 1$ and $k = 0$.   ■

Corollary 2.   The solution of

$$T(n) = \begin{cases} cT(n-1) + d & \text{for } n \geq 0 \\ d & \text{for } n = 0 \end{cases}$$

where c and d are real numbers and $c \neq 1$ is

$$T(n) = d \cdot \frac{c^{n+1} - 1}{c - 1}$$

Proof.   In the general formula, let $f(n) = d$ for $n \geq 0$ and $k = 0$. The sum becomes

$$T(n) = d \cdot \sum_{k=0}^{n} c^k$$

$$= d \cdot \frac{c^{n+1} - 1}{c - 1}$$ ■

When c is not equal to 1 in the general formula and the function $f(n)$ is a constant for all $n \geq k$ for some k, the solution becomes the sum of a finite geometric series.

In the case $f(n)$ is not a constant, a wide range of functions exist for which $\sum_{k=0}^{n} f(k)$ is known, but the answers often entail using more complicated summing techniques than those presented here. Examples using Theorem 2 and its corollaries will give some indication of the problems that can be solved using this result.

Example 1.   Solve

$$T(n) = \begin{cases} T(n-1) + n^2 & \text{for } n \geq 1 \\ 0 & \text{for } n = 0 \end{cases}$$

Solution.   In the general formula, $f(n) = n^2$ for $n \geq 0$, $c = 1$, and $k = 0$. Since $T(0) = f(0)$, by Corollary 1 the solution is

$$T(n) = \sum_{l=1}^{n} l^2 = \frac{1}{6} \cdot (2n+1) \cdot n \cdot (n+1)$$

See Theorem 9(b) in Section 7.10 for a derivation of this formula.   ■

Example 2.   Solve

$$T(n) = \begin{cases} 3T(n-1) + 4 & \text{for } n \geq 1 \\ 4 & \text{for } n = 0 \end{cases}$$

Solution.   In the general formula, $f(n) = 4$ for $n \geq 0$, $c = 3$, and $k = 0$. By Corollary 2, the solution is

$$T(n) = 4 \cdot \frac{3^{n+1} - 1}{3 - 1} = 2 \cdot (3^{n+1} - 1)$$ ■

9.3 Exercises

1. Solve $T(n) = T(n-1) + n$ for $n \geq 1$ with $T(0) = 2$.
2. Solve $T(n) = T(n-1) + n$ for $n \geq 1$ with $T(0) = 7$.
3. Solve $T(n) = T(n-1) + 2n + 1$ for $n \geq 1$ with $T(0) = 1$.
4. Solve $T(n) = T(n-1) + 5$ for $n \geq 1$ with $T(0) = 1$.
5. Solve $T(n) = T(n-1) + 2$ for $n \geq 1$ with $T(0) = 2$.
6. Solve $2T(n+1) - T(n) = 3$ for $n \geq 1$ with $T(0) = 3$.
7. Solve $T(n+1) = -T(n) + 1$ for $n \geq 1$ with $T(0) = 1$.
8. Solve $T(n) = T(n-1) + 1$ for $n \geq 1$ with $T(0) = 1$.
9. Solve $T(n) = T(n-1) + n$ for $n \geq 2$ with $T(1) = 1$.
10. Solve $T(n) = T(n-1) + n$ for $n \geq 2$ with $T(1) = 0$.
11. Solve $T(n) = T(n-1) + n^3$ for $n \geq 1$ with $T(0) = 0$.
12. Solve $T(n) = T(n-1) - n + 3$ for $n \geq 1$ with $T(0) = 2$.
13. Solve $T(n) = T(n-1) + 2$ for $n \geq 1$ with $T(0) = 1$.
14. Consider n coplanar straight lines, no two of which are parallel and no three of which pass through a common point. Find and solve the recurrence relation that describes the number of disjoint areas into which the lines divide the plane.
15. Find and solve the recurrence relation for the number of ways to arrange n distinct objects in a row.
16. Find and solve the recurrence relation that describes the number of regions created by mutually overlapping circles on a piece of paper provided no three circles have a common intersection point and each pair of circles intersects in exactly two points. Begin by drawing a picture for such a configuration when $n = 1, 2, 3$, and 4.
17. How many strings of length n formed using the alphabet $\{0,1,2,3\}$ have an even number of zeros?
18. A very old puzzle book (dated 1917) contained the following problem: A man had a basket containing n potatoes. He asked his child to place these potatoes on the ground in a straight line. The distance between the first and second potatoes was to be one yard, between the second and third potatoes three yards, between the third and fourth potatoes five yards, and so on. After placing all the potatoes as required, how far would the child walk, starting at the first potato, to pick up all n potatoes? How many potatoes would the child pick up in the first mile of walking?
19. The population of a fruit fly colony doubles every day. If a colony of fruit flies has 100 members at the start of an experiment, how many fruit flies will be present after n days? If the population triples every day, how many fruit flies will be present after n days?
20. The value of a dollar investment after n years of compounding at an interest rate of i percent per year is the future value of the investment and is denoted $FV(n)$. Find and solve a recurrence relation that gives $FV(n)$ in terms of $FV(n-1)$. Assume that interest is calculated at the end of each year. What will be the future value of $1000 after 3, 4, 5, and 10 years? Compare this to Exercise 33 in Section 1.9.
21. Find the general solution for a first-order homogeneous recurrence relation with constant coefficients. Is the restriction that the coefficients be constant necessary?
22. Two versions of the bubble sort are given below. Determine the complexity of each procedure. For the recursive version of the code, find and solve the recurrence relation

that describes the complexity of the procedure by counting the number of comparisons made between pairs of elements.

Algorithm: Iterative Bubble Sort

INPUT: An array A containing N integers
OUTPUT: The integers sorted in nondecreasing order

for $J = 1$ to $N - 1$ do
    for $I = 1$ to $N - 1$ do
        if $(A[I] > A[I + 1])$ then swap their values

Algorithm: Recursive Bubble Sort

INPUT: An array A containing N integers
OUTPUT: The integers are sorted in nondecreasing order

RecursiveBubble (N) /* The initial call */

RecursiveBubble (k) /* The recursive procedure */
    if $(k = 1)$ then
        return
    else
        for $I = 2$ to k do
            if $A[I] < A[I - 1]$ then
                swap their values
        RecursiveBubble $(k - 1)$

---

23. Let A be an array with N elements for some positive integer N. *SelectionSort* sorts the elements of A in decreasing order by swapping the largest element in $A[1], A[2], A[3], \dots, A[S - 1]$ with $A[S]$ where S ranges from N down to 1. Find and solve a recurrence relation that describes the complexity of *SelectionSort*. Let comparison of two elements be the key operation. Assume that the minimum operation on a set of size M requires $M - 1$ comparisons for any $M \in \mathbb{N}$.

> **Algorithm: Selection Sort**
>
> **INPUT:** A list of distinct values $List[1], \ldots, List[N]$
> **OUTPUT:** $List[1], \ldots, List[N]$ with values in increasing order
>
> *SelectionSort(List, N)* /* Initial call */
>
> *SelectionSort (List, M)* /* Recursive procedure */
>   if $(M = 1)$ then
>   return
>   else
> $S = $ index of maximum element of $List[1 .. M]$
> $Temp = List[S]$
> $List[S] = List[M]$
> $List[M] = Temp$
>   *SelectionSort$(A, M - 1)$*

24. For 2×2 matrices, the complexity of **matrix multiplication** can be determined using the algorithm described below where the number of arithmetic operations (additions, subtractions, and multiplications) are one measure of complexity. Let $A = \begin{pmatrix} a_{11} & a_{12} \\ a_{21} & a_{22} \end{pmatrix}$ and $B = \begin{pmatrix} b_{11} & b_{12} \\ b_{21} & b_{22} \end{pmatrix}$. Calculate

$$m_1 = (a_{11} + a_{22})(b_{11} + b_{22})$$
$$m_5 = (a_{21} + a_{22})b_{11}$$
$$m_2 = (a_{11} - a_{22})(b_{21} + b_{22})$$
$$m_6 = a_{11}(b_{12} - b_{22})$$
$$m_3 = (a_{11} - a_{21})(b_{11} + b_{12})$$
$$m_7 = a_{22}(b_{21} - b_{11})$$
$$m_4 = (a_{11} + a_{12})b_{11}$$

Now, let $C = \begin{pmatrix} c_{11} & c_{12} \\ c_{21} & c_{22} \end{pmatrix}$ where

$$c_{11} = m_1 + m_2 - m_4 + m_7$$
$$c_{12} = m_4 + m_6$$
$$c_{21} = m_5 + m_7$$
$$c_{22} = m_1 - m_3 - m_5 + m_6$$

(a) Use the algorithm described to find the matrix product for

$$\begin{pmatrix} 3 & 5 \\ 4 & 8 \end{pmatrix} \quad \begin{pmatrix} 2 & 7 \\ 5 & 9 \end{pmatrix}$$

(b) For matrices of size $2^r \times 2^r$, partition them into $2^{r-1} \times 2^{r-1}$ submatrices, and use the procedure shown for 2×2 matrices to compute the product. Carry out this process for the following matrices:

$$\begin{pmatrix} 3 & 5 & 2 & 5 \\ 4 & 8 & 7 & 3 \\ 4 & 6 & 9 & 9 \\ 3 & 6 & 9 & 7 \end{pmatrix} \quad \begin{pmatrix} 2 & 7 & 3 & 5 \\ 5 & 9 & 4 & 6 \\ 3 & 5 & 9 & 8 \\ 5 & 3 & 2 & 1 \end{pmatrix}$$

(c) Show that matrix multiplication using this algorithm satisfies the recurrence relation

$$f_r = 7f_{r-1} + 18 \cdot 2^{r-2} \text{ for } r \geq 1$$

Solve this recurrence. How does this result compare with the classical method of multiplying matrices?

25. Write a procedure to solve the general n-disk Tower of Hanoi problem using the following idea: Number the disks from smallest to largest, starting at 1. Consider the towers as being in a triangular formation with the pegs numbered counterclockwise. Describe the disk moved and the direction it should be moved where XC means move the top disk on peg X in a clockwise direction to the next peg and XA means to move the top disk on peg X in a counterclockwise direction. For example, the following moves solve the problem for $n = 3$:

$$\text{1C 2A 1C 3C 1C 2A 1C}$$

Draw pictures of the three towers and how the disks are positioned for each of the moves indicated.

9.4   Fibonacci Recurrence Relation

The **Fibonacci sequence** is the sequence of numbers 1, 1, 2, 3, 5, 8, 13, 21, 34, 55, ... This sequence is defined by a recurrence relation that gives the nth Fibonacci number as a function of the $(n-1)$-st and $(n-2)$-nd Fibonacci numbers. The recurrence relation that defines the terms of this sequence is

$$F_n = F_{n-1} + F_{n-2} \quad \text{for } n \geq 2$$
$$F_1 = 1$$
$$F_0 = 1$$

The recurrence relation has order two, since F_n is given in terms of the two preceding values of the recurrence relation. (For a fuller description of the background of the recurrence relation, see Section 1.8.3). To calculate F_n, begin by calculating all the preceding values one at a time, starting with F_2. As an example, F_5 may be calculated using this procedure as follows:

$$F_2 = F_1 + F_0 = 1 + 1 = 2$$
$$F_3 = F_2 + F_1 = 2 + 1 = 3$$
$$F_4 = F_3 + F_2 = 3 + 2 = 5$$
$$F_5 = F_4 + F_3 = 5 + 3 = 8$$

Although it is always possible to determine any value of a recurrence relation by building up values of the recurrence relation from the initial values, finding a function of n that can be evaluated to calculate directly the value of the recurrence at n is more interesting.

9.4.1   Second Order-Recurrence Relations

The definition of such notions as solution and initial values given for first-order recurrence relations generalize naturally to **second-order recurrence relations.** The method for solving second-order recurrence relations that is presented here generalizes to higher-order recurrence relations, but we will not deal with the general theory for nth-order recurrence relations. The recurrence relations to be solved are of the form

$$a_n = k_1 a_{n-1} + k_2 a_{n-2} \quad \text{for } n \geq 2$$

where k_1 and k_2 are arbitrary constants with $k_2 \neq 0$ and a_i is shorthand for $a(i)$ for $i = 0, 1, 2, \ldots$. This recurrence relation is **second order,** because the recurrence relation is defined in terms of the two preceding terms and no other terms. By writing the recurrence as

$$a_n - k_1 a_{n-1} - k_2 a_{n-2} = 0$$

we see that the function on the right-hand side is the zero function. Consequently, the recurrence relation is **homogeneous.** A second-order recurrence relation will require two boundary values or initial values to evaluate the terms needed to determine a value for a_n.

By examining a class of first-order recurrence relations, it is possible to gain an insight regarding the form of a function that will be a solution to a second-order recurrence relation. Consider the first-order recurrence relation

$$a_n = \begin{cases} ba_{n-1} & \text{for } n \geq 1 \\ 1 & \text{for } n = 0 \end{cases}$$

Using back substitution, we find the solution to be the function $a_n = b^n$. The feature to focus on in this case is that the solution for each n is of the form b^n, a constant raised to the nth power. In the case of a second-order recurrence relation, it would be nice if the solution were some combination of two different such functions, since this would be a natural generalization of the result for first-order recurrence relations. At this point, however, we have no reason to expect that this will be the case. Proceeding on the basis that the solution is of this form may lead to a contradiction. On the other hand, if we proceed on the assumption that the solution is of this form and no contradiction is found, then an insight regarding how such recurrence relations can be solved may be found.

As an example, let us suppose a solution for the Fibonacci recurrence is a function of the form $a_n = c^n$ for some c and then see what this implies. By substituting the values of a_n for $n, n-1$, and $n-2$ into the Fibonacci recurrence, more information about this possibility can be determined. For example, it may be possible to determine whether such a function can even be a solution:

$$a_n - a_{n-1} - a_{n-2} = 0$$
$$c^n - c^{n-1} - c^{n-2} = 0$$
$$c^{n-2} \cdot (c^2 - c - 1) = 0$$

Since the product of two real numbers—namely, c^{n-2} and $c^2 - c - 1$—is zero, either $c^{n-2} = 0$ or $c^2 - c - 1 = 0$. If $c^{n-2} = 0$, then $c = 0$ and $a_n = 0$ for all $n \geq 0$. This solution, called the zero function, is the **trivial solution** of the recurrence relation. For a homogeneous recurrence relation, the zero function is always a solution. Since the primary interest is in nontrivial solutions, suppose $c \neq 0$, which implies that $c^2 - c - 1 = 0$. This equation is called the **characteristic equation** of the recurrence relation. The roots of this equation are

$$c = \frac{1 + \sqrt{5}}{2}, \frac{1 - \sqrt{5}}{2}$$

This discussion began by asking if the nontrivial solutions of a second-order homogeneous recurrence relation have a form similar to the nontrivial solution of a first-order homogeneous recurrence relation. By trying such a function, we found that if a second-order homogeneous recurrence relation had such a nontrivial solution, then the two possible constants that had the property were the solutions to the characteristic equation of the recurrence relation. Now, we prove that not only these two functions but also any linear combination of two such functions are solutions of the recurrence relation. Much like the previous discussion of back substitution, a conjecture about the form of a solution has been found, and we must now prove that such functions really are solutions. It is beyond the scope of this text to prove the fundamental result that no other nontrivial solutions of a second-order homogeneous recurrence relation exist. We will therefore just state the result.

Theorem 1. Let $a_n = k_1 a_{n-1} + k_2 a_{n-2}$ for $n \geq 2$. Suppose the characteristic equation of this recurrence relation has distinct roots c_1 and c_2. Any function of the form $H(n) = Ac_1{}^n + Bc_2{}^n$, where A and B are arbitrary constants, is a solution of this recurrence relation.

Proof. To verify that a function of the form $H(n)$ is a solution, write with all terms on the left-side of the equal sign and then substitute $H(n)$ for some value of n into the recurrence relation, giving

$$a_n - k_1 a_{n-1} - k_2 a_{n-2} =$$
$$H(n) - k_1 H(n-1) - k_2 H(n-2) =$$
$$Ac_1^n + Bc_2^n - k_1(Ac_1^{n-1} + Bc_2^{n-1}) - k_2(Ac_1^{n-2} + Bc_2^{n-2}) =$$
$$A(c_1^n - k_1 c_1^{n-1} - k_2 c_2^{n-2}) + B(c_2^n - k_1 c_1^{n-1} - k_2 c_2^{n-2}) =$$
$$Ac_1^{n-2}(c_1^2 - k_1 c_1 - k_2) + Bc_2^{n-2}(c_2^2 - k_1 c_2 - k_2)$$

Since both c_1 and c_2 are roots of the characteristic equation of the recurrence $c_1^2 - k_1 c_1 - k_2 = 0$ and $c_2^2 - k_1 c_2 - k_2 = 0$. Therefore,

$$Ac_1^{n-2}(c_1^2 - k_1 c_1 - k_2) + Bc_2^{n-2}(c_2^2 - k_1 c_2 - k_2) = 0$$

and the conclusion follows.

To prove the converse, that is, every solution is of the form $Ac_1^n + Bc_2^n$ for every $n \in \mathbb{N}$, let $n_0 = 0$ and $T = \{n : \text{any solution } H \text{ has the required for } H(n)\}$.

(Base step)   Assume $H(0) = A + B$ and $H(1) = Ac_1 + Bc_2$ for some choice of A and B. Then, $0, 1 \in T$.

(Inductive step)   Now choose $n \geq 2$ and assume that $H(0), H(1), \ldots, H(n-1)$ have the required form for every solution H. We know $H(n) = k_1 H(n-1) + k_2 H(n-2)$. By the inductive hypothesis both $H(n-1)$ and $H(n-2)$ have the required form, so we substitute these terms and simplify the algebra:

$$H(n) = k_1(Ac_1^{n-1} + Bc_2^{n-1}) + k_2(Ac_1^{n-2} + Bx_c^{n-2})$$
$$= Ac^{n-2}(k_1 c_1 + k_2) + Bc_2^{n-2}(k_1 c_2 + k_2)$$

Since c_1 and c_2 are solutions of the complementary equation, we have $c_1^2 - k_1 c_1 - k_2 = 0$ and $c_2^2 - k_1 c_2 - k_2 = 0$. Substituting the values for c_1^2 and c_2^2 for $k_1 c_1 + k_2$ and $k_1 c_2 + k_2$, we get

$$H(n) = Ac_1^{n-2} c_1^2 + Bc_2^{n-2} c_2^2$$
$$= Ac_1^n + Bc_2^k$$

as required and $n \in T$.

By the Strong Form of Mathematical Induction we conclude that $T = \mathbb{N}$.   ∎

Theorem 2.   (Solution of Second-Order Recurrence Relations)   The general solution of $a_n = k_1 a_{n-1} + k_2 a_{n-2}$, where $n \geq 2$ with k_1 and k_2 as constants and $k_2 \neq 0$, is $a_n = Ac_1^n + Bc_2^n$ when the recurrence relation has distinct roots c_1 and c_2 for its characteristic equation. Each assignment of particular values to A and B gives rise to a **particular solution** of the recurrence.

In the case discussed here, the characteristic equation has two nonequal roots. The function H has a slightly different form when the two roots are equal.

9.4.2   Solving the Fibonacci Recurrence

Using Theorem 1, the general solution of the Fibonacci recurrence is given by

$$F_n = A\left(\frac{1 + \sqrt{5}}{2}\right)^n + B\left(\frac{1 - \sqrt{5}}{2}\right)^n$$

where A and B are arbitrary constants and $n \geq 0$. It remains to determine values for A and B so that F_n is given directly as a function of n. This can be done by using the two boundary values for the recurrence relation to determine two linear equations in the unknowns A and B. The boundary values of the Fibonacci recurrence are given for $n = 0, 1$. The system of equations resulting is

$$F_0 = 1 = A\left(\frac{1+\sqrt{5}}{2}\right)^0 + B\left(\frac{1-\sqrt{5}}{2}\right)^0$$

$$F_1 = 1 = A\left(\frac{1+\sqrt{5}}{2}\right)^1 + B\left(\frac{1-\sqrt{5}}{2}\right)^1$$

$$1 = A + B$$

$$1 = \frac{1+\sqrt{5}}{2}A + \frac{1-\sqrt{5}}{2}B$$

The solution of this system of equations is

$$A = \frac{1}{\sqrt{5}}\left(\frac{1+\sqrt{5}}{2}\right)$$

and

$$B = -\frac{1}{\sqrt{5}}\left(\frac{1-\sqrt{5}}{2}\right)$$

Therefore, the solution of the Fibonacci recurrence relation

$$F_n = F_{n-1} + F_{n-2}$$

with initial conditions $F_0 = 1$ and $F_1 = 1$ is

$$F_n = \frac{1}{\sqrt{5}}\left(\frac{1+\sqrt{5}}{2}\right)^{n+1} - \frac{1}{\sqrt{5}}\left(\frac{1-\sqrt{5}}{2}\right)^{n+1}$$

It is instructive to evaluate this function for some small values of n to see that this is, indeed, the solution as well as to provide an additional check that all the arithmetic is correct:

$$F_0 = \frac{1}{\sqrt{5}}\left(\frac{1+\sqrt{5}}{2}\right)^1 - \frac{1}{\sqrt{5}}\left(\frac{1-\sqrt{5}}{2}\right)^1$$

$$= \frac{1}{2\sqrt{5}} + \frac{1}{2} - \frac{1}{2\sqrt{5}} + \frac{1}{2}$$

$$= 1$$

$$F_4 = \frac{1}{\sqrt{5}}\left(\frac{1+\sqrt{5}}{2}\right)^5 - \frac{1}{\sqrt{5}}\left(\frac{1-\sqrt{5}}{2}\right)^5$$

$$= \frac{1}{\sqrt{5}}\left(\frac{1+5\sqrt{5}+50+50\sqrt{5}+125+25\sqrt{5}}{32}\right)$$

$$\qquad - \frac{1}{\sqrt{5}}\left(\frac{1-5\sqrt{5}+50-50\sqrt{5}+125-25\sqrt{5})}{32}\right)$$

$$= \frac{1}{\sqrt{5}} \cdot \frac{160\sqrt{5}}{32} = 5$$

For small values of n, it often is much easier to use the recurrence relation directly to calculate values of the recurrence relation than to evaluate a complex function. The point is that a closed form of the function can be used to calculate any term of the sequence directly. The solution also gives a tool for directly studying properties of the recurrence relation for large n.

9.4.3 Rules for Solving Second-Order Recurrence Relations

The procedure used to solve the Fibonacci recurrence is an application of one method for solving a large class of second-order recurrence relations. Before doing another example, we summarize the procedure.

Solving Second-Order Homogeneous Recurrence Relations with Constant Coefficients Using the Complementary Equation with Distinct Real Roots
$$H(n) + AH(n-1) + BH(n-2) = 0,$$
$$H(n_1) = D, \text{ and } H(n_2) = E.$$

STEP 1: Assume $f(n) = c^n$ is a solution, and substitute for $H(n)$, yielding the characteristic equation

$$c^2 + Ac + B = 0$$

STEP 2: Find the roots of the characteristic equation: c_1 and c_2. Use the quadratic formula if the equation does not factor. If $c_1 \neq c_2$, then the general solution is

$$S(n) = Ac_1^n + Bc_2^n$$

STEP 3: Use the initial conditions to form the system of equations

$$H(n_1) = D = Ac_1^{n_1} + Bc_2^{n_2}$$
$$H(n_2) = E = Ac_1^{n_2} + Bc_2^{n_2}$$

STEP 4: Solve the system of equations found in step 3, getting A_0 and B_0 as the two solutions. Form the particular solution

$$H(n) = A_0c_1^n + B_0c_2^n$$

Quite often, the boundary conditions will be the values of the recurrence at 0 and 1. All that is required of the boundary values is that they be values of the recurrence relation given for two different natural numbers. In the case where the boundary values are not given for consecutive natural numbers, the domain of the recurrence relation must be specified to include all the integer values greater than or equal to the smaller integer for which a boundary value is given.

Example 1. Solve the recurrence relation $a_n - 6a_{n-1} - 7a_{n-2} = 0$ for $n \geq 5$ where $a_3 = 344$ and $a_4 = 2400$.

Solution. Form the characteristic equation and then factor it:

$$c^2 - 6c - 7 = 0$$

$$c = 7, -1$$

Form the general solution of the recurrence relation $a_n = A7^n + B(-1)^n$, and solve the system of equations determined by the boundary values $a_3 = 344$ and $a_4 = 2400$ to get the particular solution:

$$a_3 = A7^3 + B(-1)^3$$
$$a_4 = A7^4 + B(-1)^4$$

Now, substituting 344 and 2400 for a_3 and a_4 gives

$$344 = 343A - B$$
$$2400 = 2401A + B$$

Adding the two equations gives

$$2744 = 2744A$$
$$1 = A$$

It follows that $B = -1$. Therefore, $a_n = 7^n + (-1)^{n+1}$ for $n \geq 3$ is the particular solution. ∎

9.5 Exercises

Solve Exercises 1 through 5 using the characteristic equation method.

1. $a_n = 2a_{n-1} + 3a_{n-2}$ for $n \geq 2$ where $a_0 = 2$ and $a_1 = 2$.
2. $a_n = 4a_{n-1} + 21a_{n-2}$ for $n \geq 2$ where $a_0 = 3$ and $a_1 = 7$.
3. $a_n + 8a_{n-1} + 12a_{n-2} = 0$ for $n \geq 2$ where $a_0 = -2$ and $a_1 = 6$.
4. $a_n - 8a_{n-1} + 12a_{n-2} = 0$ for $n \geq 2$ where $a_0 = 3$ and $a_1 = -1$.
5. $a_n = 2a_{n-1} + 15a_{n-2}$ for $n \geq 2$ where $a_0 = 3$ and $a_1 = -1$.
6. How many binary sequences with no two consecutive 0's are there of length k where $k \geq 0$? Exhibit all such sequences for length less than or equal to six. Determine the number of ways a coin can be flipped n times in which no two consecutive heads occur.
7. Let $m, n \in \mathbb{N}$. Using m colors, find and solve a recurrence relation that gives the number of ways to color a disk with n sectors such that no pair of neighboring sectors receive the same color.
8. Find the number of arrangements of $1, 2, \ldots, n$ such that no integer is more than one place removed from its position in the natural order $1 - 2 - \cdots - n$.
9. For students who have been introduced to the Backus-Naur forms, how many valid expressions can be formed using exactly k symbols from the set $\{0, 1, 2, 3, 4, 5, 6, 7, 8, 9, +, -, /, *\}$, each with whatever repetition is desired. The normal form for an expression is

$$expression >::=< expression >< digit > \mid < expression >< sign >< digit >$$

10. How many ways can an athlete run up a set of stairs if either one or two steps are taken at a time?

11. Let $f(n)$ be the number of strings of n symbols formed using the symbols 0, 1, and 2 such that no two consecutive 0's occur. Show that $f(n) = 2f(n-1) + 2f(n-2)$. Solve the recurrence, and enumerate all such strings of length four.

9.6 Divide and Conquer Paradigm

One aspect of the study of algorithms is to recognize how algorithms designed to solve quite different problems are actually similar. One effective strategy for algorithm design, called **divide and conquer,** suggests splitting a problem of size n (number of inputs or size of input) into a number of instances of the same problem but each of a smaller size. The strategy then proposes solving the smaller problems separately before putting the solutions of the smaller problems together to solve the original problem. We will show how the divide-and-conquer strategy is used to develop effective algorithms for searching, sorting, and multiplying "large" integer values.

Suppose the solution of the original problem of size n can be found by solving a number of subproblems of size n/d for some d. There are three types of divide-and-conquer algorithms. The first solves fewer than d subproblems of size n/d to solve the original problem. An algorithm of this type is the binary search algorithm. The second solves d subproblems of size n/d to solve the original problem. An algorithm of this type is the merge sort algorithm. The third solves more than d subproblems of size n/d to solve the original problem. An algorithm of this type is the extended precision multiplication algorithm for integers. A general recurrence relation by which all these algorithms have their complexity represented will be solved, and this solution will be interpreted in terms of the algorithms that are presented. Finally, the complexity of each type of divide-and-conquer algorithm will be determined.

9.7 Binary Search

When searching for an element in a set of elements from a linearly ordered set, the procedure of examining each of the elements one at a time could be used. This procedure may require that every element be examined before the search can conclude. If the application involves repeating this search procedure many times on a large set of elements, then it becomes very time-consuming. To reduce the time needed to search for a single element, a divide-and-conquer approach can be used provided the set has its elements stored in sorted order. Proceed by dividing the sorted set into two equal—or almost equal—parts. Let one part contain all the elements less than the middle element of the set stored in order; let the other part contain all the elements greater than or equal to the middle element, also stored in order. Determine which part can possibly contain the element sought by comparing it to the middle element of the set. The part of the set that cannot contain the element sought is then eliminated from further consideration. The procedure is repeated on the set of elements that can still possibly contain the element being sought. This set of elements is about half the size of the previously examined set. The process of dividing the set in half and then ignoring half of the elements in the next step of the algorithm is continued until it is determined that the element is or is not in the set.

Algorithm: Binary Search

INPUT: Array $A[1 .. n]$ with n elements in increasing order and an item X to be sought in A
OUTPUT: If X is in A, its location is given in *Found;* otherwise, *Found* $= 0$

BinSrch $(A, X, 1, n)$ /* The initial call */

BinSrch $(A,$ *Item, Left, Right*) /* The recursive procedure */
1. if (*Left* > *Right*) then
        return 0
    else
2.   *Midpoint* $= \lfloor (Left + Right)/2 \rfloor$
3.   if (*Item* $=$ $A[Midpoint]$) then
4.   return *Midpoint*
        else
5.   if (*Item* $<$ $A[Midpoint]$) then
6.   *BinSrch*$(A, Item, Left, Midpoint - 1)$
            else
7.   *BinSrch*$(A, Item, Midpoint + 1, Right)$

This is the same algorithm as the one used previously to find a *Name* in a phone directory (see Section 1.10.4).

9.7.1   Correctness

To show that *BinSrch* terminates and returns the correct answer, a proof by induction can be given. The proof of the inductive step is based on the following observations. First, notice that the values of *Left* and *Right* have the property that for any execution of *BinSrch* where *Left* $\leq$ *Right,* either X is found (line 3) in $A(Left), \ldots, A(Right)$, the position of X in A is returned, and the procedure terminates, or the value of one of *Right* and *Left* is changed. For any execution of lines 1 through 4, in which X is not found in the middle of the elements of A being considered, either (line 3) decreases *Right* by one or (line 4) increases *Left* by one. Therefore, if the element X is not found, the values of *Left* and *Right* will eventually satisfy the condition *Left* > *Right* in line 1, which will cause a value of zero to be returned and the procedure to terminate. The correctness will follow from an induction on n, the size of the ordered set being searched.

Figure 9.4 represents the execution of *BinSrch* in looking for 9 in the set

$$\{1, 2, 4, 6, 9, 13, 15, 18, 20, 21, 23, 25\}$$

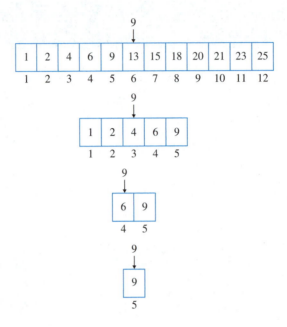

Figure 9.4   Using a binary search.

9.7.2   Complexity

For *BinSrch*, define $T(n)$ to be the number of times that *BinSrch* will be executed when A has n elements. The key operation in determining the complexity of the binary search is the comparison of the element you are searching for with an element of the array. In the case of *BinSrch*, one comparison between X and the middle element of $A[Left] \ldots A[Right]$ is made to see if X is at that position. If X is not at that middle position, then the algorithm is executed on the half of the elements of A that could still contain X. We can describe the complexity of this procedure as

$$T(n) \leq \begin{cases} T(\lfloor n/2 \rfloor) + 1 & \text{for } n \geq 2 \\ 1 & \text{for } n = 1 \end{cases}$$

The function is an upper bound, because X can be found at some position causing the search to terminate without making all the comparisons that would be necessary if X were not in A. (The solution of this recurrence relation will follow as a corollary to the solution of the recurrence relations that describe the complexity of the class of divide-and-conquer algorithms presented later.)

9.8 Merge Sort

Sorting algorithms not only are studied in a variety of computer science courses but also are used as an important processing step in many applications. The merge sort algorithm examined here uses a divide-and-conquer strategy. The idea is to divide the original set of elements in half, complete the sorting process for each half of the set, and then merge these two smaller sets, forming a sorted version of the original set. The order of worst-case complexity of this sorting algorithm is as good as that of any sorting algorithm that only uses comparisons between pairs of elements as a measure of complexity. After presenting an algorithm for this process and showing that it is correct, we find a recurrence relation that describes the complexity of the algorithm. Finally, an example will be given that shows how the algorithm works.

Algorithm: Merge Sort

INPUT: An array A with N elements
OUTPUT: The elements of A are sorted in nondecreasing order

MergeSort $(A, 1, N)$ /* The initial call */

MergeSort $(A, Left, Right)$ /* The recursive procedure */
    if $(Left < Right)$ then
        $Midway = \lceil (Left + Right - 1)/2 \rceil$
        MergeSort $(A, Left, Midway)$
        MergeSort$(A, Midway + 1, Right)$
        Merge$(A, Left, Midway, Right)$

9.8.1 Correctness

A proof by induction shows that this algorithm is correct. For an array with one element, the condition $Left < Right$ is false, so the procedure terminates. Certainly the single element is returned in sorted order. Now, assume the procedure is correct for all arrays with fewer than N elements, and show that the procedure is correct for an array with N elements. Since $Left < Right$ is true, the procedure is executed on two arrays of size less than N. By induction, the procedure is correct on such arrays. Now, assume *Merge* is correct, so the two sorted sublists will be merged to form a sorted version of the original array of N elements. Therefore, by induction, *MergeSort* is correct for arrays of any size N.

9.8.2   Example

Figure 9.5 shows the algorithm applied to an array with eight elements. The recursive calls to *MergeSort* are shown in the upper half of the diagram, and the calls to *Merge* are shown in the lower half.

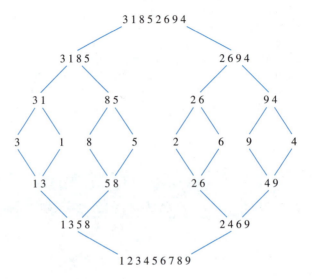

Figure 9.5   Using a merge sort.

9.8.3   Complexity

Let the complexity of *MergeSort* on an array with n elements be $T(n)$. This function counts the two kinds of commands that are executed in each call to *MergeSort*. To complete a call to *MergeSort* with an n-element array where n is even requires that *MergeSort* be executed twice for arrays of size $n/2$, followed by the merging of two sorted arrays of size $n/2$. The complexity of merging two lists of size $n/2$ will be proportional to the time that is required to make at most n comparisons. (See Exercise 7 in Section 1.12.4.) A slight modification of this argument to handle the case of N being odd leads to the following recurrence relation:

$$T(n) \leq \begin{cases} T(\lfloor n/2 \rfloor) + T(\lceil n/2 \rceil) + n & \text{for } n > 1 \\ 1 & \text{for } n = 1 \end{cases}$$

This is an upper bound, because you may not require n comparisons to merge the two lists. (The solution of this recurrence relation will follow as a corollary to the solution of the recurrence relations that describe the complexity of the class of divide-and-conquer algorithms presented later.)

9.9   Multiplication of *n*-Bit Numbers

Because the design of a computer includes a decision about the size of the unit of storage for holding integer values, the exact multiplication of integers with an arbitrarily large number of digits usually cannot be done using a computer's multiplication operation. Since exact multiplication of large numbers cannot be accomplished using the hardware structure of a computer, a software approach is needed. In a software approach, a number is assumed to be represented by its binary expansion consisting of *n* bits for *n* as large as needed. The exact **multiplication of *n*-bit numbers** when *n* is large can be accomplished using a divide-and-conquer strategy.

The example that follows shows how two 8-bit numbers are decomposed so that the multiplication of these two numbers involves multiplication of only 4-bit numbers:

$$X = 01101100, Y = 10011001,$$
$$A = 0110, B = 1100, C = 1001, D = 1001,$$
$$X = A2^4 + B, \qquad Y = C2^4 + D,$$
$$XY = (A2^4 + B)(C2^4 + D)$$
$$= AC\ 2^8 + BC2^4 + AD2^4 + BD$$

The multiplications involve 4-bit digits even though X and Y are 8-bit digits. The multiplications by 2^4 and 2^8 can be carried out simply as a shift operation and not as a multiplication. Assume, now, that X and Y are *n*-bit numbers with $n = 2^k$ for some k. The restriction to the case $n = 2^k$ will make the calculation easier to present without losing any essential information about this method.

Let X and Y be *n*-bit numbers with $A(C)$ the first $n/2$ bits of $X(Y)$ and $B(D)$ the remaining bits. The decompositions of X and Y are as follows:

Calculation I

$$X = A2^{n/2} + B$$
$$Y = C2^{n/2} + D$$
$$XY = A \cdot C2^n + (A \cdot D + B \cdot C)2^{n/2} + B \cdot D$$

This method for calculating XY involves four multiplications using pairs of numbers of size $n/2$. Repeat the divide-and-conquer strategy to calculate each of these $n/2$-bit products using $n/4$-bit numbers. Now, let $T(n)$ be the measure of complexity for multiplying two *n*-bit numbers where the key operation measured is the multiplication of two *n*-digit numbers. The complexity of this method is described by the following recurrence relation:

$$T(n) = 4T(n/2) + m_1 n$$

where m_1 is a constant and $n = 2^k$.

The term $4T(n/2)$ accounts for the $n/2$-digit multiplications while $m_1 \cdot n$ accounts for the additions and shifting operations (multiplications by a power of 2) required. Later, it will be proved that the order of the complexity of this method is n^2, which is the same order of complexity as the usual method of multiplication. The usual method of multiplication is shown for two 4-bit numbers in Figure 9.6.

$$
\begin{array}{r}
1\,0\,1\,1 \\
0\,1\,1\,1 \\
\hline
1\,0\,1\,1 \\
1\,0\,1\,1 \\
1\,0\,1\,1 \\
0\,0\,0\,0 \\
\hline
1\,0\,0\,1\,1\,0\,1
\end{array}
$$

Figure 9.6   Usual method of multiplication.

Consider the following sequence of operations where X and Y are n-bit numbers (for convenience, $n = 2^k$ for some k) that are decomposed as before:

Calculation II

Decompose:

$$X = A \cdot 2^{\frac{n}{2}} + B \qquad Y = C \cdot 2^{\frac{n}{2}} + D$$

Perform:

$$
\begin{aligned}
U &\leftarrow (A + B) \cdot (C + D) \\
V &\leftarrow A \cdot C \\
W &\leftarrow B \cdot D \\
Z &\leftarrow V \cdot 2^n + (U - V - W) \cdot 2^{\frac{n}{2}} + W
\end{aligned}
$$

By expanding $U - V - W$, we see that Z is another way of calculating XY:

$$
\begin{aligned}
Z &= V \cdot 2^n + (U - V - W) \cdot 2^{\frac{n}{2}} + W \\
&= A \cdot C \cdot 2^n + ((A + B) \cdot (C + D) - A \cdot C - B \cdot D) \cdot 2^{\frac{n}{2}} + B \cdot D \\
&= A \cdot C \cdot 2^n + (A \cdot D + B \cdot C) \cdot 2^{\frac{n}{2}} + B \cdot D \\
&= XY
\end{aligned}
$$

For the moment, assume that no carry bits are generated by the additions used in finding U. What we observe is that this method uses more additions and subtractions to find XY than in Calculation I but has fewer multiplications. Since multiplications are the more complex operation, this may be an acceptable trade-off. The recurrence relation that describes this calculation of XY is

$$T(n) = 3T(n/2) + m_2 n$$

where m_2 is a constant and $n = 2^k$. The term $m_2 \cdot n$ represents the additions, subtractions, and shifting operations required in computing XY. Later, we will prove that the order of complexity of this method is approximately $n^{1.59}$. This procedure not only gives a method to overcome the usual representation limitations of actual computers but also allows the computation to be done more efficiently. Although the value of n is restricted to 2^k for some k, $T(n)$ gives an upper bound for the complexity of $T(i)$ for any i. It is always possible to embed a problem of size i in a problem of size 2^j, where $2^{j-1} < i < 2^j$ by adding zeros to the left of the most significant digit of n.

One last detail to handle is how to deal with the product $(A + B) \cdot (C + D)$ if the additions generate an $(n/2 + 1)$-bit number. In case $A + B$ and/or $C + D$ are $(n/2 + 1)$-bit numbers, proceed as follows:

Carry Calculation

$A + B = A_1 \, 2^{n/2} + B_1$ A_1 is the leading bit of A and B_1 the remaining bits

$C + D = C_1 \, 2^{n/2} + D_1$ C_1 is the leading bit of C and D_1 the remaining bits

$(A + B) \cdot (C + D) = A_1 \cdot C_1 2^n + (A_1 \cdot D_1 + B_1 \cdot C_1)2^{n/2} + B_1 \cdot D_1$

Since B_1 and D_1 are $n/2$-bit numbers, the term $B_1 \cdot D_1$ can be computed using the recursive method. The other operations can be computed directly, since they involve 1-bit numbers or they can be accomplished by simple shifting operations.

Figure 9.7 shows how two 4-bit numbers would be multiplied using this procedure. The reader should try an example in which the carry digit comes into play.

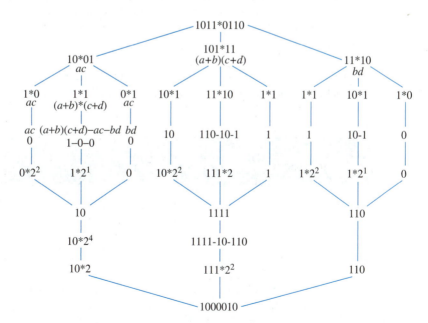

Figure 9.7   Multiplication of two 4-bit numbers.

9.10　Divide-and-Conquer Recurrence Relations

To focus attention on the kinds of recurrence relations that represent the divide-and-conquer algorithms, we have examined three different algorithms. To avoid some of the detailed analysis that is needed to handle problems of size n for any n, we make a simplifying assumption about the form of the integers for which a **divide-and-conquer recurrence relation** is defined. The divide and conquer algorithms examined are restricted to those that satisfy a recurrence relation of the form

$$T(n) = \begin{cases} CT\left(\frac{n}{d}\right) + f(n) & \text{for } n = d^k \ (k > 0) \\ f(1) & \text{for } n = 1 \ (k = 0) \end{cases}$$

where C is a constant and $k \geq 0$. A solution for this recurrence would only give an exact or closed-form answer for the case that $n = d^k$ for some k. This is often enough information about the recurrence to determine the complexity for all n. In the case of the functions that are normally encountered, it is reasonable to assume that if m lies in the interval

$$d^k \leq m \leq d^{k+1}$$

then

$$T\left(d^k\right) \leq T(m) \leq T\left(d^{k+1}\right)$$

One certainly expects that $T\left(d^{k+1}\right)$ is an upper bound for $T(m)$ whenever $m \leq d^{k+1}$, since solving a problem for m objects normally should not be more complicated than solving the same problem for a larger number of elements. With the aid of this assumption about n, the computations are considerably simplified. This simplification allows removal of the floor and ceiling functions from the recurrence used to describe the complexity of the binary search and merge sort procedures, since $\lceil n/d \rceil = \lfloor n/d \rfloor = n/d$ when $n = d^k$ for some $k \geq 0$.

After solving the divide-and-conquer recurrence in terms of the parameters C, n, and the function f, this solution will lead to a description of the behavior of the algorithms examined earlier:

Algorithm:	Value of Parameters		
Binary search	$C = 1$,	$d = 2$,	$f(n) = 1$
Merge sort	$C = 2$,	$d = 2$,	$f(n) = n$
n-Bit multiplication	$C = 3$,	$d = 2$,	$f(n) = cn$, c a constant

The method of solution is the same as that used with the first-order recurrence relations—that is, use back substitution until a pattern is recognized and then a proof by induction that this pattern leads to the solution.

Theorem 1. **(Solution of Divide-and-Conquer Recurrences)** The solution of

$$T(n) = \begin{cases} CT(n/d) + f(n) & \text{for } n = d^k, \ k \geq 1 \\ f(1) & \text{for } n = 1 \end{cases}$$

is the function $T(n) = \sum_{j=0}^{k} C^j f(n/d^j)$.

Motivation for Proof.

(Base step) For $k = 0$,

$$\sum_{j=0}^{m} C^j f(n/d^j) = f(1) = T(1)$$

(Inductive step) Now, suppose that the formula holds for $m > 0$. That is,

$$T(m) = \sum_{j=0}^{m} C^j f(n/d^j)$$

Now, prove that $T(m+1)$ is also given by this expression. In this case, prove that

$$T(m+1) = \sum_{j=0}^{m+1} C^j f(d^{m+1}/d^j)$$

Back substitutions will give

$$\begin{aligned} T(n) &= CT(n/d) + f(n) \\ &= C\{CT(n/d^2) + f(n/d)\} + f(n) \\ &= C^2 T(n/d^2) + Cf(n/d) + f(n) \\ &= C^2\{CT(n/d^3) + f(n/d^2)\} + Cf(n/d) + f(n) \\ &= C^3 \cdot T(n/d^3) + C^2 \cdot f(n/d^2) + Cf(n/d) + f(n) \end{aligned}$$

The pattern is

$$T(n) = \sum_{j=0}^{k} C^j f(n/d^j)$$

Proof. Now, verify that this is the solution using induction on k where $n = d^k$. For $k = 0$,

$$\sum_{j=0}^{0} C^j f(1/d^j) = f(1) = T(1)$$

Now, suppose that the formula holds for $m = k$ where $k < 0$, and show that it holds for $m = k + 1$:

$$T(d^{k+1}) = CT(d^k) + f(d^{k+1})$$

$$= C\sum_{j=0}^{k} C^j \cdot f(d^k/d^j) + f(d^{k+1}) \quad \text{(Induction hypothesis)}$$

$$= \sum_{j=0}^{k} C^{j+1} \cdot f(d^{k+1}/d^{j+1}) + f(d^{k+1}) \quad \left(\frac{d^k}{d^j} = \frac{d^{k+1}}{d^{j+1}} \right)$$

$$= \sum_{j=0}^{k+1} C^j \cdot f(d^{k+1}/d^j)$$

as required.

Corollary 1.   The solution to the binary search recurrence relation

$$T(n) \leq \begin{cases} T(n/2) + 1 & \text{for } n = 2^k \ k \geq 1 \\ 1 & \text{for } n = 1 \end{cases}$$

is $T(n) \leq \log_2(n+1)$.

Proof.   Let $C = 1$, $d = 2$, and $f(n) = 1$. The theorem gives

$$T(n) \leq \sum_{j=0}^{k} 1 = k + 1$$

Since $k = \log_2(n)$, $T(n) = 1 + \log_2(n)$.

Corollary 2.   The solution to the merge sort recurrence relation

$$T(n) \leq \begin{cases} 2T(n/2) + n & \text{for } n = 2^k, \ k \geq 1 \\ 1 & \text{for } n = 1, \end{cases}$$

is $T(n) \leq n + n\log_2(n)$.

Proof.   Let $C = 2$, $d = 2$, and $f(n) = n$. The theorem gives

$$T(n) \leq \sum_{j=0}^{k} 2^j (n/2d^j)$$
$$= (k+1)n \quad (k = \log_2(n))$$
$$= n + n\log_2(n)$$

Corollary 3.   The solution to the n-bit multiplication recurrence relation

$$T(n) = \begin{cases} 3T(n/2) + mn & \text{for } n = 2^k, \ k \geq 1, \ m \text{ a constant} \\ m & \text{for } n = 1 \end{cases}$$

is $T(n) = 3mn^{\log_2 3} - 2mn$.

Proof.   Let $C = 3$, $d = 2$, and $f(n) = mn$. The theorem gives

$$T(n) = \sum_{j=0}^{k} 3^j \cdot \frac{mn}{2^j}$$
$$= mn \sum_{j=0}^{k} \left(\frac{3}{2} \right)^j$$
$$= mn \frac{(3/2)^{k+1} - 1}{(3/2) - 1}$$
$$= mn \frac{\frac{3^{k+1} - 2^{k+1}}{2^{k+1}}}{\frac{3-2}{2}}$$

$$= \frac{mn}{2^k} \cdot (3^{k+1} - 2^{k+1}) \quad (n = 2^k)$$

$$= 3m \cdot 3^k - 2m \cdot 2^k$$

The following identity can be used to simplify this expression:

$$3^k = (2^{\log_2(3)})^k = (2^k)^{\log_2(3)} = n^{\log_2(3)}$$

Therefore, the solution is $T(n) = 3m \cdot n^{\log_2(3)} - 2mn$. ∎

Corollary 3 gives a solution to the recurrence for multiplying two n-bit numbers as described earlier. Since $\log_2(3) = 1.59$ and the term $2mn$ is subtracted from $3m \cdot n^{\log_2(3)}$, the complexity of this algorithm is of the order $n^{1.59}$. The recurrence for the usual multiplication algorithm has a term $4T(n/2)$. By replacing three by four in Corollary 3, it is seen that the order of complexity of the usual multiplication is $\mathbf{O}(n^2)$.

9.10.1 Complexity of Divide-and-Conquer Recurrence Relations

Often, the more important information about an algorithm is its complexity rather than an exact formula to describe its behavior for a particular case. For the three types of divide and conquer algorithms with the function $f(n) = k$ where k is a constant, Table 9.1 shows both the solution and the complexity of the solution for each of the types.

Table 9.1   Divide and Conquer Complexity

$T(n) = CT(n/d) + k$		$T(1) = k$	
		$T(n)$	$T(n)$ is $\mathbf{O}(*)$
$C = 1$		$\log_d(n) + 1$	$\log_d(n)$
$C = d$		$(dn - 1)/(d - 1)f(1)$	n
$C \neq d$	$C \neq 1$	$(Cn^{\log_d(C)} - 1)/(C - 1)f(1)$	$n^{\log_d(C)}$

Using Theorem 1 to determine the values in Table 9.1 is often referred to as the **Master Theorem.**

9.11 Exercises

1. Show that the recurrence relation $a_n = a_{n-1}/(1 + a_{n-1})$ with $a_0 = C$ has the function $S(n) = C/(1 + Cn)$ as a solution.
2. Find solutions for the following recurrences using back substitution, and then verify the correctness of these solutions by induction on k where $n = d^k$.

   (a)
   $$a_n = \begin{cases} a_{n/d} + e & \text{for } k > 0, n = d^k, d \neq 1 \\ e & \text{for } n = 1 \end{cases}$$

   (b)
   $$a_n = \begin{cases} Ca_{n/d} + e & \text{for } C \neq d, C \neq 1, k > 0, n = d^k \\ e & \text{for } n = 1 \end{cases}$$

3. Solve these recurrences using back substitution. Verify the solutions are correct by induction.

(a)
$$a_n = \begin{cases} 3a_{n/2} + 4 & \text{for } k > 0, \ n = 2^k \\ 7 & \text{for } n = 1 \end{cases}$$

(b)
$$a_n = \begin{cases} 5a_{n/5} + 7 & \text{for } k > 0, \ n = 5^k \\ 12 & \text{for } n = 1 \end{cases}$$

(c)
$$a_n = \begin{cases} 2a_{n/3} + 5 & \text{for } k > 0, \ n = 3^k \\ 7 & \text{for } n = 1 \end{cases}$$

(d)
$$a_n = \begin{cases} 7a_{n/4} + 3 & \text{for } k > 0, \ n = 4^k \\ 1 & \text{for } n = 1 \end{cases}$$

4. Find two sorted lists of length five that require nine comparisons to merge into one sorted list of size 10.
5. Write a divide-and-conquer algorithm to find the largest and smallest element in a set with 2^k elements for some $k > 0$. Determine the complexity of your algorithm.
6. If $n \geq 0$, show that
$$\left\lfloor \frac{n}{2} \right\rfloor + \left\lceil \frac{n}{2} \right\rceil = n$$

(*Hint:* Consider the cases n odd and n even separately.)
7. Determine the difference in performance between the methods described for calculating the Fibonacci numbers:
(a) Directly implement the relation $F_n = F_{n-1} + F_{n-2}$.
(b) Define a recursive function that has F_{n-1} and F_{n-2} passed as parameters when calculating F_n.
(c) Use the relations
$$F_{2n} = (F_n + 2F_{n-1}) \cdot F_n$$
$$F_{2n-1} = F_n^2 + F_{n-1}^2$$
$$F_{2n-2} = (2F_n - F_{n-1}) \cdot F_{n-1}$$

to generate pairs of Fibonacci numbers.

9.12   Chapter Review

A function often can be defined either recursively or iteratively. Programs also can be written in either way. This chapter is a brief introduction to the mathematics that can be used to compute the complexity of some recursive algorithms. Recursive algorithms can have their complexity described by recurrence relations. The problem we deal with here is to find a closed form for these recurrence relations. A recurrence relation with a value at n that depends on the value of the function at $n - 1$ is called a first-order recurrence relation. The method of back substitution is used to solve such recurrences. A recursive function with a value at n that depends on the value of the function at both $n - 1$ and $n - 2$ is called a second-order recurrence relation. The method of complementary equations is used to solve such recurrences. Finally, divide-and-conquer recurrence relations determine the value of the function at n in terms of the value of the function at n/d where n is normally a power of d. The solution of this class of recurrence relations leads to a classification of divide-and-conquer recurrence relations.

    The algorithm for the Tower of Hanoi problem exemplifies first-order recurrence relations. The Fibonacci sequence recurrence relation exemplifies second-order recurrence relations. The different types of divide-and-conquer recurrence relations are exemplified by binary search, merge sort, and n-bit multiplication.

9.12.1   Terms, Theorems, Algorithms

9.1–9.3   Summary

TERMS

back substitution

boundary value

first order

first-order recurrence relation

initial value

matrix multiplication

nonhomogeneous

solution

Tower of Hanoi

ALGORITHMS

Iterative Bubble Sort

Recursive Bubble Sort

Selection Sort

Tower of Hanoi

THEOREM

Solution of First-Order Recurrence Relations

9.4   Summary

TERMS

boundary values

characteristic equation

Fibonacci sequence

homogeneous

initial values

particular solution

second order

second-order recurrence relation

trivial solution

THEOREM

Solution of Second-Order Recurrence Relations

9.6–9.10   Summary

TERMS

divide and conquer

divide-and-conquer relation

multiplication of n-bit numbers

ALGORITHM

Binary Search

Merge Sort

THEOREMS

Master Theorem

Solution of Divide-and-Conquer Recurrences

9.12.2   Starting to Review

1. For $n = 0, 1, 2$, and 3, show that $y_n = 2 \cdot 3^n$ is a solution for $y_0 = 2$ and $y_n = 3y_{n-1}$ for $n > 0$.

2. For $n = 0, 1, 2$, and 3, show that $y_n = 1/n!$ is a solution for $y_0 = 1$ and $y_n = y_{n-1}/n$ for $n > 0$.

3. Find the value of a_n for $n = 0, 1, 2, 3, 4$, and 5 where $a_n = 3a_{n-1} - 2a_{n-2}$ for $n > 2$ where $a_0 = 2$ and $a_1 = -3$.

4. For $n = 2, 3$, and 4, determine if $F_n = 3 \cdot 2^n + 5 \cdot 3^n$ is a solution for $H_n = 5H_{n-1} - 6H_{n-2}$ where $H_0 = 8$ and $H_1 = 21$.

5. Find the characteristic equation for the recurrence relation $F_n = F_{n-1} + 3F_{n-2} - 6F_{n-3}$.

6. Let $x^2 - 4x - 5 = 0$ be the characteristic equation for a second-order recurrence relation $F(n)$ for $n \in \mathbb{N}$. Find the general solution for this recurrence relation.

7. Let $a_n = A2^n + B3^n$ be the general solution of a recurrence relation with initial conditions $a_0 = 0$ and $a_1 = 1$. Find the particular solution that satisfies these initial conditions.

8. Let $a_n = a_{n-1} + n$ where $a_0 = 0$. Solve for a_n.

9. Solve the Lucas recurrence relation: $L_n = L_{n-1} + L_{n-2}$ where $L_0 = 2$ and $L_1 = 1$. (A definition of Lucas numbers is given in Exercise 25 found in Section 1.9.)

10. Solve $a_n = a_{n/2} + 1$ where $a_1 = 1$ and $n = 2^k$ for $k \in \mathbb{N}$.

9.12.3   Review Questions

1. Prove that the sequence $b_n = C$ where C is a constant and $n \in \mathbb{N}$ is a solution to the recurrence relation $a_n - a_{n-1} = 0$ for $n \geq 1$ where $a_0 = C$.

2. Find the unique solution to the recurrence relation in Exercise 1 where $a_0 = 5$.

3. Prove that the sequence $b_n = C_1 + C_2 2^n$ where C_1 and C_2 are constants is a solution to the recurrence relation $a_n - 3a_{n-1} + 2a_{n-2} = 0$.

4. Find the unique solution to the recurrence relation in Exercise 3 where $a_0 = 5$ and $a_1 = 6$.

5. Conjecture and prove a formula for the sum of the elements of the Lucas sequence given

$$L_0 + L_1 + L_2 + \cdots + L_n$$

where $n \in \mathbb{N}$.

6. Find a recurrence relation for the number of ways to arrange flags on a flagpole that is n-feet tall using red flags two-feet high and white flags one-foot high. Suppose there are three times as many white flags as red flags on any flagpole. Flags of the same color may occur next to each other.

7. Solve $a_n = a_{n-1} + n(n-1)$ for $n \geq 1$ where $a_0 = 1$.

8. Solve $a_n - 3a_{n-1} - 4a_{n-2} = 0$ for $n \geq 2$ where $a_0 = a_1 = 1$.

9. Verify by induction that a function of the form $a_n = A_1 n + A_2$ where $A_1, A_2 \in \mathbb{R}$ is a solution to $a_n = da_{n/d} + e$ where $n = d^k$ for $k \in \mathbb{N}$.

10. Solve $a_n = 7a_{n/3} + 5$ where $a_1 = 5$ and $n = 3^k$ for $k \in \mathbb{N}$.

11. Find a recurrence relation for the number of ways to make a pile of n chips using garnet, gold, red, white, and blue chips such that no two gold chips are together.

9.12.4   Using Discrete Mathematics in Computer Science

1. Find a recurrence relation for the sum of the first n natural numbers for any $n \in \mathbb{N}$. Solve the recurrence relation. (This closed form represents the complexity of **selection sort** that was described in Section 1.7.1.)

2. (a) Find a recurrence relation to describe the complexity of the recursive code to compute a^n for any positive real number a and any natural number n:

Algorithm: Computing a^n Recursively

INPUT: $a \in \mathbb{R}$ with $a > 0$ and $n \in \mathbb{N}$
OUTPUT: a^n

$recursivePower(a, n)$
if $n = 0$ then
    $recursivePower(a, n) = 1$
else
    $recursivePower(a, n) = a \cdot recursivePower(a, n - 1)$

   (b) Find a closed form for the function found in part (a).

3. (a) Find a recurrence relation to describe the complexity of the recursive code for computing the nth Fibonacci number:

Algorithm: Compute F_n

INPUT: $n \in \mathbb{N}$
OUTPUT: F_n

$recursiveFibonacci(n)$
if $n = 0$ then
    $recursiveFibonacci(0) = 1$
else
    if $n = 1$ then
        $recursiveFibonacci(1) = 1$
    else
        $recursiveFibonacci(n) = recursiveFibonacci(n - 1)$
                     $+ recursiveFibonacci(n - 2)$

   (b) Find a closed form for the function found in part (a).

4. (a) Find a recurrence relation to describe the complexity of the recursive code for carrying out a linear search of n elements $a_1, a_2, \ldots, a_n$:

Algorithm: Search for x in $a_1, a_2, \ldots, a_n$

INPUT: $x \in \mathbb{R}$ and sequence $a_1, a_2, \ldots, a_n$
OUTPUT: *TRUE* if x is in the list and *FALSE* if it is not

recursiveLinearSearch$(x, 1, n)$

recursiveLinearSearch(x, i, n)
if $x = a_i$ then
    print *TRUE*
else
    if $i = n$ then
        return *FALSE*
    else
        recursiveLinearSearch$(x, i + 1, n)$

(b) Find a closed form for the function found in part (a).

5. Find and solve a recurrence relation that describes the complexity of each of the blocks of code shown. The blocks of code evaluate a polynomial at some point. Compare your answers to what you found for Exercise 6 in Section 5.6.4

(a)

Algorithm: Evaluate $P(x) = a_0 + a_1 x + a_2 x^2 + \cdots + a_n x^n$ at $x = x_0$

INPUT: $n \in \mathbb{N}$, the coefficients of P stored in $a[0 \ldots n]$, and a real number x_0
OUTPUT: $P(x_0)$

$Poly = a_0$
$x = 1$
for $i = 1$ to n
    $x = x_0 \cdot x$
    $Poly = a_i \cdot x + Poly$
print *Poly*

(b)

Algorithm: Evaluate $P(x) = a_0 + a_1x + a_2x^2 + \cdots + a_nx^n$ at $x = x_0$

INPUT: $n \in \mathbb{N}$, the coefficients of P stored in $a[0 \ldots n]$, and a real number x_0
OUTPUT: $P(x_0)$

$Poly = a[0]$
for $i = 0$ to n
    $x = x_0$
    for $j = 1$ to i
        $x = x \cdot x_0$
    $Poly = Poly + a[i] \cdot x$
print $Poly$

(c) The code shown implements Horner's algorithm.

Algorithm: Evaluate $P(x) = a_0 + a_1x + a_2x^2 + \cdots + a_nx^n$ at $x = x_0$

INPUT: $n \in \mathbb{N}$, the coefficients of P stored in $a[0 \ldots n]$, and a real number x_0
OUTPUT: $P(x_0)$

$Poly = x_0 \cdot a[n] + a[n-1]$
/* down to means to subtract 1 each time through the loop until $i < 0$ */
for $i = n - 2$ down to 0
    $Poly = Poly \cdot x_0 + a[i]$
print $Poly$

---

6. For the two versions of the sorting routine called *INSORT*, determine the complexity of the code presented:

    (a)   function *INSORT*$[X, n]$
          if $n \leq 1$ then
              return
          INSORT$[X, n - 1]$
          $T = X[n]$
          for $i = n - 1$ down to 1
              if $X[i] \leq T$ then
                  $X[i + 1] = X[i]$
                  $X[i] = T$
                  $T = X[i]$

    (b)   function *INSORT*$[X, n]$
          if $n \leq 1$ then
              return
          for $j = 2$ to n
              $T = X[j]$
              for $i = j - 1$ down to 1
                  $X[i] \geq T$ then
                  $X[i + 1] = X[i]$
                  $X[i] = T$
                  $T = X[i]$

7. (a) Find a recurrence relation for the number of binary decision trees on a set of n distinct numbers. (*Hint:* See Figure 6.42).
   (b) Use the recurrence relation found in part (a) to determine the number of binary search trees for 5, 7, 9, and 10 items.

Appendix A

Languages and Regular Sets

A particularly important application of finite sequences is languages. The words of a language are finite sequences of symbols or **strings** from some finite set of symbols, called an **alphabet.** For example, <u>begin</u> or <u>wherefore</u> or <u>a</u> or <u>string</u> or <u>doesn't</u> or <u>cooperation</u> are words. In this context, a sequence is written in the form <u>begin</u>, not the form <u>b e g i n</u>. The symbols in the words themselves will be underlined in this section to distinguish between talking about a word (underlined) and using a word (not underlined). The operation of concatenation defined for two words x and y, denoted as $x \cdot y$, lists the second word immediately after the first word. Concatenation is shown by just running the sequences together—for example, <u>book</u> · <u>keeper</u> = <u>bookkeeper</u>. The examples, so far, all use the characters of the English alphabet, including apostrophes. One could just as well look at any other alphabet, such as the Spanish alphabet, which is slightly different, or all the keys on a computer keyboard that print characters on the screen except for space. In general, we can define words relative to any alphabet. More generally, for any two sets of words A and B, we let

$$A \cdot B = \{x \cdot y : x \in A \text{ and } y \in B\}$$

Definition 1.   An *alphabet* is a finite set.

Definition 2.   Let Σ be an alphabet. Then, Σ^* is the set of all finite sequences of elements of Σ. The elements of Σ^* are *strings* or *words* over Σ. The *empty string,* the string with no symbols, is denoted Λ or ε.

Example 1.

(a) Let $\Sigma = \{\underline{a}\}$ (a set with one element). Then, the set of words over Σ is $\{\Lambda, \underline{a}, \underline{aa}, \underline{aaa},$ $\underline{aaaa}, \underline{aaaaa}, \underline{aaaaaa}, \ldots \}$.
(b) Let $\Sigma = \{a, b\}$. Then, the set of words over Σ is $\{\Lambda, \underline{a}, \underline{b}, \underline{aa}, \underline{ab}, \underline{ba}, \underline{bb}, \underline{aaa}, \underline{aab}, \underline{aba},$ $\underline{abb}, \underline{baa}, \underline{bab}, \underline{bba}, \underline{bbb}, \ldots\}$

Definition 3.   A *language* over an alphabet Σ is a set of strings over Σ. Alternatively, a language is a subset of Σ^*.

One way to study language is to define a **language** to be the set of words in the latest edition of the language's standard dictionary. So, by this definition, <u>cooperation</u> $\in$ *English* and <u>psymathastrygronomy</u> $\notin$ *English.* Similarly, to study a computer language, you define the set of words to be the set of all legal **tokens** of the language where a token in a computer language corresponds roughly to a word or a punctuation mark of a human language. For example, the program

> program *Programmer* (input, output);
> begin
>   write (3 <= 4)
> end.

contains the tokens

<u>program</u>   <u>*Programmer*</u>   <u>(</u>   <u>input</u>   <u>,</u>   <u>output</u>   <u>)</u>   <u>;</u>
<u>begin</u>   <u>write</u>   <u>(</u>   <u>3</u>   <u><=</u>   <u>4</u>   <u>)</u>   <u>end</u>   <u>.</u>

The traditional compiler has a procedure called a **scanner** that divides the program into tokens. Here, as noted above, the alphabet Σ is the set of all characters that can be typed onto a computer screen except for space, which only divides tokens.

Another very different way to look at a language is to define it to be the set of grammatically correct sentences. Now, the alphabet is the set of words of the language plus the set of punctuation marks, and the strings are called **sentences.** Words are separated by spaces. Punctuation marks also may need to be separated by spaces, depending on context. So, the following string contains five tokens:

<u>Wherefore</u> <u>art</u> <u>thou</u> <u>Romeo</u> <u>?</u>

with the last token being the question mark. Grammatical correctness is determined by the grammar, or **syntax,** of the language.

Similarly, a computer language can be defined to be the set of syntactically correct programs in the language. The alphabet is the set of (grammatical categories of) tokens. The program *Programmer* is a string of 10 (categories of) tokens. The traditional compiler has a procedure called a **parser** that does most of the grammatical checking.

We have described two very different ways of looking at a language, neither of which has mentioned meaning! English could also be defined to be the set of grammatically correct and *meaningful* sentences of English. For example, Chomsky's example <u>Colorless</u> <u>green</u> <u>ideas</u> <u>sleep</u> <u>furiously</u> is in the set of grammatically correct sentences but is not in the set of meaningful sentences.

As an illustration of a mathematical formalism, we return to the question of how you describe the words in a language. Identifying an approximation to the words of English is relatively easy: Just specify English as the set of all words listed in the *Oxford English Dictionary.* For a computer language, this identification becomes harder. A computer program called a **compiler** must read a program in one language and translate it into another (a machine language). A part of the compiler divides a computer program into tokens. The rules for what is a token must be clear. In theory, the set of possible tokens in most computer languages is infinite. Although the number of actual tokens allowed by almost any actual compiler is finite (for example, variable names of 5 million characters usually are not acceptable), having a dictionary of all possible tokens and looking up candidate tokens in the dictionary would be prohibitively expensive. Fortunately, there is a theoretical concept—that of a regular set—that elegantly describes the tokens of many languages.

Definition 4.   Let x be a string over a finite alphabet Σ. For, $n \in \mathbb{N}$, define $x^0 = \Lambda$ and $x^{n+1} = x \cdot x^n$.

Definition 5.   Let A and B be two sets of words:

(a) $AB = \{x \cdot y : x \in A \text{ and } y \in B\}$
(b) $A^0 = \{\Lambda\}$, and for all $n \in \mathbb{N}$, $A^{n+1} = A \cdot A^n$
(c) $A^* = A^0 \cup A^1 \cup A^2 \cup \cdots \cup A^n \cup \cdots$
(d) $A^+ = A^1 \cup A^2 \cup \cdots \cup A^n \cup \cdots$

$A \cdot B$ consists of all words formed by concatenating a word from A and a word from B. So, A^n consists of all words formed by concatenating n words from A together (possibly all the same word, possibly all different). A^* consists of all words formed by concatenating any number (possibly 0) of words from A together. (The $*$ notation is called the Kleene star, after the mathematician Stephen Kleene [1909–1994, b. United States].) A^+ consists of all words formed by concatenating one or more words from A together. For any set A, $A^+ = AA^*$ (why?), so the operator $^+$ is used only to make definitions simpler.

Example 2.   Let $A = \{\underline{aa}, \underline{b}\}$, and let $B = \{\underline{bb}, \underline{cc}, \Lambda\}$. Then:
$A \cdot B = \{\underline{aabb}, \underline{aacc}, \underline{aa}, \underline{bbb}, \underline{bcc}, \underline{b}\}$
$A^3 \quad = \{\underline{aaaaaa}, \underline{aaaab}, \underline{aabaa}, \underline{aabb}, \underline{baaaa}, \underline{baab}, \underline{bbaa}, \underline{bbb}\}$
A^* contains the following strings and infinitely many others:
  Λ (the string in A^0),
  $\underline{aa}, \underline{b}$ (the strings in A^1)
  $\underline{aaaa}, \underline{aab}, \underline{baa}, \underline{bb}$ (the strings in A^2)
  $\underline{aaaaaa}, \underline{aaaab}, \underline{aabaa}, \underline{aabb}, \underline{baaaa}, \underline{baab}, \underline{bbaa}, \underline{bbb}$ (the strings in A^3)

Since the number of tokens that can be generated using Definition 4 is infinite, a program must have some way to recognize correctly formed tokens. For this reason, the set of possible identifiers in most computer languages is required to be a **regular set.** This gives the programmer broad latitude in choosing names while still making the scanning algorithm relatively straightforward. Like the definition of the Fibonacci numbers, the definition of the term *regular set* is recursive. The pattern is like a proof by induction, with a **Base case** replacing the **Base step** and a **Closure rule** replacing the **Inductive step.** (See Section 2.1.1 for a similar notion with formulas.)

Definition 6.   Let Σ be an alphabet.

(Base case)   Any finite subset (including $\emptyset$) of Σ^* is a regular set.

(Closure rule)   If A and B are regular sets, so are $A \cup B$, $A \cdot B$, and A^*.

Example 3.   An identifier such as a program name or a variable name in a computer language consists of a letter followed by a string of letters and digits, such as $\underline{x}$ or $\underline{abcadabra123testing0}$. The set of identifiers is a regular set.
  Let

$$Letter = \{\underline{A}, \underline{B}, \underline{C}, \dots, \underline{Z}, \underline{a}, \underline{b}, \underline{c}, \dots, \underline{z}\}$$
$$Digit = \{\underline{0}, \underline{1}, \underline{2}, \underline{3}, \underline{4}, \underline{5}, \underline{6}, \underline{7}, \underline{8}, \underline{9}\}$$

Letter and *Digit* are both finite and, hence, regular. So, *Letter* $\cup$ *Digit* is also regular; it is the union of two regular sets. So, (*Letter* $\cup$ *Digit*)* is regular, since it is the Kleenestar of a regular set. (*Letter* $\cup$ *Digit*)* consists of all finite strings of letters and digits. Finally, *Letter*((*Letter* $\cup$ *Digit*)*) is also regular, consisting of all strings formed from a single letter followed by a finite string of letters and digits—that is, all identifiers.

Definition 6 allows identifiers to be of any finite length. Although a compiler will normally have a maximum allowable length, this part of the theory works out much more nicely if no maximum is assumed. Actually, a few strings of letters are not legal identifiers because of their restricted use in a language. These words are called **reserved words,** and they may only be used as intended, not as identifiers. The definition above can be modified to show that the actual set of identifiers is a regular set.

Example 4.

(a) An integer constant consists of a sequence of one or more digits, possibly preceded by a $+$ or $-$ sign, such as $\underline{3}$ or $\underline{+9876543210.}$ The set of integer constants is also regular. Let

$$UnsignedInteger = (Digit)^+$$

which is regular, since $Digit$ is regular. Then,

$$Integer = \{\underline{+}, \underline{-}, \Lambda\}(UnsignedInteger^+)$$

Since $\{\underline{+}, \underline{-}, \Lambda\}$ is finite, and, thus, regular, $Integer$ is regular.

(b) One way to write a real number constant uses an optional sign ($+$ or $-$), followed by one or more digits, followed by a decimal point, followed by one or more digits, such as in $\underline{-3.1415926535}$. The set of strings of one or more digits was called $UnsignedInteger$ above, and the set of real number constants of this form is

$$SimpleReal = Integer.UnsignedInteger$$

which is regular.

(c) There are two other ways to write a real constant, both of which use scientific notation. The number 6.02×10^{23} would be expressed as $\underline{6.02E23}$, with the E preceding the power of 10. It could also be written as $\underline{6.02E+23}$. This form has what was called a $SimpleReal$ followed by $\underline{E}$, followed by an integer. So, the set of all constants of this form is $SimpleReal\underline{E}Integer$, which is regular. The set of strings of the third form is $Integer\underline{E}Integer$, which is also regular. Finally, the set of reals is

$$(SimpleReal) \cup (SimpleReal\underline{E}Integer) \cup (Integer\underline{E}Integer)$$

which is also regular.

How is all this used in compiling a computer program? The language designer for each type of token in the language must define what its (regular) set of possible values are, often using something called **regular set notation.** The first phase of a compiler is called a **scanner** or **lexical analyzer.** It scans the input program for tokens. At each step, it takes the longest string of symbols that is in any of those regular sets. Suppose it has come to a line in the program $\underline{Jones} <= \underline{Smith;}$. J, Jo, Jon, and Jones are all legal variable names, and no token may include a blank. So, the scanner identifies $\underline{Jones}$ as a variable name—actually, it identifies $\underline{Jones}$ as an identifier. Another part of the compiler decides whether this identifier names a variable, a subprogram, or whatever else. The symbols $<=$ and $<$ are both legal comparison operations, so it chooses $<=$, the longer, as its next token. Finally, $\underline{Smith}$ is a legal token, but $\underline{Smith;}$ is not. So, it chooses $\underline{Smith}$ and $\underline{;}$ as its final tokens.

Appendix B

Finite Automata

A fundamental problem in computer science is to decide whether a word is in a language. For the case of regular sets or regular languages, the problem is solved using a very special sort of computer, called a **finite automaton.**

Definition 1.   A **finite automaton** consists of five elements:

(a) An alphabet Σ. Our input strings will be elements in Σ^*.
(b) A finite set Q, the elements of which are called **states.**
(c) A designated **start state** in Q called q_0.
(d) Some subset of Q is designated as the set of **accepting states.**
(e) A **transition function** δ, which indicates what state should be entered as a result of processing the next letter in the input word. By convention $\delta(x\omega, q) = \delta(\omega, \delta(x, q))$ for any word ω and any element $x \in \Sigma$. A word $\omega \in \Sigma^*$ is **accepted** or *is in the language* if $\delta(\omega, q_0)$ is in the set of accepting states.

Example 1.   Let $\Sigma = \{0, 1\}$ and $Q = \{q_0, q_1, q_2\}$. q_0 is the start state, and $\{q_2\}$ is the set of accepting states. The function δ is defined as

$$\delta(0, q_0) = q_2 \quad \delta(0, q_1) = q_2 \quad \delta(0, q_2) = q_0$$
$$\delta(1, q_0) = q_1 \quad \delta(1, q_1) = q_1 \quad \delta(1, q_2) = q_1$$

The word 11001 is not in the language. The word 10010 is in the language.

Solution.   We must compute $\delta(11001, q_0)$:

$$\delta(11001, q_0) = \delta(1001, \delta(1, q_0)) = \delta(1001, q_1)$$
$$\delta(1001, q_1) = \delta(001, \delta(1, q_1)) = \delta(001, q_1)$$
$$\delta(001, q_1) = \delta(0, (0, q_1)) = \delta(01, q_2)$$
$$\delta(01, q_2) = \delta(1, \delta(0, q_2)) = \delta(1, q_0)$$
$$\delta(1, q_0) = q_1$$

Since 11001 does not lead to an accepting state, this word is not in the language. We must now compute $\delta(10010, q_0)$:

$$\delta(10010, q_0) = \delta(0010, \delta(1, q_0)) = \delta(0010, q_1)$$
$$\delta(0010, q_1) = \delta(010, \delta(0, q_1)) = \delta(010, q_2)$$

$$\delta(010, q_2) = \delta(10, \delta(0, q_2)) = \delta(10, q_0)$$
$$\delta(10, q_2) = \delta(0, \delta(1, q_0)) = \delta(0, q_1)$$
$$\delta(0, q_1) = q_2$$

Since 10010 leads to an accepting state, this word is in the language.   ■

Definition 2.   The language **accepted** by a finite automaton or automata is the set of strings in Σ^* accepted by the finite automaton.

Example 2.   The finite automaton shown in Figure B.1 accepts the language of binary words that represent positive odd integers. States with double-circle boundaries are accept states. The symbol on a directed edge indicates what the next state is for that symbol. The figure is called a state diagram for the finite automaton.

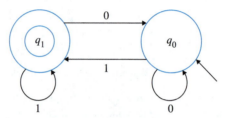

Figure B.1   Automaton to accept odd integers.

Example 3.   The finite automaton shown in Figure B.2 accepts the language of binary words that represent even positive integers.

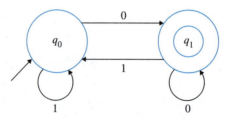

Figure B.2   Automaton to accept even integers.

B.1   Exercises

1. Let $\Sigma = \{0, 1\}$ and $Q = \{q_0, q_1, q_2\}$. q_0 is the start state, and $\{q_1\}$ is the set of accepting states. The function δ is defined as

δ	q_0	q_1	q_2
0	q_1	q_1	q_2
1	q_2	q_1	q_0

Draw the state diagram for the automata. Determine which of the following words are in the language:

(a) 0 0 1 1 0 1 1

(b) 1 1 0 1 1 0 1

(c) 1 1 1 1 0 0 0 0 1 1 1 1 0

2. Let $\Sigma = \{0, 1\}$ and $Q = \{q_0, q_1, q_2\}$. q_0 is the start state, and $\{q_1\}$ is the set of accepting states. The function δ is defined as

δ	q_0	q_1	q_2
0	q_1	q_1	q_2
1	q_2	q_2	q_0

Draw the state diagram for the automata. Determine which of the following words are in the language:

(a) 0 0 1 1 0 1 1

(b) 1 1 0 1 1 0 1

(c) 1 1 1 1 0 0 0 0 1 1 1 1 0

3. Let $\Sigma = \{a, b, c\}$ and $Q = \{q_0, q_1, q_2, q_3\}$. q_0 is the start state, and $\{q_1\}$ is the set of accepting states. The function δ is defined as

δ	q_0	q_1	q_2	q_3
a	q_1	q_2	q_3	q_0
b	q_2	q_3	q_0	q_1
c	q_1	q_0	q_3	q_2

Draw the state diagram for the automata. Determine which of the following words are in the language:

(a) *acbaccabb*

(b) *babacacbcca*

(c) *cbaccbbababc*

4. Let $\Sigma = \{a, b, c\}$ and $Q = \{q_0, q_1, q_2, q_3\}$. q_0 is the start state, and $\{q_1, q_2\}$ is the set of accepting states. The function δ is defined as

δ	q_0	q_1	q_2	q_3
a	q_1	q_2	q_2	q_0
b	q_2	q_1	q_0	q_1
c	q_1	q_0	q_3	q_3

Draw the state diagram for the automata. Determine which of the following words are in the language:

(a) *acbacbbcabb*

(b) *cbababcacabcca*

(c) *bbbcbaccbbababc*

5. In the transmission of a string of binary digits, three consecutive 1's being received is an indication of an error in the transmission. Define an automata that will detect this error condition. Draw the state diagram for the automata you define.

6. The standard way to describe the individual *tokens* ("words") of a computer language is with *regular expressions* or *finite state automata*. We gave examples in Appendix A; here, we just consider a simple example. The regular expression $(a|bb)^*$ means that a string may be built up by any number of occurrences of a's or bb's—for example, *abbaaabb* consists of one a, followed by one bb, followed by three a's, followed by one bb. The symbol | means exclusive or; the symbol * (called the *Kleene star*) means any number of occurrences—including 0. The language *denoted* by the regular expression is the set of all strings that can be built that way:

$$\{\Lambda, a, aa, bb, aaa, abb, bba, aaaa, aabb, abba, bbaa, bbbb, \ldots\}$$

where the symbol Λ just names the empty string—the string consisting of 0 symbols.

(a) Previously, we listed all strings of length less than or equal to four in the language denoted by $(a|bb)^*$. Assume there are four strings of length four in the language. Find, and solve a recurrence relation for the number of strings of length n in the language for arbitrary non-negative integer n.

(b) Find and solve a recurrence relation for the number of strings of length n in the language denoted by $(a|bb|c)^*$.

(c) Find and solve a recurrence relation for the number of strings of length n in the language denoted by $(a|bb|cc)^*$.

(d) How many strings of length n are there in the language denoted by $(a|bb|c)^*|(a|bb|cc)^*$? Why?

Index